THE IMPORTANCE OF BIOLOGICAL INTERACTIONS IN THE STUDY OF BIODIVERSITY

Edited by **Jordi López-Pujol**

The Importance of Biological Interactions in the Study of Biodiversity
Edited by Jordi López-Pujol

Published by InTech

Janeza Trdine 9, 51000 Rijeka, Croatia

Publishing Process Manager Dragana Manestar

Technical Editor Teodora Smiljanic

Cover Designer InTech Design Team

Additional hard copies can be obtained from orders@intechopen.com

The Importance of Biological Interactions in the Study of Biodiversity,
Edited by Jordi López-Pujol
p. cm.
ISBN 978-953-307-751-2

For EU product safety concerns:
IN TECH d.o.o., Prolaz Marije Krucifikse Kozulić 3, 51000 Rijeka, Croatia,
info@intechopen.com or visit our website at intechopen.com.

Meet the editor

Dr. Jordi López-Pujol is a full-time researcher at the Botanic Institute of Barcelona (CSIC-ICUB). His research interests include the conservation biology (he is a specialist in conservation genetics of rare and endangered species) and biogeography of plants, particularly in the Mediterranean Basin. He is also deeply interested in the conservation and the evolution of the flora of China (he did his postdoctoral stay at the Institute of Botany of Beijing, Chinese Academy of Sciences). Having authored some 80 scientific papers (including 24 articles in SCI journals), he serves regularly as a reviewer in numerous journals, and is recipient of several awards including the Burgen Scholar Award of the Academia Europaea (2007) and the 'Font i Quer' Botany Award of the Institute for Catalan Studies (2008).

Contents

Preface

The term 'biodiversity' was coined in the middle 1980s but became popularized in 1992 at the United Nations Conference on Environment and Development (held in Rio de Janeiro). According to the Convention on Biological Diversity (CBD) which came into force one year and half after the Rio summit, biodiversity is defined as *"the variability among living organisms from all sources including, inter alia, terrestrial, marine and other aquatic ecosystems and the ecological complexes of which they are part; this includes diversity within species, between species and of ecosystems"* (UNEP, 1992). This definition, as also occurs with many other older or more modern, clearly states that the term biodiversity comprises all the variety of life, in all its manifestations, at all its levels of organization, and including their complex **interactions**.

The **biological interactions** are, thus, a central aspect of the biological diversity. For example, it has no much sense to study a single species without taking into account the rest of the species occurring in that habitat and how they interact. However, interactions should be studied in its broadest sense, i.e. considering not only the relationships between living organisms but also those between living organisms and the abiotic elements of the environment (e.g. soils, water, climate). Following an amplified definition of biodiversity (e.g. Harrison et al., 2002), the concept of interactions should also be expanded to include those which occurred in the past (which are sometimes traceable in the fossil record) and also those between any living organism and the humans.

The biological interactions are extremely complex and varied by definition. Besides their beneficial effects (mainly the so-called 'ecosystem services', e.g. pollination, water purification, soil formation), some interactions, in contrast, can have extremely pervasive effects on biodiversity itself. For example, the interactions between biodiversity and humans often produce severe losses to the former (e.g. by habitat fragmentation or climate change). Another kind of interactions, those occurring between the native living organisms and the alien ones, are considered by many as the second most serious threat to biodiversity (after habitat fragmentation), which in turn produce serious economic losses and negative impacts on the human health.

This volume contains several contributions that illustrate the state of the art of the academic research in the field of biological interactions in its widest sense. The book has been divided in three sections correponding to the three main kinds of biological interactions: interactions between living organisms (Section 1), interactions between living organisms and humans (Section 2), and interactions between living organisms and abiotic factors (Section 3). We should be aware, however, that this arrangement of chapters follows criteria of practicality, as the three kinds of interactions are often intermingled; for example, it is acknowledged that most of the interactions between native an alien invasive species have their origin in (or are enhanced by) the human activities.

Finally, I would like to express my gratitude to Natalia Reinić and Dragana Manestar for their invaluable technical assistance in book publishing. I also thank to all the authors for their contributions. I hope that the book will be an useful tool for students, reseachers, natural resources managers or, simply, those interested in biodiversity.

Jordi López-Pujol
Botanic Institute of Barcelona (CSIC-ICUB)
Barcelona,
Spain

References

Harrison, I.J.; Laverty, M.F. & Sterling, E.J. (2002). What is biodiversity?, In: *Life on Earth – An encyclopedia of biodiversity, ecology, and evolution*, N. Eldredge (Ed.), pp. 1-30, ABC-CLIO, Santa Barbara

UNEP (United Nations Environment Programme) (1992). *Convention on biological diversity*, United Nations Environment Programme, Environmental Law and Institutions Program Activity Centre, Nairobi

Part 1

Interactions Between Living Organisms

1

Bacterial Biodiversity in Natural Environments

Ricardo C. Guerrero-Ferreira and Michele K. Nishiguchi
Emory University. School of Medicine. Department of Pediatrics. Division of Pediatric Infectious Diseases. Atlanta, GA
New Mexico State University. Department of Biology. Las Cruces, NM
United States of America

1. Introduction

Increased accessibility to the technologies for high-throughput sequencing has revealed the diversity and dynamism of bacterial genomes. It is now known that variation in gene content between bacterial strains may encompass 30–35% of the genes in the genome. Because this genetic diversity and genome variability triggers the emergence of pathogens as well as novel metabolic capabilities in the newly originated strains, there are implicit consequences to human health and the economy. Equally, genomic flexibility is understandably an impacting factor for bacterial populations because of the important role in their evolution and speciation. Conversely, in natural environments, bacteria species are constantly exposed to chemical, physical, and trophic gradients, as well as intra- and inter-specific interactions that may play an additional role in shaping bacterial biodiversity.

More specifically in interactions between bacteria and hosts, it is well accepted that the bacterial counterpart are highly susceptible to genetic changes. They usually have increased generation times when compared to eukaryotic organisms, and are genetically more diverse (Steinert et al., 2000). These aspects, in addition to the production of extremely large populations, allow bacterial species to be efficient at acquiring novel metabolic traits that facilitate their success in colonizing new environments.

Highly controlled processes of genetic regulation and genetic diversity are responsible for the ability of bacteria to live and survive under environmental conditions that are continuously changing. Processes that give rise to the genetic variability in bacteria are ultimately responsible for bacterial adaptation. Such processes are represented by point mutations, homologous recombination, and horizontal gene transfer events. Genetic and phenotypic variation is more frequently observed among bacteria since they are haploid organisms and are more susceptible to such changes that are not masked by recombination.

2. Horizontal/lateral gene transfer and biodiversity

Horizontal or lateral gene transfer (HGT or LGT) is one factor, if not the most important mechanism, influencing genomic variability and diversity in bacteria. New research efforts have recognized the importance of this process and aim to understand the rates of genetic exchange in bacterial species in natural settings. Whole genome analysis has corroborated that bacterial evolution may occur by horizontal gene flow between a range of species and genera. The current section briefly describes the role of gene transfer processes between various bacterial species, and whether this influences microbial biodiversity in a variety of ecological niches in natural environments.

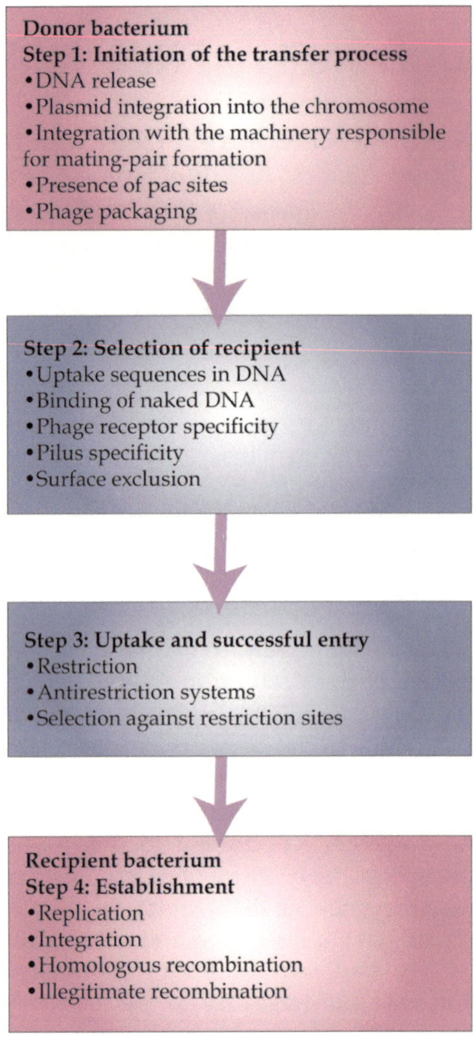

Fig. 1. Modified from Thomas & Nielsen (2005). This figure outlines the steps that take place during transfer of DNA from a donor to a recipient bacterium. The process starts with the availability of DNA from the donor cell and ends with DNA being acquired permanently by the recipient bacterium.

HGT has been observed in a wide variety of species, both in the Archaea and Bacteria domains (Smets & Barkay 2005). A number of mechanisms have accounted for the amount of transfer in specific groups, namely gene acquisition, homologous recombination, and orthologous replacement (Boucher & Stokes 2006). These processes are particularly important for changing the ability of an organism that is "clonal" and never changing, to one that has newly acquired traits that allow adaptation, speciation, and evolution to a new ecological niche. Numerous studies have documented the similarity between species of bacteria based on phylogenetic analysis of specific genes. For example, the HMG-coenzyme A reductase gene (*mvaA*), responsible for lipid metabolism, is found in a number of *Vibrio* species and was likely transferred from an archaeal donor, since mevalonate biosynthesis/degradation is an archaeal trait (Boucher & Doolittle 2000). Likewise, studies analyzing metabolic networks in *Escherichia coli* have demonstrated that particular changes are due to HGT, with very little contribution from gene duplication events (Pal et al., 2005). These changes can be linked to bacterial response to the environment, particularly when the change requires some specific metabolic capability allowing the organism to adapt more quickly to the selection imposed by the surrounding habitat. Such HGT events are usually driven by newly acquired genes that are coupled by their enzymatic pathways (*i.e.*, operons), which allow processes such as transport and degradation of external nutrients, or accommodation of an abiotic pressure (temperature, salinity). Interestingly, most HGT loci that are environment-specific are not expressed under normal laboratory conditions, demonstrating that selection of HGT loci is in part driven by adaptation to novel environments (Pal et al., 2005). This supports that HGT is a mechanism that is probably more common in natural environments than previously thought; indeed, when analyzing genes that are physiologically coupled, their functions are specific for certain environmental conditions (*i.e.*, arabinose or mannitol uptake; (Pal et al., 2005; Thomas & Nielsen 2005).

HGT has also been examined via phylogenetic reconstruction, where similar suites of genes that group together do not have a common ancestor (Gogarten & Townsend 2005). Unexpected phylogenetic distributions can therefore be explained as either HGT or an ancient gene duplication followed by differential gene loss. Oftentimes, deep-branching lineages with commonly used loci (rRNA) may also contain artifacts that are exhibited during phylogenetic reconstruction, and may provide discordance when compared to less conserved (faster adapting) molecules. This can be observed in genes that experience little or no purifying selection and are oftentimes saturated with substitutions, resulting in little phylogenetic information (Gogarten & Townsend 2005). Interestingly, examining the ratio of non-synonymous to synonymous substitutions (K_a/K_s) between *E. coli* and *Salmonella enterica* demonstrated that most horizontally acquired genes were under purifying selection, despite the K_a/K_s ratio being higher than other *E. coli* genes (0.19 vs. 0.05; (Daubin & Ochman 2004). Another example of this "neutral" selection is found in *Vibrio splendidus* (Thompson et al., 2005), where diversity at the genome level is huge compared to the sequence divergence at the 16S rRNA locus. Genome size differed between 4.5 and 5.6 Mb, eluding that there are multiple subpopulations that have unique ecological niches, despite that most of the HGT events are neutral to the recipient. If HGT events are rare, they have the probability of becoming fixed (due to selective sweeps), and are not detected under modern molecular analysis (Gogarten & Townsend 2005). Thus, in contrast to network modeling predictions, HGT may be selectively filtered against in order to deter any novel deleterious functions that may override adaptive advantages to a novel environment.

Clearly, the acquisition of genes through HGT is a much quicker and effective way for an organism to adapt to changing environments rather than their evolution via natural selection (Smets & Barkay 2005). This can be supported by observations of beneficial gene acquisition, such as antibiotic resistance, degradation pathways for xenobiotics, and bioremediation. But such observations may not be driven by environmental change alone; specific gene cassettes or mobile genetic elements may be augmented due to the increased presence of substrates that are useable by such organisms. Recently, there have been *in vitro* experiments on microbial communities to determine whether HGT events are induced by changes in environmental conditions through plasmid transfer (Sorensen et al., 2005). Such studies have allowed the detection of environmental hotspots that influence the rate of transfer via conjugation. Combining this experimental information with mathematical models (Sorensen et al., 2005) that utilize variables such as the rate of transfer, formation of new conjugants, density of donors and recipients, cell growth, and plasmid loss in homogeneous and mixed communities will be helpful in determining whether HGT is an important mechanism for driving ecological adaptations. This is particularly important in epidemics where pathogenic bacteria are more increasingly virulent. Since HGT events basically drive the evolution of bacterial "chimeras", categorizing whether a particular strain or species is genetically similar is becoming more and more difficult with modern technology (Gevers et al., 2005). The combination of both genetic background and ecological specificity will undoubtedly be the future criteria used for understanding how HGT drives microbial evolution in natural populations.

3. The role of bacterial viruses in bacterial biodiversity

In addition to the inter-specific relations that occurs within bacterial populations in nature, the association between bacteria and their viruses (bacteriophages or phages) is, quantitatively speaking, the dominant host-pathogen relationship in nature (Calendar 2006). Interactions between bacteria and phages are also expected to be particularly important, owing to the considerably fast rates of evolution of the two counterparts, the essential role bacterial communities play in ecosystem dynamics, and the emerging interest on phages as an alternative to antibiotics in the control and treatment of bacterial infections in agricultural and clinical settings (Levin & Bull 2004). More importantly, recent studies on soil bacteria and their phages have demonstrated that ecological interactions alone are not sufficient to explain the structure, population dynamics, and function of microbial communities in nature, but that rapid coevolution of bacteria and bacteriophages is also indispensable (Gómez & Buckling 2011).

Bacteriophages (also known as phages) are viruses that infect bacteria. They are widespread, with many known groups existing and found in abundance in open and coastal waters, sediments, soils, and animal tissues (Ackermann 2003). Their general life cycle (Fig. 2) varies between phage families, but generally involves adsorption, infection, and release from the host (Calendar 2006). During this cycle of phage production, the cell's metabolic machinery is reprogrammed to continually produce new phage particles with the components of the biosynthetic apparatus rerouted from basal tasks necessary for bacterial growth (Campbell 2003).

Among bacteriophage groups, infection by temperate bacteriophages often results in modification of existing properties or the acquisition of new capabilities in the bacterial host (Waldor 1998). Bacteriophages are able to integrate within the host genome during infection (a process known as lysogenic conversion), making them accountable for bacterial adaptation to new niches (Canchaya et al., 2003) and known contributors to host virulence (Rajadhyaksha & Rao 1965; Takeda & Murphy 1978; Waldor & Mekalanos 1996; Lee et al., 1999; Oakey & Owens 2000). In actuality, the process of lysogenic conversion is a key player in the evolution of Gram-positive and Gram-negative pathogens.

By definition, lysogeny is the process by which bacteriophage genome is stored in a quiescent state within the genome of a host bacterium (lysogen) (Canchaya et al., 2003). During this harboring period, transcription of the phage (temperate) genome does not take place, allowing the bacterial host to remain functional. Activation of phage transcription at this time would result in cell death (Campbell 2001). Exchange of genetic material from the virus to the bacteria can be so all-encompassing that bacteriophages have become recognized as considerable, if not the most important drivers of bacterial evolution (Krisch 2003). Temperate phages are thought of as important players in bacterial evolution because of their ability to establish long-term genetic symbioses with their host bacterium (Abedon & Lejeune 2005).

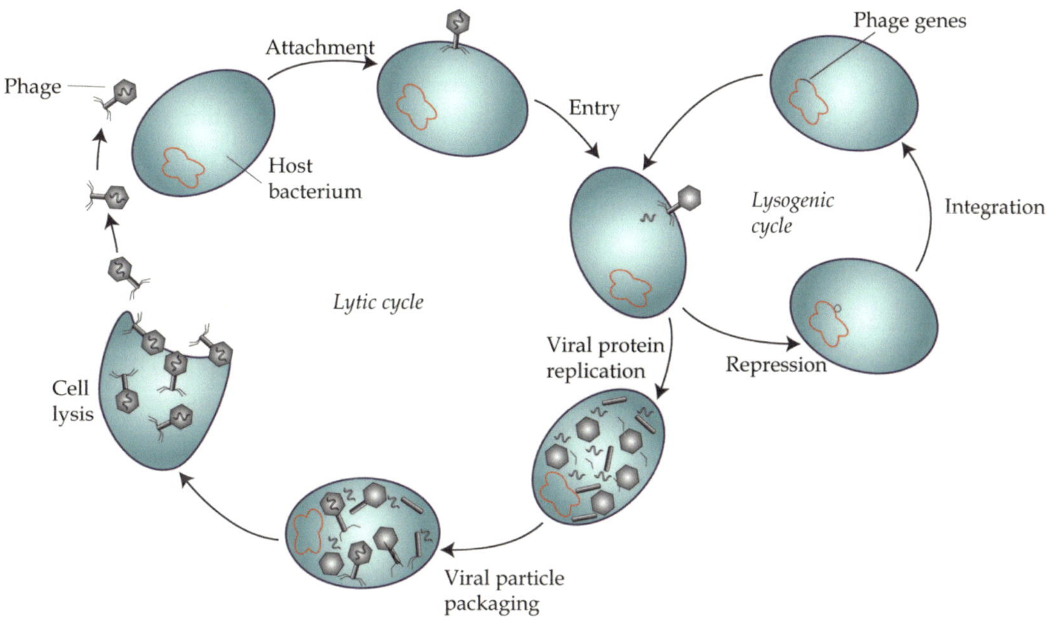

Fig. 2. Basic Phage life cycle, modified from Campbell (2003). Adsorption includes extracellular search (diffusion-mediated), random encounter between phage and host bacterium, attachment of phage to bacterium via a specific receptor, and injection of nucleic acids into the bacterial cytoplasm. This figure represents infection by a temperate phage. Phage development is temporarily repressed and phage DNA integrates into the bacterial chromosome (lysogenic cycle). Virulent phages, as well as temperate phages during their lytic cycles assemble by means of the bacterial metabolic machinery. Lastly, the cell lyses for phage progeny release.

These types of genetic associations have severe consequences in human populations owing to the variety of bacterial virulence factors that are known to be of prophage origin (Brussow et al., 2004). Among others, human diseases such as botulism, diphtheria, cholera, and *E. coli* associated conditions are virus mediated. For a more comprehensive review of prophage associated diseases please refer to (Boyd et al., 2001; Boyd & Brüssow 2002; Brussow et al., 2004).
In a recent work by Canchaya et al. (2004) it was determined that prophages are particularly abundant in the genomes of bacterial pathogens. As expected, the authors confirmed that

the presence of these prophages was in most cases responsible for encoding virulence genes and that the phenotypic characteristics that allow a strain its "uniqueness" within a bacterial consortia were contributed by the viral genome. However, this observation is not unique to pathogenic bacteria since other types of symbioses may require the bacterium to acquire particular functions to successfully colonize a host. For instance, in the gut commensal *Lactobacillus johnsonii,* it has been demonstrated that prophage derived genetic material contributes to approximately 50% of strain-specific DNA (Ventura et al., 2003).

Mechanistically, it would not be beneficial for a bacterium to fix an entire prophage genome. On the other hand, phage-derived functions that have been co-opted by the host bacterium would very likely be subjected to fixation (Casjens 2003). This makes sense considering that new ecological niches can be exploited by a bacterial species more rapidly with the acquirement of genetic material in the form of mobile DNA of phage origin. Genes of viral origin that are of no intrinsic evolutionary value to the bacterium are consequently expected to be deleted (Casjens 2003; Brussow et al., 2004). Considering that a very small amount of prophage DNA is found in the bacterial chromosome, this raises the question of why phages do not accumulate in large numbers in most cases. Campbell (2001) suggested that some genes may remain phage-borne instead of being incorporated into the bacterial genome when the host does not benefit constantly, but rather intermittently, from the product of these genes.

Prophages from bacterial pathogens that encode virulence factors have two situations that are observed (Brussow et al., 2004). Firstly, a phage-encoded toxin could be directly responsible for causing the specific disease. This is the case of *Vibrio cholerae,* Shiga toxin-producing *Escherichia coli, Corynebacterium diphtheriae,* and *Clostridium botulinum* (Abedon & Lejeune 2005). Conversely, the bacterial host may carry more than just the prophage material, and each phage-encoded factor contributes incrementally to the fitness of the host (either by direct contribution to fitness or by causing disease).

4. The role of biofilms in bacterial biodiversity

It is widely understood that most bacteria found in natural environments, as well as clinical and industrial settings, exist in biofilms. These are complex communities of microorganisms attached to surfaces or to the tissues of specific hosts, or any substrate with the adequate supply of nutrients and water (Costerton et al., 1987). These surface-associated communities are often composed of more than one species that interact with one another and their environment, and are distinct from bacteria growing in a free-living, planktonic state (Stewart & Franklin 2008).

Biofilm formation has evolved as a strategy of bacteria to establish themselves as a substrate-associated community in the environment or to become more persistent and less invasive to a host, while simultaneously taking advantage of the availability of nutrients found in those settings. The biofilm state is considered the stable period in a biological cycle that is comprised of several steps, namely initiation, maturation, maintenance, and dissolution (Fig. 3). Cells initially attach to a surface, which in most cases requires swimming motion generated by rotating flagella, and is initiated in response to specific environmental stimuli, such as nutrient availability. In most cases, the organisms undertake a series of physiological and morphological changes, transitioning from free-living, planktonic cells to non-motile, surface-attached cells. Biofilms continue to persist and grow for as long as the nutrient requirements are met. Once they are nutrient deprived, the cells separate from the surface and initiate to a free-living state (O'Toole et al., 2000).

Due to the variations in environmental conditions within the biofilm, represented by both chemical and biological heterogeneity, members of a biofilm community are subject to different selective pressures according to their location within the biofilm matrix. Therefore, bacterial cells not only express phenotypic traits that allow adaptation for growth in these surface-associated communities (as opposed to planktonic growth), but they also display phenotypic variability that allows them to thrive within a chemically heterogeneous environment.

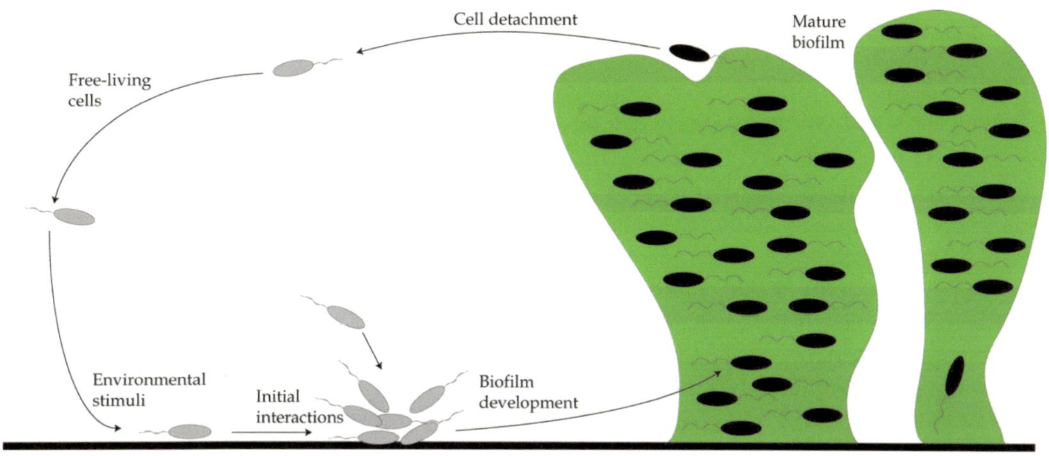

Fig. 3. Model of biofilm development. Modified from O'Toole et al. (2000). Free-living cells establish contact with other cells or with surfaces, which results in the formation of microcolonies and further maturation of the biofilm matrix. Cells from a mature biofilm can go back to a planktonic lifestyle to complete the cycle of biofilm formation.

It is expected that the chemical variability within a biofilm matrix would lead to considerable variability in the physiology of the cells that occupy the various areas within the community (de Beer et al., 1994; Xu et al., 1998). As observed in liquid cultures *in vitro,* where varying growth conditions such as temperature, aeration, and nutrient availability may impair the ability of the bacteria to grow, it is not surprising that limiting conditions within specific regions of a mature biofilm may slow or even completely stop bacterial growth and activity (Chavez-Dozal & Nishiguchi 2011). Also, metabolic waste accumulation would have an effect on the physiological state of the bacteria, mostly by changes in pH within the matrix (Stewart & Franklin 2008).

One important aspect affecting the success of a multi-species biofilm community is the ability of each member of the consortium to adapt to the presence of a second species. In a two-species community (*Acinetobacter* sp. (strain C6) and *Pseudomonas putida* (strain KT2440)), Hansen et al. (2007) demonstrated that selection in an environment such as a biofilm leads to the evolution of unequal interactions. Specific mutations in the genome of one species lead to adaptation to the presence of the other. The resulting community proved to be more successful in stability and productivity, than the ancestral community. This indicates that simple mutations due to the interactions in the biofilm generated a more intimate and specialized association.

Biofilms are ideal for the exchange of genetic material of various origins (bacterial or viral). Several studies have also demonstrated that bacterial conjugation (horizontal transfer of genetic material between two cells by physical contact) occurs within biofilms (Christensen et al., 1998; Hausner & Wuertz 1999) and this process is known to have a high impact in the evolution of bacterial lineages (Ochman et al., 2000). In a study of *E. coli* K12 biofilms, Ghigo (2001) studied how conjugative plasmids directly contribute to the ability of a bacterial cell to establish a biofilm. In this study, the author demonstrated that natural conjugative plasmids expressed factors that promote the transition of the bacteria to a biofilm forming state from a free-living state, and argued that this process supports the infectious transfer of the plasmid. Considering that antibiotic resistance is carried by bacteria through conjugative plasmids, the use of antibiotics and biocides in clinical and agricultural settings may have promoted the selection for resistant strains (bearing specific plasmids) that are more likely to form a biofilm.

It is clear that in order to be successful in the environment, a bacterial community needs to be efficient in growth and reproduction. However, it is equally important to be able to avoid, tolerate, and defend themselves against natural predators. Most studies on bacterial predation have looked at the strategies they use to increase their survival under grazing pressure by protozoans (Matz & Kjelleberg 2005). Among these adaptive traits, cell surface properties (Wildschutte et al., 2004), motility (Matz & Jurgens 2005), microcolony establishment, and quorum sensing (Matz et al., 2004) are the most studied and their results suggest that grazing by protozoans is an important contributor to bacterial diversification and to the selection of specific adaptations to defend themselves against predators. Biofilm formation has therefore emerged as an adaptive response to flagellate predation. Previous results (Matz et al., 2004) have demonstrated that *Pseudomonas aeruginosa* cells transition into a microcolony forming state upon encounter with a predator. These cell conglomerates reach a size that is beyond the prey size of the protozoan. In addition, mature biofilms build up acute toxicity to the flagellate predator via quorum sensing-mediated up-regulation of lethal compounds.

Bacterial-host interactions during mutualistic symbiosis are another, well studied example of associations in which bacteria utilize adaptive strategies of survival and reproduction in order to fight the normal defense mechanisms of the host (McFall-Ngai 1994; McFall-Ngai 1998). Similar to virulence determinants in bacteria which are regulated in their expression by both environmental and host factors (Heithoff et al., 1997; Soto et al., 2009), many novel genes are selectively expressed during the establishment and persistence of a mutualistic association (Jones & Nishiguchi 2006; Guerrero-Ferreira & Nishiguchi 2010). An example of this type of association is the mutualistic interaction between *Vibrio fischeri* and the bobtail squid *Euprymna scolopes*. It is understood that the bacterial symbionts are able to establish themselves within the host tissue by forming biofilm in the epithelium-lined crypts of the squid light organ. This was demonstrated by Ariyakumar & Nishiguchi (2009), where *V. fischeri* mutants with a reduced ability to form biofilm in vitro were unable to successfully colonize squid light organs and were not detected in any section of the crypt region.

Biofilms are the leading cause of contamination of medical devices and in industrial and agricultural settings. The initial adhesion and further colonization of bacteria onto solid surfaces is essential for biofilm formation, and therefore is the cause of infections of material of biological or medical use (Shemesh et al., 2010). Formation of microcolonies within a biofilm facilitates genetic exchange, favors genetic diversity, and promotes phenotypic variability within bacterial communities. Further understanding of these phenomena is necessary to understand the mechanisms bacterial communities utilize to infect and persist in humans and other organisms and surfaces. Deciphering the factors that control bacterial diversity will not only permit a more vigorous model of bacterial evolution and speciation but also a more comprehensive analysis of the likelihood of emergence of new biofilm-forming infectious agents.

5. Conclusion

Bacterial diversity in natural populations is continually being revitalized and revisited due to the availability of whole genomes, *in situ* measurements of HGT, and manipulation of regulatory genes that are influenced by changes in the natural environment. It is especially important to consider the diversity of bacteria, and what selection pressures have driven the evolution of species or strains that can accommodate such a wide ecological breadth. Combining phylogenetics, metabolic networks, models of HGT, and phenotypic characterization of ecotypes, will help provide meaningful explanations of how bacteria can adapt so quickly to specific abiotic and biotic factors, and what forces are important to create the diversity of microbes we observe today.

6. Acknowledgements

The authors would like to thank the editors of this book for including this interesting topic in bacterial diversity. R.C.G-F. is supported by grants from the National Institutes of Health. M.K.N. is supported by grants from the National Science Foundation and the National Institutes of Health.

7. References

Abedon, S. T., & Lejeune, J. T. (2005). Why bacteriophage encode exotoxins and other virulence factors. *Evolutionary Bioinformatics Online*, Vol. 1, No. 1, pp. 97-110.

Ackermann, H. W. (2003). Bacteriophage observations and evolution. *Research in Microbiology*, Vol. 154, No. 4, pp. 245-251.

Ariyakumar, D. S., & Nishiguchi, M. K. (2009). Characterization of two host-specific genes, mannose-sensitive hemagglutinin (mshA) and uridyl phosphate dehydrogenase (UDPDH) that are involved in the *Vibrio fischeri–Euprymna tasmanica* mutualism. *FEMS Microbiology Letters*, Vol. 299, No. 1, pp. 65-73.

Boucher, Y., & Doolittle, W. F. (2000). The role of lateral gene transfer in the evolution of isoprenoid biosynthesis pathways. *Molecular Microbiology*, Vol. 37, No. 4, pp. 703-716.

Boucher, Y., & Stokes, H. W. (2006). The roles of lateral gene transfer and vertical descent in Vibrio evolution, In: *The Biology of Vibrios*. F. L. Thompson, B. Austin & J. Swings, pp. 84-94. ASM Press, Washington, D.C.

Boyd, E. F., & Brüssow, H. (2002). Common themes among bacteriophage-encoded virulence factors and diversity among the bacteriophages involved. *Trends in Microbiology*, Vol. 10, No. 11, pp. 521-529.

Boyd, E. F., Davis, B. M., & Hochhut, B. (2001). Bacteriophage-bacteriophage interactions in the evolution of pathogenic bacteria. *Trends in Microbiology*, Vol. 9, No. 3, pp. 137-144.

Brussow, H., Canchaya, C., & Hardt, W. D. (2004). Phages and the evolution of bacterial pathogens: from genomic rearrangements to lysogenic conversion. *Microbiology and Molecular Biology Reviews*, Vol. 68, No. 3, pp. 560-602.

Calendar, R. (2006). *The Bacteriophages*, Oxford University Press, Inc., New York, NY.

Campbell, A. (2003). The future of bacteriophage biology. *Nature Reviews Microbiology*, Vol. 4, No. 6, pp. 471-477.

Campbell, A. M. (2001). *Lysogeny*, John Wiley & Sons, Ltd.

Canchaya, C., Fournous, G., & Brüssow, H. (2004). The impact of prophages on bacterial chromosomes. *Molecular Microbiology*, Vol. 53, No. 1, pp. 9-18.

Canchaya, C., Fournous, G., Chibani-Chennoufi, S., Dillmann, M. L., & Brussow, H. (2003). Phage as agents of lateral gene transfer. *Current Opinion in Microbiology*, Vol. 6, No. 4, pp. 417-424.

Canchaya, C., Proux, C., Fournous, G., Bruttin, A., & Brussow, H. (2003). Prophage genomics. *Microbiology and Molecular Biology Reviews*, Vol. 67, No. 2, pp. 238-276.

Casjens, S. (2003). Prophages and bacterial genomics: what have we learned so far? *Molecular Microbiology*, Vol. 49, No. 2, pp. 277-300.

Chavez-Dozal, A., & Nishiguchi, M. K. (2011). Variation in biofilm formation among symbiotic and free-living strains of *Vibrio fischeri*. *Journal of Basic Microbiology*, Vol. 51, pp. 1-7.

Christensen, B. B., Sternberg, C., Andersen, J. B., Eberl, L., Moller, S., Givskov, M., & Molin, S. (1998). Establishment of new genetic traits in a microbial biofilm community. *Applied and Environmental Microbiology*, Vol. 64, No. 6, pp. 2247-2255.

Costerton, J. W., Cheng, K.-J., Geesey, G. G., Ladd, T. I., Nickel, J. C., Dasgupta, M., & Marrie, T. J. (1987). Bacterial biofilms in nature and disease. *Annual Review of Microbiology*, Vol. 41, No. 1, pp. 435-464.

Daubin, V., & Ochman, H. (2004). Bacterial genomes as new gene homes: the genealogy of ORFans in E. coli. *Genome Research*, Vol. 14, No. 6, pp. 1036-1042.

de Beer, D., Stoodley, P., Roe, F., & Lewandowski, Z. (1994). Effects of biofilm structures on oxygen distribution and mass transport. *Biotechnology and Bioengineering*, Vol. 43, No. 11, pp. 1131-1138.

Gevers, D., Cohan, F. M., Lawrence, J. G., Spratt, B. G., Coenye, T., Feil, E. J., Stackebrandt, E., Van de Peer, Y., Vandamme, P., Thompson, F. L., & Swings, J. (2005). Opinion: Re-evaluating prokaryotic species. *Nature Reviews Microbiology*, Vol. 3, No. 9, pp. 733-739.

Ghigo, J. M. (2001). Natural conjugative plasmids induce bacterial biofilm development. *Nature*, Vol. 412, No. 6845, pp. 442-445.

Gogarten, J. P., & Townsend, J. P. (2005). Horizontal gene transfer, genome innovation and evolution. *Nature Reviews Microbiology*, Vol. 3, No. 9, pp. 679-687.

Gómez, P., & Buckling, A. (2011). Bacteria-Phage Antagonistic Coevolution in Soil. *Science*, Vol. 332, No. 6025, pp. 106-109.

Guerrero-Ferreira, R. C., & Nishiguchi, M. K. (2010). Differential gene expression in bacterial symbionts from loliginid squids demonstrates variation between mutualistic and environmental niches. *Environmental Microbiology Reports*, Vol. 2, No. 4, pp. 514-523.

Hansen, S. K., Rainey, P. B., Haagensen, J. A., & Molin, S. (2007). Evolution of species interactions in a biofilm community. *Nature*, Vol. 445, No. 7127, pp. 533-536.

Hausner, M., & Wuertz, S. (1999). High rates of conjugation in bacterial biofilms as determined by quantitative in situ analysis. *Applied and Environmental Microbiology*, Vol. 65, No. 8, pp. 3710-3713.

Heithoff, D. M., Conner, C. P., Hanna, P. C., Julio, S. M., Hentschel, U., & Mahan, M. J. (1997). Bacterial infection as assessed by in vivo gene expression. *Proceedings of the National Academy of Sciences*, Vol. 94, No. 3, pp. 934-939.

Jones, B. W., & Nishiguchi, M. K. (2006). Differentially expressed genes reveal adaptations between free-living and symbiotic niches of *Vibrio fischeri* in a fully established mutualism. *Canadian Journal of Microbiology*, Vol. 52, No. 12, pp. 1218-1227.

Krisch, H. M. (2003). The view from Les Treilles on the origins, evolution and diversity of viruses. *Research in Microbiology*, Vol. 154, No. 4, pp. 227-229.

Lee, S. H., Hava, D. L., Waldor, M. K., & Camilli, A. (1999). Regulation and temporal expression patterns of *Vibrio cholerae* virulence genes during infection. *Cell*, Vol. 99, No. 6, pp. 625-634.

Levin, B. R., & Bull, J. J. (2004). Population and evolutionary dynamics of phage therapy. *Nature Reviews Microbiology*, Vol. 2, No. 2, pp. 166.

Matz, C., Bergfeld, T., Rice, S. A., & Kjelleberg, S. (2004). Microcolonies, quorum sensing and cytotoxicity determine the survival of Pseudomonas aeruginosa biofilms exposed to protozoan grazing. *Environmental Microbiology*, Vol. 6, No. 3, pp. 218-226.

Matz, C., & Jurgens, K. (2005). High motility reduces grazing mortality of planktonic bacteria. *Applied and Environmental Microbiology*, Vol. 71, No. 2, pp. 921-929.

Matz, C., & Kjelleberg, S. (2005). Off the hook--how bacteria survive protozoan grazing. *Trends in Microbiology*, Vol. 13, No. 7, pp. 302-307.

McFall-Ngai, M. J. (1994). Animal-Bacterial Interactions in the early life history of marine invertebrates: The *Euprymna scolopes/Vibrio fischeri* symbiosis. *American Zoologist*, Vol. 34, No. 4, pp. 554-561.

McFall-Ngai, M. J. (1998). The development of cooperative associations between animals and bacteria: establishing detente among domains. *American Zoologist*, Vol. 38, No. 4, pp. 593-608.

O'Toole, G., Kaplan, H. B., & Kolter, R. (2000). Biofilm formation as microbial development. *Annual Review of Microbiology*, Vol. 54, No. 1, pp. 49-79.

Oakey, H. J., & Owens, L. (2000). A new bacteriophage, VHML, isolated from a toxin-producing strain of *Vibrio harveyi* in tropical Australia. *Journal of Applied Microbiology*, Vol. 89, No. 4, pp. 702-709.

Ochman, H., Lawrence, J. G., & Groisman, E. A. (2000). Lateral gene transfer and the nature of bacterial innovation. *Nature*, Vol. 405, No. 1, pp. 299-304.

Pal, C., Papp, B., & Lercher, M. J. (2005). Adaptive evolution of bacterial metabolic networks by horizontal gene transfer. *Nature Genetics*, Vol. 37, No. 12, pp. 1372-1375.

Rajadhyaksha, A. B., & Rao, S. S. (1965). The role of phage in the transduction of the toxinogenic factor in *Corynebacterium diphtheriae*. *Journal of General Microbiology*, Vol. 40, No. 3, pp. 421-429.

Shemesh, M., Tam, A., Aharoni, R., & Steinberg, D. (2010). Genetic adaptation of Streptococcus mutans during biofilm formation on different types of surfaces. *BMC Microbiology*, Vol. 10, No. 1, pp. 51.

Smets, B. F., & Barkay, T. (2005). Horizontal gene transfer: perspectives at a crossroads of scientific disciplines. *Nature Reviews Microbiology*, Vol. 3, No. 9, pp. 675-678.

Sorensen, S. J., Bailey, M., Hansen, L. H., Kroer, N., & Wuertz, S. (2005). Studying plasmid horizontal transfer in situ: a critical review. *Nature Reviews Microbiology*, Vol. 3, No. 9, pp. 700-710.

Soto, W., Gutierrez, J., Remmenga, M. D., & Nishiguchi, M. K. (2009). Salinity and temperature effects on physiological responses of Vibrio fischeri from diverse ecological niches. *Microbial Ecology*, Vol. 57, No. 1, pp. 140-150.

Steinert, M., Hentschel, U., & Hacker, J. (2000). Symbiosis and pathogenesis: evolution of the microbe-host interaction. *Naturwissenschaften*, Vol. 87, No. 1, pp. 1-11.

Stewart, P. S., & Franklin, M. J. (2008). Physiological heterogeneity in biofilms. *Nature Reviews Microbiology*, Vol. 6, No. 3, pp. 199-210.

Takeda, Y., & Murphy, J. R. (1978). Bacteriophage conversion of heat-labile enterotoxin in *Escherichia coli*. *Journal of Bacteriology*, Vol. 133, No. 1, pp. 172-177.

Thomas, C. M., & Nielsen, K. M. (2005). Mechanisms of, and barriers to, horizontal gene transfer between bacteria. *Nature Reviews Microbiology*, Vol. 3, No. 9, pp. 711-721.

Thompson, J. R., Pacocha, S., Pharino, C., Klepac-Ceraj, V., Hunt, D. E., Benoit, J., Sarma-Rupavtarm, R., Distel, D. L., & Polz, M. F. (2005). Genotypic diversity within a natural coastal bacterioplankton population. *Science*, Vol. 307, No. 5713, pp. 1311-1313.

Ventura, M., Canchaya, C., Pridmore, D., Berger, B., & Brussow, H. (2003). Integration and distribution of Lactobacillus johnsonii prophages. *Journal of Bacteriology*, Vol. 185, No. 15, pp. 4603-4608.

Waldor, M. K. (1998). Bacteriophage biology and bacterial virulence. *Trends in Microbiology*, Vol. 6, No. 8, pp. 295-297.

Waldor, M. K., & Mekalanos, J. J. (1996). Lysogenic conversion by a filamentous phage encoding cholera toxin. *Science*, Vol. 272, No. 5270, pp. 1910-1914.

Wildschutte, H., Wolfe, D. M., Tamewitz, A., & Lawrence, J. G. (2004). Protozoan predation, diversifying selection, and the evolution of antigenic diversity in Salmonella. *Proceedings of the National Academy of Sciences of the United States of America*, Vol. 101, No. 29, pp. 10644-10649.

Xu, K. D., Stewart, P. S., Xia, F., Huang, C. T., & McFeters, G. A. (1998). Spatial physiological heterogeneity in Pseudomonas aeruginosa biofilm is determined by oxygen availability. *Applied and Environmental Microbiology*, Vol. 64, No. 10, pp. 4035-4039.

2

Host-Plant Specialisation and Diurnal Dynamics of the Arthropod Community Within *Muhlenbergia robusta* (Poaceae)

Víctor López-Gómez and Zenón Cano-Santana
Facultad de Ciencias, Universidad Nacional Autónoma de México
Mexico

1. Introduction

It is well-known that herbivorous insects are very specialised in terms of their food. It has been reported that these insects feed on only one or a few genera of plants, even within a single family (Bernays & Graham, 1988; Schoonhoven et al., 2005). Certain factors have been found to be decisive in determining the range of hosts of herbivorous insects. Among the most important are (1) the secondary compounds in the plants, (2) the presence of predators, and (3) the insects' mating behaviours. These factors are discussed below.

Secondary compounds are one of the most effective strategies that plants use to avoid predation by herbivores, for example, as toxins or in feeding deterrents that kill insects or slow their rates of development (Lill & Marquis, 2001; Schowalter, 2006). The noxious effects of secondary compounds on insects are crucial to the preferences of feeding insects, and therefore, the ranges of the host plants of phytophagous insects (Bernays & Graham, 1988; Cates, 1980).

Natural enemies can influence the host ranges of phytophagous specialists. Moreover, it has been proposed that species seek out enemy-free spaces to reduce their mortality (Gilbert & Singer, 1975; Lawton, 1978). In fact, Price et al. (1980) recorded insect herbivores that changed their host plant to a new toxic plant that provided protection against enemies.

The literature describes some phytophagous insects that restrict their host range to comply with patterns of mate-finding behaviour. This is true even in plants that do not have a relationship with the food preferences of insects (Labeyrie, 1978).

Conversely, arthropod predators are generalised in their food selection (Sabelis, 1992). For this reason, habitat selection by arthropods depends on the services that the habitat provides to increase their chances of survival. It has been reported that the abundance of arthropod predators within plant communities is related to habitats offering (1) abundant prey; (2) refuge from predation, e.g., cannibalism and intraguild predation; (3) easier and more effective spotting and capture of prey; (4) a more favourable microclimate; and (5) access to alternative resources (Langellotto & Denno, 2004).

In spite of the knowledge gathered about host-plant specialisations within several arthropod species (Bernays & Graham, 1988; Cates, 1980; Feeny, 1976), there is little information about the level of arthropod specialisation in the use of their habitats at the community level (i.e., species that carry out all of their activities on the host plant). Descriptions of arthropod communities on host plants assume that all species have the same level of specialisation in the use of their habitat. To address this theoretical problem, only the most abundant taxa of the community have been studied. Previous studies about the ratio of arthropods with high specialisation in the use of their habitats are difficult to find. This kind of research could provide important data about the dynamics of the arthropod community on their host plant and about the possible main flows of matter and energy within arthropod-plant ecosystems.

The main goals of this study are (1) to determine the ratio of species in the arthropod community in a grass ecosystem (*Muhlenbergia robusta,* Poaceae) with high levels of specialisation in the use of the host plant (i.e., species that carry out all their activities on the host plant), by studying arthropod communities with similar habitats (i.e., herbaceous patches and litter) at four different times throughout the day, and (2) to determine the diurnal variation of the arthropod community structure (richness, abundance, index of diversity and composition) in three different herbaceous habitats (*M. robusta,* herbaceous patches, and litter).

2. Methods

2.1 Area of study

This study was carried out in the Reserva Ecológica del Pedregal de San Ángel (REPSA) (19°19′N, 99°11′W), which is located on the main campus of the Universidad Nacional Autónoma de México, southwest of Mexico City. This ecological reserve has an area of 237 ha and an elevation of 2300 m. The vegetation of the reserve can be characterised as a xerophilous scrubland, and the area has a sub-humid climate. This site has an annual mean temperature of 16.1 °C, and its annual mean rainfall is 835 mm (César-García, 2002). The reserve has a wet season between May and October. The area is located over a basaltic substratum that was deposited 1650 to 2000 years ago during the eruption of the Xitle volcano (Carrillo, 1995). Most plant species are herbaceous or shrub-like; however, there are a few small trees from 3 to 7 m in height.

2.2 Study system

Muhlenbergia robusta (Fourn.) Hitchc. (Poaceae) is a perennial grass 1 to 2 m tall. This plant accounts for approximately 15% of the aboveground net primary productivity in the REPSA (Cano-Santana, 1994). This plant flowers between June and August and bears fruit between September and June (César-García, 2002). It has a distribution between 2250 and 3200 m in elevation (Rzedowski & Rzedowski, 2001).

2.3 Collection

With each collection effort, we had the intention to trap as many arthropods as possible in each of the study habitats; unfortunately, the heterogeneous geomorphology in the REPSA did not allow for the use of the same trapping technique in each habitat. For this reason, the most suitable technique for each kind of habitat (*M. robusta,* herbaceous patches and litter) was used.

Twenty-four *M. robusta* plants —which showed approximately 48 to 73 cm of diameter at ground level—were collected at random at 4 different times of day (0100 to 0300 h, 0700 to 0900 h, 1300 to 1500 h and 1900 to 2100 h). Six grasses were collected during each time period. The collection took place in July 2006 in a large site in the nuclear zone of the REPSA

with the presence of some trees. Each selected grass plant was completely wrapped and protected with a plastic bag in the field and was later extracted using a pick and shovel.
To obtain an authentic epiphytic arthropod community from the herbaceous patches of each grass, an entomological net was struck ten times in the four nearest patches where the herbaceous patches were dominant and *M. robusta* was not present.
To acquire the arthropods associated with a litter habitat, for each plant, the litter of the four nearest patches without vegetation was collected using a 24 cm diameter circle as a sampling unit.
On the same day of the collection, the three habitat samples (*M. robusta*, herbaceous patches, and litter) were taken to the laboratory, where arthropods were manually extracted from them. Only organisms ≥ 3 mm in corporal length were considered. Extracted fauna were initially sorted into morphospecies, a common practice in biodiversity studies that does not compromise scientific accuracy (Oliver & Beattie, 1996) and has some clear advantages when expertise in all taxonomic groups is not available (Gaston, 1996). The morphospecies were identified and then sent to several taxonomists for species identification. The community attributes of each sample were recorded considering richness per plant, abundance per plant, and diversity. Diversity was recorded using the Shannon-Wiener index with a natural logarithm (H').
The aboveground dry weight of each plant was obtained by drying the plant in an electric oven at 50°C to a constant weight, and the plants were then weighed using an analytical balance (Ohaus AV812, ± 0.005 g).
In August 2007, the relative coverages of the principal landscape elements in the site were determined (*i.e.*, *M. robusta*, herbaceous patches, litter, exposed rock, and shrubbery and arboreal plants) using Canfield´s method with two lines of 8 m that traversed the site.

2.4 Statistical analysis

To determine the effects of the sampling schedule (0100, 0700, 1300, and 1900 h) and the type of habitat (grass, herbaceous patches and litter) on the community attributes (richness, abundance, and H'), two-way factorial ANOVA were calculated (Zar, 2010). Tukey's multiple comparison tests were then done on significant ANOVA tests. Richness and abundance were transformed using the equation

$$X' = \sqrt{0.5 + X} \qquad (1)$$

because they are discrete variables (Zar, 2010). Statistical analyses were conducted with Statistica software (StatSoft, 2007).
To determine the effect of the kind of habitat or the schedules of collection on the composition of arthropod communities, a principal component analysis (PCA) was calculated with Prime software (PRIMER-E, 2001).
To determine the similarity of species composition among the different communities, Jaccard's index of similarity was applied, considering the twelve treatments (four schedules × three habitats).

3. Results

On the 24 grasses, a total of 139 arthropod taxa and 1529 individuals were found; the herbaceous patch sampling registered 150 arthropod taxa and 1594 individuals; the litter sampling found 60 arthropod taxa and 248 individuals.

Two-way factorial ANOVA tests showed a significant effect of habitat type, the hour of sample collection, and the type of habitat × the hour of collection on richness ($F_{3,\ 60}$=8.1, P=0.001; $F_{2,\ 60}$=84.1, P<0.001; $F_{6,\ 60}$=7.6, P<0.001, respectively), abundance ($F_{3,\ 60}$=3.1, P=0.03; $F_{2,\ 60}$=41.9, P<0.001; $F_{6,\ 60}$=4.1, P=0.001, respectively), and index of diversity ($F_{3,\ 60}$=7.5, P<0.001; $F_{2,\ 60}$=40.4, P<0.001; $F_{6,\ 60}$=4.0, P<0.001, respectively).

Different schedules did not result in significant changes in arthropod mean richness or abundance or in *H'* in *M. robusta* (Figs. 1A, B, C). In contrast, the lowest richness and abundance averages on the herbaceous patch habitats were at 0100 h. At 0700 h, they showed a sudden increase, and at 1300 h and 1900 h, they showed a slight decrease (Figs. 1A, B). The mean of the arthropod index of diversity was unchanged at different times in the herbaceous habitat patches (Fig. 1C). In the litter habitat, the lowest arthropod richness and diversity averages were at 1300 h, and the highest averages were at 0700 h (Figs. 1A, C). Abundance averages were constant at different times (Fig. 1B).

In the *M. robusta* habitat, the highest average abundance of Formicidae was at 1300 h. Diplopoda, Coleoptera, Hemiptera, Blattodea, and Araneae did not show a clear peak average abundance (Fig. 2A). Orthoptera, Homoptera, Coleoptera, and Diptera showed their lowest average abundance at 0100 h in the herbaceous patch habitat. Later, these taxonomic groups increased their average abundances between 0700 h and 1300 h, and then they showed a slight decrease at 1900 h (Fig. 2B). There was a clear peak of abundance at 0700 h for Coleoptera, Araneae, Formicidae, and Diplopoda in the litter habitat, whereas Chilopoda did not appear (Fig. 2C).

There were six taxa that appeared only in *M. robusta*, four of which were registered at all times (Thomisidae 10: Araneae, *Phlegyas* sp.: Hemiptera, Armadillidiidae 2: Isopoda, *Blatta* sp.: Blattodea), and two of which were registered at three times (*Novalene* sp.: Araneae and *Dinocheirus tenoch* Chamberlin 1929: Pseudoscorpiones). There were three exclusive taxa for the herbaceous stratum habitat, of which only Cicadellidae 10 (Homoptera) was present at all times. In the litter habitat, Chrysomelidae 12 (Coleoptera) was present at almost all times, except at 1300 h.

Sphenarium purpurascens Charpentier 1842 (Orthoptera), *Crematogaster* sp. (Formicidae), and Melyridae 14 (Coleoptera) were registered on *M. robusta* and in herbaceous patch habitats. Polydesmida 3 (Diplopoda), *Paratrechina* sp. (Formicidae) and Coleoptera 41 were registered on *M. robusta* and in litter habitats. There were no taxa found both on herbaceous patches and in litter habitats; there were no fauna that used all three kinds of habitats.

PCA shows that the arthropod communities have more similarity in their assemblage by the type of habitat (*M. robusta*, herbaceous stratum, and litter) (Fig. 3) than by the time of collection (0100 to 0300 h, 0700 to 0900 h, 1300 to 1500 h and 1900 to 2100 h) (Fig. 4).

The highest Jaccard similarity index among arthropod communities within *M. robusta* at different local times was between 0700 h and 1700 h, and the lowest was between 0100 h and 1300 h (Table 1). The average of all similarity indices was 0.59 (± 0.04 SE).

	Local solar time (h)		
	0100	**0700**	**1300**
0700	0.603		
1300	0.448	0.566	
1900	0.673	0.717	0.518

Table 1. Jaccard similarity indices among arthropod communities within *M. robusta* at different local times solar hours (0100, 0700, 1300, 1900). Collection: July 2006, Reserva Ecológica del Pedregal de San Ángel, Mexico City.

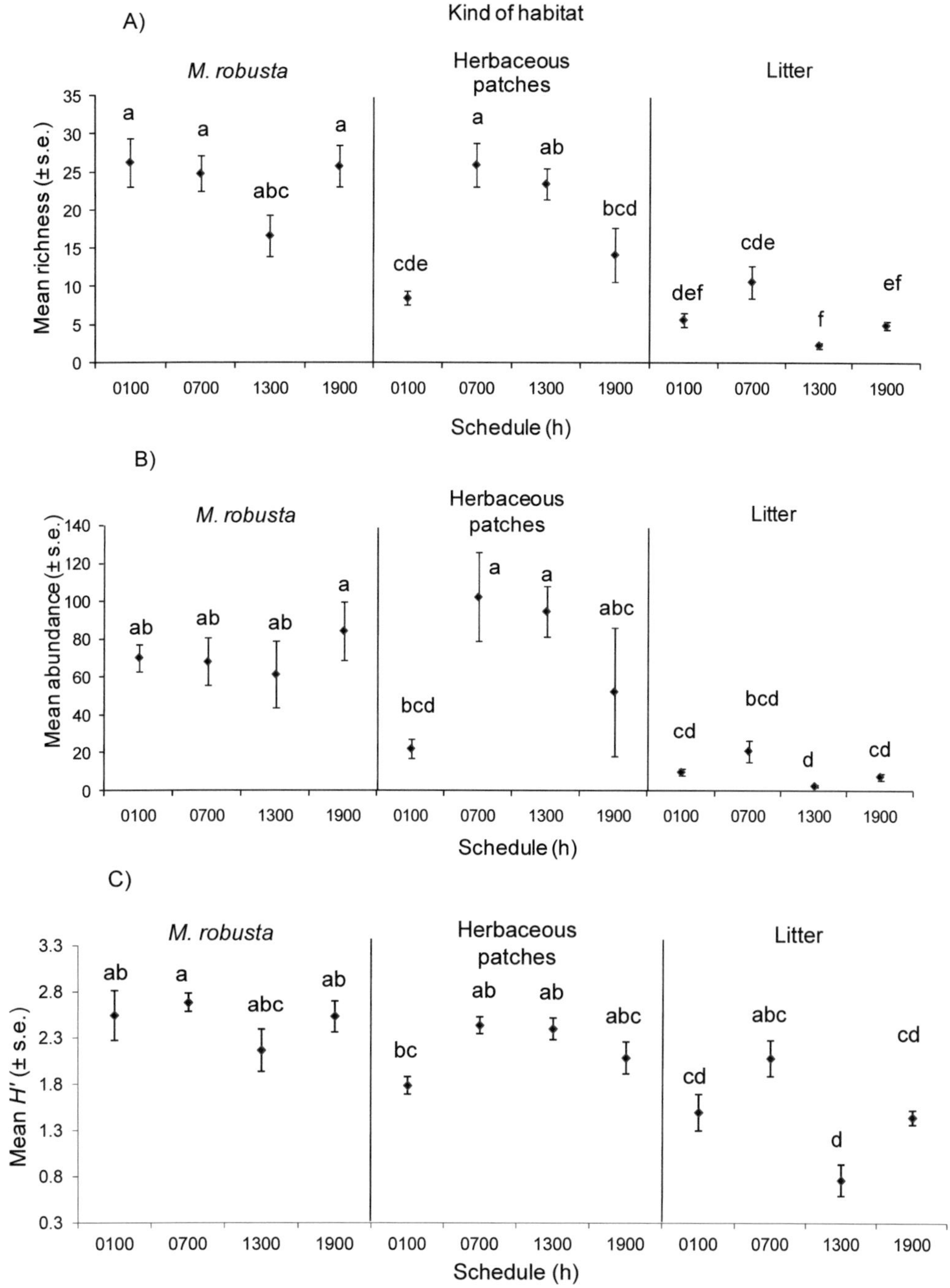

Fig. 1. Arthropod-fauna average richness (A), abundance (B), and index of diversity (C) (H') in three different kinds of habitats (*M. robusta*, herbaceous patches and litter) during four different sampling times (0100, 0700, 1300, 1900 h). Collection: July 2006, Reserva Ecológica del Pedregal de San Ángel, Mexico City. Letters denote significant differences (α=0.05). Values are means ± SE

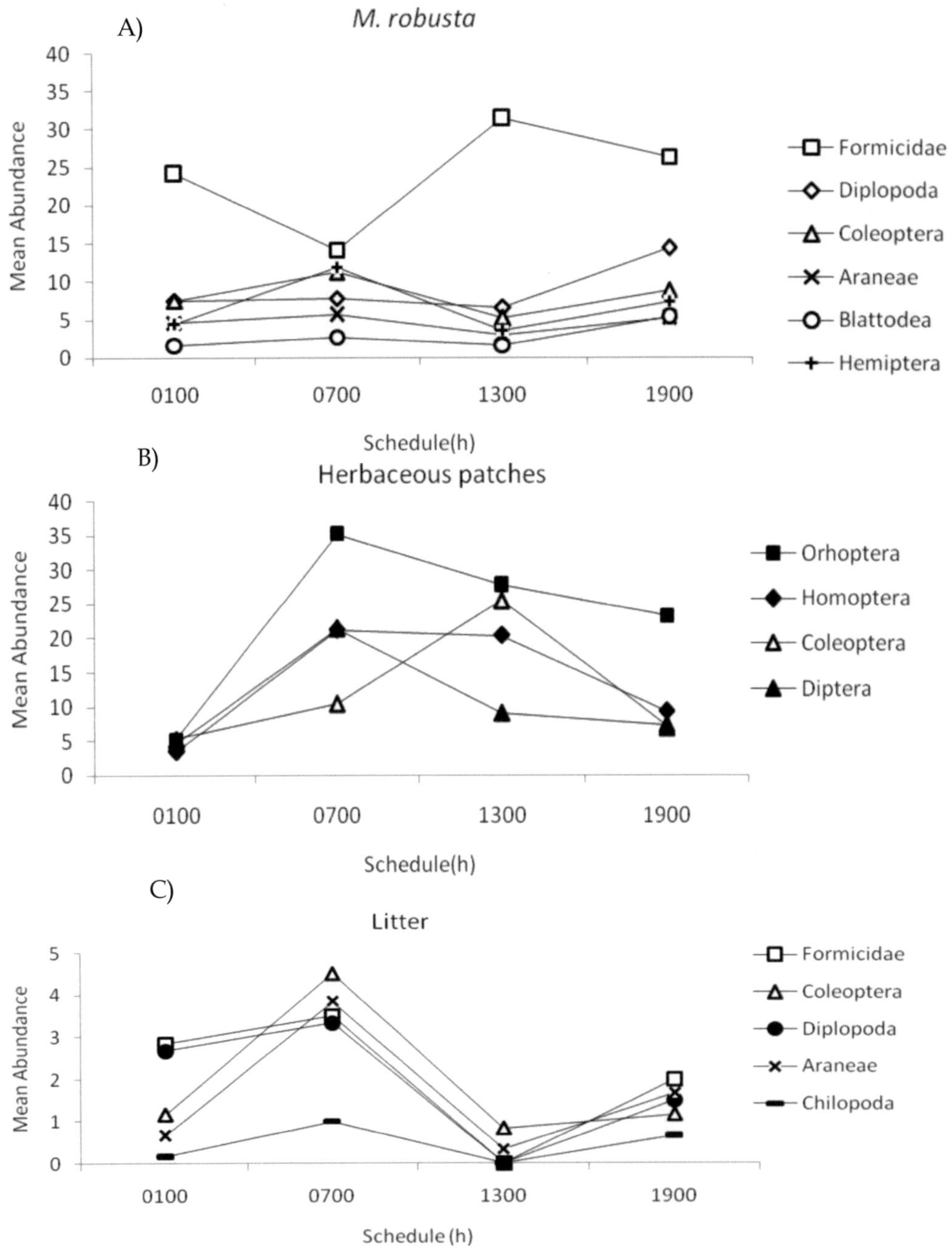

Fig. 2. Taxonomic group mean abundances during four different schedules (0100, 0700, 1300, 1900 h) in three different habitats: *Muhlenbergia robusta* (A), herbaceous patches (B), and litter (C). For the *M. robusta* and herbaceous patches habitats, only the taxonomic groups with an average abundance of greater than 5 are shown. Collection: July 2006, Reserva Ecológica del Pedregal de San Ángel, Mexico City

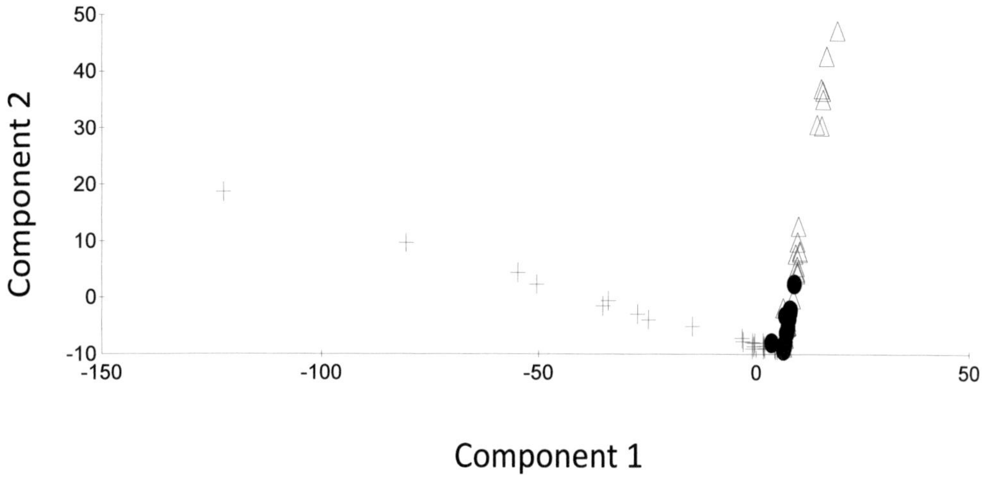

Fig. 3. PCA diagram that shows the arthropod communities of three kinds of habitats: *M. robusta* (+), herbaceous patches (Δ) and litter (•). Collection: July 2006, Reserva Ecológica del Pedregal de San Ángel, Mexico City

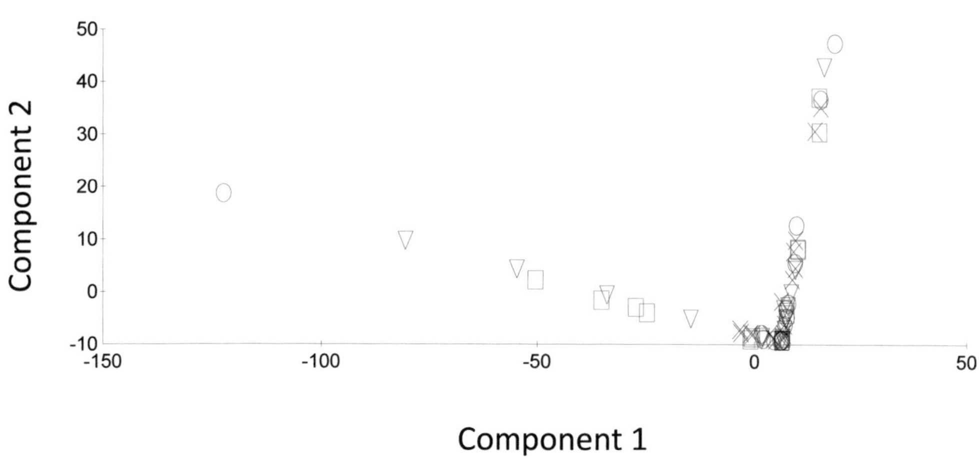

Fig. 4. PCA diagram that shows the arthropod communities at four times throughout the day: 0100 to 0300 h (×), 0700 to 0900 h (∇), 1300 to 1500 h (□) and 1900 to 2100 h (○). Collection: July 2006, Reserva Ecológica del Pedregal de San Ángel, Mexico City

Canfield's method showed that *M. robusta*'s presence was the most dominant at the site of the study. Following that, in order of importance, were the herbaceous patches, litter, exposed rock, and finally, shrubbery and trees (Fig. 5).

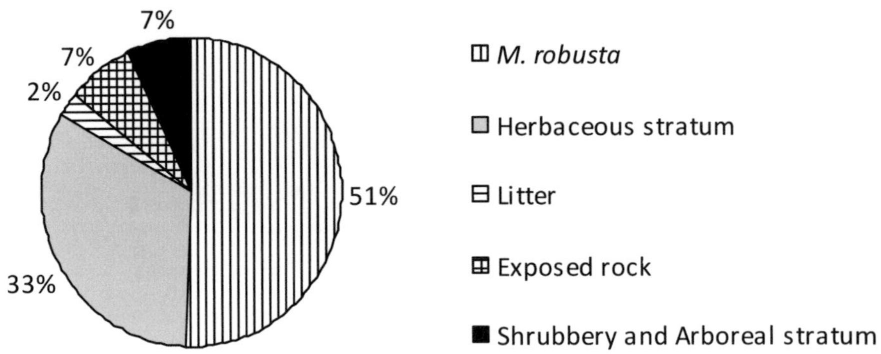

Fig. 5. Relative coverage of *M. robusta* and landscape elements (herbaceous patches, litter, exposed rock, and shrubbery and tree plants) in a sunny location at the Reserva Ecológica del Pedregal de San Ángel, Mexico City. Collection, August 2007

4. Discussion

Only 4.3% of the arthropod taxa (six morphospecies) were specialised on *M. robusta* in the use of habitat, which suggests that they carry out most of their activities (foraging, hiding and meeting) within this grass. These specialist arthropods showed the main functional groups in an ecosystem: herbivorous (*Phlegyas* sp.: Hemiptera), saprophagous (Armadillidiidae 2: Isopoda and *Blatta* sp.: Blattodea) and predatory (Thomisidae 10: Araneae, *Novalene* sp.: Araneae and *Dinocheirus tenoch*: Pseudoscorpiones). This suggests that grass conditions offer most of the requirements of these taxa in a microhabitat, i.e., alternative prey or food resources and refuge from predation.

Study results indicate that *Phlegyas* sp. (Hemiptera) could be a probable phytophagous specialist feeding on this grass. As the literature has reported, herbivorous insects are very specialised in the selection of their food (Bernays & Graham, 1988). The three predators with significant habitat specialisation on grass (Thomisidae 10: Araneae, *Novalene* sp.: Araneae and *Dinocheirus tenoch*: Pseudoscorpiones) show signs that the *M. robusta* structure facilitates their hunting strategies and provides suitable refuge for avoiding predation (Langellotto & Denno, 2004). It was observed that the habitat structure of the host plant can influence a community of spiders in plants. This was shown through a robust pattern of growth in the natural enemies of arthropods (hemipterans, mites, parasitoids and spiders) in complex structural habitats. These complex habitats provide a broad range of favourable conditions that attract natural enemies and decrease the need to move in search of more suitable conditions (Sunderland & Samu, 2000). In the same way, the two saprophagous taxa specialists on *M. robusta* (Armadillidiidae 2 and *Blatta* sp.) indicate that the layer of dead organic matter typical on *M. robusta* (located in its base at ground level) could be an appropriate source of food and protection against predators (Jabin et al., 2004; Schmidt et al., 2005).

Most of the arthropod community taxa within *M. robusta* (i.e., 133 morphospecies) were generalised in their use of the different available herbaceous habitats. This could be attributed to the great variety of life forms and requirements that are characteristic of the Phylum Arthropoda. These organisms can be categorised as (1) taxa with a regular association with *M. robusta* and (2) taxa that use *M. robusta* and other herbaceous habitats.

One example of a taxon with a regular association with this grass could be *Sphenarium purpurascens* (Orthoptera), a grasshopper that eats the pollen and fruit of *M. robusta* (Mendoza & Tovar-Sánchez, 1996). Results show that this orthopteran was found in herbaceous patches at all times but was recorded in *M. robusta* only at 1300 h. This grasshopper likely forages on the reproductive structures of the grass only at this specific hour of the day because of favourable environmental conditions, as has been recorded for other floral visitors in this ecological reserve (Figueroa-Castro & Cano-Santana, 2004).

As an example of a taxon that uses *M. robusta* along with other herbaceous habitats, Polydesmida 3 (Diplopoda) was registered in the grass at all times and in the litter habitat at three times. This can be interpreted to mean that saprophagous insects use these two habitats simultaneously because they offer food and refuge against adverse conditions. Other studies have also registered a direct relationship between saprophagous abundance and the amount of litter available (Jabin et al., 2004; Schmidt et al., 2005).

Apparently, there is no taxon that uses all three kinds of habitats. However, there are arthropods that likely use all of the described habitats. Of these, most are probably fliers. Unfortunately, their numbers could not be recorded because of their high mobility and the limitations of our sampling techniques.

The *M. robusta* habitat had the greatest coverage of all the landscape types (51%), which explains the richness and abundance of the arthropods (139 taxa and 1529 individuals) found within this habitat. This landscape provides a greater quantity and variety of habitats as well as resources for the fauna. Similarly, species-area relationship (SAR) has described a direct link between the richness of arthropods and the extension of their host plant distribution (Lawton, 1978; Marshall & Storer, 2006; Ozanne et al., 2000; Southwood et al., 1982).

Despite the low coverage (33%) of the herbaceous patches, this habitat shows the highest arthropod richness (150 taxa) in comparison with the other two kinds of habitats. This could be because the herbaceous patches habitat comprises many species of plants that offer a greater variety of habitats and food for the arthropod community; this permits the establishment of more species with diverse requirements (Symstad et al., 2000).

Results show that the structure of the arthropod community within *M. robusta* is constant throughout the day, based on (1) the richness and abundance per plant and the diversity (H') and (2) the stable abundances of the principal taxonomic groups within the grass throughout the day. However, Jaccard indices and PCA indicate that arthropod communities' assemblages change throughout the day (Fig. 4). This suggests that all of the available habitats for arthropods in *M. robusta* are fully occupied all day long and that arthropod communities within *M. robusta* have a particular assemblage with a remarkable turnover of species (41%).

PCA indicates that those arthropod communities within *M. robusta* present a remarkably different species composition compared to herbaceous patches and litter habitats. This could be explained because *M. robusta* offers diverse (1) microclimatic conditions, (2) types of resources and (3) interactions with other species. These factors are decisive in determining the establishment of species (Begon et al., 2006).

In the herbaceous patches habitat, arthropod richness and abundance—and the abundance of the principal taxonomic groups—showed a sudden increase at 0700 h. Following that, the recorded numbers decreased gradually. This indicates that arthropods experience a peak of activity at 0700 h in this habitat. These results agree with a study of arthropod floral visitor activity of four Compositae plants (*Eupatorium petiolare, Dahlia coccinea, Tagetes lunulata* and *Verbesina virgata*) in the REPSA (Figueroa-Castro & Cano-Santana, 2004). These authors found that the highest frequency of visits of anthophilous arthropods was between 0845 and 1645 h. The number of arthropod visitors on flowers was related to higher temperatures and lower relative humidity levels, which is directly related to arthropods' physiological responses to the environment.

In comparison to other habitats, the litter habitat showed the lowest richness, abundance, and diversity. This may be true because, for the majority of the arthropod community, this habitat is used only as a pass-through location for dispersion; the results show that the peak of arthropod mobility is at 0700 h. Moreover, this habitat represents an exposed location to predators because of the absence of vegetation; nevertheless, records indicate that it could be an appropriate habitat for saprophagous arthropods. Another reason could be the differences in the sample techniques for arthropod collection. The sample sizes for the three techniques were designed to achieve equality between them, but the lower records for the litter habitat could signify that the sample size should have been bigger for this habitat.

We are conscious that our results have limitations in their interpretation because of the difficulty in comparing these arthropod communities from different habitats when different trapping techniques were used. However, this study provides an approach to determining the level of specialisation of the arthropod community to a host plant and shows the diurnal dynamics of the whole arthropod community within a plant; both of these aspects of arthropod ecology have been little studied. For future studies that will try to corroborate our records, it may be appropriate to use an identical trapping technique on each of the treatments, if possible.

5. Conclusions

We conclude that the ratio of arthropod species with a high level of specialisation in the use of the *M. robusta* host plant was very low (4.3%). Furthermore, the structure of the arthropod community (richness, abundance, and index of diversity) in the grass was constant throughout the day, although the diurnal variation in the community assemblage shows a remarkable change (41%).

6. Acknowledgements

We thank Santiago Zaragoza Caballero, Harry Brailowsky, Rafael Gaviño Rojas, Iván Castellanos Vargas, and Cristina Mayorga Martínez for the identification of specimens. Thanks to Marco Romero-Romero for technical laboratory support. We thank Susana

Alejandre Grimaldo and Erick Daniel Villamil for field assistance. VLG received a scholarship from CONACYT-Mexico within the Posgrado en Ciencias Biológicas.

7. References

Begon, M., Townsend, C. R. & Harper, J. L. (2006). *Ecology: from individuals to ecosystems.* (4th), Blackwell Science, 978-1-4051-1117-1, Oxford.

Bernays, E. & Graham, M. (1988). On the evolution of host specificity in phytophagous arthropods. *Ecology*, Vol. 69, No. 4, (August 1988), pp. (886-892), 0012-9658.

Cano-Santana, Z. (1994). *Flujo de energía a través de Sphenarium purpuracens (Orthoptera: Acrididae) y productividad primaria neta aérea en una comunidad xerófila.* Doctoral thesis, Universidad Nacional Autónoma de México, México.

Carrillo, C. (1995). *El Pedregal de San Ángel.* (1st), UNAM, 968-36-4786-3, México.

Cates, R. G. (1980). Feeding patterns of monophagous, oligophagous and polyphagous insect herbivores: the effect of resource abundance and plant chemistry. *Oecologia*, Vol. 46, No. 1, (july 1980), pp. (22-31), 1432-1939.

César-García, S. B. (2002). *Análisis de algunos factores que afectan la fenología reproductiva de la comunidad vegetal de la Reserva del Pedregal de San Ángel.* Bachelor thesis, Universidad Nacional Autónoma de México, México.

Feeny, P. P. (1976). Plant apparency and chemical defense, In: *Biochemical interactions between plants and insects*, Wallace, J. W. & Mansell, R. L., pp. (1-40), Plenum Press, 0-306-34710-5, New York.

Figueroa-Castro, D. M. & Cano-Santana, Z. (2004). Floral visitor guilds of five allochronic flowering Asteraceous species in a xeric community in Central Mexico. *Environmental Entomology*, Vol. 33, No. 2, (April 2004), pp. (297-309), 1938-2936.

Gaston, K. J. (1996). *Biodiversity: A Biology of Numbers and Difference.* Blackwell, Cambridge.

Gilbert, L. E. & Singer, M. C. (1975). Butterfly ecology. *Annual Review of Ecology and Systematics*, Vol. 6, No. 1, (November 1975), pp. (365-397), 0066-4162.

Jabin, M., Mohr, D., Kappes, H. & Topp, W. (2004). Influence of deadwood on density of soil macro-arthropods in a managed oak-beech forest. *Forest Ecology and Management*, Vol. 194, No. 1-3, (June 2004), pp. (61-69), 0378-1127.

Labeyrie, V. (1978). Reproduction of insects and co-evolution of insects and plants. *Entomologia Experimentalis et Applicata*, Vol. 24, No. 3, (November 1978), pp. (496-504), 1570-7458.

Langellotto, G. A. & Denno, R. F. (2004). Responses of invertebrate natural enemies to complex-structured habitats: a meta-analytical synthesis. *Oecologia*, Vol. 139, No. 1, (February 2004), pp. (1-10), 0029-8549.

Lawton, J. H. (1978). Host-plant influences on insect diversity: the effects of space and time, In: *Symposia of the Royal Entomological Society of London 9, Diversity of insects faunas*, Mound, L. A. & Waloff, N., pp. (105-125), Blackwell Scientific Publications, 0-12-713750-5, Oxford.

Lill, J. T. & Marquis, R. J. (2001). The effects of leaf quality on herbivore performance and attack from natural enemies. *Oecologia*, Vol. 126, No. 3, (February 2001), pp. (418-428), 0029-8549.

Marshall, J. M. & Storer, A. J. (2006). Influence of *Centaurea biebersteinii* patch size on *Urophora quadrifasciata* (Dipt. Tephritidae) in Michigan, USA. *Journal of Applied Entomology*, Vol. 130, No. 2, (March 2006), pp. (91-95), 1439-0418.

Mendoza, C. P. & Tovar-Sánchez, E. (1996). *Ecología de forrajeo de Sphenarium purpurascens (Orthoptera: Acrididae) en la Reserva del Pedregal de San Ángel, D.F. México*. Bachelor thesis, Universidad Nacional Autónoma de México, México.

Oliver, I. & Beattie, A. J. (1996). Invertebrate morphospecies as surrogates for species: a case study. *Conservation Biology*, Vol. 10, No. 1, (February 1996), pp. (99-109),

Ozanne, C. M. P., Speight, M. R., Hambler, C. & Evans, H. F. (2000). Isolated trees and forest patches: patterns in canopy arthropod abundance and diversity in *Pinus sylvestris* (Scots Pine). *Forest Ecology and Management*, Vol. 137, No. 1-3, (October 2000), pp. (53-63), 0378-1127.

Price, P. W., Bouton, C. E., Gross, P., McPheron, B. A., Thompson, J. N. & Weis, A. E. (1980). Interactions among three trophic levels: influence of plants on interactions between insect herbivores and natural enemies. *Annual Review of Ecology and Systematics*, Vol. 11, No. 1, (November 1980), pp. (41-65), 0066-4162.

PRIMER-E. 2001. PRIMER 5. version 5.2.8.

Rzedowski, G. C. & Rzedowski, J. (2001). *Flora fanerogámica del Valle de México*. (2nd), Instituto de Ecología, 970-9000-17-9, Pátzcuaro.

Sabelis, M. W. (1992). Predatory arthropods, In: *Natural enemies: the population biology of predators, parasites and diseases*, Crawley, M. J., pp. (225-264), Blackwell Scientific, 0-632-02698-7, Oxford.

Schmidt, M. H., Lefebvre, G., Poulin, B. & Tscharntke, T. (2005). Reed cutting affects athropod communities, potentially reducing food for passerine birds. *Biological Conservation*, Vol. 121, No. 2, (January 2005), pp. (157-166), 0006-3207.

Schoonhoven, L. M., van Loon, J. J. A. & Dicke, M. (2005). *Insect-plant Biology*. (2nd), Oxford University Press, 978-0-19-852594-3, Oxford.

Schowalter, T. D. (2006). *Insect Ecology: An Ecosystem Approach*. (2nd), Academic, 978-0-12-088772-9, Amsterdam.

Southwood, T. R. E., Moran, V. C. & Kennedy, C. E. J. (1982). The richness, abundance and biomass of the arthropod communities on trees. *The Journal of Animal Ecology*, Vol. 51, No. 2, (June 1982), pp. (635-649), 1365-2656.

StatSoft. 2007. Statistica (data analysis software system). version 8.0.

Sunderland, K. & Samu, F. (2000). Effects of agricultural diversification on the abundance, distribution, and pest control potential of spiders: a review. *Entomologia Experimentalis et Applicata*, Vol. 95, No. 1, (April 2000), pp. (1-13), 1570-7458.

Symstad, A. J., Siemann, E. & Haarstad, J. (2000). An experimental test of the effect of plant functional group diversity on arthropod diversity. *Oikos*, Vol. 89, No. 2, (April 2003), pp. (243-253), 1600-0706.

Zar, J. H. (2010). *Biostatistical Analysis*. (4th), Prentice Hall, 0-13-081542-X, Upper Saddle River.

3

Global Impact of Mosquito Biodiversity, Human Vector-Borne Diseases and Environmental Change

Sylvie Manguin[1] and Christophe Boëte[2]
[1]*Institut de Recherche pour le Développement (IRD), Faculté de Pharmacie (UM1), Montpellier*
[2]*American University of Paris, Department of Computer Science, Mathematics and Science, Paris*
France

1. Introduction

More than 3,400 species of mosquito have been recorded worldwide. They include 37 genera, all within a single family, the Culicidae, itself divided into 3 sub-families, Toxorhynchitinae, Anophelinae, and Culicinae. They occur in tropical and temperate zones, even above the Arctic Circle but are absent in the Antarctic. They are found as high as 6,000 m (above sea level) in mountainous regions and as deep as 1,250 m (below sea level) in caves and mines (Lane & Crosskey, 1993).

If there is obviously an academic interest in the description and understanding of mosquito biodiversity, its study is also a major issue because of the risk associated with invasive species and the emergence and spread of vector-borne diseases. The efficiency, speed and reach of modern transport networks put indeed people at risk from the emergence of new strains of familiar diseases or completely new diseases (Guimerà et al., 2005). The global growth of economic activity, tourism, and human migration is leading to ever more cases of the movement of both diseases vectors and the pathogens they carry (Tatem et al., 2006b), increasing at the same time the biodiversity of mosquitoes around the world. In the current context of global warming, the expansion of areas suitable for the major disease vectors is becoming a threat and even a reality for some regions with several species being classified as invasive. To temperate this point, it is important to mention that among all mosquito species recorded, only 10% are regarded as efficient vectors of infectious agents having a considerable impact on human welfare and health. However, this small fraction is responsible for some of the worst scourges of humankind and the most important vectors belong essentially to three genera including *Anopheles*, *Aedes*, and *Culex* among which some species have been quite successful of wide-scale invasions.

An invasive species is defined as a species having a great impact on its new environment and spreading with measurable growth and distance dispersed (Daehler, 2001). Three major biological characteristics are found among the most invasive mosquito species, their close association with humans, egg resistance to desiccation (genus *Aedes*), and small larval habitats such as a wide variety of water holding containers among which man-made ones are totally suitable. Dispersal of mosquitoes can be active or passive. In the former case, mosquitoes are able to extent actively their range by flying from one habitat to another, but their flight capacity being fairly reduced (few kilometers for most of them), this natural dispersal will not allow them to quickly travel long distances. In the latter case, passive dispersal allows long-range transportation that can be either due to natural conditions (wind) or human-assisted (population movements), increasing the risk of spread of mosquitoes and vector-borne diseases. Under exceptional conditions, for instance strong winds, transport of mosquitoes is occurring on long distances into new areas. The case of the arrival of *An. gambiae* in Reunion Island (200 km from Mauritius) is attributed to the passage of a cyclone, and led to the first malaria epidemics in 1868 (Julvez et al., 1990). However, most of mosquito invasions are due to human-assisted transportation. For centuries, ship-borne transportation allowed man to travel long distances bringing with him immature stages of mosquitoes able to cope with the transport constraints. This is especially true for *Aedes aegypti* and *Culex quinquefasciatus* that have the capacity to develop in ship water storage (Mouchet et al., 1995). Mosquito larvae that occupy small water bodies, such as natural or artificial containers, are easily transported and when the conditions are favourable, mosquitoes establish and invade new territories. Compared to other mosquito genera, *Aedes* eggs have the property to resist to desiccation for several months that facilitates the spread at a global scale of some species, in particular *Ae. aegypti* (Figs. 1A, 2) and *Ae. albopictus* (Figs. 1B, 3). These two *Aedes* species are sylvatic mosquitoes that have become closely associated with the peridomestic environment and have been transported worldwide with goods and people.

In this chapter, mosquito biodiversity is considered with a special attention to species invasions at a global or regional scale with the risk of spreading vector-borne diseases. The factors influencing mosquito invasion are examined and environmental changes are discussed. Finally, mosquito vector control strategies are exposed in relation with the question of biodiversity.

2. The Mosquito family: The usual suspects

Mosquitoes as vectors

Among the vector-borne diseases, malaria is probably the the most famous one, responsible for the biggest burden in terms of mortality despite the existence of methods for prevention and treatment. It is only transmitted by about 70-80 species, all of them from the genus *Anopheles* (Bruce-Chwatt, 1980; Manguin et al., 2008a; Mouchet et al., 2004). Less studied but still important, some of these species can also transmit other parasites such as lymphatic microfilariae (*Wuchereria bancrofti, Brugia malayi, Brugia timori*) (Buck, 1991; Mak, 1987; Manguin et al., 2010). It is noteworthy to mention that these microfilarial parasites are not only transmitted by *Anopheles* in rural areas, but also by *Aedes* and *Culex* species in urban areas. Those latter ones are also well known for their role in the transmission of various arboviruses respectively dengue, yellow fever, chikungunya for *Aedes* and West Nile, Japanese Encephalitis, Rift Valley for *Culex* species (Lane & Crosskey, 1993).

About the complexity of classification

Numerous studies have shown that most of the vectors belong to complexes in which sibling species cannot be distinguished based on morphological characteristics alone, highlighting the importance of biodiversity in the world of mosquito vectors. Then, recent advances in molecular systematics have provided simple and reliable methods for unambiguous species identification and as a result increased the mosquito biodiversity with new species described based on molecular approaches (Besansky, 1999; Manguin et al., 2008b; Paskewitz et al., 1993; Smith & Fonseca, 2004). These molecular assays are crucial to identify the correct vector species before implementing efficient control programs. Examples of misidentification due to overlapping morphological characters may have important consequences as non-vectors are targeted hampering vector control programs (Singh et al., 2010; Van Bortel et al., 2001).

Are mosquitoes just pests?

If most studies dealing with mosquitoes are related with their role in the spread of infectious diseases and the will to reduce their vectorial capacity, it appears fundamental to consider their place in the biosphere, especially as part of the food chain and the possible positive impacts they can have. Thus, some of them, *Aedes communis* and *Aedes canadensis*, are said to be involved in the pollination of orchids (*Habenaria* species) in North American swamps (Thien, 1969). There are however very few data on this point and this even reveals the cruel lack of ecological data on mosquitoes and especially their role in the ecosystem. This is particularly true for the ones that are not considered as public health problems. It is interesting to notice that among them, the pitcher-plant mosquito, *Wyeomyia smithii* has the privilege to have been the subject of investigations that have revealed that this mosquito species has evolved in response to global warming (Bradshaw & Holzapfel, 2001). The particularity of this mosquito is that it is found only in the phytotelma of the purple pitcher plant, *Sarracenia purpurea* where it is the top-predator controlling the bacterial diversity in this small ecosystem (Peterson et al., 2008). Such an interest for the ecology and evolution of mosquito species is usually uncommon for species that are not considered as vector or nuisance. The development of studies in this field appears today highly needed to better understand the role of mosquitoes in community ecology, in species interactions and thus in ecosystem functioning.

2.1 *Aedes aegypti*

The yellow fever mosquito *Aedes aegypti* (Fig. 1A) is the principal urban vector of dengue (Fig. 2), chikungunya and yellow fever. It is an African species closely associated with humans: it feeds on human blood (anthropophilic), although not exclusively, and breeds in the peridomestic environment in artificial containers. Maritime dispersal in the Mediterranean region—in ships carrying infested water storage vessels—may have begun as early as 1000 B.C. From the 16th century onwards the species invaded the Americas, carried from Africa in slave ships (Tabachnick & Walter, 1991). The transatlantic passage generally took six to eight weeks, during which water storage containers must have been a prolific breeding site and the slaves and crew a ready source of blood. In consequence, it was not unusual for ships to arrive with an active yellow fever epidemic onboard (Rodhain, 1996), hence the yellow flag that still signifies a ship in quarantine.

The species was quick to colonise the New World, and this was accompanied by repeated and devastating yellow fever epidemics in seaports as far north as Boston (MA) and in inland cities such as Memphis (TN) and St Louis (MO). *Aedes aegypti* is now widespread in tropical Americas and has been recorded in 21 states of the United States (Gubler & Clark, 1995; Reiter, 2010b). From the mid-19th century onwards, it became widespread and common in much of India and Southeast Asia, probably accelerated by an increase in seaborne trade. Curiously, yellow fever has never been documented in the region, but dengue and chikungunya quickly became endemic and are now a major health problem. In the Polynesian islands, dispersal of *Ae. aegypti* between 1924 and 1986 has been attributed in large part to the construction of new airports (Mouchet et al., 1995; Séchan et al., 1993). Although well-established, often common in sub-tropical and many temperate regions, it disappeared from the entire Mediterranean region after World War II, perhaps because of widespread use of DDT for malaria eradication. In the past two decades it has become scarce in many regions of the United States; this has been attributed to the spread of a second introduced species, the Asian Tiger mosquito *Ae. albopictus* (Hobbs et al., 1991; Lounibos et al., 2010; O'Meara et al., 1995; Reiter, 2010b).

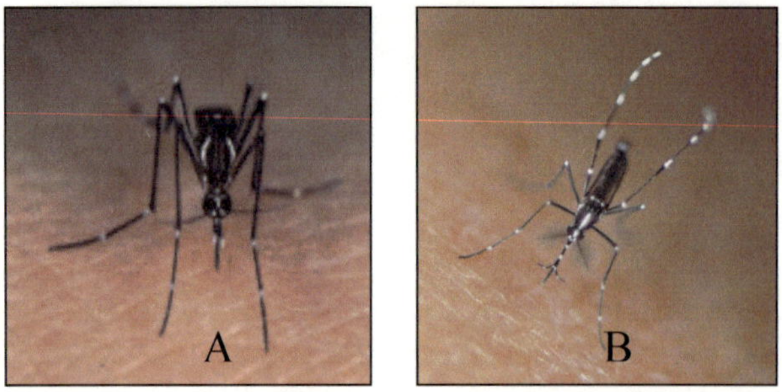

Fig. 1. A, Adult female *Aedes (Stegomyia) aegypti* (Photo © IRD/Jean-Pierre Hervy) and B, Adult male *Aedes (S.) albopictus* (Photo © IRD/Michel Dukhan). These species are the principal urban vectors of dengue and chikungunya throughout the world

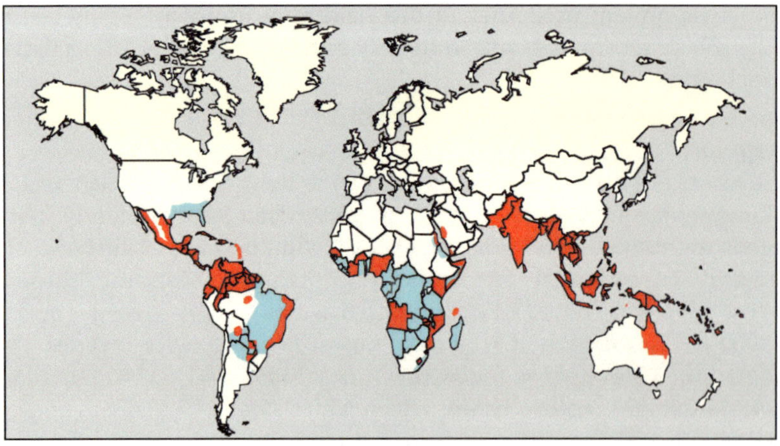

Fig. 2. Worldwide distribution of *Aedes aegypti* and Dengue fever in 2006; in blue, areas infested with *Ae. aegypti*; in red, areas with *Ae. aegypti* and recent epidemic dengue fever (Map by G.G. Clark, Public Domain)

2.2 *Aedes albopictus*

Although *Aedes albopictus*, the Asian Tiger mosquito (Fig. 1B), is widespread in urban areas, it is not restricted to the peri-domestic environment and feeds readily on many species of vertebrates. Until the mid-20th century, the only established populations outside Asia were in the Hawaiian Islands, Madagascar, several small islands in the south-western Indian Ocean and various islands in Polynesia and other regions of the Pacific, but in the 1970s a new infestation appeared in Albania and the species quickly became widespread and abundant. In 1983, the appearance of a single specimen in Memphis (TN, United States) was attributed to container transportation (Reiter & Darsie, 1983) and this was confirmed in 1985 when widespread infestations throughout the southern United States were traced to a global trade in used tires, with millions of tires exported from various countries in Asia, particularly Japan (Hawley et al., 1987; Reiter & Sprenger, 1987). Laboratory studies revealed that strains from northern Asia are cold-hardy and exhibit a winter diapause that is progressively absent at lower latitudes (Hawley et al., 1987); these mechanisms presumably limit potential range in newly infested countries. In the United States, the species is now established as far north as Nebraska and Illinois where temperatures below -20°C are not uncommon. In 1988, *Ae. albopictus* reached Mexico (probably exported from the United States) and spread into all countries of Central America. The species was already present in Brazil (1985) and is now occurring throughout South America, except Chile (Forattini, 1986; Reiter, 2010b). It has spread rapidly in Africa, with infestations identified in southeast Nigeria (1991), Cameroon (2000), Equatorial Guinea (2003), Gabon (2007), Central African Republic (2010), Algeria (2010) (Diallo et al., 2010; Fontenille & Toto, 2001; Gubler, 2003; Izri et al., 2010; Pages et al., 2009; Toto et al., 2003) and is also reported from Lebanon, Syria and Israel (Reiter, 2010a). Likewise in Europe: it is now present and well established in at least 14 countries, including Italy (1991), France (1999),

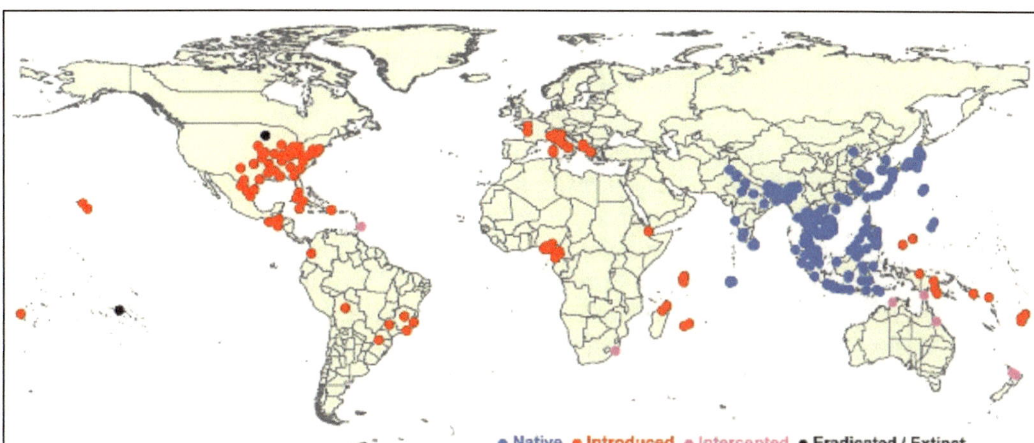

Fig. 3. Worldwide distribution of *Aedes albopictus* in 2004; in blue: native distribution; in red: invaded areas; in pink: intercepted specimens; in black: species eradicated or extinct (Map courtesy of Landcare Research New Zealand Limited 2005, available on www.landcareresearch.co.nz). More countries have now been infested (see text above)

Spain (2006), as well as Switzerland, several countries in the Balkans and Malta (Aranda et al., 2006; Dalla Pozza & Majori, 1992; Reiter, 2010b; Schaffner & Karch, 2000; Scholte & Schaffner, 2007). Compared to *Ae. aegypti*, its vectorial capacity is generally held to be low because it is not host specific (Gratz, 2004; Gubler, 2003) and virus dissemination from midgut into salivary glands occurs at lower rates (Lambrechts et al., 2010). However, recent outbreaks of dengue (Hawaii, 2001-02) and chikungunya (Reunion Island, 2005-06; Italy

2007; Gabon 2007) confirm that *Ae. albopictus* can function as an epidemic vector, at least under certain conditions (Effler et al., 2005; Pages et al., 2009; Renault et al., 2007). The accelerating expansion of global air travel for business and tourism, and the rapid rise in the prevalence and incidence of urban transmission in the tropics guarantee that such outbreaks are likely to become increasingly common. This growing risk shows the need for implementation of regular monitoring of populations as the vectorial status is a dynamic process that may evolve towards greater efficiency (Lambrechts et al., 2010). The potential role of *Ae. albopictus* acting as bridge vector of more arboviruses, such as yellow fever (Reiter, 2010a), to be brought into peridomestic environments, increasing the risk of human infection, is feared and calls for a close surveillance.

2.3 Other *Aedes* or *Ochlerotatus* species

Other species of *Aedes* or *Ochlerotatus,* two genera of the Aedini tribe (Reinert et al., 2004), have invaded new territories, including *Ochlerotatus* (*Ae.*) *japonicus* in North America and Hawaiian Islands (Fonseca et al., 2010; Hughes et al., 2008; Larish & Savage, 2005) and Central and Western Europe (Schaffner et al., 2004; 2009), *Ae. bahamensis* in south Florida (O'Meara et al., 1989), *Oc.* (*Ae.*) *atropalpus* in Italy (Romi et al., 1997) and in the Netherlands (Scholte et al., 2009), and *Ae. togoi* in the Pacific Northwest (Belton & Belton, 1990) and Malaysia (Ramalingam, 1969).

2.4 *Culex pipiens* complex

Mosquitoes in the *Culex pipiens* complex are ubiquitous in tropical, subtropical and temperate regions in urban, suburban and rural habitats worldwide (Smith & Fonseca, 2004). They are important as vectors of a number of arboviruses in temperate and tropical regions (e.g. West Nile Virus, Rift Valley, St. Louis encephalitis, Japanese encephalitis) and parasites including Bancroftian filariasis and a large number of avian protozoans.

In the tropics and sub-tropics *Culex pipiens quinquefasciatus* (Figs. 4, 5) is the classic species in water highly contaminated with organic material, even in latrines and septic tanks, but it also breeds in cleaner water in artificial containers, often sympatric with *Ae. aegypti*. The pathways of global expansion of this species appear to have followed those of *Ae. aegypti*. There is evidence that it invaded Australia between the end of the 18th century and early 19th century (Marks, 1972) and New Zealand at some time prior to 1848 (Weinstein et al., 1997). Its appearance on Maui, in the Hawaii Islands (the first mosquito to do so) is attributed to the arrival of ships from Mexico around 1826 (Dine, 1904), but this was undoubtedly followed by multiple additional introductions (Fonseca et al., 2000). The presence of the species, in conjunction with the arrival the European House Sparrow (*Passer domesticus*) infected with the avian malaria parasite *Plasmodium relictum* had a catastrophic impact on the unique local avifauna; many species became extinct and others only survive at altitudes too high for mosquito-borne transmission (Fonseca et al., 2000; Van Riper et al., 1986). Colonisation of the Pacific islands appears to have been linked to military movements during World War II as well as marine trade with Australia (Belkin, 1962). Introduction of *Cx. p. quinquefasciatus* in the New World also appears to have been recent but the pathway is unclear. On the African and Asian continents, populations present a high allelic richness that suggests multiple introductions from different sources (Fonseca et al., 2006). The common House Mosquito, *Culex p. pipiens,* is probably of European origin. Winter diapause in the adult stage, and the ability to survive in sub-zero temperatures have enabled it to colonise temperate regions of north and south America (Lane & Crosskey, 1993; Vinagradova, 2000).

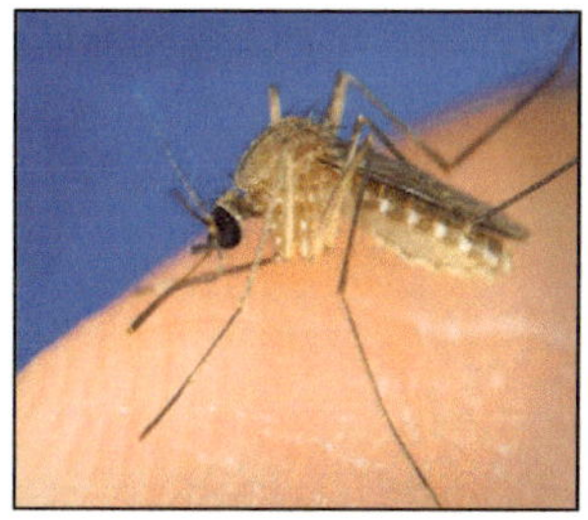

Fig. 4. Adult of *Culex quinquefasciatus* (Photo courtesy CDC/James Gathany)

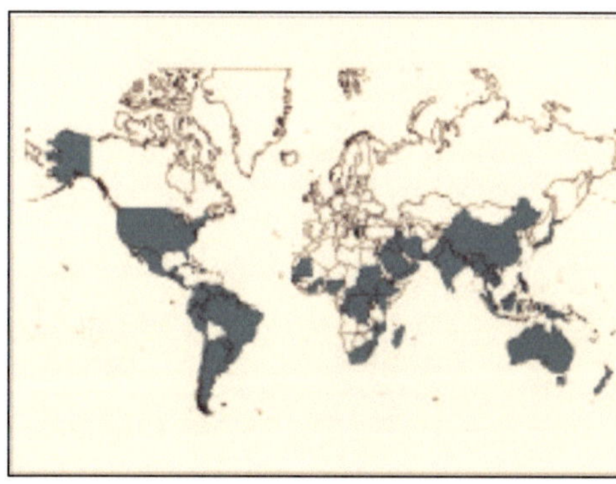

Fig. 5. Worldwide distribution map of *Culex quinquefasciatus* in 2010 (Map courtesy of WRBU, USA, www.wrbu.org)

2.5 *Anopheles* species

The reduction of travel time with faster ships or aircrafts allowed the transportation of more mosquitoes, as well as other genera that require specific environment for reproduction such as *Anopheles*. This major change in travel velocity has provoked cases of invasion of *Anopheles* vectors. In 1864, the opening of the Tamatave-Port-Louis steam boat line was immediately followed by the invasion of *An. gambiae* s.l. (Fig. 6), major vector of *Plasmodium falciparum* malaria in Africa (Fig. 7), from Madagascar to Mauritius, a malaria-free island prior to this introduction. One year later, the first malaria epidemics burst in the harbour of Mauritius and propagate all over the island (Mouchet et al., 1995).

Fig. 6. *Anopheles gambiae* (Photo © IRD /Jean-Pierre Hervy)

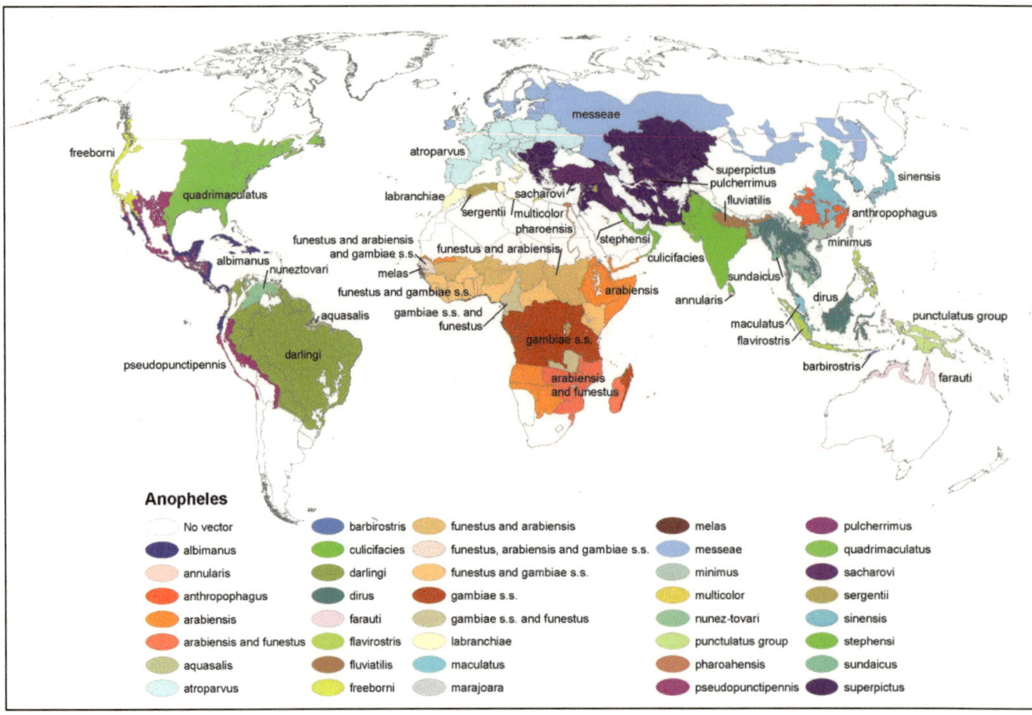

Fig. 7. *Anopheles* mosquito range map (U.S. CDC, Public Domain)

However, the most famous *Anopheles* vector invasion concerns the devastating introduction of *An. gambiae* s.l., specifically *An. arabiensis* (Parmakelis et al., 2008) which is one species of the Gambiae Complex, from Africa to north-eastern Brazil in 1930 (Soper & Wilson, 1943). *Anopheles* females were most likely transported from Dakar (Senegal) to Natal (Brazil) in a fast ship (sloop) which was regularly crossing the Atlantic Ocean (Rodhain, 1996). The environmental conditions found in Natal were close to the ones of the native area; therefore the species gradually invaded the north-eastern part of the country to cover 54,000 km^2 after 10 years, spreading malaria and killing 16,000 persons (Killeen et al., 2002). Malaria was already endemic in Brazil but the local vectors were not as efficient as the African vector, *An. arabiensis*, in disseminating falciparum malaria. One might also wonder about the role played by the interaction between African *Anopheles* and South American malaria parasites. Although costly, the military organisation of the vector control program, based on the systematic suppression of larval habitats and the use of Paris Green in larval sites that could not be eliminated, was successful as the species was eradicated by 1940 from Brazil (Soper & Wilson, 1943).

In 1942, another invasion by *An. gambiae* s.l., presumably *An. arabiensis* again, was reported in Upper Egypt, spreading 850 km along the Nile River. The vector was most likely imported by boats from Sudan as transportation was greatly increased during this period of war. This invasion was followed by a major malaria epidemic responsible for an estimated 12,000 deaths (Shousha, 1948). Based on the same energetic vector control methods than in Brazil (Paris Green against larvae and insecticide spraying against adults), *An. gambiae* s.l. was eradicated by 1945 (Shousha, 1948).

With the era of aircraft transportation, the travel time has been reduced so much that in less than two days an insect could reach almost any point of the world, as well as an infected person. The phenomenon called "airport malaria" is due to the introduction of *Anopheles* vectors from an endemic area to a non-endemic one where people, living near international airports but who never travelled to a malaria-risk area, can be infected. The highest risk comes from western and central Africa as most of the air traffic includes stopover in west European airports. Among the numerous *Anopheles* mosquitoes found in these international airports, *An. gambiae* s.l. (*An. gambiae* or *An. arabiensis*), the most efficient malaria vector in the world, is frequent during the hot summer period (July-August) as climatic conditions coincide with the region of origin (Mouchet, 2000). However, in the case of airport malaria, the infected tropical *Anopheles* mosquito has very poor chance of development as temperate climate is unfavourable to its establishment. Even the growth of travel back and forth to endemic areas presents a negligible risk of invasion of *Anopheles* vectors in temperate regions but this risk increases in tropical and subtropical areas where the climate barrier is absent.

The constant increase in sea and air traffic raises the risk of introducing new vectors, especially through shipment of goods and small aircrafts that introduced numerous mosquito species worldwide contributing locally to an increase of the biodiversity. Reports show for instance that all five mosquito species found in Hawaii are non-indigenous, as well as 17 of Guam's 24 species (only 7 are endemic) (Lounibos, 2002). However, most vector invasions, especially in regions where resources for correct entomological surveys are lacking, go unnoticed because of a lack of surveillance. Only few regions in the world are controlling invasive mosquitoes and a future universal reporting system as well as qualified personal need to be implemented. Multiple introductions of the same invasive species seem to be a fundamental evolutionary factor that may also play a role on changes in pathogen virulence which impacts important epidemiological parameters as shown in Hawaii with the increase in the altitudinal range of *Cx. quinquefasciatus* and the differences in vector competence to avian malaria after the second introduction of the vector (Fonseca et al., 2006).

3. Factors influencing mosquito invasion

Most successful mosquito invaders have arrived by ships (Fig. 8) (Calder & Laird, 1994). However, this mode of transportation implies that the imported mosquitoes are able to cope with travel conditions and the length of journey that can last months for sailing vessels. Desiccation-resistant eggs are strongly associated with becoming an introduced species (Juliano & Lounibos, 2005). Thus *Aedes* species can travel in large numbers at the egg or immature stages which enhance survival in non-native environments and greatly increase their probability of population establishment (Grevstad, 1999). For aircraft transportation, mosquitoes mainly travel at the adult stage and typically in low numbers. Although, this type of transportation is quick allowing any mosquito to reach its destination in several hours, the success of invasion is poor because the probability of population establishment decreases as lower number of specimens is released. However, some successful invasions after aircraft transportation have been reported, such as *Ae. albopictus* in Hawaii (Kilpatrick et al., 2004) and *Ae. aegypti* in Polynesian islands (Mouchet et al., 1995; Séchan et al., 1993).

Fig. 8. Cargo loaded with containers passing the Panama Canal in 2010 (Photo S. Manguin)

Global village … also for mosquitoes

The risk of vector invasion is now constantly present and even increasing as transport networks continue to expand in reach, speed of travel, number of travellers and goods carried. This expansion of the global traffic is associated with an increase of biological invasions which may cause nuisance and threats to human health (Tatem et al., 2006a). The vectors and the pathogens they carry can now move further, faster and in greater numbers than ever before. The current main road of mosquito invaders goes from Asia to Americas which corresponds to the predominant direction of tire shipments (Reiter, 1998). The combination of traffic volumes and climatic appropriateness can explain a great deal of invasion pathways of most cosmopolitan vectors, as shown for *Ae. albopictus* (Tatem et al., 2006a). The establishment of a vector in a new area requires several conditions, which vary from a species to another. The environment and the type of larval habitats are major components that will allow either the global invasion of new vector species (e.g. *Ae. albopictus*), or its establishment in a new area due to environmental changes (e.g. global warming), or will restrict it to specific areas. The immature stages of *Anopheles* species tend to develop in more natural larval habitats, although anthropogenic sites have been reported such as wells, domestic tanks, etc. For *Aedes* and *Culex*, their habitats are highly anthropogenic, and thus suitable for invasion and extension of their geographic distribution. Although most recorded invasions concern *Aedes* species and to a lesser extent *Culex* species, the establishment and spread of *Anopheles* species, in particular *An. gambiae* or its sister species *An. arabiensis*, could be devastating as proved by the few examples related above. The former species occurs in hotter and wetter climates of sub-Saharan Africa compared to the latter one that prefers drier environments as found during the invasions of Northeastern Brazil or Upper Egypt. This specificity linked to each species led Tatem et al. (2006a) to speculate on the most favourable regions that could be invaded outside Africa, such as Southeast Asia and Central/South America for *An. gambiae* and Middle East and Canary Islands for *An. arabiensis*, although it was this latter species that invaded the north-eastern part of Brazil in 1930 (Parmakelis et al., 2008). These high-risk regions of invasion by the most efficient malaria vectors in the world have been defined based on the combination of local climate and volume of sea traffic. However, this risk is relatively small as the sea traffic volumes from sub-Saharan Africa are still quite low compared to the Asian traffic. To the level of traffic volume and the appropriate environment, it is important to add biological traits that favour mosquito transportation. *Anopheles* mosquitoes are a lot less adapted to man-made larval habitats and to urban areas than *Aedes* or *Culex* species. Therefore, their transportation is mainly occurring at the adult stage and in low numbers and this greatly reduces the success of invasion. As a consequence, the distribution of the *Anopheles* species is either regional or local (Fig. 7), but no species has a cosmopolitan distribution such as *Aedes* (Figs. 2, 3) or *Culex* (Fig. 5) species (Hay et al., 2010; Lounibos, 2002; Mouchet, 1999).

Mosquito genus	*Anopheles*	*Aedes*	*Culex*
Main invasive phase	Terrestrial phase (adults)	Aquatic phase (eggs, larvae)	Aquatic phase (eggs, larvae)
Main human-assisted introduction mode	Airplanes, speed boats, cars (few hours to days)	Small containers (domestic water jars, tires) in ships (few days to several months)	Small containers transported by ships (few days to several months)
Biological characteristics		Eggs resistant to desiccation	
Major invasive vectors (origin to invaded regions)	*An. arabiensis* (Africa to Mauritius, Brazil, Upper Egypt)	*Ae. aegypti* (Africa to cosmopolitan), *Ae. albopictus* (Asia to Americas, Europe, Africa)	*C. pipiens* (Old World to North America), *C. quinquefasciatus* (Africa to Americas, Asia, New Zealand, southern Europe)
Major associated vector-borne diseases	Malaria	Dengue, yellow fever, chikungunya	West Nile, Japanese Encephalitis, Rift Valley

Table 1. Characteristics of invasion by the main mosquito genera

Other barriers exist such as competition from local mosquitoes that represents an obstacle for the development of the alien species (Hay et al., 2005). In addition, equilibrium with competitors, predators and local pathogenic agents must be reached in order to have a productive larval development in new ecosystems. In Mauritius, as both *An. arabiensis* and *An. funestus* were imported, the latter species was eradicated after insecticide vector control while the other one persisted (Mouchet et al., 1995). The invasion of *An. arabiensis* in Brazil was stopped and the species eradicated by larvicide use (Soper & Wilson, 1943), whether this vector control is not efficient in Africa. This is most likely due to the precarious adaptation of this imported species into a new environment (Mouchet et al., 1995). More cases of imported *Anopheles* species eradication or spontaneous disappearance have been reported. The same occurred for some cases of imported *Aedes* vectors, for instance with the elimination of *Ae. aegypti* from the Mediterranean Basin and *Ae. albopictus* from a Polynesian atoll (Taiaro Island, Tuamotu Archipelago) where it was intentionally introduced to take over from *Ae. polynesiensis* which is the local vector of lymphatic filariasis. The non-indigenous species failed to displace the native vector and within 1-4 years disappeared from the island (Rosen et al., 1976).

4. How is the mosquito biodiversity distributed around the world?

As stated in the introduction, mosquitoes are occurring worldwide, except in the Antarctic. They occur heterogeneously around the world. *Anopheles* mosquitoes for instance are absent from islands such as Greenland, Iceland, the Seychelles, New Caledonia and islands of central and eastern Polynesia (Fig. 7) (Mouchet et al., 2004; Robert et al., 2011). Several studies on mosquito biogeography have shown that the highest biodiversity of mosquitoes occurs in Southeast Asia and the Neotropics (Gaston & Hudson, 1994). There is a latitudinal biodiversity gradient with species richness increasing toward the equator (Foley et al., 2007) and several Asian countries, including Indonesia, Malaysia, and Thailand, along with Brazil have the highest numbers of mosquito species. Based on the 24 *Anopheles* species complexes currently recognized (Carvalho-Pinto & Lourenço-de-Oliveira, 2004; Coetzee et al., 1999; Harbach, 2004; Krzywinski et al., 2011; Manguin et al., 2008b; Narang et al., 1993), 12 occur in Asia (50%) while 5, 3, 3, and 1 are respectively distributed in Americas (21%), Africa (13%), Australia-Pacific (13%) and Europe (4%) (Table 2). In addition, *Anopheles* species

Anopheles species Complex (Subgenus)	Distribution
Albitarsis (*Nyssorhynchus*)	Americas
Annularis (*Cellia*)	Asia
Annulipes (*Cellia*)	Australia-Pacific
Barbirostris (*Anopheles*)	Asia
Benarocchi (*Nyssorhynchus*)	Americas
Claviger (*Anopheles*)	Europe
Crucians (*Anopheles*)	Americas
Cruzii (*Kerteszia*)	Americas
Culicifacies (*Cellia*)	Asia
Dirus (*Cellia*)	Asia
Farauti (*Cellia*)	Australia-Pacific
Fluviatilis (*Cellia*)	Asia
Gambiae (*Cellia*)	Africa
Gigas (*Anopheles*)	Asia
Leucosphyrus (*Cellia*)	Asia
Lindesayi (*Anopheles*)	Asia
Lungae (*Cellia*)	Australia-Pacific
Marshallii (*Cellia*)	Africa
Minimus (*Cellia*)	Asia
Nili (*Cellia*)	Africa
Nivipes (*Cellia*)	Asia
Nuneztovari (*Nyssorhynchus*)	Americas
Subpictus (*Cellia*)	Asia
Sundaicus (*Cellia*)	Asia
Total: 24	Asia (12), Americas (5), Africa (3), Australia-Pacific (3), Europe (1)

Table 2. *Anopheles* species complexes in relation to their geographic distribution

richness in collecting sites is higher in Asia, with more than 10 species per site, often reaching 12 to 15 species and up to 20 species in some localities (Coosemans et al., 2006; Garros et al., 2008), compared to generally less than 10 species per site, even less than 5 species, on the American and African continents (based on the combination of different collecting methods) (S. Manguin, unpublished data).

5. Mosquito biodiversity, human health and environmental change

Introduced vectors may have a serious impact on human health either by simultaneously introducing a novel pathogen (e.g. *Ae. aegypti* and yellow fever/dengue in Americas, airport malaria cases), by spreading a native pathogen (e.g. *An. arabiensis* and *P. falciparum* malaria in Brazil, *Ae. aegypti* and dengue in Asia), or by the independent introductions of a non-native vector and novel pathogen (e.g. *C. pipiens* and West Nile virus in America, *Ae. albopictus* and chikungunya epidemics in Indian Ocean islands and outbreaks in Europe) (Juliano & Lounibos, 2005). Climate is obviously an important factor for the spread of most invasive mosquitoes. In addition to transportation, global climate changes could indeed increase the risk of vector invasion and the spread of vector-borne diseases even under temperate climate (Rogers et al., 2001). This adds to the constant increase in international transportation that is a major issue in the dispersion of vectors and pathogens. This needs to be monitored on a regular basis and at a global scale in order to prevent from emerging disease epidemics. In a more general way, travel is a potent force in vector invasion along with disease emergence and spread, and the continued growth in global air travel and ship-borne trade is increasing the risk of disease vector invasions (Tatem et al., 2006b). As transport links are becoming busier and further between spatially distant, but climatically similar regions of the world, global transportation by seaborne or through the worldwide airline network increases the risks of deliberate or accidental movements and establishment of climatically sensitive exotic organisms in new areas (Tatem, 2009). With traffic levels continuing to rise and climates changing regionally, these risks will vary, both seasonally and year-by-year.

Beside invasion of some mosquito species, there is a confirmed loss of biodiversity from anthropogenic origins and this may greatly affect human health. Variation in the diversity of vertebrate hosts may indeed have an influence on the risk of human exposure to vector-borne diseases. High biodiversity can protect human health by reducing the risk of disease transmission due to the diversity of hosts, also called the "dilution effect" (Ostfeld & Keesing, 2000). On the contrary, reduced biodiversity can increase the risk of disease transmission by concentrating the source pool on few available and competent hosts, unless the biodiversity loss causes the reduction in competent and amplifying hosts (Keesing et al., 2010). This is valid under four conditions: (1) the vector is generalist, able to blood feed on several hosts, such as *Ae. albopictus*. However, this condition is not verified for the main mosquito vectors of human diseases as they generally present an anthropophilic behaviour, such as *An. gambiae* or *Ae. aegypti*. In this case, the dilution effect is not reducing disease transmission as the vectors feed almost exclusively on humans. (2) The vector acquires pathogens from its hosts through its biting behaviour. Although some *Aedes* and *Culex* species may get infected by a transovarial transmission, the large majority of mosquito vectors contract pathogens after an infected bite. (3) The reservoir host is competent for transmitting pathogens. (4) Most competent reservoir host tends to be a community dominant. When all four conditions are met, high diversity in the community of hosts strongly reduces the transmission of vector-borne diseases. Ostfeld (2009) has well documented the effect of biodiversity loss on the increase of vector-borne zoonotic disease transmission; on the contrary, the preservation of intact ecosystems and their endemic biodiversity should reduce the prevalence of some infectious diseases, in particular transmission of zoonotic pathogens such as West Nile Virus and *Culex* vectors (Keesing et al., 2010). Therefore, in certain cases, biodiversity may play a dual role in the emergence and transmission of vector-borne diseases with high diversity providing a larger potential source of novel pathogens and vectors, and in the meantime biodiversity can reduce pathogen transmission.

Other studies showed that, in the case of malaria vectors, changes in aquatic predator diversity can have a major impact on larval population and therefore on infectious disease risk. Larval habitat disturbance markedly reduced the predator diversity which had for consequence to increase the density of *An. gambiae* (Carlson et al., 2009). As human population increases, urbanization and environmental changes due to an expansion of land cultivation is occurring at a faster pace. Ecological disturbance and development of irrigation is participating in the increase of malaria, as demonstrated in Africa (Minakawa et al., 2006; Mouchet et al., 1998). Deforestation is obviously one of these types of disturbance. A forest is a closed environment that favours main vectors such as *An. dirus* in Asia, but forest clearing for cultivation has resulted in the development of more heliophilic vectors, such as *An. harrisoni* in Southeast Asia (Manguin et al., 2008b). A similar situation also occurred in Africa with *An. gambiae* (Livadas et al., 1958). Dam, irrigation ditches, rice fields, and borrow pits made for bricks used in the construction of houses, constitute favourable larval habitats for *An. gambiae* and *An. arabiensis* and in the latter case the mosquito find next to each other suitable larval habitats and human blood source (Mouchet & Brengues, 1990). In central African highlands, the natural marshes of papyrus, that produces a layer of oil at the water surface inhibiting the anopheline larval development, has been widely replaced by cultivation of valley bottom for food crops and fish ponds offering suitable mosquito habitats. This land use change led to the rapid development of *An. gambiae* and, in 1994, a deadly malaria epidemic burst in a region of Uganda located above 1,500 meters of altitude and previously considered nearly malaria-free (Mouchet et al., 1998). In association to the environmental change due to the demographic pressure, the climate with abnormal rainfall, more than 2-fold a normal year, was also a determinant factor that induced this epidemic. Finally, it is noteworthy that zones of unstable malaria are more sensitive to environmental changes and climatic variations than stable malaria regions (Mouchet et al., 1998).

6. Vector control strategies

While *Anopheles* species are distributed into distinct biogeographical zones, the main *Aedes* and *Culex* vector species have a worldwide distribution with some invasive species, such as *Aedes albopictus* being the current example of a rapid and efficient global spread responsible for epidemics of dengue and chikungunya. Each continent has its main malaria vectors (Hay et al., 2010; Mouchet et al., 2004), among them *Anopheles gambiae, An. arabiensis* and *An. funestus* being the most efficient ones in sub-Saharan Africa (Sinka et al., 2010a); *An. darlingi*, the main vector in Latin America (Sinka et al., 2010b); *An. culicifacies* s.l., *An. dirus* s.l., *An. fluviatilis* s.l., *An. maculatus, An. minimus* s.l., and *An. stephensi* in Asia; and *An. farauti* s.l., *An. koliensis* and *An punctulatus* in the Australian-Pacific region (Sinka et al., 2011). In comparison, the distribution of species such as *Aedes aegypti* covers the tropical belt across all continents (Fig. 2) or is cosmopolitan as for *Culex quinquefasciatus* (Fig. 5) (WHO, 1989). This fundamental difference has important consequences on vector control strategies with global control programs that may be applied at a continental scale in the latter case, when regional strategies need to be developed in the case of *Anopheles* mosquitoes.

Vector control strategies are mainly based on three fundamental methods: (1) physical control such as the destruction of larval habitats by environmental modifications, (2) chemical control based on insecticide use, (3) biological control by use of biolarvicides (Rozendaal, 1997).

6.1 Physical control

Physical control is possible by removing larval sites (containers, water deposits), especially efficient against *Aedes* or *Culex* species. Larval habitats can also be altered such as drainage operations, intermittent irrigation, change in river discharge or configuration, removal water plants or algae, making habitats unsuitable. Few interesting results against *Anopheles* mosquitoes have been reported but they require massive engineering efforts (Mouchet et al., 2004).

6.2 Chemical control

The chemical control is based on insecticide impregnation of material, as well as regular spray of insecticides/larvicides by terrestrial or airways according to the nature of the habitats and their size. Vector control programs against *Anopheles* mosquitoes are mainly based on the use of insecticide-treated nets (ITN), long-lasting insecticidal nets (LLINs), or indoor residual spraying (IRS) which remain effective methods of control when used properly. Larviciding can also be implemented in special situations (well known breeding sites mainly man-made). The use of different insecticides in rotation or mosaic can also be recommended in order to avoid (or at least delay) the problems of resistance (Hemingway et al., 1997; Lenormand & Raymond, 1998). In association to larval habitat removal, vector control program against *Aedes* mosquitoes based on insecticidal spraying (organophosphates, pyrethroids) using hand-held apparatus, road vehicle or aircraft presents some efficacy. However, for *Ae. aegypti* which tends to rest inside houses, indoor spraying is more appropriate. For *Ae. albopictus*, vector control appears to be more difficult due to its ability to breed away from humans, but its outdoor resting behaviour makes outdoor spraying an effective strategy of control. Two pyrethroid insecticides, alpha-cypermethrin and lambda-cyhalothrin, applied uniformly to the inside surfaces of either dry tires that were partially filled with water 24 h after spraying, or wet tires partially filled with water prior to spraying showed their efficacy to prevent mosquito larval colonization in disused car tires. Both insecticides using either application method show great potential for the prevention of invasion against *Ae. aegypti* and *Ae. albopictus* (Pettit et al., 2010).

6.3 Biological control

Some organisms have proved to be effective against mosquito larvae such as larvivorous fish (*Gambusia, Poecilia*), predatory insect larvae (dragonfly, *Toxorhynchites* mosquitoes), copepods, nematode worms, fungi, and bacterial larvicides. The latter ones have concentrated much attention. Thus, *Bacillus sphaericus* is widely used against *Culex* species (Skovmand et al., 2009), while megadoses of dry formulations of *Bacillus thuringiensis* subspecies *israelensis* (Bti) proved to be efficient for residual control of *Ae. aegypti* in small containers (Ritchie et al., 2010). Recent studies have also shown that the use of entomopathogenic fungi could provide an interesting tool having the advantage to be evolution-proof (Blanford et al., 2005; Read et al., 2009; Scholte et al., 2005).

6.4 What about keeping mosquitoes and replacing the vectors

It is interesting here to mention that the current methods are aiming at reducing (or eliminating) mosquito populations and vector species. A recent debate has taken place on the question "What would happen if there were no more mosquitoes?" Despite the collateral damage of eradicating mosquitoes that would result in the loss of a group of pollinators and a primary food source for many species, a world without mosquitoes could be seen as more secure to us, in particular the elimination of *Anopheles* would be quite beneficial for mankind (Fang, 2010; Wickson, 2010). However, the limitations in mosquito control strategies make a world without mosquitoes unlikely (Fang, 2010). In addition, the real impact on ecosystems worldwide is difficult to apprehend, especially as discussions undergone illustrate the nadir in which mosquito ecology lies. In relation with the role of mosquitoes in the ecosystem, it is interesting to notice that among the current effort in developing a transgenic approach against vector-borne disease, one envisaged method is to replace populations of vectors by non-vectors. Even if the relevance of this is highly debated (Boëte, 2006), one might consider that it would have the advantage to leave mosquitoes in the food chain but the balance between this advantage and the potential adverse effects needs to be properly evaluated.

Nowadays and in the current efforts to reduce the burden of vector-borne diseases, the implementation of integrated vector management (IVM) activities that combine the use of different control strategies in relation to local vector bionomics appear the best strategy to move forward (Chanda et al., 2008).

7. Conclusions

Aedes aegypti and *Ae. albopictus* are forest mosquitoes. In their original habitat they breed in tree-holes, rock holes, fruit husks and other small collections of water. In the peridomestic environment they exploit man-made articles that substitute for these breeding sites. In the modern urban environment the abundance of such articles is the key factor in the abundance of these species and of the burgeoning problem of dengue, chikungunya and other viral infections that they transmit. Modern transportation technology has encouraged widespread commerce in items that can be infested; classic examples are used tires, live plants, and exotic fishes. Accelerating global trade in such items will increase the probability of introduction of such species into new regions. The advent of tens of millions of tightly packed, locked shipping containers (Fig. 8), the onus on speed in their handling and transportation, and above all their delivery unopened to their ultimate destination, have made conventional inspection in harbour redundant (Reiter, 2010a). In addition, the exponential increase in cheap air travel and transport has facilitated the movement of human and animal pathogens. In the past three decades, this mobility has sparked outbreaks of diseases such as dengue and chikungunya in many places around the world. Recent events include epidemics of chikungunya in La Reunion Island and northern Italy and of dengue in the Cape Verde Islands. Spectacular examples of other exotic pathogens, such as the explosive panzootic of West Nile virus in the New World are here to remind us the vital importance of a better knowledge and good understanding of the fine interactions between pathogens, mosquitoes and environmental changes, and the need to clearly admit and evaluate our role and responsibility in those phenomenons.

8. Acknowledgment

We would like to thank Jean Mouchet for his contribution in the revision of the manuscript. We are also grateful to Paul Reiter for many helpful suggestions.

9. References

Aranda, C.; Eritja, R. & Roiz, D. (2006). First record and establishment of the mosquito *Aedes albopictus* in Spain. *Medical Veterinary Entomology,* Vol.20, No.1, pp. 150-152

Belkin, J.N. (1962). *The mosquitoes of the South Pacific (Diptera: Culicidae).* University of California Press, Berkeley, USA

Belton, P. & Belton, O.C. (1990). *Aedes togoi* comes aboard. *Journal of the American Mosquito Control Association,* Vol.6, No.2, pp. 328-329

Besansky, N.J. (1999). Complexities in the analysis of cryptic taxa within the genus *Anopheles*. *Parassitologia,* Vol.41, pp. 97-100

Blanford, S.; Chan, B.H., Jenkins, N., Sim, D., Turner, R.J., Read, A.F. & Thomas, M.B. (2005). Fungal pathogen reduces potential for malaria transmission. *Science,* Vol.308, No.5728, pp. 1638-1641

Boëte, C. (2006). *Genetically modified mosquitoes for malaria control.* Eurekah/Landes Bioscience, Georgetown, USA

Bradshaw, W.E. & Holzapfel, C.M. (2001). Genetic shift in photoperiodic response correlated with global warming. *Proceedings of the National Academy of Sciences,* Vol.98, No.25, pp. 14509-14511

Bruce-Chwatt, L.J. (1980). *Essential malariology.* William Heinemann Med Books Ltd, London, UK

Buck, A. (1991). *Filariasis, In: Strickland, T. G., Hunter's Tropical Medicine,* 7th Edition. W.B. Saunders Company, Baltimore, USA

Calder, L. & Laird, M. (1994). Mosquito travelers, arbovirus vectors and the used tire trade. *Travel Medicine International,* Vol.12, pp. 3-12

Carlson, J.C.; Dyer, L.A., Omlin, F.X. & Beier, J.C. (2009). Diversity cascades and malaria vectors. *Journal of Medical Entomology,* Vol.46, No.3, pp. 460-464

Carvalho-Pinto, C.J. & Lourenço-de-Oliveira, R. (2004). Isoenzimatic analysis of four *Anopheles* (*Kerteszia*) *cruzii* (Diptera: Culicidae) populations of Brazil. *Memórias do Instituto Oswaldo Cruz,* Vol.99, No.5, pp. 471-475

Chanda, E.; Masaninga, F., Coleman, M., Sikaala, C., Katebe, C., Macdonald, M., Baboo, K.S., Govere, J. & Manga, L. (2008). Integrated vector management: the Zambian experience. *Malaria Journal,* Vol.7, p. 164

Coetzee, M.; Estrada-Franco, J.G., Wunderlich, C.A. & Hunt, R.H. (1999). Cytogenetic evidence for a species complex within *Anopheles pseudopunctipennis* theobald (Diptera: Culicidae). *American Journal of Tropical Medicine & Hygiene,* Vol.60, No.4, pp. 649-653

Coosemans, M.; Thuan, L.K., Socheat, D., Phompida, S., Baimai, V., Manguin, S., Harbach, R.E. & Hemingway, J. (2006). Monitoring insecticide resistance and mapping malaria vectors in Southeast Asia: a prerequisite for sustainable malaria vector control. INCO report, IC4-CT-2002-10041, 202 p

Daehler, C.C. (2001). Two ways to be an invader, but one is more suitable for ecology. *Bulletin of the Ecological Society of America*, Vol.81, pp. 101-102

Dalla Pozza, G. & Majori, G. (1992). First record of *Aedes albopictus* establishment in Italy. *Journal of the American Mosquito Control Association*, Vol.8, No.3, pp. 318-320

Diallo, M.; Laganier, R. & Nangouma, A. (2010). First record of *Ae. albopictus* (Skuse 1894), in Central African Republic. *Tropical Medicine & International Health*, Vol.15, No.10, pp. 1185-1189

Dine, D.L.V. (1904). Mosquitoes in Hawaii. *Bulletin of the Hawaii Agricultural Experimental Station*, Vol.6, p. 30

Effler, P.V.; Pang, L., Kitsutani, P., Vorndam, V., Nakata, M., Ayers, T., Elm, J., Tom, T., Reiter, P., Rigau-Perez, J.G., Hayes, J.M., Mills, K., Napier, M., Clark, G.G. & Gubler, D.J. (2005). Dengue fever, Hawaii, 2001-2002. *Emerging Infectious Diseases*, Vol.11, No.5, pp. 742-749

Fang, J. (2010). Ecology: A world without mosquitoes. *Nature*, Vol.466, No.7305, pp. 432-434

Foley, D.H.; Rueda, L.M. & Wilkerson, R.C. (2007). Insight into global mosquito biogeography from country species records. *Journal of Medical Entomology*, Vol.44, No.4, pp. 554-567

Fonseca, D.M.; LaPointe, D.A. & Fleischer, R.C. (2000). Bottlenecks and multiple introductions: population genetics of the vector of avian malaria in Hawaii. *Molecular Ecology*, Vol.9, No.11, pp. 1803-1814

Fonseca, D.M.; Smith, J.L., Wilkerson, R.C. & Fleischer, R.C. (2006). Pathways of expansion and multiple introductions illustrated by large genetic differentiation among worldwide populations of the southern house mosquito. *American Journal of Tropical Medicine & Hygiene*, Vol.74, No.2, pp. 284-289

Fonseca, D.M.; Widdel, A.K., Hutchinson, M., Spichiger, S.E. & Kramer, L.D. (2010). Fine-scale spatial and temporal population genetics of *Aedes japonicus*, a new US mosquito, reveal multiple introductions. *Molecular Ecology*, Vol.19, No.8, pp. 1559-1572

Fontenille, D. & Toto, J.C. (2001). *Aedes* (*Stegomyia*) *albopictus* (Skuse), a potential new Dengue vector in southern Cameroon. *Emerging Infectious Diseases*, Vol.7, No.6, pp. 1066-1067

Forattini, O.P. (1986). Identification of *Aedes* (*Stegomyia*) *albopictus* (Skuse) in Brazil. *Revista de Saúde Pública*, Vol.20, No.3, pp. 244-245

Garros, C.; Van Nguyen, C., Trung, H.D., Van Bortel, W., Coosemans, M. & Manguin, S. (2008). Distribution of *Anopheles* in Vietnam, with particular attention to malaria vectors of the *Anopheles minimus* complex. *Malaria Journal*, Vol.7, p. 11

Gaston, K.J. & Hudson, E. (1994). Regional patterns of diversity and estimates of global insect species richness. *Biodiversity and Conservation*, Vol.3, pp. 493-500

Gratz, N.G. (2004). Critical review of the vector status of *Aedes albopictus*. *Medical & Veterinary Entomology*, Vol.18, No.3, pp. 215-227

Grevstad, F.S. (1999). Experimental invasions using biological control introductions: the influence of release size on the chance of population establishment. *Biological Invasions*, Vol.1, pp. 313-323

Gubler, D.J. & Clark, G.G. (1995). Dengue/dengue hemorrhagic fever: the emergence of a global health problem. *Emerging Infectious Diseases*, Vol.1, No.2, pp. 55-57

Gubler, D.J. (2003). *Aedes albopictus* in Africa. *Lancet Infectious Diseases*, Vol.3, No.12, pp. 751-752

Guimerà, R.; Mossa, S., Turtschi, A. & Amaral, L.A. (2005). The worldwide air transportation network: Anomalous centrality, community structure, and cities' global roles. *Proceedings of the National Academy of Sciences*, Vol.102, No.22, pp. 7794-7799

Harbach, R.E. (2004). The classification of genus *Anopheles* (Diptera: Culicidae): a working hypothesis of phylogenetic relationships. *Bulletin of Entomological Research*, Vol.94, pp. 537-553

Hawley, W.A.; Reiter, P., Copeland, R.S., Pumpuni, C.B. & Craig, G.B., Jr. (1987). *Aedes albopictus* in North America: probable introduction in used tires from northern Asia. *Science*, Vol.236, No.4805, pp. 1114-1116

Hay, S.I.; Guerra, C.A., Tatem, A.J., Atkinson, P.M. & Snow, R.W. (2005). Urbanization, malaria transmission and disease burden in Africa. *Nature Reviews Microbiology*, Vol.3, No.1, pp. 81-90

Hay, S.I.; Sinka, M.E., Okara, R.M., Kabaria, C.W., Mbithi, P.M., Tago, C.C., Benz, D., Gething, P.W., Howes, R.E., Patil, A.P., Temperley, W.H., Bangs, M.J., Chareonviriyaphap, T., Elyazar, I.R., Harbach, R.E., Hemingway, J., Manguin, S., Mbogo, C.M., Rubio-Palis, Y. & Godfray, H.C. (2010). Developing global maps of the dominant *Anopheles* vectors of human malaria. *PLoS Medicine*, Vol.7, No.2, p. e1000209

Hemingway, J.; Penilla, R.P., Rodriguez, A.D., James, B.M., Edge, W., Rogers, H. & Rodriguez, M.H. (1997). Resistance management strategies in malaria vector mosquito control. A large-scale field trial in Southern Mexico. *Pesticide Science*, Vol.51, pp. 375-382

Hobbs, J.H.; Hughes, E.A. & Eichold, B.H., 2nd. (1991). Replacement of *Aedes aegypti* by *Aedes albopictus* in Mobile, Alabama. *Journal of the American Mosquito Control Association*, Vol.7, No.3, pp. 488-489

Hughes, T.H.; Irwin, P.M., Kaufman, A., Sage, H., Pagac, B.B., Jr. & Paskewitz, S.M. (2008). First records of *Aedes japonicus japonicus* in Wisconsin. *Journal of the American Mosquito Control Association*, Vol.24, No.4, pp. 583-584

Izri, A.; Bitam, I. & Charrel, R.N. (2010). First entomological documentation of *Aedes (Stegomyia) albopictus* (Skuse, 1984) in Algeria. *Clinical Microbiology and Infection*, DOI: 10.1111/j.1469-0691.2010.03443.x

Juliano, S.A. & Lounibos, L.P. (2005). Ecology of invasive mosquitoes: effects on resident species and on human health. *Ecology Letters*, Vol.8, No.5, pp. 558-574

Julvez, J.; Mouchet, J. & Ragavoodoo, C. (1990). [Historical epidemiology of malaria in the archipelago of the Mascarenes (Indian Ocean)]. *Annales de la Société belge de Médecine Tropicale*, Vol.70, No.4, pp. 249-261

Keesing, F.; Belden, L.K., Daszak, P., Dobson, A., Harvell, C.D., Holt, R.D., Hudson, P., Jolles, A., Jones, K.E., Mitchell, C.E., Myers, S.S., Bogich, T. & Ostfeld, R.S. (2010). Impacts of biodiversity on the emergence and transmission of infectious diseases. *Nature*, Vol.468, No.7324, pp. 647-652

Killeen, G.F.; Fillinger, U., Kiche, I., Gouagna, L.C. & Knols, B.G. (2002). Eradication of *Anopheles gambiae* from Brazil: lessons for malaria control in Africa? *Lancet Infectious Diseases,* Vol.2, No.10, pp. 618-627

Kilpatrick, A.M.; Gluzberg, Y., Burgett, J. & Daszak, P. (2004). Threat of West Nile virus to Hawaii. *Ecohealth,* Vol.1, pp. 205-209

Krzywinski, J.; Li, C., Morris, M., Conn, J.E., Lima, J.B., Povoa, M.M. & Wilkerson, R.C. (2011). Analysis of the evolutionary forces shaping mitochondrial genomes of a Neotropical malaria vector complex. *Molecular Phylogenetics and Evolution,* Vol.58, No.3, pp. 469-477

Lambrechts, L.; Scott, T.W. & Gubler, D.J. (2010). Consequences of the expanding global distribution of *Aedes albopictus* for dengue virus transmission. *PLoS Neglected Tropical Diseases,* Vol.4, No.5, p. e646

Lane, R.P. & Crosskey, R.W. (1993). *Medical Insects and Arachnids*, British Museum Edition. Chapman & Hall, London, UK

Larish, L.B. & Savage, H.M. (2005). Introduction and establishment of *Aedes* (*Finlaya*) *japonicus japonicus* (Theobald) on the island of Hawaii: implications for arbovirus transmission. *Journal of the American Mosquito Control Association,* Vol.21, No.3, pp. 318-321

Lenormand, T. & Raymond, M. (1998). Resistance management: the stable zone strategy. *Proceeding of the Royal Society of London,* Vol.265, pp. 1985-1990

Livadas, G.; Mouchet, J., Gariou, J. & Chastang, R. (1958). [Can one foresee the eradication of malaria in wooded areas of South Cameroun]. *Rivista di Malariologia* Vol.37, No.4-6, pp. 229-256

Lounibos, L.P. (2002). Invasions by insect vectors of human disease. *Annual Review of Entomology,* Vol.47, pp. 233-266

Lounibos, L.P.; O'Meara, G.F., Juliano, S.A., Nishimura, N., Escher, R.L., Reiskind, M.H., Cutwa, M. & Greene, K. (2010). Differential Survivorship of Invasive Mosquito Species in South Florida Cemeteries: Do Site-Specific Microclimates Explain Patterns of Coexistence and Exclusion? *Annals of the Entomological Society of America,* Vol.103, No.5, pp. 757-770

Mak, J.W. (1987). Epidemiology of lymphatic filariasis. *Ciba Foundation Symposium,* Vol.127, pp. 5-14

Manguin, S.; Carnevale, P., Mouchet, J., Coosemans, M., Julvez, J., Richard-Lenoble, D. & Sircoulon, J. (2008a). *Biodiversity of malaria in the world.* John Libbey Eurotext, Paris, France

Manguin, S.; Garros, C., Dusfour, I., Harbach, R.E. & Coosemans, M. (2008b). Bionomics, taxonomy, and distribution of the major malaria vector taxa of *Anopheles* subgenus *Cellia* in Southeast Asia: An updated review. *Infection, Genetics and Evolution,* Vol.8, No.4, pp. 489-503

Manguin, S.; Bangs, M.J., Pothikasikorn, J. & Chareonviriyaphap, T. (2010). Review on global co-transmission of human *Plasmodium* species and *Wuchereria bancrofti* by *Anopheles* mosquitoes. *Infection, Genetics and Evolution,* Vol.10, No.2, pp. 159-177

Marks, E.N. (1972). Mosquitoes (Culicidae) in the changing Australian environment. *Queensland Naturalist,* Vol.20, pp. 101-116

Minakawa, N.; Omukunda, E., Zhou, G., Githeko, A. & Yan, G. (2006). Malaria vector productivity in relation to the highland environment in Kenya. *American Journal of Tropical Medicine and Hygiene,* Vol.75, No.3, pp. 448-453

Mouchet, J. & Brengues, J. (1990). [Agriculture-health interface in the field of epidemiology of vector-borne diseases and the control of vectors]. *Bulletin de la Société Pathologique Exotique,* Vol.83, No.3, pp. 376-393

Mouchet, J.; Giacomini, T. & Julvez, J. (1995). [Human diffusion of arthropod disease vectors throughout the world]. *Cahiers Santé,* Vol.5, No.5, pp. 293-298

Mouchet, J.; Manguin, S., Sircoulon, J., Laventure, S., Faye, O., Onapa, A.W., Carnevale, P., Julvez, J. & Fontenille, D. (1998). Evolution of malaria in Africa for the past 40 years: impact of climatic and human factors. *Journal of the American Mosquito Control Association,* Vol.14, No.2, pp. 121-130

Mouchet, J. (1999). [Vectors and environmental factors in malaria]. *Transfusion Clinique et Biologique,* Vol.6, No.1, pp. 35-43

Mouchet, J. (2000). Airport malaria : a rare disease still poorly understood. *Eurosurveillance,* Vol.5, No.7, pp. 75-76

Mouchet, J.; Carnevale, P., Coosemans, M., Julvez, J., Manguin, S., Richard-Lenoble, D. & Sircoulon, J. (2004). *Biodiversité du paludisme dans le monde.* John Libbey Eurotext, Paris, France

Narang, S.K.; Klein, T.A., Perera, O.P., Lima, J.B. & Tang, A.T. (1993). Genetic evidence for the existence of cryptic species in the *Anopheles albitarsis* complex in Brazil: allozymes and mitochondrial DNA restriction fragment length polymorphisms. *Biochemical Genetics,* Vol.31, No.1-2, pp. 97-112

O'Meara, G.F.; Larson, V.L., Mook, D.H. & Latham, M.D. (1989). *Aedes bahamensis*: its invasion of south Florida and association with *Aedes aegypti. Journal of the American Mosquito Control Association,* Vol.5, No.1, pp. 1-5

O'Meara, G.F.; Evans, L.F., Jr., Gettman, A.D. & Cuda, J.P. (1995). Spread of *Aedes albopictus* and decline of *Ae. aegypti* (Diptera: Culicidae) in Florida. *Journal of Medical Entomology,* Vol.32, No.4, pp. 554-562

Ostfeld, R.S. & Keesing, F. (2000). The function of biodiversity in the ecology of vector-borne zoonotic diseases. *Canadian Journal of Zoology,* Vol.78, No.12, pp. 2061-2078

Ostfeld, R.S. (2009). Biodiversity loss and the rise of zoonotic pathogens. *Clinical Microbiology Infection,* Vol.15 Suppl 1, pp. 40-43

Pages, F.; Peyrefitte, C.N., Mve, M.T., Jarjaval, F., Brisse, S., Iteman, I., Gravier, P., Tolou, H., Nkoghe, D. & Grandadam, M. (2009). *Aedes albopictus* mosquito: the main vector of the 2007 Chikungunya outbreak in Gabon. *PLoS One,* Vol.4, No.3, p. e4691

Parmakelis, A.; Russello, M.A., Caccone, A., Marcondes, C.B., Costa, J., Forattini, O.P., Sallum, M.A., Wilkerson, R.C. & Powell, J.R. (2008). Historical analysis of a near disaster: *Anopheles gambiae* in Brazil. *American Journal of Tropical Medicine & Hygiene,* Vol.78, No.1, pp. 176-178

Paskewitz, S.M.; Wesson, D.M. & Collins, F.H. (1993). The internal transcribed spacers of ribosomal DNA in five members of the *Anopheles gambiae* species complex. *Insect Molecular Biology,* Vol.2, pp. 247-257

Peterson, C.N.; Day, S., Wolfe, B.E., Ellison, A.M., Kolter, R. & Pringle, A. (2008). A keystone predator controls bacterial diversity in the pitcher-plant (*Sarracenia purpurea*) microecosystem. *Environmental Microbiology,* Vol.10, No.9, pp. 2257-2266

Pettit, W.J.; Whelan, P.I., McDonnell, J. & Jacups, S.P. (2010). Efficacy of alpha-cypermethrin and lambda-cyhalothrin applications to prevent *Aedes* breeding in tires. *Journal of the American Mosquito Control Association,* Vol.26, No.4, pp. 387-397

Ramalingam, S. (1969). New record of *Aedes* (*finlaya*) *togoi* (Theobold) in West Malaysia. *Medical Journal of Malaya,* Vol.23, No.4, pp. 288-292

Read, A.F.; Lynch, P.A. & Thomas, M.B. (2009). How to make evolution-proof insecticides for malaria control. *PLoS Biology,* Vol.7, No.4, p. e1000058

Reinert, J.F.; Harbach, R.E. & Kitching, I.J. (2004). Phylogeny and classification of Aedini (Diptera: Culicidae), based on morphological characters of all life stages. *Zoological Journal of the Linnean Society,* Vol.142, pp. 289-368

Reiter, P. & Sprenger, D. (1987). The used tire trade: a mechanism for the worldwide dispersal of container breeding mosquitoes. *Journal of the American Mosquito Control Association,* Vol.3, No.3, pp. 494-501

Reiter, P. (1998). *Aedes albopictus* and the world trade in used tires, 1988-1995: the shape of things to come? *Journal of the American Mosquito Control Association,* Vol.14, No.1, pp. 83-94

Reiter, P. (2010a). A mollusc on the leg of a beetle: Human activities and the global dispersal of vectors and vector-borne pathogens, In: *Infectious Disease Movement in a Borderless World: Workshop Summary,* David A. Relman, E.R.C., and Alison Mack, Rapporteurs; Forum on Microbial Threats; Institute of Medicine (Ed.), pp. 324, The National Academies Press, Washington, D.C., USA

Reiter, P. (2010b). Yellow fever and dengue: a threat to Europe? *Eurosurveillance,* Vol.15, No.10, p. 19509

Renault, P.; Solet, J.L., Sissoko, D., Balleydier, E., Larrieu, S., Filleul, L., Lassalle, C., Thiria, J., Rachou, E., de Valk, H., Ilef, D., Ledrans, M., Quatresous, I., Quenel, P. & Pierre, V. (2007). A major epidemic of chikungunya virus infection on Reunion Island, France, 2005-2006. *American Journal of Tropical Medicine & Hygiene,* Vol.77, No.4, pp. 727-731

Ritchie, S.A.; Rapley, L.P. & Benjamin, S. (2010). *Bacillus thuringiensis* var. *israelensis* (Bti) provides residual control of *Aedes aegypti* in small containers. *American Journal of Tropical Medicine & Hygiene,* Vol.82, No.6, pp. 1053-1059

Robert, V.; Rocamora, G., Julienne, S. & Goodman, S.M. (2011). Why are anopheline mosquitoes not present in the Seychelles? *Malaria Journal,* Vol.10, No.1, p. 31

Rodhain, F. (1996). [Insects know no border]. *Médecine et Maladies Infectieuses,* Vol.26 Suppl 3, pp. 408-414

Rogers, D.J.; Randolph, S.E., Lindsay, S. & Thomas, C. (2001). *Health effects of climate change in the U.K.* U.K. Department of Health, London, UK

Romi, R.; Sabatinelli, G., Savelli, L.G., Raris, M., Zago, M. & Malatesta, R. (1997). Identification of a North American mosquito species, *Aedes atropalpus* (Diptera: Culicidae), in Italy. *Journal of the American Mosquito Control Association,* Vol.13, No.3, pp. 245-246

Rosen, L.; Rozeboom, L.E., Reeves, W.C., Saugrain, J. & Gubler, D.J. (1976). A field trial of competitive displacement of *Aedes polynesiensis* by *Aedes albopictus* on a Pacific atoll. *American Journal of Tropical Medicine & Hygiene,* Vol.25, No.6, pp. 906-913

Rozendaal, J.A. (1997). *Vector control: methods for use by individuals and communities.* WHO, Geneva, Switzerland

Schaffner, F. & Karch, S. (2000). [First report of *Aedes albopictus* (Skuse, 1984) in metropolitan France]. *Comptes Rendus de l'Académie des Sciences - Série III - Sciences de la Vie,* Vol.323, No.4, pp. 373-375

Schaffner, F.; Van Bortel, W. & Coosemans, M. (2004). First record of *Aedes* (*Stegomyia*) *albopictus* in Belgium. *Journal of the American Mosquito Control Association,* Vol.20, No.2, pp. 201-203

Schaffner, F.; Kaufmann, C., Hegglin, D. & Mathis, A. (2009). The invasive mosquito *Aedes japonicus* in Central Europe. *Medical & Veterinary Entomology,* Vol.23, No.4, pp. 448-451

Scholte, E. & Schaffner, F. (2007). *Waiting for the tiger: Establishment and spread of the* Aedes albopictus *mosquito in Europe.* Wageningen Academic Publishers, Wageningen, The Netherlands

Scholte, E.J.; Ng'habi, K., Kihonda, J., Takken, W., Paaijmans, K., Abdulla, S., Killeen, G.F. & Knols, B.G. (2005). An entomopathogenic fungus for control of adult African malaria mosquitoes. *Science,* Vol.308, No.5728, pp. 1641-1642

Scholte, E.J.; Den Hartog, W., Braks, M., Reusken, C., Dik, M. & Hessels, A. (2009). First report of a North American invasive mosquito species *Ochlerotatus atropalpus* (Coquillett) in the Netherlands, 2009. *Eurosurveillance,* Vol.14, No.45,

Séchan, Y.; Lardeux, F., Loncle, S., Rivière, F. & Mouchet, J. (1993). *Les arthropodes vecteurs de maladies et agents de nuisances en Polynésie Française.* Atlas de la Polynésie Française, Orstom, France

Shousha, A.T. (1948). Species-eradication: The eradication of *Anopheles gambiae* from Upper Egypt, 1942-1945. *Bulletin of the World Health Organization,* Vol.1, No.2, pp. 309-352

Singh, O.P.; Nanda, N., Dev, V., Bali, P., Sohail, M., Mehrunnisa, A., Adak, T. & Dash, A.P. (2010). Molecular evidence of misidentification of *Anopheles minimus* as *Anopheles fluviatilis* in Assam (India). *Acta Tropica,* Vol.113, No.3, pp. 241-244

Sinka, M.E.; Bangs, M.J., Manguin, S., Coetzee, M., Mbogo, C.M., Hemingway, J., Patil, A.P., Temperley, W.H., Gething, P.W., Kabaria, C.W., Okara, R.M., Van Boeckel, T., Godfray, H.C., Harbach, R.E. & Hay, S.I. (2010a). The dominant *Anopheles* vectors of human malaria in Africa, Europe and the Middle East: occurrence data, distribution maps and bionomic précis. *Parasites & Vectors,* Vol.3, p. 117

Sinka, M.E.; Rubio-Palis, Y., Manguin, S., Patil, A.P., Temperley, W.H., Gething, P.W., Van Boeckel, T., Kabaria, C.W., Harbach, R.E. & Hay, S.I. (2010b). The dominant *Anopheles* vectors of human malaria in the Americas: occurrence data, distribution maps and bionomic précis. *Parasites & Vectors,* Vol.3, p. 72

Sinka, M.E.; Bangs, M.J., Manguin, S., Chareonviriyaphap, T., Patil, A.P., Temperley, W.H., Gething, P.W., Elyazar, I.R., Kabaria, C.W., Harbach, R.E. & Hay, S.I. (2011). The dominant *Anopheles* vectors of human malaria in the Asia-Pacific region: occurrence data, distribution maps and bionomic précis. *Parasites & Vectors,* Vol.4, No.1, p. 89

Skovmand, O.; Ouedraogo, T.D., Sanogo, E., Samuelsen, H., Toe, L.P. & Baldet, T. (2009). Impact of slow-release *Bacillus sphaericus* granules on mosquito populations followed in a tropical urban environment. *Journal of Medical Entomology,* Vol.46, No.1, pp. 67-76

Smith, J.L. & Fonseca, D.M. (2004). Rapid assays for identification of members of the *Culex* (*Culex*) *pipiens* complex, their hybrids, and other sibling species (Diptera: culicidae). *American Journal of Tropical Medicine & Hygiene,* Vol.70, No.4, pp. 339-345

Soper, F.L. & Wilson, D.B. (1943). *Anopheles gambiae* in Brasil, 1930-1940. (New York, USA, The Rockefeller Foundation), p. 262

Tabachnick, W.J. & Walter, J. (1991). Evolutionary genetics of arthropod-borne diseases. The yellow fever mosquito. *American Entomologist,* Vol.37, pp. 14-26

Tatem, A.J.; Hay, S.I. & Rogers, D.J. (2006a). Global traffic and disease vector dispersal. *Proceedings of National Academy of Sciences of the USA,* Vol.103, No.16, pp. 6242-6247

Tatem, A.J.; Rogers, D.J. & Hay, S.I. (2006b). Global transport networks and infectious disease spread. *Advances in Parasitology,* Vol.62, pp. 293-343

Tatem, A.J. (2009). The worldwide airline network and the dispersal of exotic species: 2007-2010. *Ecography (Cop.),* Vol.32, No.1, pp. 94-102

Thien, L.B. (1969). Mosquito pollination of *Habenaria obtusata* (Orchidaceae). *American Journal of Botany,* Vol.56, No.2, pp. 232-237

Toto, J.C.; Abaga, S., Carnevale, P. & Simard, F. (2003). First report of the oriental mosquito *Aedes albopictus* on the West African island of Bioko, Equatorial Guinea. *Medical & Veterinary Entomology,* Vol.17, No.3, pp. 343-346

Van Bortel, W.; Harbach, R.E., Trung, H.D., Roelants, P., Backeljau, T. & Coosemans, M. (2001). Confirmation of *Anopheles varuna* in vietnam, previously misidentified and mistargeted as the malaria vector *Anopheles minimus. American Journal of Tropical Medicine & Hygiene,* Vol.65, No.6, pp. 729-732

Van Riper, C.; Van Riper, S.G., Goff, M.L. & Laird, M. (1986). The epizootiology and ecological significance of malaria in Hawaiian land birds. *Ecological Monographs,* Vol.56, pp. 327-344

Vinagradova, E.B. (2000). Culex pipiens pipiens *mosquitoes: taxonomy, distribution, physiology, genetics, applied importance and control.* Pensoft, Moscow, Russia

Weinstein, P.; Laird, M. & Browne, G. (1997). Exotic and endemic mosquitoes in New Zealand as potential arbovirus vectors (Wellington, UK, Ministry of Health), p. 16

WHO. (1989). Geographical distribution of arthropod-borne disease and their principal vectors. In: *WHO/VBC/89.967,* WHO (Ed.), Geneva, Switzerland

Wickson, F. (2010). Mosquitoes: just how much biodiversity does humanity need? *Nature,* Vol.466, No.7310, p. 1041

4

Exotic Insects in Italy: An Overview on Their Environmental Impact

Costanza Jucker and Daniela Lupi
University of Milan/ DiPSA
Italy

1. Introduction

Hundreds of alien arthropods, native to different continents or introduced from other parts of Europe, have been recorded in Italy in the last few decades (Jucker et al., 2009; Pellizzari & Dalla Montà, 1997; Pellizzari et al., 2005). Italy is at high risk of introduction of exotic insects, particularly because of the country's climatic conditions, which support many subtropical species, and its position in the middle of the Mediterranean Sea, which makes Italy a commercial and tourist crossroad. Invasion is a normal ecological process, but it can be incentivized by human activities with processes far more rapid than those that occur in purely native systems. Biological invasion by exotic species is considered one of the main threats to the loss of biodiversity; it can cause enormous damage in terms of socio-economic costs, harm human health, and result in ecological losses (Kenis et al., 2009). The rise in economic costs can be attributed to yield losses, in agriculture and horticulture, or to increases in the production costs for pest management. The severe economic impact of these species is evident: the crudest estimate of the total known monetary impact of alien species in Europe is close to 10 billion euros annually (COM, 2008). Moreover, the damage to human health leads to major costs connected to control programs and public health measures. Damage is due to irritation and disease transmission by hematophagous species, or to allergies, and other maladies by other groups of insects.
The introduction of a species affects the ecosystem of the new habitat in different ways: the invasive species can be a herbivore, predator, or parasitoid, and can compete with indigenous species via different interactions (competition, disease transmission, hybridization). The major pathways of arthropod migration are indeed accidental, represented by different means of transports, the trade in ornamental plants and other goods, and tourism; natural climatic events can also be responsible for the movement of arthropods. We likewise have to consider that sometimes the introduction of the invasive species is voluntary, and even if the insect itself is helpful, can become a serious threat to a country's biodiversity. The deliberate introduction into the new environment can be for productive purposes or as a control agent (the classical biological control, based on importation of natural enemies from the country of origin of the exotic pest). *Rodolia cardinalis* (Mulsant), imported in California in the late 19th century to control cottony cushion scale, has ever since been a textbook example illustrating the accomplishments of biological control.

2. Factors influencing alien species establishment and spread

Only some alien arthropods become significant as an outcome of invasion; whether this happens depends both on the species and on the susceptibility of the invaded ecosystem. Moreover, once the species has settled, damage can be heavy only when the population is sufficiently numerous. Activities such as agriculture, logging, and grazing enhance the establishment of exotics by creating disturbed sites for colonization (Sakai et al., 2001). However, we consider the climate, the hosts, the presence of natural enemies, the reproduction strategies, and habitat fragmentation as the factors that strongly influence the settlement of the alien species in the new environment and make it invasive.

2.1 Climate

Among the factors that influence the settlement and the invasiveness of a species, one of the most important is the climate and, consequently, the synchronization of the lifecycle in the new habitat. The climate effect is more evident in arthropods because they are poikilothermic, which means that temperature strongly influences their development.

A species may arrive in a new potential habitat, but the differences between the season and the instar of development of the insect may not allow the species to survive and settle. There have been many interceptions of dangerous pests that would not have even survived in the new habitat. *Leucinodes orbonalis* Guen. (Lepidoptera, Pyralidae), which was included in the European and Mediterranean Plant Protection Organization (EPPO) Alert List after more than 120 interceptions of infested *Solanum* fruits imported from Asia and Africa were made by several EPPO member countries (EPPO, 2011), can be reported as an example of this lack of synchronization in Italy. In our country the species was detected as larvae during phytosanitary control on easter eggplant fruits at Milan Airport in winter (Jucker et al., 2007), a period not suitable for insect development. We cannot exclude the possibility, however, that arrival in a more favorable season would allow this species to settle.

Sometimes a species of tropical or sub-tropical origin that cannot survive winter in a temperate climate can overcome this difficulty by entering greenhouses or houses. *Trialeurodes vaporariorum* (Westwood) (Hemiptera, Aleyrodidae), *Leptoglossus occidentalis* Heidemann (Hemiptera, Coreidae), and *Harmonia axyridis* Pallas (Coleoptera, Coccinellidae) are just a few examples of the application of this strategy in Italy. *T. vaporariorum,* accidentally introduced into Europe and Italy in the middle of the 19th century, is originally a sub-tropical species; although it can withstand slight frost for a short time, it overwinters in glasshouses in areas of colder climate. *L. occidentalis,* discovered near Vicenza in Northern Italy in autumn 1999 (Tescari, 2001), and *H. axyridis,* reared and introduced as a biocontrol agent (Pervez & Omkar, 2006), can survive in unfavorable low temperatures by entering houses at the onset of cold weather in search of a protected site to spend the winter.

Global warming can also influence alien species settlement (Perrings et al., 2010). Climate change, in fact, is expected to alter biodiversity, causing variation in phenology, genetic composition, and species ranges, and affecting species interactions and ecosystem processes both in autochthonous and in exotic species (Sutherst, 2000). Global warming could provide new opportunities for introduction to areas where, until recently, introduced species were not able to survive. The Mediterranean fruit fly *Ceratitis capitata* Wiedemann (Diptera, Tephritidae), a highly polyphagous quarantine pest that probably originates from sub-Saharan Africa, has limited its distribution to areas under the 41st parallel. In northern Italy, with its normal climate conditions, the insect is not able to overwinter in the field at

preimaginal stages, although adults can survive indoors and at temperatures higher than 10-12°C for prolonged periods (Rigamonti, 2004). However, global warming may allow stable colonization of *C. capitata* populations at higher latitudes than at present.
Moreover, a country's climatic situation can be favorable to a major or minor number of generations per year; it can influence the oviposition period, number of eggs, preimaginal development, and adult survival, and it can also make a species more or less aggressive. An example is *Lissorhoptrus oryzophilus* Kuschel (Coleoptera, Erirhinidae), one of the major insect pests in rice-growing regions of the United States, China, Japan, and Italy. This species is univoltine in many areas, including Italy, but it can be bivoltine or multivoltine in regions where the climate is warmer (Chen et al., 2005; Lupi et al., 2007; Shang et al., 2004).

2.2 Host

There are numerous examples of successful adaptation to Italian environments because of the presence of valuable hosts. We have to consider that many invading species are polyphagous, so they have a higher potential of establishment than do monophagous or oligophagous species.
Sometimes in the country of origin the insect coevolves with its host, and the development on other hosts in different countries (either varieties or species) leads to major damage. One of the first invasive agricultural insect pests that hit the Italian economy in the past was the American vine phylloxera, *Viteus (=Daktulosphaira) vitifoliae* (Fitch) (Hemiptera, Phylloxeridae) (Strapazzon & Girolami, 1983). Detected in Europe in the late 19th century, it is an example of the complexity of the host-plant relationship. Within the natural range of diffusion of phylloxera and its host, *Vitis vinifera* plants show very little damage from the aphid (Gullan & Cranston, 2010). The situation was different upon its arrival in Europe, where plants had no resistance to the aphid: on *V. vinifera* European species, the insect produces nodose and tuberose galls on the subapices of young roots, causing incalculable economic damage and social consequences (Kenis & Branco, 2010). The problem was largely solved by replanting European cultivars grafted onto resistant American rootstocks. The Asian black hornet *Vespa velutina* Lepeletier (Hymenoptera, Vespidae) is an example of predator-prey coevolution. This wasp is endemic to southeast Asia, where it is a predator of social wasps and bees, including both the native *Apis cerana* F. and the introduced *Apis mellifera* L. (Hymenoptera, Apidae). While *A. cerana* colonies have evolved defense strategies against *V. velutina*, the European honeybee sustains significantly greater losses (Tan et al., 2010). Even if the insect is not yet arrived in Italy, our country is at high risk of introduction because the insect has already settled in southwestern France (Haxaire, 2006).
Sometimes the species becomes invasive only after a host shift. An example is the rice water weevil *L. oryzophilus*, which was originally confined to spontaneous gramineous and cyperaceous plants in North America. When rice was introduced in the area where the insect lived, there was a host shift to it and the insect invasion started (Chen et al., 2005; Lupi et al., 2010). However, rice is not enough for insect development as spontaneous plants are still necessary in the spring, when the adults emerge from overwintering sites and rice is not yet emerged (Lupi et al., 2009; Tindall & Stout, 2003).

2.3 Natural enemies

When an exotic pest colonizes a new habitat, the absence of natural control agents can contribute to its settlement and to the consequent damage. Sometimes research in the country of origin can be useful in finding natural enemies as an alternative to chemical control. In fact, in many situations biological control could represent a valid and stable help to manage both autochthonous and exotic pests.

Among the recent examples of classical biological methods applied to introduced pests in Italy, we have to mention *Neodryinus typhlocybae* (Ashmead) (Hymenoptera, Dryinidae) and *Torymus sinensis* Kamijo (Hymenoptera, Torymidae). In 1987, *N. typhlocybae* was introduced in Italy into the Veneto region for the biological control of *Metcalfa pruinosa* (Say) (Hemiptera, Flatidae). First detected in Italy in 1979, this planthopper native of the United States rapidly spread because of its polyphagy and lack of natural enemies (Strauss, 2009; Zangheri & Donadini, 1980). Among the parasitoids in North America, *N. typhlocybae* was considered the most promising control agent. For this reason, a biological control project started in Italy at the end of the 1980s to import the parasitoid. Larvae develop as ectoparasitoids of *M. pruinosa* immatures, and adults also feed on the host, thus augmenting pest mortality. Over the past years, this insect has been successfully released in many urban and agricultural areas of Italy, and this has resulted in its establishment (Alma et al., 2005).

The classical biological method based on the use of the parasitoid *T. sinensis* seems to be the most effective in reducing the populations of *Dryocosmus kuriphilus* Yasumatsu (Hymenoptera, Cynipidae), a species of Chinese origin that attacks chestnut trees and that invaded Japan, Korea, Nepal, the United States, and Europe (Brussino et al., 2002; Moriya et al., 2003; Quacchia et al., 2009). Removing twigs or protecting them with nets proved to be very labor intensive and produced little results. Chemicals were also inefficient as the immature stages are protected within the gall.

New interactions among recently introduced species and endemic natural enemies are also very important. Again, *D. kuriphilus* and *H. axyridis* can be used as examples. It has been reported in many countries and apparently in Italy as well that *D. kuriphilus,* when established, rapidly recruits parasitoids attacking oak gall wasps locally (Aebi et al., 2007) and that *H. axyridis* can be parasitized by *Dinocampus coccinellae* (Shrank) (Hymenoptera, Braconidae), a general parasitoid of Coccinellidae present in Italy (Dindo et al., 2011). The relationship among alien pests and autochthonous useful species is particularly remarkable when the origin of a species is unknown or few data are available on natural enemies in the native country. We refer to the horse chestnut leaf miner *Cameraria ohridella* Deschka and Dimic (Lepidoptera, Gracillariidae), described as a new species in Macedonia and whose origin is still under debate (Deschka & Dimic, 1986; Valade et al., 2009). Detected in Italy in the 1990s, it spread rapidly and caused a great deal of aesthetic damage to the plants it attacked. Studies on this species showed that numerous native parasitoids adapted to *C. ohridella* with, sometimes, a very high parasitization percentage (Lupi, 2005; Gröbler & Lewis, 2008; Grabenweger et al., 2009). We also refer to *Anoplophora chinensis* Forster (Coleoptera, Cerambycidae), the citrus longhorned beetle, for which information on parasitoids in the native country is lacking. Studies carried out in Italy led to the detection of some polyphagous endemic parasitoids and the Asiatic egg parasitoid, *Aprostocetus anoplophorae* Delvare (Hymenoptera, Eulophidae) , probably introduced with *A. chinensis* and the efficacy of which is still under study (Delvare et al., 2004; Hérard et al., 2004). Studies concerning the presence of natural enemies of *Cacyreus marshali* Butler (Lepidoptera, Lycaenidae) in Italy gave evidence to the presence of the oophagous *Trichogramma evanescens* (Westwood) (Hymenoptera, Trichogrammatidae), already detected in other European countries, and of the larval parasitoid *Aplomya confinis* Fallen (Diptera, Tachinidae) (Vicidomini & Dindo, 2006).

2.4 Reproduction strategies

Among the traits that promote successful invasion after initial colonization, there is the establishment of a self-sustaining population. Reproduction strategies that feature the ability to reproduce asexually or to shift between r- and K-selected strategies seem to play a really important role in this process (Sakai et al., 2001). Among insects reproducing sexually at low densities in the founder population, the efficiency in mate finding can influence insect settlement. The ability to reproduce asexually overcomes this difficulty and facilitates the settlement. Because the founding population of an invading species often suffers from a lack of genetic diversity, organisms that reproduce sexually could be more prone to extinction because of inbreeding (Liebhold & Tobin, 2008). Hoffmann et al. (2011), evaluating two databases from North America and Italy, found that parthenogenetic species cover 45% (North America) or 48% (Italy) of pest species derived from genera where parthenogenesis occurred. They concluded that several factors may contribute to the high incidence of parthenogenesis among pests. The most likely factor is a stable and uniform agricultural environment with an abundance of resources. In such a habitat, the same genotype may be continuously favored by selection, leading to a selective advantage of some parthenogenetic lineages over a sexual population. Among the successful parthenogenetic species in Italy we have to mention many Hemiptera. Moreover, *Lissorhoptrus oryzophilus* is a parthenogenetic species in nearly all colonized areas and in Italy (Lupi et al., 2007).

The lack of genetic diversity has a notable exception in the Argentine ant *Linepithema humile* (Mayr) (Hymenoptera, Formicidae), for which reduced genetic variability likely enhanced the invasion success (Suarez et al., 2008). Native to South America, it is now a cosmopolitan pest. In its area of origin the species has a social system with multicoloniality and systematic aggression between workers from different nests, while in invaded countries, it exhibits an extraordinary social organization, called unicoloniality, in which individuals mix freely among physically separated nests because of a loss of aggression between colonies having the same recognition alleles. From 33 Argentine ant populations collected along the Mediterranean and Atlantic coasts from Portugal to Italy, it was established that in this area there are only two supercolonies and that aggression never occurred between individuals from the same supercolony, even when taken from very distant nests (Giraud et al., 2002).

2.5 Habitat fragmentation

Both long-distance dispersal and stratified diffusion can greatly increase invasion rates: short-range expansion is mainly due to insect adult active dispersal flying, crawling, and swimming, whereas long-range expansion is due to accidental movements caused by human transportation. The alteration of land-use patterns has resulted in the fragmentation of habitats, ecosystems, and landscapes in most parts of the world, and several theoretical explorations indicate that spread rates are affected by habitat fragmentation (Hoffmeister et al., 2005; Perrings et al., 2010). In fragmented habitats very specialized species or those species with poor dispersal ability may suffer more than generalistic and invasive species. The Italian outbreaks of *Cameraria ohridella*, *Diabrotica virgifera virgifera*, and *Lissorhoptrus oryzophilus* have expanded by stratified dispersal, involving both continuous diffusion and discontinuous long-distance dispersal (Ciosi et al., 2011; Gilbert et al., 2004; Lupi et al., 2010).

3. Secondary effects of alien species introduction

Many alien species are dangerous not only for the direct damage they can cause, but also for their secondary effects. Invasive species can act as vectors for pathogens both to plants and to humans. The impact of invasives on native species, communities, and ecosystems has been widely recognized for decades. Ecological interactions between native and invasive species may be direct (predation, herbivory, parasitism, competition, mutualism) or indirect (habitat alteration, apparent predation, cascading trophic interactions) and result in changes in the population biology of the native species.

3.1 Pathogen transmission

Alien species can behave as vectors of pathogens (e.g., nematodes, fungi, viruses, bacteria) to both plants and animals.

Hemiptera are known for their aptitude to transmit pathogens. Among them we have whiteflies, aphids, psyllid, and leafhoppers. *Scaphoideus titanus* Ball (Hemiptera, Cicadellidae), a leafhopper introduced in Europe from North America and currently spread in vineyards, is the vector of Flavescence Dorée, a quarantine grapevine disease caused by the bacterium '*Candidatus Phytoplasma vitis*' (Angelini et al., 2001).

Among thrips, *Frankliniella occidentalis* (Pergande) (Thysanoptera, Thripidae), a North American invasive species introduced in Europe in 1983 and currently widespread, is the main vector of the tomato spotted wilt virus, a serious disease in several economically important crops worldwide and in many wild plants that can behave as reservoirs for the virus in the environment. This disease was first described in Australia in 1915 (Brittlebank, 1919). In Europe the virus was present before the Second World War, but the malady was noticeable only when *F. occidentalis* arrived as the preexisting vector, *Thrips tabaci* (Thysanoptera, Thripidae), has a low transmission rate (Inoue & Sakurai, 2006).

Coleoptera are important as they can be vectors of wood nematodes. The presence of the pinewood nematode *Bursaphelenchus xylophilus* Steiner & Buhrer, a major threat to pine forests in Japan, is remarkable. This quarantine disease, whose vector is the cerambycid *Monochamus alternatus* (Coleoptera, Cermabycidae) Hope, has been found in Europe, particularly in Portugal (Mota et al., 1999). In Italy neither the insect nor the nematode has been detected so far.

Many insects can bite and transmit disease, at least potentially. An Asian mosquito species, *Stegomyia albopicta* (Skuse) (*sensu* Reinert et al., 2004) (Diptera, Culicidae), often referred to as the "tiger mosquito", is the vector of major human diseases such as dengue, yellow fever, the West Nile virus, and chikungunya (Eritja, 2005). This species began to spread worldwide in the 1970s thanks to the marine transport of tires and other goods, leading to colonization of many areas of the world. European concern rose when the species was detected in Italy, first in Genoa and one year later near Padua (Dalla Pozza & Majori, 1992). Fortunately the species has invaded many countries without carrying human pathogens. However, we cannot ignore the unexpected outbreaks of chikungunya fever in northern Italy (Angelini et al., 2007), the sudden appearance of West Nile virus in North America, and the increasing frequency of Rift Valley fever epidemics in the Arabian Peninsula.

3.2 Autochthonous species competitiveness

Many studies have documented invaders that show a superior ability to exploit local resources as compared with native residents (Bazzocchi et al., 2004; Burgio et al., 2008; Holway, 1998) or as compared with non-invading introduced species. The Argentine ant (*Linepithema humile*) is not only an agricultural and house pest but is also competitively superior to the native ant species in both interference and exploitative competition in the

riparian woodlands of northern California and in the alteration of plant community structure (Holway, 1998; Suarez et al., 2008).
The intentional introduction of natural enemies for biological control has raised some criticism (Michaud, 2002) on the potentially negative effects induced by exotic species on indigenous natural ones and, in general, on non-target native taxa (Howarth, 1991; Pearson & Callaway, 2003; Van Lenteren et al., 2003). We must consider that the more a natural control agent is specific, the less negative the impact is on the environment. Predators have a generally wide range of hosts whereas parasitoids are usually more specific and so potentially less invasive. The coccinellid *H. axyridis*, a voracious predator native to southern Siberia, China, Taiwan, Korea, and Japan, is a major example of this impact. This coccinellid has been introduced in North America and Europe to control aphid and scale populations, and has settled (Pervez & Omkar, 2006). *H. axyridis* has been used in greenhouses since the 1990s, and its negative impact was already reported in 1995 in America. Nowadays this species is considered the most invasive ladybeetle in the world: it reduces the biodiversity of the new colonized areas, directly affecting useful insects, replacing the native ladybeetles, and altering the ecosystem equilibrium (Alyokhin & Sewell, 2004).

3.3 Landscape alteration

Alien species invasion can contribute to landscape alteration. This effect can be due directly to insect feeding or indirectly to the necessity of altering the environment because of eradication programs. The devastating impact of the introduction of the red palm weevil *Rhynchophorus ferrugineus* (Olivier) (Coleoptera, Curculionidae) on landscape palms is well known (Jucker et al., 2007; Longo & Tamburino, 2005; Sacchetti et al., 2005). In many coastal areas in Italy the urban landscape is principally based on palm, and many of such palms represent a patrimony of high value. The settlement of this weevil in our country represents a serious environmental problem liable to have significant repercussions both on the landscape and on tourism. Eradication programs that consist in the removal of trees infested with *Anoplophora chinensis,* detected in northern Italy (Colombo & Limonta, 2001a), are altering the landscape of many areas, especially urban and suburban (Maspero et al., 2007).
Another example of habitat alteration is correlated to the voluntary introduction of the Ailanthus silkmoth (*Samia cynthia*), a saturniid moth, used to produce silk and introduced in Italy at the end of the 19th century as a possible alternative to *Bombyx mori*. The impact of this insect on biodiversity has been indirect because the insect itself cannot be considered an invasive insect as it does not threaten any other species, nor does it invade the niche of others. Its introduction, however, was associated with that of its preferred host, *Ailanthus altissima,* a deciduous tree indigenous to China, which is a serious threat to ecosystems in introduced areas, as the plant is very competitive and it contains allelopathic chemicals that may inhibit the growth of surrounding native plants (Heisey, 1990).

3.4 Human nuisance

Sometimes the insects do not transmit pathogens but annoy humans and cattle. Examples are *Leptoglossus occidentalis* and *Harmonia axyridis,* which do not bite, sting, feed on people and pets, or reproduce inside the house, but disturb because of their large number in houses, buildings, and food industries, and because they can give off a pungent odor when touched. The habit of *H. axyridis* to aggregate on grapes in vineyards before harvest also results in great consequences for the grape industry (Linder et al., 2009). Further cases are reported of *H. axyridis* leading to patient complaints and symptoms of rhinitis, wheezing, and urticaria after exposure to the beetles (Albright et al., 2006; Yarbrough et al., 1999). A problem, encountered in crowded parks in cities, is represented by the presence of species of the genus *Corythucha* (Hemyptera, Tingidae). The species usually feed on ornamental plants, but if they fall on people, during probing, they can bite. A historical example for Italy is the abundant presence on *Platanus* spp. of *Corythucha ciliata* Say.

4. Italian situation on invasive species

Within the Research Project of European Commission DAISIE (www.europe-aliens.org), an inventory of alien species in Europe was taken. Among introduced organisms, terrestrial invertebrates, mostly insects, represent the major cases of exotic introductions (Roques et al., 2009). Concerning Italy, 425 species of insects and 22 mites were recorded from 1492 to 2006 (Jucker et al., 2007, 2009; Pellizzari & Dalla Montà, 1997; Pellizzari & Faccoli, 2007). In our analysis we considered the species detected in Italy as exotic pests introduced by human activities or by passive transport from other countries, starting from 1945. We did not count insects that are naturally spreading in different Italian regions (e.g. as a result of the global warming process). The list was made after an examination of different material, mostly published in scientific journals and the EPPO website. We considered the species as being of agricultural, forest, or human importance for their different impact. Species introduced as biological control or species of stored products were not counted. The year of publication was considered as the same of that of detection; we have to point out, however, that in many cases the arrival could have happened some years before the official report, because most introductions occur unintentionally. A total of 291 exotic insects have been listed.

4.1 Country of origin

The native area of each species was taken from the publication of the interception. For our purposes, the world area was divided as follows: Africa (excluding Mediterranean Basin countries), Asia, Australia, North America, Central and South America, Mediterranean Basin, Europe (excluding Mediterranean Basin countries), and Pacific Islands (Australia not included). When the origin was not sure, as the case of many cosmopolitan species, we reported it as unknown (Fig. 1).

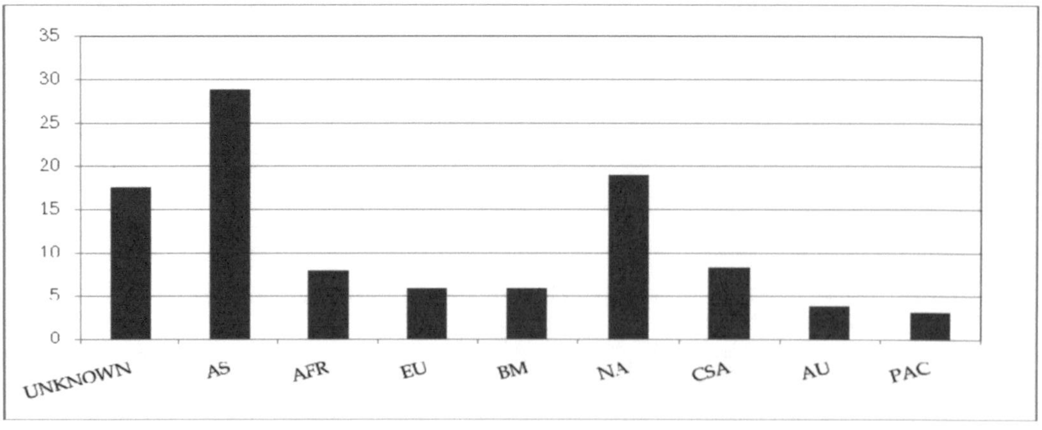

Fig. 1. Original country of the alien species in Italy. AS, Asia; AFR, Africa (excluding Mediterranean Basin countries); EU, Europe; BM, Mediterranean Basin countries; NA, North America; CSA, Central-South America; AU, Australia; PAC, Pacific Islands.

Asia accounts for the largest proportion of Italian alien species (28.9%), followed by North America (19.0%). In many cases (17.5%) the country of origin is unknown. Next, in order of decreasing importance, come the Central-South American countries (8.2%), Africa (7.9%), and European countries, in particular from the East (5.8%) and Mediterranean Basin countries (5.8%). Australia and other Pacific islands together represent 6.9%. In addition, the proportion of the species coming from Asia and North America over the last 30 years represents 47.9% of the total.

Figure 2 illustrates the temporal evolutions of the detections of non-indigenous arthropods, with the data organized in decades (apart the first 15 years 1945-1960). The findings of alien species starts to increase significantly from 1980, and the highest number of new records was in the decade 1991-2000, with 111 species. In the last 10 years the number of new introductions has only slightly decreased. Considering our data, the rate of introduction increased from an average of 2.3 species/year until 1990 to an average of 9.9 species/year in the last 20 years.

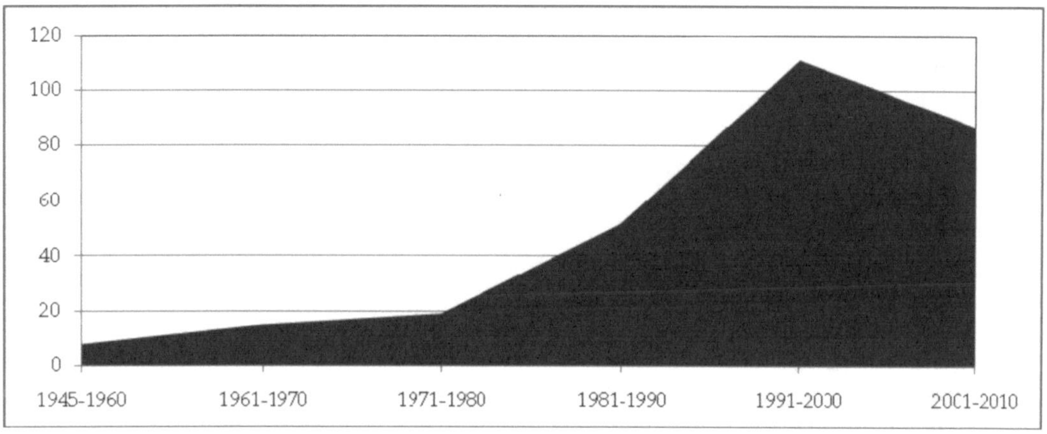

Fig. 2. Trend of non-indigenous species recorded in Italy in the recent decades.

4.2 Taxonomy

In the Figure 3 the taxonomic groups of the exotic insects in Italy are represented. Detailed descriptions of exotic species in the different orders follow.

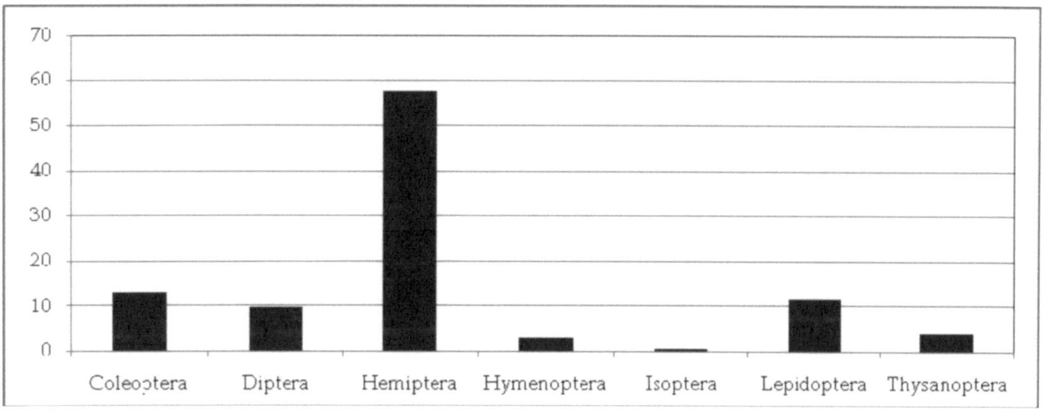

Fig. 3. Percentage of the orders introduced in Italy from 1945.

4.2.1 Hemiptera

In our list of phytophagous insects, Hemiptera alone represent more than 57.7% of the total of the recorded species: 97% are Homoptera, while only 3% belong to Heteroptera. Members of this order (particularly psyllids, scales, aphids, and whiteflies) can easily escape during the phytosanitary controls because of their small size and their position, often hidden under the foliage, inside the flowers, or on the plant roots. They mostly come from Asia and North America, and many of them are of worldwide distribution, with the country of origin unclear. Considering the host, more than 50% attack woody plants, while 34% are phytophagous of various herbaceous plants. Only in a small number of cases the pathway of introduction well known. Because a lot of species are host-plant specific, in many cases they are introduced with the trade of their host plant. The major vector of the species in the new habitat is represented by ornamental woody plants. The psyllid *Acizzia jamatonica* (Kuwayama), native to Asia, was first recorded in Europe, in Italy, in 2002 (Alma et al., 2002). It was introduced in our country with its host plant, *Albizia julibrissin*, a plant species used in streets and in parking areas. Belonging to the same genus are *A. acaciaebaileyanae* (Froggatt) and *A. uncatoides* (Ferris & Klyver), native to Australia and host of *Acacia* spp. and *Albizia* spp. Other examples of phytophagous insects on ornamental plants that are now widespread in our country are the two tyngids, *Stephanitis takeyai* Drake & Maa from Japan, and *S. pyrioides* Scott from North America, both pests of *Rhododendron* spp. (Colombo & Limonta, 2001b; Del Bene & Pluot-Sigwalt, 2005). The same plant species can host the aphid *Illinoia azaleae* (Mason) and the whitefly *Pealius azaleae* (Baker & Moles) (Del Bene et al.; 1991 Süss, 1973). The aphid *Illinoia liriodendri* (Monell) was found in Italy in 2001 (Limonta, 2001) on *Liriodendron tulipifera*, an exotic ornamental tree largely used for the color of its foliage and, in the past, for the absence of pests. Among scales affecting various ornamental woody plants mainly in gardens and urban parks, we refer to *Ceroplastes japonicus* Green (Kozar et al., 1984) and *C. ceriferus* (Fabricius) (Mori et al., 2001), both originating from Asia. *Pulvinaria hydrangeae* (Steinweden) is native to North America and its host is *Tilia* (Pellizzari, 1976). Recently, the red gum lerp psyllid *Glycaspis brimblecombei* Moore, attacking *Eucalyptus camaldulensis* in urban and rural landscapes, was reported (Laudonia & Garonna, 2010). *Corythucha ciliata* and *C. arcuata* are two Heteroptera introduced to Italy from North America and living respectively on *Platanus* spp. and on *Quercus* spp. (Bernardinelli & Zandigiacomo, 2000; Servadei, 1966). Two species with a high poliphagy are the well-known *Metcalfa pruinosa* and *Acanalonia conica* (Say) (D'Urso & Uliana, 2004). They both live on wild and cultivated trees, shrubs, and grasses belonging to different families, some of which have an economic importance (Alma et al., 2005; Wilson & McPherson, 1980).

Some Homoptera that damage forest plants are the scales *Matsucoccus feytaudi* Ducasse on *Pinus pinaster* and *Marchalina hellenica* Gennadius on pines (Arzone & Vidano, 1981; Kozar et al., 1984). We can also add *Gilletteella coweni* (Gillette) on Douglas fir and *Cinara cedri* and *C. laportei* on *Cedrus* (Covassi, 1971; Covassi & Binazzi, 1974, 1981). The western conifer seed bug *Leptoglossus occidentalis* is a serious pest of conifer seed production throughout western North America; economic damage consists in a reduction of the quantity and quality of seeds because of the trophic activity of adults and young instars sucking the seeds' endosperm. In Italy at the end of the 1990s the total weight of pine cones for pine nut production for food was about 40,000 tonnes per year. Since the introduction of the pest, production of pine nuts has rapidly decreased and in 2009 cone harvests from Italian stone pine forests declined by as much as 95% (Roversi et al., 2011).

Herbaceous plants that host Hemiptera include graminaceae and various flowers; *Tryonimus* spp. and *Antonina graminis* Maskell are just some examples (Pellizzari, 1994).

Among fruits plants, *Vitis vinifera* had been seriously attacked since the 1960s by *Scaphoideus titanus*, and more recently by the leafhopper *Erythroneura vulnerata* Fitch (Duso et al., 2005). The latter pest is widespread in the United States and Canada where it has been reported as a pest of wild and cultivated grapes, and of other host plants. Up to now its occurrence in commercial vineyards is relatively unimportant, probably because of insecticide applications carried out during summer to control other pests. Many hemipteran pests of *Citrus* cultivation are found in Italy: *Aonidiella citrina* (Coquillett), *Bemisia citricola* Gomez Menor, *Dialeurodes citri* Ashmead, *Aphis spiraecola* Patch, *Aleurothrixus floccosus* Maskell, *Chrysomphalus aonidum* L., *Unaspis yanonensis* (Kuwana), and *Aleurocanthus spiniferus* Quaintance. The last one is a quarantine species, representing a serious risk to Italy.

4.2.2 Coleoptera

Non-indigenous Coleoptera (13.1%) are mostly Cerambycidae, Chrysomelidae, Curculionidae, and Scolytidae. Asia and America have contributed the most non-indigenous Coleoptera, with more than 65% together, followed by Australia and the Pacific islands. Africa and Europe are of secondary importance.

Some alien beetles settled easily in our country, becoming agricultural pests. Omitting the well-acclimatized *Leptinotarsa decemlineata* Say, found in Italy in 1945, we want to focus on *Diabrotica virgifera virgifera* (LeConte), the northern corn rootworm from Central America, detected in 1992 near the Belgrade International Airport (Bača, 1994). The species, considered the major pest of maize in North America, settled in the new colonized area and now is present in 19 European countries. The chrysomelid arrived in Italy in 1998 and now occupies all the maize area in northern and central Italy. Its arrival in new countries can be attributed to aircraft; Miller et al. (2005) demonstrated that the pest has been introduced in Europe at least three times. Economic damage in Italy was observed in 2001 on only a few maize hectares (Boriani et al., 2002). More recently, the rice water weevil *Lissorhoptrus oryzophilus* was recorded in northern Italy (Caldara et al., 2004), but because this finding accidentally occurred during faunistic studies in the litter in Ticino Park (Lombardy), it was impossible to acquire information on the pathway of introduction of the insect (Lupi et al., 2010). Besides, until the rice water weevil detection, there were only a few arthropod species related to rice cultivation in Italy, with few outbreaks and little damage (Süss et al., 2008).

Among the exotic beetles that damage ornamental plants, some are extremely harmful and are listed as quarantine species. Among the most dangerous species of wooden plants worldwide we have to report *Anoplophora chinensis* and *A. glabripennis* (Motschulski), native to the Far East and detected in Italy in 2000 and in 2007, respectively (Colombo & Limonta, 2001a; Maspero et al., 2007). The first one has acclimatized in our country and is spreading to the north despite the eradication program. A single infestation has also been present in Rome from 2008. *A. glabripennis* seemed to be eradicated in the first area but in 2009 a new report was made in Veneto Region (EPPO, 2011). *A. chinensis* was introduced with living trees, while *A. glabripennis* was found in the backyard of a private company that imports valves, and other metallic parts in wood packaging material from China. They are both xylophagous species that damage many broad-leaf plants. The high rate of polyphagy of these species increases the possibility of the pest spreading, making eradication more difficult. Up to now in Italy for the eradication of *A. chinensis* more than 18,000 plants have been removed and the cost of the eradication program has amounted to about 12 million euros. Originating from Asia, the red palm weevil, *Rhynchophorus ferrugineus*, is now threatening palms in public and private villas, historical gardens, roads and squares with inestimable environmental and cultural value (in particular *Phoenix canariensis*, *P. dactylifera* and *Washingtonia robusta*), spreading in all Italian regions where palms are grown (Sacchetti et al., 2005). Because of the low efficacy of control methods, all infested plants are cut and removed. *Megaplatypus mutatus* Chapuis, polyphagous on broad-leaf species, can compromise the stability of plants. Coming from South America, it is considered a quarantine species. In Italy serious damage has been recorded on *Populus* sp. (Tremblay et al., 2000).

Not included among quarantine pests but considered by EPPO as a potential threat and thus included in the Alert List is the Asiatic xylophagous *Psacothea hilaris* (Pascoe), a pest of fig and mulberry trees. Detected in 2005, it has shown how dangerous it can be by killing infested plants in the area where it has settled (Jucker et al., 2006). Its bio-ethology in the new habitat is still under study. Another Asiatic exotic insect threatening *Ficus carica* is *Aclees* spp., found in 2005 in a nursery. Larvae bore tunnels at the base of the trunk and in the roots, compromising the health of the plants (Ciampolini et al., 2005). Some other xylophagous pests of ornamental and forest plants introduced into Italy are the cerambycid *Xylotrechus stebbingi* Gahan from North America, the two Scolytidae *Xylosandrus crassiusculus* (Motsculsky) from the Pacific islands and *Phloeotribus liminaris* (Harris) from North America (Dioli & Viganò, 1990; Pennacchio et al., 2003, 2004). *Phoracantha semipunctata* Fabricius, *Phoracantha recurva* (Newman), and *Gonipterus scutellatus* Gyllenhal are three pests of *Eucalyptus* native to Australia, introduced with infested wooden material (Arzone, 1976; Sama & Bocchini, 2003; Tassi, 1970). All these species, which affect different plants, have not shown a particular impact on our environment yet, but they represent further records among the numerous introductions in our country.

4.2.3 Lepidoptera

Lepidoptera represent 11.7% of the alien species in Italy; 30% come from Asia, 23% are native to America (North and Central-South), and 20% are of unknown origin. Others are European or African species.

In this order some species that cause serious economic damage are accounted. We refer, for example, to the Geranium Bronze *Cacyreus marshalli* (Butler), one of the most harmful insect pests of cultivated *Geranium* and *Pelargonium*. Native to South Africa, it was recorded in 1996 in central Italy and is now widespread (Trematerra et al., 1997; Lupi & Jucker, 2004). In 2002 *Paysandisia archon* (Burmeister), a palm tree pest native to South America, was found in southern Italy and easily settled in our country (Espinosa et al., 2003). This is the only exotic castniid known in Europe, and it is present in almost all Mediterranean coasts. In Europe the species has been found on several palms. Originally feeding on *Trithrinax campestris*, the species in Europe has switched from the host plants to *Phoenix canariensis*, *Latania* sp. and on the only native European palm, *Chamaerops humilis* (Montagud, 2004). In Italy *Chamaerops humilis*, *Trachycarpus fortunei*, *Phoenix canariensis*, and *Washingtonia* represent the hosts. The tomato leaf borer *Tuta absoluta* (Meyrick), counted in the EPPO A2 list of quarantine pests, was found in Italy in 2008 and has shown itself to be very aggressive on different Solanaceae, especially tomatoes (Viggiani et al., 2009).

Cameraria ohridella Deschka & Dimic, *Phillonorycter robiniellus* Clemens, *Parectopa robiniella* Clemens, *Phyllocnistis citrella* Stainton, and *P. vitegenella* Clemens are exotic leaf miners belonging to the family Gracillariidae, attacking woody plants. In particular, *C. ohridella* (Butin & Fuhrer, 1994) causes aesthetic damage through larval trophic activity. Host trees (*Aesculus hippocastanum*) rarely die, but can be completely defoliated in the middle of summer. Moreover, recent studies suggest that *C. ohridella* could have a potential negative impact on native leaf miners via apparent competition and could also represent a risk for *Acer pseudoplatanus* in some areas (Péré et al., 2010).

4.2.4 Diptera

Diptera are 9.6% of all the arthropods introduced in Italy, and species in the family Agromyzidae and Cecydomyiidae predominate. Horticultural and ornamental trade represents the most important pathway of introduction. Most of the species are of unknown origin (53.5%); others come essentially from North America, followed by Asia and Central-South America. Although they represent less than 10% of the introduced exotic insects, they include pest of economic importance.

The quarantine leafminers Agromizydae *Liriomyza trifolii* L. and *L. huidobrensis* Blanchard, coming respectively from North America and from Central-South America, caused much economic damage in the greenhouse cultivation of flowers and in general horticultural crops (Arzone, 1979; Süss, 1992).

Some species affect ornamental plants. The honeylocust gall midge *Obolodiplosis robiniae* (Haldeman) is strictly specific to *Robinia pseudacacia*. Detected in northern Italy in 2002, it is now present throughout the country (Bella, 2007). *Cecydomia gleditchiae* (Osten Sacken) cause galls on leaflets of *Gleditsia triacanthos*, while *Phytoliriomyza jacarandae* Steyskal & Spencer is the leafminer of *Jacaranda mimosifolia* (Bella et al., 2007; Bolchi & Volontè, 1984). Because of the species they attack and their monophagy, the importance of these pests is minimized to aesthetic damage, particularly in the area where these tree species are planted.

Rhagoletis completa Cresson represents an economically important pest of *Juglans regia* in the United States, especially in California. The species was recorded in Italy in 1991 (Duso, 1991) on some plants in the west of the country and could represent a serious threat to this tree and eventually to peach, known to be another host plant.

A species with economic importance is the newly introduced Drosophilidae *Drosophila suzukii* (Matsumura). Coming from the Far East, the species was recently recorded in North America and in Europe as noxious on strawberries, raspberries and other *Rubus*, blueberries, sweet cherries, plums, and many others fruit crops (EPPO, 2011). The dissemination of the pest over long distances is assured both by plants for planting and fruits. In Italy the species was detected in 2009 in the province of Trento (Grassi et al., 2009) and today represents a serious risk for many fruit crops with possible economic damage; *D. suzukii* was added to the Alert list in 2010 (EPPO, 2011).

Among diptera affecting humans we have to remember the mentioned *Stegomyia albopicta*.

4.2.5 Thysanoptera

Alien thrips (4.1%) are native to Asia, North and Central-South America, Australia, and the Pacific islands. Because of their size and the part of plants infested, they are mostly transferred through trade in ornamental greenhouse plants; in fact, this tropical species can easily establish inside greenhouses.

Frankliniella occidentalis (Perg.) is considered one of the major worldwide crop pests, listed within the quarantine species producing high economic damage to different crops. Detected in Italy in 1987, it has represented a serious problem in glasshouses, both for direct and indirect damage, since the 1990s (Arzone et al., 1989). In 1998 a massive infestation of *Echinothrips americanus* Morgan (Scarpelli & Bosio, 1999) was found in a heated greenhouse. This is a polyphagous species that can affect plants in glasshouses. Coming from America, it was probably introduced with ornamentals plants for planting. Another thrips on ornamentals in glasshouses is *Bradinothrips musae* (Hood), detected in 1999 in Lombardy on *Spatiphyllum* sp. and *Musa* sp. (Colombo et al., 1999). *Pezothrips kellyanus* (Bagnall), originally from Australia (Conti, 2001), was found on Citrus in 1998. This thrips can also attack *Jasmimum fruticans*, *Pittosporum tobira*, and *Lonicera* spp.

Thrips palmi Karny is a quarantine polyphagous pest, especially of Cucurbitaceae and Solanaceae, and represents a serious threat to numerous crops worldwide. Native to Asia, several interceptions have been reported, especially on cut flowers (mainly orchids) imported from Thailand, but up to now no settlement is known in Europe (Marullo, 1997).

4.2.6 Hymenoptera

Hymenoptera exotic species in Italy are mostly beneficials used as biological control agents. Pests in this order are few, representing 3.1% of the alien species. Belonging to the Eulophidae family and associated with *Eucalyptus* spp. are the species *Lectocybe invasa* Fisher & La Salle, *Ophelimus maskelli* Ashmead, and *Leprosa milga* Kim & La Salle; they are native to Australia and the Pacific islands and probably have been introduced with the host plant.

In 2002, the Hymenoptera Cinipidae *Dryocosmus kuriphilus* Yasumatzu, considered to be the most important pest of chestnut worldwide, was detected for the first time in Europe (in Piedmont, Italy) (Melika et al., 2002). From its first detection the pest rapidly expanded its colonization and now is present in quite the entire country, threatening chestnut production and *Castanea* trees, largely cultivated in some Italian regions. This quarantine species arrived with material for grafting.

Moreover, some exotic ants, mostly of tropical or subtropical origin, were recorded as new for the Italian fauna (Jucker et al., 2008). Because of their small size and their ability to nest in different materials, ants can easily be transferred by humans. Many species are found in glasshouses (e.g., *Pheidole megacephala* (Fabricius), *P. nodus* Smith, and *Tetramorium bicarinatum*) (Limonta & Colombo, 2003) or indoors in heated buildings. Among exotic ants, we include the well-known Argentine ant *Linepithema humile* (Mayr) and the pharaoh ant *Monomorium pharaonis* (Linnaeus). Considered as anthropic pests, they are mostly present in buildings. At present, if we exclude the Argentine ant, most of these non-indigenous ants seem harmless to environments and humans.

4.2.7 Isoptera

Among Isoptera (0.7%), the presence of *Cryptotermis brevis* (Walker), native to Central America, was recorded in 1994 (Tremblay & Priore, 1997). The species was first found in the South of Italy, but afterwards other findings were made in Palermo and in northern Italy (Raineri et al., 2001). Further interception was made in 2003, when an infestation of adults belonging to the *Coptotermes* genus was found on a boat in Naples harbor. These termites add to the other species of Isoptera still present in our country (*Kalotermes flavicollis* F. and *Reticulitermis lucifugus* (Rossi) (Savoldelli & Lupi, 2008).

4.3 Pathways

Beyond the natural spreading of species, most alien insects are accidentally transferred across countries through passive transport arising from commercial activities and human movement. Many interceptions occur on live wooden plants, mostly ornamental plants, considering the entire plant or just part of it. The import of bonsais has recently been receiving more and more attention for the introduction of alien insects, such as *Anoplophora chinensis* and numerous Homoptera (e.g. *Neophyllaphis podocarpi* Takahashi, *Lopholeucaspis japonica* (Cockerell), *Rhizoecus hibisci* Kawai & Takagi, and *Tinocallis ulmiparvifoliae* Matsumura). Wood packaging material and timber in general are associated with Coleoptera, in particular xylophagous, and Isoptera. Recently, some harmful cerambycids were imported on those goods, e.g. *Anoplophora glabripennis, Psacothea hilaris,* and *Megaplatypus mutatus. Anoplophora chinensis* is also known to be transported on pallets.

Notwithstanding phytosanitary regulations, these cryptic species can easily escape control because they live most of their biological cycle hidden inside the wood. Moreover, up to now effective detection methods do not exist, even though researchers are studying various techniques (e.g. acoustic methods, ultrasound, x-rays, tomography, infrared thermo imaging, detection dogs). Species are often transported inside living plants or wooden material without any observable symptoms outside. Only when the pests are in the new area and the environmental conditions are suitable do they emerge, causing much damage. Where living plants are concerned, horticultural and ornamental trade is probably the most significant pathway. Larvae of leaf miners can easily hide inside leaves or fruits (e.g. *Liriomyza* spp., *Rhagoletis completa*), as for the gall midge. Cut flowers, plant materials for propagation, and seeds are also known to carry pests (*Liriomyza trifolii, Dryocosmus kuriphilus, Frankliniella occidentalis,* etc.).

Means of transport also represent a possible pathway of entrance of exotic insects. The introduction of *Aedes albopictus* in Italy from the United States is ascribed to the trade of secondhand tires, where females lay eggs; moreover, larvae of *A. albopictus* were found inside bags watering "lucky bamboos" (*Dracaena senderiana*). Diffusion over short distances can be attributed to cars or trucks. The spreading of *Cameraria ohridella* inside Europe can also be ascribed to land transport. The arrival of *D. virgifera virgifera* in Europe, near the Belgrade airport, can be attributed to transfer by plane. Afterwards, many other new findings in different European countries were made close to airport areas by plane, as well the introduction of *Metcalfa pruinosa*.

Intentional introductions represent a high percentage of the species in some orders, like Hymenoptera and some Diptera, due to the large number of species introduced as biological control agents, which is not considered in our work.

5. Conclusion

The introduction of exotic arthropods is a dynamic non-stop process with new species reported each year. Insects probably represent the taxon with the most numerous exotic species in Italy. The history of new introductions in our country is similar to that in other parts of the world. The increasing globalization of trade and people can raise this trend with new interceptions every day, especially in industrialized countries with a high level of commercial exchanges. Moreover, the global warming phenomenon can enhance the risk of some species—particularly of tropical or subtropical origin—settling in the Mediterranean area, allowing them to spread in the new colonized area. Moreover, many Italian exotic species are listed among the 100 most invasive species in Europe (e.g. *Anoplophora chinensis, A. glabripennis, Stegomyia albopicta, Diabrotica virgifera virgifera, Harmonia axyridis*), and in many cases they have been recorded in our country for the first time in Europe, confirming that Italy is at high risk of introduction.

6. Acknowledgment

This research was supported by the University of Milan Research Project "Unimi 5permille: RISEKO" and was developed within the Research Projects financed by Lombardy Region and MiPAAF (Agricultural, Alimentary and Forestry Ministry).

The authors are grateful to all their colleagues that made these researches possible and in particular to our Professor Mario Colombo for his support, and to Prof. Lidia Limonta and F. Romana Eördegh for their useful comments.

7. References

Aebi, A.; Schönrogge, K.; Melika, G.; Quacchia, A.; Alma, A. & Stone, G.N. (2007). Native and introduced parasitoids attacking the invasive chestnut gall wasp *Dryocosmus kuriphilus*. *EPPO Bulletin*, Vol. 37, No. 1, (April 2007), pp. 166-171, ISSN 1365-2338

Albright, D.D.; Jordan-Wagner, D.; Napoli, D.C.; Parker, A.L.; Quance-Fitch, F.; Whisman, B.; Collins, J.W. & Hagan, L.L. (2006). Multicolored Asian lady beetle hypersensitivity: a case series and allergist survey. *Annals of Allergy, Asthma & Immunology*, Vol. 97, No. 4, (October 2006), pp. 521-527, ISSN 1081-1206.

Alma, A.; Tedeschi, R. & Rossi, J. (2002). *Acizzia jamatonica* (Kuwayama) nuova psilla per l'Europa (Homoptera Psillidae). *Informatore Fitopatologico*, Vol. 52, No. 4, (April 2002), pp. 64-65, ISSN 0020-0735

Alma, A.; Ferracini, C. & Burgio, G. (2005). Development of a Sequential Plan to Evaluate *Neodryinus typhlocybae* (Ashmead) (Hymenoptera: Dryinidae) Population Associated with *Metcalfa pruinosa* (Say) (Homoptera: Flatidae) Infestation in Northwestern Italy. *Environmental Entomology*, Vol. 34, No. 4, (August 2005), pp. 819-824, ISSN 0046-225X

Alyokhin, A. & Sewell, G. (2004). Changes in A Lady Beetle Community Following the Establishment of Three Alien Species. *Biological Invasions*, Vol. 6, No. 4, (December 2004), pp. 463–471, ISSN 1387-3547

Angelini, E.; Clair, D.; Borgo, M.; Bertaccini, A. & Boudonpadieu, E. (2001). Flavescence Dorée in France and Italy: occurrence of closely related Phytoplasma isolates and their near relationships to Palatinate grapevine yellows and an alder yellows phytoplasma. *Vitis*, Vol. 40, No. 2, (June 2001), pp. 79-86, ISSN 0042-7500

Angelini, R.; Finarelli, A.C.; Angelini, P.; Po, C.; Petropulacos, K.; Macini, P.; et al. (2007). An outbreak of chikungunya fever in the province of Ravenna, Italy. In: *Eurosurveillance*, Vol. 12, No. 36, Art. 3260, 17.03.2011, Available from: http://www.eurosurveillance.org

Arzone, A. (1976). Un nemico dell'Eucalipto nuovo per l'Italia. *L'Apicoltore Moderno*, Vol. 67, No. 6, (November-December 1976), pp. 173-177, ISSN 0518-1259

Arzone, A. (1979). L'agromizide neartico *Liriomyza trifolii* (Burgess) nuovo nemico di Gerbera in Italia. *Informatore Fitopatologico*, Vol. 29, No. 3, (March 1979), pp. 3–6, ISSN 0020-0735

Arzone, A. & Vidano, C. (1981). *Matsucoccus feytaudi* Due. (Homoptera, Margarodidae), fitomizo letale a *Pinus pinaster* Ait. in Italia. *Informatore Fitopatologico*, Vol. 31, No. 10, (October 1981), pp. 3-10, ISSN 0020-0735

Arzone, A.; Alma, A. & Rapetti, S. (1989). *Frankliniella occidentalis* (Perg.) (Thysanoptera Thripidae) nuovo fitomizo delle serre in Italia. *Informatore Fitopatologico*, Vol. 39, No. 10, (October 1989), pp. 43-48, ISSN 0020-0735.

Bača, F. (1994). New member of the harmful entomofauna of Yugoslavia, *Diabrotica virgifera virgifera virgifera* LeConte (Coleoptera, Chrysomelidae). *Zastita Bilja*, Vol. 45, No. 2, pp. 125-131, ISSN 0372-7866

Bazzocchi, G.; Lanzoni, A.; Accinelli, G. & Burgio, G. (2004). Overwintering, phenology and fecundity of *Harmonia axyridis* in comparison with native coccinellid species in Italy. *BioControl* 49, (June 2004), pp. 245–260, ISSN 1386-6141

Bella, S. (2007). Presenza di *Obolodiplosis robiniae* (Haldeman, 1847) in Italia centro-meridionale e in Sicilia (Diptera Cecidomyiidae). *Bollettino di Zoologia* Agraria e di Bachicoltura Ser. II, Vol. 39, No. (3), (December 2007), pp. 239–242, ISSN 0366-2403

Bella, S.; Mazzeo, G. & Süss, L. (2007). First record for the European fauna of *Phytoliriomyza jacaranda* Steyskal & Spencer, 1978 (Diptera Agromyzidae) leafminer of *Jacaranda mimosifolia* D. Don. (Bigoniaceae). *Bollettino di Zoologia agraria e di Bachicoltura. Ser. II*, Vol. 39, No. 1, (April 2007), pp. 75-78. ISSN 0366-2403

Bernardinelli, I. & Zandigiacomo, P. (2000). Prima segnalazione di *Corythucha arcuata* Say (Heteroptera: Tingidae) in Europa. *Informatore Fitopatologico,* Vol. 50, No. 12, (December 2000), pp. 47-49, ISSN 0020-0735

Bolchi, G. & Volontè, L. (1984-1985). *Dasineura gleditchiae* (Osten Sacken), cecidomide nuovo per l'Italia (Diptera Cecidomyiidae). *Bollettino di Zoologia* Agraria e di Bachicoltura *Ser. II*, Vol. 18, No. 2, (December 1985), pp. 185-189, ISSN 0366-2403

Boriani, M.; Bettoni, D. & Notarangelo, N. (2002). Primi danni da Diabrotica su mais in Italia. *L'Informatore Agrario*, Vol. 28, No. 31, (July 2002), pp. 61-62, ISSN 0020-0689

Brittlebank, C.C. (1919). Tomato diseases. *Agricultural journal of Victoria*, Vol. 27, pp. 231-235

Brussino, G.; Bosio, G.; Baudino, M.; Giordano, R.; Ramello, F. & Melika, G. (2002). Pericoloso insetto esotico per il castagno europeo. *L'Informatore agrario.* Vol. 28, No. 37, (September 2002), pp. 59-61, ISSN 0020-0689

Burgio, G.; Lanzoni, A.; Accinelli, G. & Maini, S. (2008). Estimation of mortality by entomophages on exotic *Harmonia axyridis* versus native *Adalia bipunctata* in semi-field conditions in northern Italy. *BioControl*, Vol. 53, No (1), (February, 2008), pp. 277-287, ISSN 1386-6141

Butin, H. & Fuhrer, E. (1994). The horse-chestnut miner (*Cameraria ohridella* Deschka & Dimic), a new pest of *Aesculus ippocastanum*. *Nachrichtenblatt des Deutschen pflanzenschutzdienstes*, Vol. 46, No. 5, pp. 89-91, ISSN:00277479

Caldara, R.; Diotti, L. & Regalin R., (2004). First record for Europe of the rice water weevil, *Lissorhoptrus oryzophilus* Kuschel (Coleoptera, Erirhinidae). *Bollettino di Zoologia Agraria e di Bachicoltura Ser. II*, Vol. 36, No. 1, (April 2004), pp. 165–171, ISSN 0336-2403

Chen, H.; Chen, Z.M. & Zhou, Y.S. (2005). Rice water weevil (Coleoptera: Curculionidae) in mainland China: invasion, spread and control. *Crop Protection*, Vol. (24), No. (8), (August 2005), pp. 695-702, ISSN 0261-2194

Ciampolini, M.; Regalin, R. & Peerin, H. (2005). *Aclees cribratus*, nuovo per l'Italia nocivo al fico allevato in vivaio. *L'Informatore agrario*, Vol. 61, No. 47, (December 2005), pp. 69-71, ISSN 0020-0689

Ciosi, M.; Miller, N. J.; Toepfer, S.; Estoup, A. & Guillemaud, T. (2011). Stratified dispersal and increasing genetic variation during the invasion of Central Europe by the western corn rootworm, *Diabrotica virgifera virgifera. Evolutionary Applications*, Vol. 4, No. 1, (January 2011), pp. 54–70, ISSN 1752-4571

Colombo, M.; Rigamonti, I.E. & Eördegh, F.R. (1999). Segnalazione di *Bradinothrips musae* (Hood) (Thysanoptera Thripidae) in una serra della Lombardia. *Bollettino di*

Zoologia Agraria e di Bachicoltura, Ser II, Vol. 31, No. 2, (August 1999), pp. 231-234, ISSN 0336-2403

Colombo, M. & Limonta, L. (2001)a. *Anoplophora malasiaca* Thomson (Coleoptera Cerambycidae Lamiinae Lamiini) in Europe. *Bollettino di Zoologia Agraria e di Bachicoltura, ser II,* Vol. 33, No. 2, (August 2001), pp. 65-68, ISSN 0336-2403

Colombo, M. & Limonta, L. (2001)b. Presenza di *Stephanitis takeyai* Drake e Maa (Heteroptera Tingidae) su *Pieris japonica* (Thunb.) introdotta in Italia. *Bollettino di Zoologia Agraria e di Bachicoltura, Ser II,* Vol. 33, No. 1, (April 2001), pp. 139-142, ISSN 0336-2403

COM (2008). Towards an eu strategy on invasive. In: *Communication from the commission to the council, the european parliament, the european economic and social committee and the committee of the regions. Commission of the European Communities 789 Final,* Available from: http://ec.europa.eu/environment/nature/invasivealien/docs/1_EN_ACT_part1_v6.pdf

Conti, B. (2001). Segnalazioni Faunistiche Italiane, N° 394. *Bollettino della Società Entomologica Italiana,* Vol. 133, No. 3, pp. 267-272, ISSN 0373-3491

Covassi, M. (1971). Osservazioni preliminari sulla presenza in Italia di un afide nocivo ai cedri: *Cedrobium laportei* Remaud (Homoptera, Lachnidae). *Redia,* Vol. 52, pp. 641-652, ISSN 0370-4327

Covassi, M. & Binazzi, A. (1974). Note corologiche e morfologiche sulla *Cynara cedri* Mim. in Italia (Homoptera, Lachnidae). *Redia,* Vol. 55, pp. 331-341, ISSN 0370-4327

Covassi, M. & Binazzi, A. (1981). Contributi alla conoscenza degli Afidi delle Conifere. IV, Note su alcune specie di Adelgidi reperite in Italia (Homoptera Adelgidae). *Redia,* Vol. 64, pp. 303-329, ISSN 0370-4327

Dalla Pozza, G. & Majori, G. (1992). First record of *Aedes albopictus* establishment in Italy. *Journal of the American Mosquito Control Association,* Vol. 8, No. 3, (September 1992), pp. 318–320. ISSN 8756-971X

Del Bene, G.; Gargani, E. & Landi, S. (1991). Note su *Pealius azaleae* (Baker et Moles) (Hom. Aleyrodidae) specie nuova per l'Italia. *Redia,* Vol. 74, pp. 163-175, ISSN 0370-4327

Del Bene, G. & Pluot-Sigwalt ,D. (2005). *Stephanitis pyrioides* (Scott) (Heteroptera Tingidae): a lace bug new to Italy. *Bollettino di Zoologia Agraria e di Bachicoltura Ser II,* Vol. 37, No. 1, (April 2005), pp. 71-76, ISSN 0336-2403

Delvare, G.; Bon, Mc.; Hérard, F.; Cocquempot, C.; Maspero, M. & Colombo, M. (2004). Description of *Aprostocetus anoplophorae* n. sp. (Hymenoptera: Eulophidae), a new egg parasitoid of the invasive pest *Anoplophora chinensis* (Förster) (Coleoptera: Cerambycidae). *Annales Société Entomologique de France (N. S.),* Vol. 40, No. 3-4, (August 2005), pp. 227-233, ISSN 0037-9271

Deschka, G. & Dimic., N. (1986). *Cameraria ohridella* sp. n. (Lep., Lithocolletidae) aus Mazedonien, Jugoslawien. *Acta entomologica jugoslavica,* Vol. 22, No. 1-2, pp. 11-23, ISSN 0350-5510

Dindo, M.L.; Francati, S.; Marchetti, E.; Lanzoni, A.; Maini, S.; Jucker, C. & Lupi, D. (2011). New associations between native parasitoids and two exotic insects introduced in Italy. *Proceedings of the Second Entomophagous Insects Conference,* Antibes France, 20-23 June 2011, in press

Dioli, P. & Viganò, C. (1990). Presenza in Valtellina di un cerambice nuovo per la Fauna italiana: *Xylotrechus stebbingi* Gahan, 1906. (Insecta, Coleoptera, Cerambycidae). *Il Naturalista Valtellinese,* Vol. 1, pp. 7-10, ISSN 1120-6519

D'Urso, V. & Uliana, M. (2004). First record of *Acanalonia conica* (Issidae) in Italy. *Proceeding of third European Hemiptera Congress,* Saint Petersburg, Russia, 8-11 June 2004

Duso, C. (1991). Sulla comparsa in Italia di un Tefritide neartico del noce: *Rhagoletis completa* Cresson (Diptera: Tephritidae). *Bollettino di Zoologia Agraria e di Bachicoltura, Ser II,* Vol. 23, No. 2, (August 1991), pp. 203-209, ISSN 0336-2403

Duso, C.; Bressan, A.; Mazzon, L. & Girolami, V. (2005). First record of the grape leafhopper *Erythroneura vulnerata* Fitch (Hom., Cicadellidae) in Europe. *Journal of Applied Entomology,* Vol. 129, No. 3, (April 2005), pp. 170-172, ISSN 1439-0418

EPPO (2011). European and plant protection organization. 10.03.2011, Available from: http://www.eppo.org

Eritja, R.; Escosa R.; Lucientes, J.; Marquès, E.; Molina, R.; Roiz, D. & Ruiz, S. (2005). Worldwide invasion of vector mosquitoes: present European distribution and challenges for Spain. *Biological Invasions,* Vol. 7, No. 1, (March 2005), pp. 87–97, ISSN 1387-3547

Espinosa, B.; Muccio, P. & Di Russo, G. (2003). *Paysandisia archon,* una minaccia per le nostre palme. *L'Informatore Agrario,* Vol. 59, No. 7, (February 2003), pp. 61, ISSN 0020-0689

Gilbert, M.; Grégoire, J.C.; Freise, J.F. & Heitland, W. (2004). Long-Distance Dispersal and Human Population Density Allow the Prediction of InvasivePatterns in the Horse Chestnut Leafminer *Cameraria ohridella. Journal of Animal Ecology,* Vol. 73, No. 3, (May 2004), pp. 459-46, ISSN 0021-8790

Giraud, T.; Pedersen, J.S. & Keller, L. (2002). Evolution of supercolonies: The Argentine ants of southern Europe. *PNAS,* Vol. 99, No. 9, (April 2002), pp. 6075-6079, ISSN 0027-8424

Grabenweger, G.; Kehrli, P.; ller, I. Z.; Augustin, S.; Avtzis, N.; Bacher, S.; Freise, J.; et al. (2009). Temporal and spatial variations in the parasitoid complex of the horse chestnut leafminer during its invasion of Europe. *Biological Invasions,* Vol. 12, No. 8, (August 2010), pp. 2797–2813, ISSN 1387-3547

Grassi, A.; Calmieri, L. & Giongo, L. (2009). *Drosophila (Sophophora) suzukii* (Matsumura). Nuovo fitofago per i piccoli frutti in Trentino. *Terra Trentina,* Vol. 54, No. 10, (November 2009), pp. 19-23

Gröbler, C. B. & Lewis, O.T. (2008). Response of native parasitoids to a range-expanding host. *Ecological Entomology,* Vol. 33, No. 4, (August 2008), pp. 453–463, ISSN 0307-6946

Gullan, P.J & Cranston, P.S. (2010). *The insects: An outline of entomology.* Blackwell Publishing, pp. 1- 559, ISBN 1444330365, Oxford UK

Haxaire, J.; Bouguet, J.P. & Tamisier, J.P. (2006). *Vespa velutina* Lepeletier, 1836, une redoutable nouveauté pour la faune de France (Hym., Vespidae). *Bulletin de la Société Entomologique de France,* Vol. 111, No. 2, pp. 194, ISSN 0037-928X.

Heisey, R.M. (1990). Evidence for allelopathy by tree-of-heaven (*Ailanthus altissima*). *Journal of Chemical Ecology,*Vol. 16, No. 6, (June 1990), pp. 2039-2055, ISSN 0098-0331

Hérard, F.; Cocquempot, C.; Lopez, J.; Covi, J.; Maspero, M. & Colombo, M. (2004). Field study to evaluate the egg parasitoid *Aprostocetus anoplophorae* sp. n. (Hymenoptera: Eulophidae) on two *Anoplophora* hosts. *Proceedings, XV USDA Interagency Research Forum on Gypsy Moth and Other Invasive Species,* Annapolis USA, 13-16 January 2004

Hoffmann, A.A.; Reynolds, K.T.; Nash, M.A. & Weeks, A.R. (2011). A high incidence of parthenogenesis in agricultural pests. *Proceeding Royal Society B.,* Vol. 278, (March 2011), pp. 799–800 ISSN 1364-5021

Hoffmeister, T.S.; Vet, L.E.M.; Biere, A.; Holsinger, K. & Filse, J. (2005). Ecological and Evolutionary Consequences of Biological Invasion and Habitat Fragmentation. *Ecosystems,* Vol. 8, No. 6, (September 2005), pp. 657–667, ISSN 1387-3547

Holway, D.A. (1998). Effect of Argentine ant invasions on ground-dwelling arthropods in northern California riparian woodlands. *Oecologia,* Vol. 116, No. 1-2, (August 1998), pp. 252-258, ISSN 0029-8549

Howarth, F.G. (1991). Environmental impacts of classical biological control. *Annual Review of Entomology,* Vol. 36, (January 1991), pp. 485-509, ISSN 0066-4170

Inoue, T. & Sakurai, T. (2006). Infection of *Tomato spotted wilt virus* (TSWV) shortens the life span of thelytokous *Thrips tabaci* (Thysanoptera: Thripidae). *Applied Entomology and Zoology,* Vol. 41, No. 2, (June 2006), pp. 239–246, ISSN 0003-6862

Jucker, C.; Tantardini, A. & Colombo, M. (2006). First record of *Psacothea hilaris* (Pascoe) (Coleoptera Cerambycidae Lamiinae Lamini) in Europe. *Bollettino di Zoologia agraria e Bachicoltura Ser. II,* Vol. 38, No. 2, (August 2006), pp. 187-191, ISSN 0336-2403

Jucker, C.; Maspero, M. & Colombo, M. (2007). Principali fitofagi in ambiente urbano e loro contenimento. *Informatore Fitopatologico* Vol. 57, No. 7/8, (June 2007), pp. 22-25, ISSN 0020-0735

Jucker, C.; Quacchia, A.; Colombo, M. & Alma, A. (2007). Hemiptera recently introduced into Italy. *Bulletin of Insectology,* Vol. 61, No. 1, pp. 145-146, ISSN 1586-4502

Jucker, C.; Valentini, M. & Lupi, D. (2007). Intercettazioni di insetti esotici in Lombardia - *Proceedings XXI Congresso Nazionale Italiano di Entomologia* pp.195, Campobasso Italy 11-16 June 2007

Jucker, C.; Rigato, F. & Regalin, R. (2008). Exotic ant records from Italy (Hymenoptera, Formicidae). *Bollettino di Zoologia agraria e Bachicoltura Ser. II,* Vol. 40, No. 1, pp. 99-107, ISSN 0336-2403

Jucker, C.; Barbagallo, S.; Roversi, P.F. & Colombo, M. (eds.) (2009). *Insetti esotici e tutela ambientali: morfologia, biologia, controllo e gestione,* MiPAAF, Cermenate (CO), Italy

Kenis M, Auger-Rozenberg M-A, Roques A, Timms L, Péré C, Cock MJW, Settele J, et al. (2009.) Ecological effects of invasive alien insects. Biological Invasions 11: 21–45.

Kenis, M. & Branco, M. (2010). Impact of alien terrestrial arthropods in Europe. *BioRisk,* Vol. 4, No. 1, (July 2010), pp. 51–71, ISSN 1313-2644

Kozar, F.; Tranfaglia, A. & Pellizzari, G. (1984). New records on the scale insect fauna of Italy (Homoptera: Coccoidea). *Bollettino del Laboratorio di Entomologia agraria "Filippo Silvestri",* Vol. 41, pp. 3-10, ISSN: 0304-0658

Laudonia, S. & Garonna, A.P. (2010). The red gum lerp psyllid, *Glycaspis brimblecombei,* a new exotic pest of *Eucalyptus camaldulensis* in Italy. *Bulletin of Insectology,* Vol. 63, No. 2, pp. 233-236, ISSN 1586-4502

Liebhold, A.M. & Tobin, P.C. (2008). Population Ecology of Insect Invasions and Their Management. *Annual Review of Entomology,* Vol. 53, (January 2008), pp. 387-408, ISSN 0066-4170

Limonta, L. (2001). Heavy infestation of *Illinoia liriodendri* (Monell) (Rhynchota: Aphididae) in northern Italy. *Bollettino di Zoologia Agraria e di Bachicoltura, Ser. II,* Vol. 33, No. 2, (August 2001), pp. 133-136, ISSN 0336-2403

Limonta, L. & Colombo, M. (2003). Record of *Pheidole megacephala* (F.), *Pheidole nodus* Smith and *Tetramorium bicarinatum* Nylander (Hymenoptera Formicidae), tropical species, in nursery imported plants. *Bollettino di Zoologia Agraria e di Bachicoltura, Ser. II,* Vol. 35, No. 2, (August 2003), pp. 287- 289, ISSN 0336-2403

Linder, C.; Lorenzini, F. & Kehrli, P. (2009). Potential impact of processed *Harmonia axyridis* on the taste of 'Chasselas' and 'Pinot noir' wines. *Vitis* , Vol. 48, No. 2, pp.101–102, ISSN 0042-7500

Longo, S. & Tamburino, V. (2005). Gravi infestazioni di punteruolo rosso della palma. *L'Informatore Agrario,* Vol. 61, No. 50, (December 2005), pp. 73-74, ISSN 0020-0689

Lupi, D. & Jucker, C. (2004). Il lepidottero *Cacyreus marshalli* minaccia le colture di geranio. *Clamer informa,* Vol. 29, No. 2, pp. 51-54, ISSN 0394-9435

Lupi, D. (2005). A three year field survey of the natural enemies of the horse-chestnut leaf miner *Cameraria ohridella* in Lombardy, Italy. *BioControl,* Vol. 50, No. 1, (February 2005), pp. 113-126, ISSN 1386-6141

Lupi, D.; Colombo, M.; Giudici, M.L.; Villa, B.; Sparacino, A.C. & Ranghino, F. (2007). Present status of knowledge on *Lissorhoptrus oryzophilus* Kuschel (Rice Water Weevil) in Italy. *Proceedings of the fourth temperate rice conference,* pp. 138-139, ISBN 978-88-95616-01-8, Novara, Italy, 25-28 June 2007

Lupi, D.; Cenghialta, C. & Colombo, M. (2009). Adult feeding by the rice water weevil *Lissorhoptrus oryzophilus* on different host plants. *Bulletin of Insectology,* Vol. 62, No. 2, pp. 229-236, ISSN 1586-4502

Lupi, D.; Giudici, M.L.; Cenghialta, C.; Villa, B.; Passoni, D. & Colombo, M. (2010). On the spatial spread of the Rice water weevil, *Lissorhoptrus oryzophilus* Kuskel (Coleoptera: Erirhinidae), in Italy. *Journal of Entomological and Acarological Research, Ser II,* Vol. 42, No. 2, (August 2010), pp. 81-90, ISSN 2038-324X

Marullo, R. (1997). *Thrips palmi,* un importante parassita da quarantena per l'Italia. *Informatore Fitopatologico,* Vol. 47, No. 10, (October 1997), pp. 18-22, ISSN 0020-0735

Maspero, M.; Jucker, C. & Colombo M. (2007). First record of *Anoplophora glabripennis* (Motschulsky) (Coleoptera Cerambycidae Lamiinae Lamiini) in Italy. *Bollettino di Zoologia agraria e Bachicoltura, Ser II,* Vol. 39, No. 2, (August 2007), pp. 161-164, ISSN 0336-2403

Maspero, M.; Cavalieri, G.; D'Angelo, G.; Jucker, C.; Valentini, M.; Colombo, M.; Hérard, F.; et al. (2007). *Anoplophora chinensis - Eradication programme in Lombardia (Italy),* 10.02.2011, Available from: http://www.eppo.org.

Melika G.; Brussino, G.; Bosio, G. & Csoka, G. (2002). Chestnut gall wasp (*Dryocosmus kuriphilus* Yamatzu), a new pest of chestnut in Europe. *Növényvédelem,* Vol. 39, No 2, pp. 59-63, ISSN 0133-0829

Michaud, J.P. (2002). Classical Biological Control: A Critical Review of Recent Programs Against Citrus Pests in Florida. *Annals of the Entomological Society of America,* Vol. 95, No. 5, (September 2002), pp. 531-540, ISSN 0013-8746

Miller, N.; Estoup, A.; Toepfer, S.; Bourguet, D.; Lapchin, L.; Derridj, S.; Kim, K. S.; Reynaud, P.; Furlan, L. & Guillemaud, T. (2005). Multiple transatlantic introductions of the western corn rootworm. *Science,* Vol. 310, No. 5750, (November 2005), pp. 992, ISSN 0036-8075

Montagud, S. (2004). *Paysandisia archon* (Burmeister, 1880) (Lepidoptera, Castniidae), nuevas localizaciones en la Península Ibérica y su gestión. *Boletín Sociedad Entomológica Aragonesa,* Vol. 34, (April 2004), pp. 237–246, ISSN 1134-6094

Mori, N.; Pellizzari, G. & Tosi, P. (2001). *Ceroplastes ceriferus* (Fabricius) (Hemiptera, Coccoidea): new pest of ornamentals in Europe? *Bollettino di Zoologia Agraria e di Bachicoltura, Ser II,* Vol. 33, No. 3, (December 2001), pp. 331-336, ISSN 0336-2403

Moriya, S.; Shiga, M. & Adachi, I. (2003). Classical biological control of the chestnut gall wasp in Japan, *Proceedings of the 1st International Symposium on biological control of arthropods*, pp. 407-415, Honolulu, Hawaii, 14-18 January 2002

Mota, M.M.; Braasch, H.; Bravo, M.A.;Penas, A.C.; Burgermeister, W.; Metge, K. & Sousa, E. (1999). First report of *Bursaphelenchus xylophilus* in Portugal and in Europe. *Nematology*, Vol. 1, No. 7-8, (November 1999), pp. 727-734, ISSN 1388-5545

Pearson, D.E. & Callaway, R.M. (2003). Indirect effects of host-specific biological control agents. *Trends in Ecology and Evolution*, Vol. 18, No. 9, (September 2003), pp. 456-461, ISSN 0169-5347

Pellizzari, G. (1976). Sulla presenza in Italia *dell'Eupulvinaria hydrangeae* (Steinw.) (Homoptera, Coccoidea). *Redia*, Vol. 59, pp. 59-67, ISSN 0370-4327

Pellizzari, G. (1994). Homoptera, Coccoidea new for Italy. *Bollettino di Zoologia Agraria e di Bachicoltura, Ser II*, Vo. 26, No. 2, (August 1994), pp. 271-274, ISSN 0336-2403

Pellizzari, G. & Dalla Montà, L. (1997). Gli insetti fitofagi introdotti in Italia dal 1945 al 1995. *Informatore Fitopatologico*, Vol. 47, No. 10, (October 1997), pp. 4-12, ISSN 0020-0735. 068

Pellizzari, G.; Dalla Montà, L. & Vacante, V. (2005). Alien insect and mite pests introduced to Italy in sixty years (1945-2004). *Proceeding of Plant protection and plant health in Europe: introduction and spread of invasive species*, pp. 275-276, ISBN 1 901396 819, Berlin, Germany, 9-11 June 2005

Pellizzari, G. & Faccoli, M. (2007). Checklist degli acari e insetti introdotti in Italia dopo la scoperta dell'America. *Proceeding of XXI Congresso Nazionale Italiano di Entomologia*, pp. 64, Campobasso, Italy, 11-16 June 2007

Pennacchio, F.; Roversi, P.F.; Francardi, V. & Gatti, E., (2003). *Xylosandrus crassiusculus* (Motschulsky) a bark beetle new to Europe (Coleoptera Scolytidae). *Redia*, Vol. 86, pp. 77-80, ISSN 0370-4327

Pennacchio, F.; Faggi, M.; Gatti, E.; Caronni, F.; Colombo, M. & Roversi, P.F. (2004). First record of *Phloeotribus liminaris* (Harris) from Europe (Coleoptera Scolytidae). *Redia*, Vol. 87, pp. 85-89, ISSN 0370-4327

Péré C.; Augustin S.; Turlings T. C. J. & Kenis, M. (2010). The invasive alien leaf miner *Cameraria ohridella* and the native tree *Acer pseudoplatanus*: a fatal attraction? *Agricultural and Forest Entomology*, Vol. 12, No. 2, (May 2010), pp. 151–159, ISSN 1461-9563

Perrings, C.; Mooney, H. & Williamson, M. (eds.) (2010). *Bioinvasion and globalization. Ecology economics, management*, Oxphord univ. press, ISBN 978-0-19-956015-8, New York

Pervez A. & Omkar A. (2006). Ecology and biological control application of multicoloured Asian ladybird, *Harmonia axyridis*: A review. *Biocontrol Science and Technology*, Vol. 16, No. 2, pp.111-128, ISSN 0958-3157

Quacchia, A.; Ferracini, C.; Moriya, S., & Alma, A. (2009). Italian experience in biological control of *Dryocosmus kuriphilus*. *Proceedings of the Japan-Italy joint international symposium A Global Serious Pest of Chestnut Trees, Dryocosmus kuriphilus: Yesterday, Today and Tomorrow*, pp. 14-17, Tsukuba Japan, 24-25 November 2009

Raineri, V.; Ray, A.; Marini, M. & Zaffagnini, V. (2001). Un nuovo rinvenimento di *Cryptotermes brevis* in Italia (Isoptera). *Bollettino della Società Entomologica Italiana*, Vol. 133, No. 2, pp. 99-102, ISSN 0373-3491

Reinert, J. F.; Harbach R. E. & Kitching I. J. (2004). Phylogeny and classication of Aedini (Diptera: Culicidae), based on morphological characters of all life stages. *Zoological Journal of the Linnean Society*, Vol. 142, pp. 289-368, ISSN 0024-4082

Rigamonti, I.E. (2004). Contribution to the knowledge of *Ceratitis capitata* Wied (Diptera, Tephritidae) in Nortern Italy II. Overwintering in Lombardy. *Bollettino di Zoologia agrarian e di Bachicoltura, Ser II*, Vol. 36, No. 1, (April 2004), pp. 89-100, ISSN 0373-3491

Roques, A.; Rabitsch, W.; Rasplus, J. Y.; Vaamonde, C. L.; Nentwig, W. & Kenis, M. (2009). Alien terrestrial invertebrates of Europe. In: *Handbook of Alien Species in Europe*, Drake J. A., pp. (63-79), Springer, ISBN 978-1-4020-8279-5

Roversi, P. F.; Strong, W. B.; Caleca, V.; Maltese, M.; Sabbatini Peverieri, G.; Marianelli, L.; Marziali, L. &. Strangi, A. (2011). Introduction into Italy of *Gryon pennsylvanicum* (Ashmead), an egg parasitoid of the alien invasive bug *Leptoglossus occidentalis* Heidemann. *EPPO Bulletin*, Vol. 41, No. 1, (April 2011), pp.72–75, ISSN: 1365-2338

Sacchetti, P.; Camera, A.; Granchietti, A.; Rosi, M.C. & Marzialetti, P., (2005). Prima segnalazione in Italia del curculionide delle palme, *Rhynchophorus ferrugineus*, 15.02.2011, Available from: http://www.cespevi.it/art/rhynco.htm

Sakai, A.K.; Allendorf, F.W.; Holt, J.S.; Lodge, D.M.; Molofsky, J.; With, K.A.; Baughman, S.; Cabin, R.J.; Cohen, J.E.; Ellstrand, N.C.; McCauley, D.E.; O'Neil, P.; Parker, I.M.; Thompson, J.N. & Weller, S.G., (2001). The population biology of invasive species. *Annual Review of Ecology, Evolution, and Systematics*, Vol. 32, (November 2001), pp. 305-332, ISSN 1543-592X

Sama, G. & Bocchini, R. (2003). *Phoracanta recurva* Newman, 1840 (Insecta Coleoptera Cerambycidae) prima segnalazione per la fauna italiana. *Quaderno della Società per gli Studi Naturalistici della Romagna*, Vol. 18, (December 2003), pp. 168, ISSN 1123-6787

Savoldelli, S. & Lupi, D. (2008). New breeding grounds of *Reticulitermes lucifugus* (Rossi) (Isoptera Rhinotermitidae) in Lombady (Italy). *Bollettino di Zoologia Agraria e di Bachicoltura, Ser II*, Vol. 40, No. 1, (April 2008), pp. 91-94, ISSN 0336-2403

Scarpelli, F. & Bosio, G. (1999). *Echinothrips americanus* Morgan, nuovo tisanottero delle serre. *L'Informatore Agrario*, Vol. 55, No. 2, (January 1999), pp. 59-61, ISSN 0020-0689

Servadei, A. (1966). Un tingide neartico comparso in Italia (*Corythucha ciliata* Say). *Bollettino della Società Entomologica Italiana*, Vol. 96, pp. 94-96, ISSN 0020-0689

Shang, H.; Stout, M.J.; Zhang, Z. & Cheng, J. (2004). Rice water weevil (Coleoptera: Curculionidae) population dynamics in Luoisiana. *Journal of Entomological Science*, Vol. 39, No. 4, pp. 623-642, ISSN 0749-8004

Strapazzon, A. & Girolami, V. (1983). Infestazioni fogliari di fillossera (*Viteus vitifoliae* (Fitch)) con completamento dell'olociclo su *Vitis vinifera* (L.) innestata. *Redia*, Vol. 66, pp. 179-194, ISSN 0370-4327

Strauss, G. (2009). Host range testing of the nearctic beneficial parasitoid *Neodryinus typhlocybae*. *Biocontrol*, Vol. 54, No. 2, (April 2009), pp.163–171, ISSN 1386-6141

Suarez, A.V.; Holway, D.A. & Tsutsui, N.D. (2008). Genetics and behavior of a colonizing species: The invasive argentine ant. *American Naturalist*, Vol. 172, No. 1, (July 2008), pp. 72-84, ISSN 0003-0147

Süss, L. (1973). Reperti afidologici su alcune piante in serra nel Milanese. *Bollettino di Zoologia agraria e Bachicoltura, Ser. II*, Vol. 11, No.3, pp. 177-193, ISSN 0336-2403

Süss, L., (1992). Prima segnalazione in Italia di *Liriomyza huidobrensis* (Blanchard) (Diptera: Agromyzidae). *Bollettino di Zoologia agraria e Bachicoltura Ser. II*, Vol. 23, No. 2, (August 1992), pp. 197-202, ISSN 0336-2403

Süss, L.; Lupi, D. & Savoldelli, S. (2008). Parassiti animali, In: *Il riso*, AA.VV. (2008), pp. 332-353, Bayer CropScience, Ed. Script, Bologna

Sutherst, R.W. (2000). Climate change and invasive species: a conceptual framework, In: *Invasive Species in a Changing World*, Mooney, H.A. & Hobbs, R.J., eds, pp. 211–240, Island Press, ISBN 1-55963-781-1, Washington DC

Tan, K.; Li, H.; Yang, M.X.; Hepburn, H.R. & Radloff, S.E. (2010). Wasp hawking induces endothermic heat production in guard bees. *Journal of Insect Science,* Vol. 10, (September 2010), pp. 1-6, ISSN 1536-2442

Tassi, F. (1970). Un longicorne australiano per la prima volta in Italia (Col. Cerambicidae). *Bollettino dell'Associazione Romana di Entomologia,* Vol. 24, pp. 69-71, ISSN 0004-6000

Tescari, G. (2001). *Leptoglossus occidentalis,* Coreide neartico rinvenuto in Italia (Heteroptera Coreidae). *Lavori della Società Veneziana di Scienze Naturali,* Vol. 26, pp. 3-5, ISSN 0392-9450

Tindall, K.V. & Stout, M.J. (2003). Use of common weeds of rice as hosts for the rice water weevil (Coleoptera: Curculionidae). *Environmental Entomology,* Vol. 32, No. 5, (October, 2003), pp. 1227-1233, ISSN 0046-225X

Trematerra, P.; Zilli, A.; Valentini, V. & Mazzei, P. (1997). *Cacyreus marshalli*, un lepidottero sudafricano dannoso ai gerani in Italia. *Informatore Fitopatologico,* Vol. 47, No. 7-8, (July-August. 1997), pp. 2-6, ISSN 0020-0735

Tremblay, E. & Priore, R. (1997). Termite centro-americana in via di acclimatazione in Italia. *L'Informatore Agrario,* Vol. 53, No. 47, (December 1997), pp. 76-77, ISSN 0020-0689

Tremblay, E. ; Espinosa, B.; Mancini, D. & Caprio, G. (2000). Un coleottero proveniente dal Sudamerica minaccia i pioppi. *L'Informatore Agrario,* Vol. 53, No. 48, (December 2000), pp. 89–90, ISSN 0020-0689

Valade, R.; Kenis, M.; Hernandez-Lopez, A.; Augustin, S.; Mari Mena, N.; Magnoux, E.; Rougerie, R.; Lakatos, F.; Roques, A. & Lopez-Vaamonde, C. (2009). Mitochondrial and microsatellite DNA markers reveal a Balkan origin for the highly invasive horse-chestnut leaf miner *Cameraria ohridella* (Lepidoptera, Gracillariidae). *Molecular Ecology,* Vol. 18, No. 16, (August 2009), pp. 3458–3470, ISSN 0962-1083

Van Lenteren, J.C.; Babandrier, D.; Bigler, F.; Burgio, G.; Hokkanen, H.M.T.; Kuske, S.; Loomans, A.J.M.; Menzler-Hokkanen, I.; et al. (2003). Environmental risk assessment of exotic natural enemies used in inundative biological control. *Biocontrol,* Vol. 48, No 1, (February 2003), pp. 3-38, ISSN 1386-6141

Vicidomini, S. & Dindo, M.L. (2006). Prima segnalazione europea di parassitizzazione di *Cacyreus marshalli* (Butler) da parte di un tachinide indigeno. A*nnali del museo civico di Rovereto*, Vol. 22, pp. 213-218, ISSN 1720-9161

Viggiani, G.; Filella, F.; Delrio, G.; Ramassini, W. & Foxi, C. (2009). *Tuta absoluta,* nuovo lepidottero segnalato anche in Italia. *L'Informatore Agrario,* Vol. 65, No. 2, (January 2009), pp. 66, ISSN 0020-0689

Wilson, S.W. & McPherson, J.E. (1980). A list of the host plants of the Illinois Acanaloniidae and Flatidae (Homoptera: Fulgoroidea). *Transactions of the Illinois State Academy of Science,* Vol. 73, No. 4, pp. 21-29, ISSN 0019-2252

Yarbrough, J.A.; Armstrong, J.L.; Blumberg, M.Z.; Phillips, A.E.; McGahee, E. & Dolen, W.K. (1999). Allergic rhinoconjunctivitis caused by *Harmonia axyridis* (Asian lady beetle, Japanese lady beetle, or lady bug). *Journal of Allergy and Clinical Immunology,* Vol. 104, No. 3, (September 1999), pp. 704-705, ISSN 0091-6749

Zangheri, S. & Donadini, P. (1980). Comparsa nel Veneto di un omottero neartico: *Metcalfa pruinosa* Say (Homoptera, Flatidae). *Redia,* Vol. 63, pp. 301- 305, ISSN 0370-4327

5

Invasion, Evenness, and Species Diversity in Human-Dominated Ecosystems

Eyal Shochat[1] and Ofer Ovadia[2]
[1]*The Israel Ornithological Center, the Society for Protection of Nature in Israel*
Tel Aviv
[2]*Departemnt of Life Sciences, Ben-Gurion University of the Negev*
Beer Sheva
Israel

1. Introduction

Biological invasions have become a major ecological problem globally (Vitousek et al., 1997; Mooney & Hobbs, 2000) for several non-mutually exclusive reasons. From the perspective of economics, many invasive alien species have become pests, sometimes causing severe damages to crops (Mack et al., 2000; Pimenel et al., 2000). From the ecological point of view, invasive species can inflict drastic changes in the invaded community altering its structure and function (Simberloff, 1997; Mooney & Cleland, 2001), and leading to biodiversity loss (Herbold & Moyle, 1986; Clavero & García-Berthou, 2005). The battle against invasions of alien species requires a great deal of efforts and resources, and yet is not always successful (Kaiser, 1999; Zavaleta et al., 2001). Whereas complete eradication of successfully established invasive species is in most cases impossible, controlling their population growth may decrease economical damages, and prevent biodiversity loss in the invaded community. Furthermore, successful eradication of one alien species may result in a worse alternative stable state (Beisner et al., 2003), where another, potentially more harmful invasive species dominates the community (e.g., Zavaleta et al., 2001). Therefore, fighting invasions should be done carefully and wisely, while taking into account multiple biotic and abiotic factors and outcomes. More importantly, such actions should emerge from a deep understanding of the evolutionary and ecological mechanisms that operate at different biological hierarchies to allow invasions, i.e., starting from the individual traits characterizing the invasive species (genetic background, morphology, physiology, behavior and life history), through its population parameters (demography, fluctuations, growth rate), and up to patterns and processes characterizing the invaded community (composition, structure, and age).

The demography of many invasive species has been well documented, and their damages were explored in details (Pimentel et al., 2000). However, the ways such species enter new ecosystems and communities, and their exact effect on community structure and biodiversity are not yet fully understood (Gurevitch & Padilla, 2004). Our goal is to review the evolutionary and ecological mechanisms of bio-invasions at the species (e.g., Facon et al., 2006) and community levels (e.g., Shea & Chesson, 2002), and present some of the general effects and outcomes of invasions at both levels. We discuss various mechanisms by which invasive species manage to establish viable populations and even dominate new ecosystems, how their unique biological traits allow rapid population growth, and the extent to which their interactions with native species trigger changes in community structure and function (i.e., community resilience).

Biological invasions are controlled by several factors. Invasive species must have good dispersal abilities, or otherwise they can only immigrate to remote places with the help of humans (Facon et al., 2006). Once the invader has found its way to the new environment, it must be reasonably adapted to the local conditions. If not, changes must occur within a short time, either in the species or in the environment (Facon et al., 2006). Rapid evolutionary changes have been shown to occur in invertebrates and plants shortly after immigration to the new environment (reviewed by Whitney & Gabler, 2008). Next, the invader's population must rapidly increase in size, reaching a critical threshold above which drift and disturbances are less likely to drive the invader into extinction (reviewed by Sakai et al., 2001). Once the invader is successfully established, it has the potential to modify community structure and function. For example, even if no changes in species composition occur, the relative abundance of the different species making up the community may change, resulting in few and highly abundant species dominating many low abundance species (Shochat et al., 2010). Below we discuss the potential negative effect of such reduction in evenness on community resilience.

2. The population/species level

2.1 Evolutionary mechanisms promoting invasion

Facon et al. (2006) discussed three potential scenarios for species to invade remote new environments. In the simple case, the only restriction is the species dispersal abilities, whereas the new environment is perfectly suitable for the species. If the species manages to arrive to the new environment (normally with the help of humans) it will establish successfully within a short time (assuming that it can overcome the negative interactions with the local species). The other two cases are more complex and interesting, because they deal with nearby species that either go through genetic or plastic changes (e.g., Maron et al., 2004), or take advantage of changes in the new environment (e.g., Seabloom et al., 2003).

In the second scenario the invader takes advantage of changes in the new environment. Such scenarios are probably common in nature, where some species may even manage to invade new environments within short temporal opportunities, such as disturbances in the new habitat. In such cases, the species may enter the environment during a periodical disturbance, quickly establish a self-sustained population, small as it may be, and remain in the new environment even if conditions return to the initial ones. For example, Seabloom et al. (2003) suggested that in California replacement of perennial grasses by invasive annual grasses across vast areas occurred as a result of disturbance that reduced water and nitrogen levels, and not because the perennial grasses were weak competitors. Such cases imply that proper management regimes may facilitate restoration of native biota in parts of the ecosystem.

Once established, the alien species may find a restricted niche, and remain rare and localized, or it may find perfect conditions, thrive, and even dominate the local community. Yet, even for such successful invasions, timing is also an important factor, because environmental factors change in both time and space. For several successful invasions of species that became abundant throughout countries and continents, it took more than one introduction and several failure cases before the population has established (e.g., Pimm, 1991; Veltman et al., 1996).

Both the first and the second scenarios suggest that a key trait for any invasive species is a high genetic variation (Mooney & Cleland, 2001). On the one hand, genetic variation serves as the raw material upon which natural selection can operate (Endler, 1986). On the other hand, genetic variation is positively correlated with population fitness (Reed & Frankham, 2003). The combined effect of both increases the survival probability of the species in the new environment (Facon et al., 2006). It is likely that selection will favor those individuals better adapted to the new environment. Assuming that their fitness advantage is large, their adaptive genes will rapidly increase in frequency, and despite the loss of genetic diversity, the invader's population will become highly adapted to the environment and successfully establish.

In the third scenario the species, that is originally not adapted to the new environment, has to change genetically or epigenetically [i.e., heritable changes in gene expression and function that cannot be explained by changes in DNA sequence (Richards, 2006)] in order to invade successfully. Environmentally induced epigenetic changes, usually occurring over a relatively short period of time, are often inherited by future generations (Richards, 2006). As such, they may increase the evolutionary potential of invasive species in response to challenges and stressors in novel environments (Bossdorf et al., 2008). Although epigenetic processes may play an important role in biological invasions, to date they have not been explored in that context. Future research on invasive species may benefit from including this aspect as a potential mechanism facilitating the establishment of alien species in novel environments.

2.2 Ecological mechanisms promoting invasion

Although species must be reasonably adapted to the new environment on the first place, lack of adaptations is not necessarily the most parsimonious explanation for invasion failure (Pimm, 1991). In other words, caution should be taken before arguing that an introduced species failed to establish in a new environment because it lacks physiological or genetic adaptation necessary to survive and reproduce in this environment (Pimm, 1991). For example, if adaptations are important, one would expect that prior experience of a habitat would increase invasion success. However, no support for this prediction was found by Moulton & Pimm (1986) when analyzing bird species that were introduced to Hawaii. Specifically, invasion success of birds originating from tropical habitats was not higher than that of birds originating from temperate habitats. Similarly, Crawley (1987) could not find evidence for the idea that "climate matching" between the habitat of origin and new habitat improves the invasion success of insects.

Both studies, however, found that widespread species were more successful invaders than species with small geographical range. The most intuitive explanation for this pattern is that widespread species experience a wide range of environmental conditions and thus have an advantage over restricted species, which are less likely to find suitable conditions in the new environment (Pimm, 1991). Clearly, there are many other possible explanations for this pattern such as that widespread species better cope with competitors and predators than the more geographically restricted species (Pimm, 1991).

A key feature for a successful invasion is an efficient utilization of food resources (i.e., competitive ability), which in turn can increase both individual and population growth rate (Sakai et al., 2001). For example, in plants, Grotkopp et al. (2002) found that the seedling relative growth rate of invasive pine species was higher than in non-invasive species. In addition to this characteristic, the studied pines were characterized by short generation time and low seed mass. Yet, high population growth rate alone is not necessarily sufficient for successful invasions, since in many cases increased growth rate is also associated with increased fluctuations in population size. In such cases species may reach high population density within a short time, but also decrease soon after to low population densities that may drive the population into extinction (Pimm, 1991).

Efficient utilization of resources may also relate to behavioral aspects. In central Arizona synanthropic bird species, including several exotic ones, were found to be more efficient foragers than native species (Shochat et al., 2004a). Yet, under certain conditions, species that have such a potential to become invasive may remain in low profile in wild habitats, where harsh conditions favor native species. Agricultural and urban developments may remove some key hurdles (e.g. specific predators) or add essential components to the environments (e.g. water, breeding sites or specific food resources) that allow 'dormant invasive species' to change their mode and become invasive.

For example, extremely dry deserts favor rodent species that do not drink, but gain water from food resources, including dry seeds. Such seeds are not available to birds as they are not digestive without water. However, agricultural development in such arid zones, that adds water holes and irrigation systems to the ecosystem, not only allows the establishment of seed-eating birds, but totally removes the restriction on their fast individual and population growth rates. This extreme change in the rules of the evolutionary game creates a new environment in which very few resources remain for the nocturnal rodents in the end of the day (Shochat et al., 2004a).

3. The community level

The relationship between species diversity of the local community and the probability of successful invasions has been widely discussed in the ecological literature (e.g., Elton, 1958; Kennedy et al., 2002). Both theory (Case, 1990) and small scale experimental studies (McGrady-Steed et al., 1997; Tilman, 1997; Levine & D'Antonio, 1999; Naeem et al. 2000; Kennedy et al. 2002) indicate that diverse communities better resist invasions [but see Planty-Tabacchi et al., (1996) and Stohlgren et al., (1999) for the opposite pattern detected at the regional scale]. In a field experiment, Kennedy et al. (2002) showed that species diversity in small grassland plots enhanced invasion resistance by increasing crowding and species richness in the local plant neighborhood. Both the number of invaders and success of invading plants were reduced. These results strongly suggest that local biodiversity represents an important line of defense against the spread of invaders.

In less diverse communities, invasive species may change community organization and break assembly rules. In California, invasion of the argentine ant (*Linepithema humile*) led to disassembly of local ant community. Where the Argentine ant was present, ant communities appeared random and weakly aggregated in species co-occurrence (Sanders et al. 2003).

How extreme can the impact of invasion on the local community be? Although alien species have been argued to drive local extinctions either indirectly through exploitation competition (Byers, 2000), or directly either via interference competition (Human & Gordon, 1996; Lach, 2005) or hybridization (Rhymer & Simberloff, 1996), the overall view of extinction as a result of invasions has been long criticized and is still under debate. In a review of ten studies covering 850 plant and animal introductions, Simberloff (1981) argued that invasive species rarely influence the invaded communities. This view has been

criticized by Herbold & Moyle (1986) who questioned Simberloff's methodology and also argued that substantial density reductions can be as important as local extinctions. Davis (2003) suggested that in cases where invasion leads to biodiversity loss, it is more likely to be driven by habitat loss or predation rather than by competition. Gurevitch & Padilla (2004) also questioned the idea that invasive species are an important driver of native species extinctions. Their view has been criticized by Clavero and García-Berthou (2005) who provided a list of case studies supportive of extinction following bio-invasions [e.g., European birds (Birdlife International, 2000) and North American fish (Miller et al., 1989)].
Bio-invasions may also be facilitated if food webs are arranged in compartments (distinct food chains in which species interact more frequently among themselves than with species pertaining to other food chains) (Pimm, 1991). Such communities are easier to invade, because alien species can locate themselves among two or more food chains, relying on resources from several chains (Fig. 1). May (1982) suggested that food webs are organized in

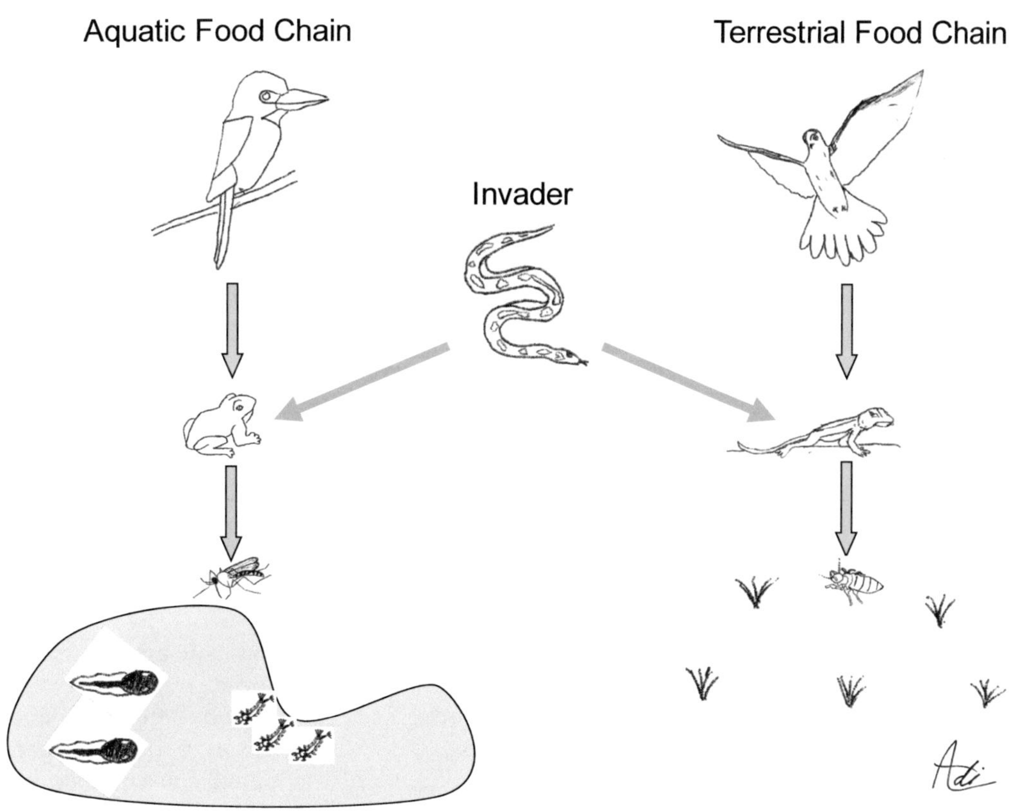

Fig. 1. Consequences of compartmentalization in food webs on biological invasions. An example showing a food web comprising of two distinct food chains occurring across two major habitat boundaries (aquatic and terrestrial) in which species only interact within chains. Such an organization can facilitate the invasion of the snake, which can consume prey from both chains. In the alternative scenario, the snake enters only one of the chains and thus suffers strong competition and possibly high predation pressure from the top predators.

compartments, but Pimm & Lawton (1980) could not find an empirical support for such organization. Recent studies (Krause et al., 2003; Stouffer & Bascompte, 2011), however, provide new evidence that food webs are indeed compartmented. Furthermore, Stouffer & Bascompte (2011) demonstrated how compartments in food webs directly buffer extinctions, while also increasing the persistence of species making up the community. Thus, if compartmented communities are indeed easier to invade, invasive species can indirectly lead to extinction, by disordering the food-web structure within the community.

In the case of hybridization the picture is much clearer, since the process of introgression should ultimately lead to elimination of native species ancestry, and eventually to its complete replacement by the invasive species (Hedrick, 2005). Examples include introgression by the introduced American Ruddy Duck (*Oxyura jamaicensis*) into the European White headed Duck (*Oxyura leucocephala*) (Green & Hughes, 2001), and introgression by domesticated Mallards (*Anas Platyrhynchos*) into the Florida Mottled Duck (*Anas fulvigula*) (Mazourek and Gray, 1994). In some cases hybridization may result in the formation of new species, either by speciation through recombination or allopolyploidy (Coyne & Orr, 2004).

In summary, since extinctions happen across evolutionary time scales, and because invasions involve many other changes in the environment, it is difficult to link extinctions and invasions. Altogether, it is agreed that the various negative effects of invasive species on community structure and biodiversity can lead, in extreme cases, to local extinctions. Below we discuss possible mechanisms that may lead to such scenarios.

3.1 The influence of invasion on community patterns and biodiversity

Elton (1958) suggested that disturbed communities are more susceptible to invasion because they comprised of mainly inferior species. This view has been revisited by Simberloff (1986) who associated disturbed communities with young and human-produced environments. Thus, the association of human-managed habitats with disturbances on one hand, and with invasive species on the other hands, implies that disturbances and invasions should be positively correlated. Furthermore, animal and plant communities in human-managed habitats are relatively new and therefore may not be fully saturated in terms of species richness. Notably, even in recently assembled, species-rich communities, the community resistance to invasion increases with its ecological age, as well as with the increase in the number of interspecific interactions (i.e., community connectedness, Post & Pimm, 1983; Pimm, 1991).

From the newer, global change perspective, the alteration of wildlands into agricultural and urban ecosystems by humans includes not only structural changes, but also involves changes in productivity, microclimate, and many other ecological variables that combine to change community structure (Shochat et al., 2004b, 2006). Yet, changes in productivity per se (bottom-up control) might be the major driver shaping plant and animal communities, with changes in predation (top-down control) acting as a secondary, yet important driver. In other words, whereas physical changes in the structure of environment are responsible for changes in community composition, changes in community structure, evenness, and the overall loss of diversity in human-managed ecosystems may be more related to changes in energy flow and resources, which in turn affect competitive interactions (Shochat et al., 2010).

Shochat et al. (2010) showed how changes related to land transformation lead to rapid population growth of a few species that become invasive. On the species rank distribution axis, such species move leftwards (i.e., they become the most abundant species). A case study on spiders in central Arizona demonstrated how dramatic such changes can be: wolf spiders that account for 7-8% of the whole spider community in desert and xeric yards, became the most abundant family in mesic yards and agricultural sites, accounting for up to 80% of the whole community (Shochat et al., 2004b, 2010). Although wolf spiders are not alien and reach these novel moist habitats from nearby arid ones, they can be treated as invasive species, as they respond to human-induced changes in productivity in the environment. Spider diversity in arid lands and xeric urban yards in Arizona is higher than in agricultural fields and mesic urban yards (Shochat et al., 2004b, 2010). Where invasive species become much more abundant than the dominant species in the original community, the overall outcome is that urban and agrarian ecosystem communities turn from fairly even to uneven communities monopolized by a few abundant species.

Evenness has been argued to enhance community resistance to total density changes (King & Pimm, 1983), an important indicator of community stability. Recent empirical studies illustrate that such stability is also linked to community resistance to invaders (Wilsey & Polley, 2002; Tracy & Sanderson, 2004). Experimentally reducing evenness of grassland species resulted in increased invasion of dicot plants, as well as in elevated levels of spittle-bug infestation (Wilsey & Polley, 2002). Similarly, Tracy & Sanderson (2004) found that maintaining pasture community productivity and evenness can effectively reduce weed invasion. Other findings, however, were less supportive of this view (e.g., Emery & Gross, 2007; Mattingly et al., 2007). Mattingly et al. (2007) found that evenness increased community productivity, but had no effect on resistance to invasion. Moreover, Emery & Gross (2007) showed that resistance to invasion was driven by the identity of the dominant species rather than by the reduction in evenness. In summary, although not always the major factor decreasing invasibility, evenness appears to play an important role in community stability and resistance.

Following the above findings on community evenness, Shochat et al. (2010) suggested a mechanism for the loss of diversity, based on foraging efficiency. They found that in the desert, species body mass was negatively correlated with foraging efficiency. That is, the most dominant species were the least efficient foragers. Such a pattern concurs with the "temporal partitioning" mechanism of species coexistence, where subordinate species may be able to find sufficient amount of food resources after dominant species quit foraging (Kotler & Brown, 1988; Ziv et al., 1993). In the urban environment, however, body size was positively correlated with foraging efficiency. Some of the larger, very efficient foragers, such as the Inca Dove (*Scardafella inca*), were completely absent from desert and xeric habitats in central Arizona, but thrived in mesic environments such as parks, lawns and agricultural fields. These species dominated artificial food patches, and obviously depleted food resources to extremely low levels, meaning that subordinate species are not likely to coexist with dove species through "temporal partitioning" (Shochat et al., 2010). Such uneven communities, where the "temporal partitioning" mechanism of coexistence collapses, must lose species, because most resources are now consumed by the few dominant species, leaving very little for the many others. Furthermore, when the populations of native species dramatically decrease, their genetic variation also decreases. Both these processes increase the probability of random extinction (i.e., drift and disturbance) (Lande, 1988).

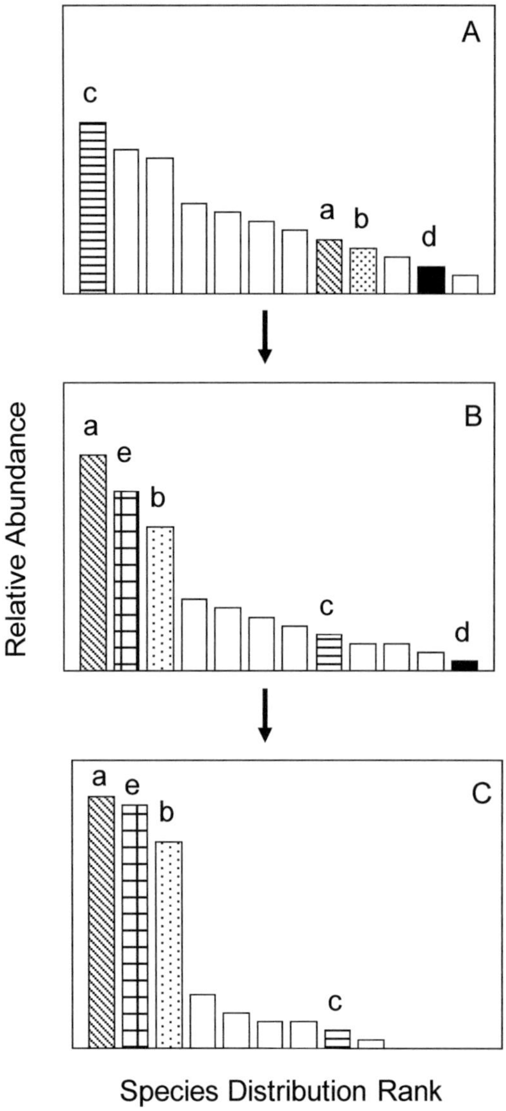

Fig. 2. Changes in community structure following anthropogenic disturbances. (A) Wildland communities are relatively even, with the most dominant species (three in this case) accounting for relatively low proportion of the whole community. Such equilibrium in wild ecosystems with low resource predictability is maintained in the long-term by high variation in productivity (temperature, rainfall, food and water resources), high predation pressure, and relatively low food abundance. (B) Human activities often dampen environmental variability by increasing overall available water and food, buffering seasonal and annual variations in resource abundance, and removing key predators. In such relatively constant and predictable environments, where key hurdles are removed, a few either dormant or alien species may rapidly spread. This includes species a and *b* that were originally 'minor players' in the community, and species *e*, as an example for an alien species that did not exist in the original community. The wildland community 'major players' (such as species *c*)

move right on the species distribution rank axis. The profile of the whole community becomes significantly less even. (C) The invasive species eventually reach their maximum population size, accounting for a relatively high proportion of the whole community, leaving much less space and resources for many of the local subordinate species. The subordinate species populations are now small, and some, like species *d*, vanish. This extinction can be the result of competition, loss of genetic variability, or stochastic events. The figure was adopted from Shochat et al. (2010).

The loss of diversity in human-managed ecosystems may therefore be the result of competition with some invasive species, though this competition may not necessarily be direct and short, but rather a long process with several stages (Fig. 2). Furthermore, it is important to note that not all invasive species may have a negative effect on local native species, and that some manage to invade by occupying vacant niches, or by "affecting many native species slightly and evenly". The goal of conservation ecologists is to learn to distinguish between invasive species that have a potential to cause severe ecological and economical damages and those that find empty niches and cause negligible damages, and to find ways to reduce, as much as possible, the negative influence of the first group's more harmful invasive species.

To understand possible ways for management of rich communities with minimized negative impact of invasive species, Shochat et al. (2010) studied the competition between an alien, aggressive species, the House Sparrow (*Passer domesticus*), and a native species, the Lesser Goldfinch (*Carduelis psaltria*). Feeders were designed to exclude sparrows, and the abundance and behavior of the goldfinch were compared between two situations: the presence of sparrow proof, and the presence of sparrow friendly feeders in urban yards. When sparrows had access to the feeders and dominated them, the number of aggressive encounters increased on the short term (1-2 hours), until the goldfinches avoided the sparrow-friendly feeders. On a longer term, Goldfinch abundance in the yards decreased, and in extreme cases reached "local extinction". In contrast, the use of "sparrow proof" feeders forced the sparrows to forage on the ground, among doves and pigeons. This turned the sparrows subordinate, as they suffered aggressive encounters from the larger birds, especially Mourning Doves (*Zenaida macroura*). At the same time, the use of sparrow proof feeders allowed the return of goldfinches to the yards within a fairly short period (1-2 weeks), and their numbers peaked under this treatment. When both feeder types were offered, Goldfinches avoided as much as possible the sparrow friendly feeders. A third granivorous passerine species, the House Finch (*Carpodacus mexicanus*), neither appeared to be suffering aggressive interactions from the similar-sized alien House Sparrow, nor to hassle the smaller Lesser Goldfinch.

These results illustrate the different extent of negative effects of alien species on native species, and how community structure and organization change depending on realized niche dimension of the alien species. Some species may be hardly affected, while other species suffer greatly from invasions. From the pure ecological perspective, this experiment shows that although some subordinate species manage to persist regionally, locally they experience a strong interference competition from the invasive species, and are thus forced to turn to alternative activities. From the applicable perspective, preventing the access of invasive species to resource patches by simple, yet clever, manipulations can largely reduce the pressure on the subordinate species. A classic example is the manipulation of nest boxes to include hole-diameter which is too small for the invasive European Starling (*Sturnus vulgaris*) in North America, that allows smaller, native cavity nesting species, to persist and breed successfully, with no obvious effect on starling densities (Newton, 1994). Creating such new niche opportunities for subordinate species can facilitate their rapid population growth as a result of immigration. This is important, because it suggests that the battle against harmful invasive species, whose goal is to sustain high diversity in human managed ecosystems, does not necessarily require great budget and effort, neither should it aim to eradicate such species. Rather, understanding the ecology and evolution of species may allow us to turn the advantages of invasive species to disadvantages, by re-opening niche opportunities for subordinate, native species.

4. Conclusions

For half a century ecologists have studied biological invasions by focusing either on the invasive species, or on characteristics of the invaded community (resources, natural enemies, or species richness) (Sakai et al., 2001; Shea & Chesson, 2002). Early studies attempted to generalize characteristics that facilitate invasions. Although some trends were identified, generalizations were limited and no clear patterns have emerged (Sakai et al., 2001). This approach has failed because the different modalities of biological invasions considered are ecosystem-specific and are also dependent on each other. Thus, understanding biological invasions requires adopting an integrated research approach, considering patterns and process that operate at different biological hierarchies jointly (Shea & Chesson, 2002). Among the many factors that should be considered, we suggest that future research should consider epigenetic processes. Such environmentally-induced epigenetic processes, often inherited by future generations, usually occur over relatively short time periods, increasing the evolutionary potential of invasive species in response to challenges in novel environments (Bossdorf et al., 2008).

Because the effects of bio-invasions on the local species pool may be complex and slow, loss of biodiversity may not always be evident over short ecological time scales. This has triggered a long-running debate among ecologists regarding the extent of the negative impact of invasive species on biodiversity (Pimm, 1991). Current views tend to agree about the remarkable negative influence of bio-invasions on both biodiversity and economy, even if no immediate loss of native species is detected (Vitousek et al., 1997; Mooney & Hobbs, 2000; Lockwood et al., 2007). We demonstrate that, for several reasons, changes in species abundance within invaded communities may be as crucial as extinctions. Small populations may suffer loss of genetic diversity, and become vulnerable to extinction by drift and disturbances. Furthermore, even if such small populations manage to persist, the reduction in genetic diversity dictates that their ability to adapt to future changes in the environment will be limited. Inflation in invasive species densities and reduction in local species abundance severely reduces the invaded community evenness, a process that may further facilitate invasions by reducing the community ability to resist invasions (Wilsey & Polley, 2002; Tracy & Sanderson, 2004).

Finally, because complete eradication of invasive species may be costly and in most cases impossible, more applicable solutions that allow re-opening niche opportunities for native species should be adopted. For example, manipulations of artificial nest sites (Newton, 1994) or feeders (Shochat et al., 2010) that exclude dominant invaders without affecting their densities yet allow restoring densities of subordinate local species.

5. References

Beisner, B.E., Haydon, D.T. & Cuddington, K. (2003). Alternative stable states in ecology. *Frontiers in Ecology and the Environment*, Vol. 1, pp. 376-382

Birdlife International (2000). *Threatened Birds of the World*, Lynx Editions and BirdLife International, ISBN 8487334288, Cambridge, UK

Bossdorf O., Richards C.L. & Pigliucci, M. (2008) Epigenetics for ecologists. *Ecology Letters*, Vol. 11, pp. 106–115

Byers J.E. (2000). Competition Between Two Estuarine Snails: Implications For Invasions Of Exotic Species. *Ecology*, Vol. 81, pp. 1225–1239

Clavero, M. & García-Berthou, E. (2005). Invasive species are a leading cause of animal extinctions. *Trends in Ecology and Evolution*, Vol. 20, p. 110

Coyne, J.A. & Orr, H.A. (2004). *Speciation*. Sinauer Associates, Sunderland, MA

Davis, M.A. 2003. Biotic Globalization: Does Competition from Introduced Species Threaten Biodiversity? *BioScience*, Vol. 53, pp. 481-489

Elton, C.S. (1958). *The ecology of invasions by animals and plants*, ISBN 9780226206387, Methuen, London

Emery, S.M. & Gross K.L. (2007). Dominant species identity, not community evenness, regulates invasion in experimental grassland plant communities. *Ecology*, Vol. 88, pp. 954-964

Endler, J.A. (1986) *Natural selection in the wild*, ISBN 9780691083872, Princeton University Press, Princeton.

Facon, B., Genton, B.J., Shykoff, J., Jarne. P., Estoup, A. & David P. (2006). A general eco-evolutionary framework for understanding bioinvasions. *Trends in Ecology and Evolution*, Vol. 21, pp. 130-135

Green, A.J. & Hughes, B. (2001) *Oxyura leucocephala. White-headed duck*. BWP Update, Vol. 3, pp. 79–90

Gurevitch, J. & Padilla, D.K. (2004). Are invasive species a major cause of extinctions? *Trends in Ecology & Evolution*, Vol. 19, pp. 470-474

Grotkopp, E., Rejmanek, M. & Rost, T.L. (2002). Toward a causal explanation of plant invasiveness: Seedling growth and life-history strategies of 29 pine (*Pinus*) species. *American Naturalist*, Vol. 159, pp. 396-419

Hedrick, P.W. 2005. *Genetics of populations* (3rd Edition), ISBN 9780763757373, Jones and Bartlett Publishers, Sudbury MA.

Herbold, B. & Moyle, P.B. (1986). Introduced Species and Vacant Niches. *American Naturalist*, Vol. 128, pp.751-760

Human, K.G. & Gordon, D.M. (1996). Exploitation and interference competition between the invasive Argentine ant, *Linepithema humile*, and native ant species. *Oecologia*, Vol. 105, pp. 405-412

Kaiser, J. (1999). Stemming the Tide of Invading Species. *Science*, Vol. 285, pp. 1836-1841

Kotler, B.P. & Brown, J.S. (1988). Environmental heterogeneity and the coexistence of desert rodents. *Annual review of ecology and systematics*, Vol. 19, pp. 281-307

King, A.W. & Pimm, S.L. (1983). Complexity and stability: a reconciliation of theoretical and experimental results. *American Naturalist*, Vol. 122, pp. 229-239

Krause, A.E., Frank, K.A., Mason, D.M., Ulanowicz R.E. &Taylor W.W. (2003). Compartments revealed in food-web structure. *Nature*, Vol. 426, pp. 282-285.

Lach, L. (2005). Interference and exploitation competition of three nectar-thieving invasive ant species. *Insect Sociaux,* Vol. 52, pp. 257-262.

Lande, R. (1988). Genetics and demography in biological conservation. *Science*, Vol. 241, pp. 1455-1460

Lockwood, J.L., Hoopes, M.F. & Marchetti M.P. (2007). *Invasion Ecology*, ISBN 9781405114189, Blackwell Publishing, Malden MA.

Mack, R.N., Simberloff D., Lonsdale W.M., Evans H., Clout M. and Bazzaz F.A. (2000). Biotic invasions: causes, epidemiology, global consequences, and control. *Ecological Applications*, Vol. 10, pp. 689–710

Maron, J.L., Vil, M., Bommarco, R., Elmendorf, S. & Beardsley, P. (2004). Rapid evolution of an invasive plant. *Ecological Monographs*, Vol. 74, pp. 261-280

Mattingly, W.B., Hewlate, R., & Reynolds, H.L. (2007). Species evenness and invasion resistance of experimental grassland communities. *Oikos*, Vol. 116, pp. 1164-1170

May, R.M. (1972). Will a large complex system be stable? *Nature*, Vol. 238, pp. 413–414.

Mazourek, J.C. & Gray, P.N. (1994). The Florida duck or the Mallard. *Florida Wildlife*, Vol. 48, pp. 29-31

Miller, R.R. Williams MJ, Williams JE. (1989) Extinctions of North American fishes during the past century. *Fisheries*, Vol. 14, pp. 22–38

Mooney, H.A. & Hobbs, R.J. (2000) *Invasive Species in a Changing World*, ISBN 9781559637824, Island Press, Washington DC.

Mooney, H.A. & Cleland, E.E. (2001). The evolutionary impact of invasive species. *Proceedings of the National Academy of Sciences*, Vol. 98, pp. 5446–5451

Moulton, M.P. & Pimm, S.L. (1986) The extent of competition in shaping an introduced avifauna. In: *Community Ecology*, J. Diamond and T. Case, (Eds.), Harper & Row, ISBN 006041202X, New York.

Newton, I. (1994). The role of nest sites in limiting the numbers of hole-nesting birds: A review. *Biological Conservation*, Vol. 70, pp. 265-276

Planty-Tabacchi, A.M., Tabacchi, E., Naiman, R.J., Deferrari, C. & Decamps, H. (1996). Invasibility of species-rich communities in riparian zones. *Conservation Biology*, Vol. 10, pp. 598–607

Pimentel, D., Lach, L., Zuniga, R. & Morrison D. (2000). Environmental and economic costs of nonindigenous species in the United States. *BioScience*, Vol. 50, pp. 53-65

Pimm, S.L. (1991). *The balance of nature? Ecological issues in the conservation of species and communities*. University of Chicago Press, Chicago, Illinois, 434 p.

Pimm, S.L. & Lawton J.H. (1980). Are food webs compartmented? *Journal of Animal Ecology*, Vol. 49, pp.879-898

Post, W.M. & Pimm, S.L. (1983). Community assembly and food web stability. *Mathematical Biosciences*, Vol. 64, pp. 169-192

Reed, D.H. & Frankham, R. (2003). Correlation between fitness and genetic diversity. *Conservation Biology*, Vol. 17, pp. 230–237

Rhymer, J. & Simberloff, D. (1996). Extinction by hybridization and introgression. *Annual Review of Ecology and Systematics*, Vol. 27, pp. 83-109

Richards, E.J. (2006). Inherited epigenetic variation - revisiting soft inheritance. *Nature Reviews Genetics*, Vol. 7, pp. 395–401

Sakai, A.K, Allendorf, F.W, Holt, J.S., Lodge, D.M., Molofsky, J., With, K.A., Baughman, S., Cabin, R.J., Cohen, J.E., Ellstrand, N.C., McCauley, D.E., O'Neil, P., Parker, I.M.,

Thompson, J.N. & Weller S.G. (2001). The population biology of invasive species. *Annual Review of Ecology and Systematics*, Vol. 32, pp. 305-332

Seabloom, W.E., Harpole, S.W., Reichman, O.J. & Tilman, D. (2003). Invasion, competitive dominance, and resource use by exotic and native California grassland species. *Proceedings of the National Academy of Sciences*, Vol. 100, pp. 13384-13389

Shea, K. & Chesson, P. (2002). Community ecology theory as a framework for biological invasions. *Trends in Ecology and Evolution*, Vol. 17, pp. 170-176.

Shochat, E., Lerman, SB, Katti, M & Lewis, D. (2004a). Linking optimal foraging behavior to bird community structure in an urban-desert landscape: field experiments with artificial food patches. *American Naturalist*, Vol. 164, pp. 232-243

Shochat, E., Stefanov, W.L., Whitehouse, M.E.A. & Faeth, S.H. (2004b). Urbanization and Spider Diversity: Influences of Human Modification of Habitat Structure and Productivity. *Ecological Applications*, Vol. 14, pp. 268-280

Shochat, E., Warren, P.S., Faeth, S.H., McIntyre, N.E. & Hope D. (2006). From patterns to emerging processes in mechanistic urban ecology. *Trends in Ecology & Evolution*, Vol. 21, pp. 186-191

Shochat, E., Lerman, S.B., Anderies, J.M., Warren, P.S., Faeth, S.H. & Nilon, C.H. (2010). Invasion, Competition, and Biodiversity Loss in Urban Ecosystems. *BioScience*, Vol. 60, pp. 199-208

Simberloff, D. (1981). Community effects of introduced species. In: *Biotic crisis in ecological and evolutionary time*. H. Nitecki Ed. 53-81, Academic press, ISBN 0125196407, New York.

Simberloff, D. (1986). Introduced species: a biogeographic and systematic perspective. In: *Ecology of biological invasions of North America and Hawaii*. H.A. Mooney and J.A. Drake (Eds.), pp. 3-26 Springer-Verlag, NY.

Simberloff, D. (1997). The biology of invasions. In: *Strangers in Paradise: impact and management od nonindigenous species in Florida*. Simbeloff, D., Schmitz, D.C. & Brown, T.C. (Eds.), pp. 3-17, Island Press, ISBN-10: 038797153X, Washington DC.

Stohlgren, T.J., Binkley, D., Chong, G.W., Kalkhan, M.A., Schell, L.D., Bull, K.A., Otsuki, Y., Newman, G., Bashkin, M. & Son, Y. (1999) Exotic plant species invade hot spots of native plant diversity. *Ecological Monographs*, Vol. 69, pp. 25–46.

Stouffer, D.B. & Bascompte, J. (2011). Compartmentalization increases food-web persistence. *Proceedings of the National Academy of Sciences*, Vol. 108, pp. 3648-3652

Tracy, B.F., & Sanderson, M.A. (2004). Forage productivity, species evenness and weed invasion in pasture communities. *Agriculture, Ecosystems and Environment*, Vol. 102, pp. 175–183

Veltman, C.J., Nee, S. & Crawley, M.J. (1996)Correlates of introduction success in exotic New Zealand birds. *American Naturalist*, Vol. 147, pp. 542-557

Vitousek, P.M., D'Antonio, C.M., Loop, L.D., Rejmánek, M. & Westbrooks, R. (1997). Introduced species: A significant component of human-caused global change. *New Zealand Journal of Ecology*, Vol. 21, pp. 1-16

Whitney, K.D. & Gabler, C.A. (2008). Rapid evolution in introduced species, 'invasive traits' and recipient communities: challenges for predicting invasive potential. *Diversity and Distributions*, Vol. 14, pp. 569–580

Wilsey, B.J., & and Polley, H.W. (2002). Reductions in grassland species evenness increase dicot seedling invasion and spittle bug infestation. *Ecology Letters,* Vol. 5, pp. 676–684

Yoshida, T., Goka, K., Ishihama, F., Ishihara, M. & Kudo, S. (2007). Biological invasion as a natural experiment of the evolutionary processes: Introduction of the special feature. *Ecological Research,* Vol. 22, pp. 849-854

Zavaleta, E.S., Hobbs, R.J. & Mooney, H.A. (2001). Viewing invasive species removal in a whole-ecosystem context. *Trends in Ecology and Evolution* Vol 16, pp. 454-459

Part 2

Interactions Between Living Organisms and Humans

6

Change in Bacterial Diversity After Oil Spill in Argentina

Graciela Pucci, María Cecilia Tiedemann, Adrián Acuña and Oscar Pucci
Universidad Nacional de la Patagonia San Juan Bosco
Argentina

1. Introduction

During the past few decades, the impact and threat of oil pollution have resulted in extensive research. The anthropogenic origins of petroleum pollution (particularly via leak of coastal oil refineries) have raised interest among the scientific community in oil pollution distribution and its effects in the environment, mainly the marine environment. Tanker accidents are the major cause of oil pollution in marine environments. On December 26th, 2007, a tanker spilled approximately 100 m^3 of crude oil into the sea, which impacted on Caleta Córdova coast, Argentina. Being part of Comodoro Rivadavia city, Caleta Córdova is a small village, where the community's local economy is based on artisanal fishing. Over the course of two months, local oil companies collaborated to manually clean the coast and hydrocarbon controls were carried out to monitor the hydrocarbon concentration. The province environmental authorities prohibited the use of chemical products to disperse hydrocarbons and the use of any other chemical product, including nitrogen and phosphorus salts to help biodegradation. On the day of the spill, there was an extraordinary high tide (6.4 m), which resulted in the definition of three areas of impact: the high area, which was the most difficult to clean up; the middle area, which was cleaned up with minimal problem and from where the samples in this study were taken; and the low tide line area.

Petroleum hydrocarbon can be degraded by microorganisms such as bacteria, fungi, yeast, and microalgae (e.g.,Atlas, 1981) Bacteria are considered to represent the predominant agents of hydrocarbon degradation in the environment and hydrocarbon-degrading bacteria are ubiquitous (Pucci et al 2009a, b). Since crude oil is made of a mixture of compounds, and since individual microorganisms metabolize only a limited range of hydrocarbon substrates (Britton, 1984), biodegradation of crude oil requires a mixture of different bacterial groups or consortia functioning to degrade a wider range of hydrocarbons (Bordenave et al., 2007; Cagnon et al., 2011). Contaminated marine environments are inhabited by a range of selected microorganisms able to tolerate and remediate pollutants that impacted the environment, leading to the dominance of pollutant-tolerant bacteria. Hence, bacterial communities in contaminated sites are typically less diverse than those in nonstressed systems (Harayama et al., 2004). Another characteristic of marine environments is that the vast majority of bacteria (90–99%) are uncultivable (Amann et al., 1995); hence, the analysis of microbial communities that contribute to in situ hydrocarbon biodegradation activities has been a challenge to microbiologists (Rollins & Colwell, 1986; Wilkinson 1988).

	Genus	%	Genus	%	Genus	%
Caleta Córdova						
	Acinetobacter	4.0	*Kurthia*	0.8	*Psychrobacter*	12.9
	Aeromonas	4.0	*Lysobacter*	0.8	*Rhodobacter*	2.4
	Arthrobacter	0.8	*Microbacterium*	1.6	*Rhodococcus*	0.8
	Bacillus	16.9	*Micrococcus*	1.6	*Shewanella*	5.6
	Brevibacillus	1.6	*Ochrobactrum*	0.8	*Sphingopyxis*	0.8
	Brevundimonas	1.6	*Paracoccus*	1.6	*Staphylococcus*	2.4
	Burkholderia	0.8	*Photobacterium*	1.6	*Vibrio*	7.3
	Halomonas	2.4	*Pseudoalteromonas*	21.0	*Zobellia*	0.8
	Kocuria	1.6	*Pseudomonas*	3.2		
Comodoro Rivadavia						
	Acinetobacter	8.12	*Microbacterium*	0.85	*Rhodobacter*	5.56
	Aeromonas	4.27	*Micrococcus*	1.71	*Rhodococcus*	1.28
	Bacillus	3.85	*Moraxella*	0.85	*Rhodovolum*	1.71
	Brevundimonas	2.99	*Ocrhobactrum*	4.27	*Roseomonas*	0.43
	Chryseomonas	0.85	*Paenibacillus*	0.43	*Salmonellas*	5.56
	Curtobacterium	0.85	*Pectobacterium*	0.43	*Shewanella*	1.28
	Enterobacter	0.85	*Photobacterium*	2.99	*Staphylococcus*	3.42
	Enterococcus	0.43	*Photorhabdus*	0.43	*Stenotrophomonas*	0.43
	Escherichia	3.42	*Proteus*	0.43	*Vibrios*	5.56
	Flavimonas	1.71	*Pseudoaltermona*	16.24	*Virgibacillus*	0.43
	Gluconobacter	1.71	*Pseudomona*	6.84	*Xanthobacter*	0.43
	Grimontia	0.85	*Pseudoxantomonas*	1.28	*Yersinia*	0.43
	Kluyveria	1.28	*Psychrobacter*	0.43	*Zobellia*	0.43
	Kokuria	3.85	*Rhizobium*	0.85		
Rada Tilly						
	Acinetobacter	1.7	*Grimontia*	0.6	*Psychrobacter*	1.1
	Aeromonas	7.4	*Kocuria*	4.0	*Rhodobacter*	4.5
	Arthrobacter	4.0	*Microbacterium*	1.1	*Rhodovulum*	0.6
	Bacillus	6.8	*Micrococcus*	11.4	*Roseomonas*	1.1
	Bradyrhizobium	1.7	*Neisseria-*	2.3	*Rothia*	0.6

	Genus	%	Genus	%	Genus	%
	Brevibacillus	2.3	*Nesterenkonia*	0.6	*Sphingomonas*	0.6
	Brevundimonas	3.4	*Ochrobactrum*	3.4	*Sporosarcina*	0.6
	Cellulomonas	0.6	*Paenibacillus*	1.7	*Staphylococcus*	10.2
	Chryseomonas	0.6	*Paracoccus*	0.6	*Stenotrophomonas*	0.6
	Corynebacterium	0.6	*Paucimonas*	0.6	*Vibrio*	1.7
	Dietzia	2.3	*Photobacterium*	1.7	*Virgibacillus*	0.6
	Escherichia	0.6	*Pseudoalteromonas*	11.9	*Zobellia*	2.3
	Flavimonas	1.1	*Pseudomonas*	2.8		
Caleta Olivia	**Genus**	**%**	**Genus**	**%**	**Genus**	**%**
	Acinetobacter	0.96	*Kocuria*	1.91	*Psychrobacter*	4.31
	Arthrobacter	0.48	*Lechevalieria*	0.48	*Rhodobacter*	3.35
	Bacillus	4.78	*Methylobacterium*	0.48	*Rhodovulum*	0.96
	Brevibacterium	1.44	'*Microbacterium*	3.35	*Salmonella*	0.48
	Brevundimonas	0.96	*Micrococcus*	3.35	*Shewanella*	2.39
	Clavibacter	0.48	*Neisseria*	1.44	*Sphingomonas*	3.83
	Corynebacterium	0.48	*Nocardioides*	0.48	*Staphylococcus*	4.31
	Curtobacterium	0.96	*Paenibacillus*	0.96	*Vibrio*	1.91
	Dietzi	0.48	*Photobacterium*	3.83	*Xanthobacter*	0.48
	Escherichia	0.48	*Pseudoalteromonas*	28.71	*Zobellia*	1.91
	Flavobacterium	0.48	*Pseudomonas*	19.14		

Table 1. Bacterial diversity taxas in four locations of San Jorge Gulf in 2006.

Since the pioneering work on marine bacteria by C.E. ZoBell, many bacterial strains have been isolated from the coastal and oceanic environments; these bacteria, including the genera *Alteromonas* (Beckman et al., 2008; Ivanova et al., 2004), *Aeromonas* (Stabili & Cavallo 2004), *Alcanovorax* (Head et al., 2006;Purkrtova et al., 2010), *Bacillus* (Oguntoyinbo, 2006), *Cycloclasticus* (Kasai et al., 2002), *Chromobacterium, Flavobacterium, Marinobacter spp., Shewanella* (Holt et al., 2005), *Microscilla* (Lennon, 2007), *Micrococcus* (Süss et al., 2004), *Rhodococcus* (Süss et al., 2004), *Photobacterium, Planacoccus, Pseudomonas, Pseudoalteromonas,* and *Vibrios*, among other genera, have been considered to be representative of marine bacteria. Many of them have the capacity to use different hydrocarbons (Head et al., 2006, Yu et al., 2005a, b).

Comodoro Rivadavia, known as Argentina's oil capital, is a city with an intense oil activity. However, and contrary to belief, there are no oil refineries in town, which means that the crude oil needs to be transported by ship on a daily basis; hence, the possibility of marine and coastal pollution is an ongoing risk. Caleta Córdova is located approximately 20 km north of Comodoro Rivadavia city in the San Jorge Gulf. The beaches of Caleta Córdova are characterized by being used for different purposes. In the south of Caleta Córdova, crude oil

is loaded into transport vessels by means of a buoy located about 4 km offshore. The local port is located in the town area of the beach, where activities related to fishing and shellfish harvesting are carried out. The northern area is used for recreation during the summer due to its large size and absence of industrial activity.
Bacterial diversity in four geographical points of San Jorge Gulf (caleta Olivia, Rada Tilly, Comodoro Rivadavia and Caleta Córdova) was studied before the oil spill (Table 1). The method used for the bacterial identification was fatty acids methyl ester (FAME).
The current study examined sediments collected from one spill located in the Caleta Córdova beach in Comodoro Rivadavia city, and assessed the temporal trends in distributions and concentrations of petroleum hydrocarbons from December 2007 to June 2008. The sediments were analyzed for total petroleum hydrocarbons (TPHs) and mineralization of crude oil and distilled oil. Bacterial counts were carried out and strain were isolated and identified. The analysis of the community growth on liquid medium was also included in the study.

2. Material and methods

2.1 Sampling

Sediment samples were collected 1 year before the spill and on six occasions during the study between December 2007 and June 2008 (on days 4, 20, 40, 60, 120, and 164 after the spill) to determine the temporal changes and spatial distribution of the petroleum in the sediment on the same coordinate. The samples of intertidal sediment were collected in Caleta Córdova from the coast, from the top layer (0–30 cm). The temperature at the sediment surface was around 14 °C. The fuel-oil sample containing 48.9% of aliphatic hydrocarbon, 35.9% of aromatic hydrocarbon, and 15.1% of resins and asphaltenes hydrocarbon, was obtained from the contaminated site after the spill.

2.2 Chemical analysis

Gas Chromatography (GC). Two grams of each individual sample were dissolved in 5 ml of pentane, phase separated, and percolated through 2 g of silicalgel. One mL of the eluate was carefully evaporated until dryness to determine the fuel oil content of the sample. The fractions were analyzed and quantified by gas chromatography utilizing Varian 3800 GC, equipped with a split/splitless injector, a flame ionization detector, and a capillary column VF-5ms (30 m, 0.25 mm, 0.25 µm). The injector and detector temperatures were maintained at 200 °C and 340 °C respectively. The Sample (1 µL) was injected in split mode and the column temperature was raised from 45 to 100°C at a rate of 5 °C/min and a second ramp from 100 to 275 °C at a rate of 8 °C/min with the final temperature of 275 °C, held for 5 minutes.
Hydrocarbons by Infrared determination. EPA Method 418.1. Two grams of each individual sample were dissolved in 10 mL of carbon tetrachloride, phase separated, and percolated through 2 g of silicalgel and the absorbance was measured at 2930 cm^{-1}.
Soxhlet determination. Hydrocarbon concentration was determined by Soxhlet extractor using trichlorinethane as the extraction solvent. The extracted hydrocarbons were quantified on a mass difference basis (Pucci & Pucci, 2003) and separated into class fractions by silica gel column chromatography. Aliphatic, aromatic and polar oil fractions were eluted with

250 mL of hexane, 150 mL of benzene and 150 mL of chloroform-methanol 1:1 respectively (Acuña et al., 2008).

2.3 Respiration

Basal respiration was measured by monitoring CO_2 evolution using Na OH to capture it. 5 g of sediment were poured into a brown bottle with 50 mL of sea water and 50 µL of mineral medium hydrocarbon degrading bacteria (HDB) (K_2HPO_4 100 g, $(NH_4)_2PO_4$ 200 g, distilled water 1000 mL, pH 7) and 0.1% of gasoil, diesel oil, kerosene, lubricant oil and crude oil and intrinsic microcosms without hydrocarbons and nutrient presence were used for rest of the values obtained on hydrocarbons mineralization. The NaOH was titrated by HCl 0.1N. (Bartha et al., 1972).

2.4 Enumeration and isolation of aerobic bacteria

The culturable bacteria in sediments were counted using standard plate dilution method. One gram (wet weight) of sediments was suspended in 9 mL of sterile seawater (pH 7.2) and vortexed for 1 min at low speed. Aliquots of 100 µL of undiluted samples, and 10^{-1} to 10^{-6} dilutions were spread on BBR, BRN and MBM-PGO agar plates (Pucci et al., 2009), incubated at 28 °C for up to 21 days. The media were BBR medium (g L^{-1}: Tripteine 0.5, yeast extract 0.5, K_2HPO_4 1; $(NH_4)_2SO_4$ 2; agar-agar 15, sterile marine water 700 mL, sterile water 300 mL, pH 7.2), BRN medium (g L^{-1}: Tripteine 5, yeast extract 1, K_2HPO_4 1; $(NH_4)_2SO_4$ 2; agar-agar 15, sterile marine water 700 mL, sterile water 300 mL, pH 7.2) and MBM-PG medium (20 mL of MM: NaCl 5 g, K_2PO_4H 0.5 g, $NH_4PO_4H_2$ 0.5 g, $(NH_4)_2SO_4$ 1 g, Mg SO_4 0.2 g, KNO_3 3 g, $FeSO_4$ 0.05 g, distilled water 1000 mL), which was distributed in a Petri plate and after solidification, 30 µL of a mixture 1:1 of petroleum-diesel oil were spread on the surface (Pucci & Pucci, 2003). Grown discrete colonies were isolated and purified.

2.5 Chemotaxonomic analysis

The diversity of cultured sediment bacteria was determined by fatty acids methyl ester (FAMEs) analysis of the samples taken after summer before the spill and 6 times after the spill. FAMEs analysis allows the characterization of individual bacterial colonies. Fatty acids are extracted and compared against a database, to identify isolated bacteria. From each culture plate containing between 30 and 300 colonies, individual colonies were randomly isolated and incubated on tryptic soy broth agar for 24h. The FAMEs were extracted and analyzed as described by MIDI (MIDI Newark, Del., USA).

2.6 Communities FAMEs analysis

10 g of sediments were incubated on the medium HDB (g L^{-1}: K $_2HPO_4$ 1; $(NH)_4$ SO_4)with 0.1% phenanthrene, 0.1% hexadecane and on BBR medium. After 10 days of incubation at 28°C, the samples were centrifuged at 4000 rpm for 30 min. The FAMEs were extracted and analyzed as described by MIDI (MIDI Newark, Del., USA).

2.7 GC parameters

The MIDI microbial identification system (Microbial ID, Inc, Newark, NJ) was applied to separate fatty acid methyl ester using a gas chromatograph (HP 6890) equipped with a split/splitless injector, a flame ionization detector, a capillary column Ultra 2 (25 m, 0.2 mm, 0.33 μm); an automatic sampler; an integrator; and a program which identifies the fatty acids (Microbial ID 6.0 version). The injector and detector temperatures were maintained at 250 °C and 300 °C respectively. The Sample (2 μL) was injected in split mode and the column temperature was raised from 170 to 270 °C at a rate of 5 °C/min.

2.8 Statistical analysis

The mean and standard deviation of three replicates were calculated. The mean values were compared by ANOVA test BIOM (Applied Biostatistics Inc., 3 Heritage, Setauket, NY 117ll USA). Differences were considered significant when $P<0.05$. To identify possible similarity between FAMEs profiles, the data were subjected to analysis of variance using the PAST (Hammer & Harper 2005) and Sherlock (Microbial ID 6.0 version).

3. Results

3.1 Bacterial count

The bacterial count increased after the spill and the values did not show a significant difference ($P<0.05$) during the rest of the time (Table 2). The hydrocarbons under study on the rest of the beach were 2.2% but on the location subject to study, the TPH was only 0.22% before the incident and 4 days afterwards. On the rest of the beach, the TPH was 2.2% and there were three different areas of impact depending on the slope and the tide movement. However, the cleaning up of the oil took longer on the high area because the tide on December 26th was an extraordinary one of 6.05 meters; the ordinary tide is from 0.88 to 5.6 meters. The middle area was cleaned up in a shorter period of time (Iantanos et al., 2008).

Days	1 year before	4	20	40	60	102	164
BBR (cfu g^{-1})	3.2×10^2	1.1×10^4	9.9×10^3	2.63×10^4	1.06×10^6	2.0×10^3	5.9×10^4
BRN(cfu g^{-1})	1.7×10^2	1.36×10^6	1.24×10^5	1.87×10^5	6.0×10^4	1.2×10^4	4.7×10^3
MBM-PGO(cfu g^{-1})	4×10^1	7.5×10^5	1.29×10^6	1.79×10^5	2.12×10^6	7.2×10^4	1.74×10^5
Shannon_H	2.339	2.26	1.759	1.685	1.721	1.839	1.688
HC(%)	-	0.22	0.07	0.02	0.02	0.02	0.02
TPH -IR	-	1569.25	381.7	370	313.55	385	280.55
GC total	-	1712.150	551.510	200	187.180	154.346	153.717

Table 2. Heterotrophic (BRN), oligotrophic (BBR) and hydrocarbon degrading bacteria (MBM-PGO) counts determined 1 year before the spill and 6 times afterwards. Total hydrocarbons (TPH) determined by soxhlet (HC%), infra red (IR) and gas chromatographic (GC). (-) TPH not detected. Shannon H: Shannon diversity index.

3.2 Effects of bioremediation on oil degradation

Oil degradation was assessed by measuring carbon dioxide evolution and determining changes in oil sediment samples composition (Table 3). Carbon dioxide evolution was monitored intensively during the experience and afterwards, delta mgCO_2 days was done, which showed an important change in the values on day 4 and these values decreased with the time. The spill was crude oil, but the bacterial communities had a good performance in the mineralization of all the hydrocarbons tested. In all of the three cases (gas oil, diesel and lubricant oil), on day 4, the values of mineralization rate doubled the previous values as determined before the spill; however, on day 164 the rate values were similar to those values determined 1 year before the incident. The bacterial communities were able to use poliaromatic hydrocarbon, which is the most problematic for the human health (Table 4). Degradation on the beach by the bacterial communities was studied by monitoring the disappearance of TPH measuring by three forms through the determination of their biodegradation percentages, and confirmed by GC chromatographic analysis of the residual crude oil on the six occasions (Table 4). Detailed chromatogram analysis showed visible degradation of n-alkanes ranging from n-C16 to n-C24 during the first 40 days and n-C5 to n-C16 were detected in a small quantity. Branched alkanes such as pristine and phytane experienced degradation during the first 40 days as well.

Hydrocarbon	**1 year before**	**4**	**20**	**40**	**60**	**102**	**164**
Gas oil	13.02	34.14	19.07	21.18	20.88	15.61	16.74
Kerosene	20.92	22.71	22.9	23.7	29.28	24.01	30.86
Diesel oil	14.91	31.01	24.97	29.24	25.75	22.43	19.81
Crude oil	18.62	22.48	24.57	19.59	26.58	20.52	18.59
Lubricant oil	14.22	30.62	31.66	27.05	19.19	28.6	11.68

Table 3. Delta Mineralization rate of hydrocarbons in mg CO_2 day.

Hydrocarbons ppm.	**4**	**20**	**40**	**60**	**102**	**164**
Naphthalene	1.213	0.5782	0.09	0	0	0
2-Metylnaphthalene	1.696	1.9188	1.1	0.322	0.197	0.187
Acenaphthalene	4.033	0.698	0.349	0	0	0
Fluorene	0.185	0	0	0	0	0
Pristane	0.126	0.234	0.125	0	0	0
Phytane	59.655	5.136	0.968	0	0	0
Phenanthrene	4,.14	1.38	0.37	0	0	0
Anthracene	0.617	0.685	0	0	0	0
Fluoranthrene	4.36	2.922	0.604	0	0	0
Pyrene	0.057	0	0	0	0	0
Benzo(k)fluorantreno	4.47	1.386	2.175	0	0	0
C6-C10	1.954	0.8878	0.8354	0.536	0.481	0.469
C11-C15	14.949	10.2774	3.726	2.282	1.461	1.292
C16-C20	259.714	103.638	9.49	0	0	0
C20-C25	468.095	210.45	85.028	0	0	0
C20-C25	468.095	210.45	85.028	0	0	0
C26-C29	46.156	29.738	21.383	0	0	0
PTH total	1383.543	334.406	133.163	3.463	2.952	3.35
PAH total	22.198	10.461	5.188	0.322	0.197	0.187
n-Alkanes total	917.557	411.1	126.883	3.141	2.214	2.074
Pristane/C17	0.004	0.019	0.034	0	0	0
Phytano/C18	2.346	0.393	0.851	0	0	0

Table 4. GC-Hydrocarbon (ppm) on the samples.

3.3 Communities

In the principal component analysis (PCA), Fig. 1a, the bacterial communities, which grew in alkaline peptone, proved to be, on day 164, near the bacterial community development analyzed in samples taken a year before the spill, specifically in the seasons of autumn and summer. On the other hand, in those samples taken 4, 20 and 40 days after the spill, the bacterial community experienced a change but it eventually returned to its original community state. Most fatty acids included the same proportion of *Vibrio* fatty acids (Hoffmann et al., 2010), whereas other fatty acids were in less proportion. There were cis fatty acids and absence of trans fatty acids; however, there was a major presence of unsaturated fatty acids. After 60 days, the concentration of hydrocarbons fell, and approximately 100 days after the spill, the bacterial communities proved to have similar characteristic to the pre- spill bacterial communities. No cyclo fatty acids were identified in these communities before or after the spill; values were 1.1 to 0.6% on day 4 and on day 164. On the contrary, methyl fatty acids were present only in the first three samples taken after the spill.

On BBR medium (Fig. 1b), in those cases where the beach samples contained Hydrocarbons, the sample points (days 4, 40 and 60) in the PCA were located in the left upper quadrant, but when the samples did not contain hydrocarbons, the sample points were located in the opposite quadrant, close to those samples taken before the spill, in autumn and summer. Similar to the case of the communities grown in alkaline peptone, no cyclo fatty acids were identified in these communities before the spill and afterwards; the values were 1.2% on day 4, but the value of methyl fatty acids was 1.9% which remained stable for the rest of the time. The percentage of hydroxy fatty acids gradually decreased and on day 164, their value was only 2% higher than their value 1 year before the spill. In addition, the cis and iso fatty acids increased until day 60, which indicates the presence of Gram positive bacteria.

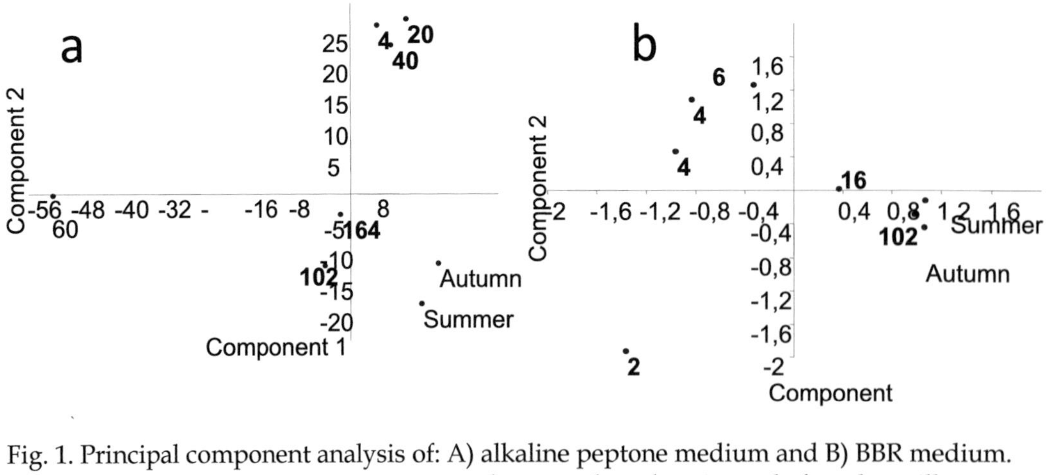

Fig. 1. Principal component analysis of: A) alkaline peptone medium and B) BBR medium. The summer and autumn points correspond to samples taken 1 year before the spill.

The medium with phenanthrene as carbon and energy source, showed that in the PCA graph, points 4, 20 and 40 are all located in the same quadrant whereas those samples taken

on day 102 have a tendency to be near the sample points referred to as autumn and summer (samples taken before the spill) (Fig. 2a). There was one exception with those samples taken on day 164 which presented a strange point with different fatty acids. Saturated fatty acids in pre-spill samples were around 40.5% in summer and 43.4% in autumn, after the spill this value changed to 73.2% and the value on day 164 was 60.5%. On the other hand, the value of iso fatty acids was 5.3% in summer and 8.45% in autumn after the spill whereas there was no significant difference in anteiso fatty acids (P>0.5). However, there was no presence of cyclo, trans and methyl fatty acids.
Hexadecane is usually used for degradation of bacterial communities and isolated strains; it does not present difficulty to be degraded and is less toxic. Principal component analysis showed a non-predictive performance (Fig. 2b). Anteiso, cyclo and methyl fatty acids were not present; and iso fatty acids were identified only when hydrocarbons were present on the beach. In addition, the percentage values of hydroxy fatty acids were around 6.8% to 15.3% on day 20 and the saturated fatty acids increased until day 60. The saturated fatty acids increased with the time and the hydroxy fatty acids did not show a pattern.

3.4 Bacteria identification

There are 34 genera and in total we found 39 different strains. Most of the species (25 out of 39) were found together in group A (*Aerococcus viridians, Bacillus atrophaeus, Bacillus pumilus Brevibacillus centrosporus, Curtobacterium flaccumfaciens, Flavobacterium ferrugineum, Halomonas aquamarina, Kocuria rosea, Kurthia gibsonii, Lysobacter enzymogenes enzymogenes, Micrococcus luteus, Microbacterium sp, Nocardia transvalensis, Pseudoalteromonas nigrifaciens, Pseudomonas aeruginosa, Psychrobacter immobilis, Rhodococcus, Rhodobacter, Roseomonas genomospecies, Shewanella putrefaciens, Streptoverticillium reticulum, Staphylococcus, Vibrio* sp., *Vibrio fischeri, Pseudoxanthomonas* sp.;). This group contained strains with the capacity to use hydrocarbon as *Rhodococcus, Pseudomonas, Micrococcus,* and *Kocuria* among other strains. This group is close to pre- spill samples and the samples taken on days 60, 102 and 164. Other small group was the group B with 9 species.

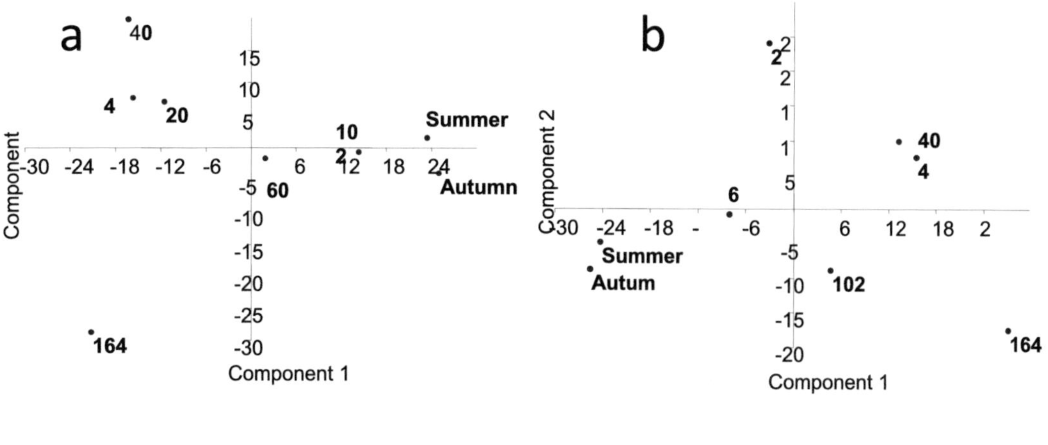

Fig. 2. Principal component analysis of a) phenanthrene and b) hexadacane as contaminants. The summer and autumn points correspond to samples taken 1 year before the spill.

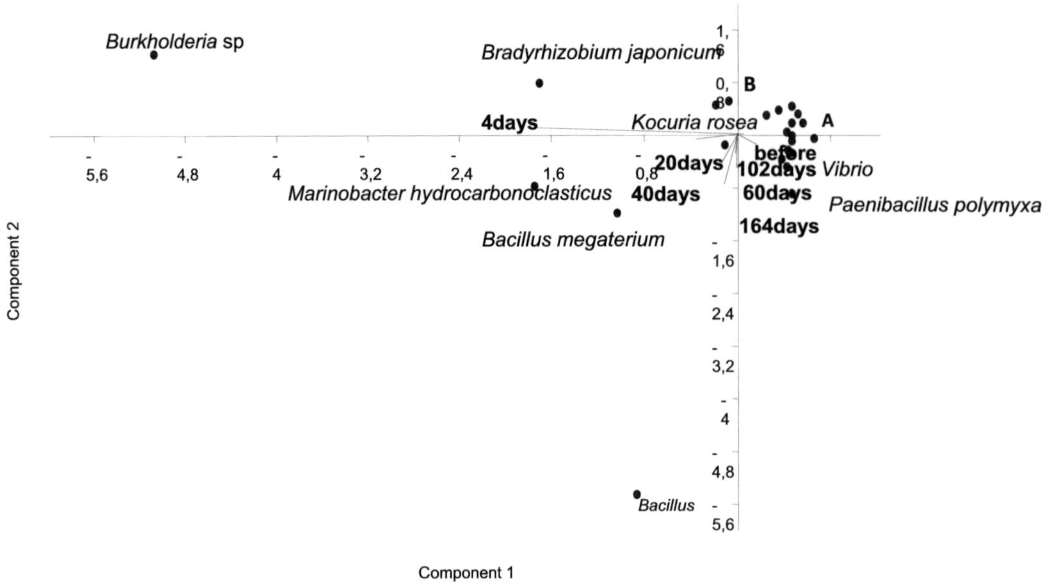

Fig. 3. Principal component analysis of bacteria and time. Group A Aerococcus viridians, Bacillus atrophaeus, Bacillus pumilus, Brevibacillus centrosporus, Curtobacterium flaccumfaciens, Flavobacterium ferrugineum, Halomonas aquamarina, Kocuria rosea, Kurthia gibsonii, Lysobacter enzymogenes enzymogenes, Micrococcus luteus, Microbacterium sp, Nocardia transvalensis, Pseudoalteromonas nigrifaciens, Pseudomonas aeruginosa, Psychrobacter immobilis, Rhodococcus, Rhodobacter, Roseomonas genomospecies, Shewanella putrefaciens, Streptoverticillium reticulum, Staphylococcus, Vibrio, Vibrio fischeri, Pseudoxanthomonas sp.

Shanon index gradually decreased, except on day 102, when it slightly increased (Table 2). Group B *Arthrobacter, Burkholderia cenocepacia, Brevundimonas vesicularis, Phyllobacterium myrsinacearum, Ochrobactrum anthropi, Pseudoalteromonas, Paucimonas lemoignei, Micrococcus sp, Zobellia uliginosa.*

4. Discussion

Oil degradation activities in natural environments are regulated by various physicochemical parameters such as water temperature, dissolved oxygen, nutrient and salt concentrations (Leahy, 1990). Seasonal variation in temperature can be a limiting factor for oil biodegradation in seawater (Minas 1999; Pucci et al., 2009a,b) and sediments (Shiaris 1989; Pucci et al., 2009a,b). The spill took place in Argentina during the summer and this was determinant in the fast decrease of the oil accumulated on the beach together with a good and fast job of a cleanup company. The variation on CO_2 accumulated changed significantly during summer and spring (Pucci et al., 2010) and this was always followed by the number of bacterial count.

The spill aimed at stimulating pollutant-degrading microorganisms to speed the recovery of contaminated ecosystems to a pre-pollution state in terms of biodiversity and ecosystem function. In this study, changes in the predominant bacterial populations occurred on day 4. Oiling and especially bioremediation led to a strong decrease in bacterial community diversity on day 20, but a rapid recovery to near pre-oiling levels of diversity did not occur subsequently. Still, despite having a similar level of biodiversity, the component organisms contributing to that diversity were somewhat different from the original community, as revealed by FAMEs analysis, which agrees with Röling et al. (2002). Kaplan & Kitts (2004), reported an increase in diversity during a bioremediation of crude oil-contaminated soil. In the present study, although diversity decreased on the time in both treated and control plots, no significant differences were found. Similarly, in a recent study on bioremediation of creosote contaminated soil, we reported depletion of diversity during the maximum biodegradation period; diversity remained low until the end of the experiment (Viñas et al., 2005; Jimenez et al., 2007).

Caleta Córdova beach had a good microbiological population, which included hydrocarbon degrading bacteria. The presence of these bacteria can be attributed to old spills, which were not officially registered but remembered by elder people. The fast bacterial growth in summer, proved by a recount and mineralization rate, confirmed that there was an old spill in the zone. Hydrocarbon degraders may comprise less than 0.1% of the microbial community in unpolluted environments but can constitute up to 100% of the viable microorganisms in oil-polluted ecosystems (Atlas, 1981). Effective biodegradation requires an appropriate population of degraders that have adequate tolerance to environmental changes (Mosbech, 2002), and the environment should be conducive to potential active microorganisms. On the beach, the biodegradation could be a minor process, but the capacity to use hydrocarbon was showed in the mineralization microcosms. The highest values of mineralization rate were not found in crude oil, probably because of the presence of polar hydrocarbons in Patagonia oil. Diesel and lubricant oil are alkane hydrocarbons which are not toxic for the bacteria whereas gas oil has a toxic compound and its mineralization rate showed high values. The length of time that degraders become acclimated and enriched before the degradation provides an approximate clue of how rapidly the microbial populations in different cold environments respond, and thus helps in making cleanup decisions. It is also found that formerly polluted frozen soils are commonly enriched in hydrocarbon-utilizing organisms (Margesin & Schinner, 1997, 1999, 2001). However, the time for different microbial species to recover and degrade petroleum contaminants varies markedly with sites.

The capacity of bacteria, especially *Pseudomonas* and *Rhodococcus*, to metabolize aerobically oil or aliphatic hydrocarbons has been well known for a long time in Patagonia (Acuña et al., 2010; Pucci & Pucci, 2003). The n-alkanes are the most biodegradable of the petroleum hydrocarbons and are attacked by more microbial species than aromatic or naphthenic compounds. However, normal alkanes in the C5-C10 range are inhibitory to any hydrocarbon degraders at high concentrations; this is because, as solvents, they disrupt lipid membrane. On the other hand, the cycloalkanes (alicyclic hydrocarbons) are less degradable that their straight-chain parent, the alkanes, but more degradable than the polycyclic aromatics (PAHs) (Jeffrey, 1980; Bacosa et al., 2010). From some years, the biodegradation of alkanes correlated with in relation to denitrification, sulfate reduction, and metanogenesis (Aeckersberg et al., 1998; Rueter et al., 1994). Moreover, aliphatic hydrocarbons are not fermentable, and only in the presence of O_2, significant aliphatic hydrocarbon degradation occurs. The initial degradation of petroleum hydrocarbons often requires the action of oxygenase enzymes and, thus, it is dependent on the presence of molecular oxygen (Brock, 1970; Atlas, 1991). Aerobic conditions are therefore necessary for the initial breakdown of petroleum hydrocarbons and in subsequent steps, nitrate or sulfate may serve as a terminal electron acceptor (Bartha, 1986), but oxygen is most commonly used. In the field of chemistry research after the 1989 Exxon Valdez oil spill accident in Alaska, Bragg et al. (1994) demonstrated that in situ oil biodegradation rates along oil polluted coasts were stimulated by fertilizer applications, suggesting that the rates depended mainly on the nitrogen concentration along the shoreline. The Maruyama et al. (2003) study also indicates that the application of inorganic nutrients (such as nitrate, ammonium, and phosphate) would have accelerated oil cleanup processes. Much more stimulation can be achieved by supplementation with organic nutrients such as yeast extracts. That was not possible in the case of Caleta Córdova spill because the environmental minister did not allow their use, but many bacteria in Patagonia soil had nitrogenases and the capacity to use hydrocarbons (Acuña et al., 2010); however, these genes were not searched during this study.

For this study, such a measure may have been of little value; in the plot, the bacterial community structure after 164 days was comparable to the pre-spill community structure. It has been suggested that restoration of the bacterial community structure to a state similar to that present prior to pollution could be used as a parameter for determination of the ecological end point of bioremediation (Stephen et al., 1999). Another explanation for this similarity would be the recovery of indigenous microbial populations present before the accident.

The use of diversity index in microbial ecology studies have been described previously by several authors (Hill et al., 2003; Hughes et al., 2001; Zrafi-Nouira et al., 2009; Pedrós-Alió, 2006). Our results are in agreement with Zrafi-Nouira et al. (2009) and Brakstad & Lødeng (2004), which indicate that the reduction of microbial diversity is observed early, from day 4 to day 164 after the spill. In fact, the crude oil spill can dramatically reduce bacterial diversity (Röling et al., 2002). Shannon diversity indexes shown in Table 2 seem to be low when compared to the results obtained in other studies, which is around 4 to 5 (Schloss & Handelsman 2006). Furthermore, lower values of the Shannon index were previously found in aged crude oil impact into the coast as indicated by Saul et al. (2005), Kaplan & Kitts (2004) and Popp the present study we found that bacterial diversity is gradually reduced during the biodegradation process, but Fig. 3 showed that the identified microorganisms, on day 164, were closer to the pre-spill community structure.

5. Conclusions

The four geographical points studied in the San Jorge Gulf share similar biodiversity and they can probably have a similar response to a crude oil spill impact. Previous to the spill, Caleta Córdova beach point had a bacterial population with the capacity to use hydrocarbons. After 60 days, there was not TPH and the communities returned to a state similar to that present before the spill. It can be concluded that the presence of hydrocarbonoclastics population is normal in the San Jorge Gulf coast and that it can rapidly respond to a natural disaster. The Shannon index was lower than that reported in other studies, and this needs further research especially of microorganisms that could grow in nitrogen absence, which is an important nutrient to biodegradation.

6. Acknowledgment

We acknowledge the financial support of CEIMA and UNPSJB. We sincerely thank Mirta Leiva, Miriam Roble and technical support of CEIMA for their invaluable and advice on technical work.

7. References

Acuña, A.J.; Pucci, O.H. & Pucci, G.N. (2008). Caracterization of a hydrocarbons biorremediation process in nitrogen deficit in a soil from Patagonia Argentina. *Ecosistema,* Vol. 17, No 2 (May, 2008), pp. 85-93, ISSN 1697-2473

Acuña, A.J.; Pucci, O.H. & Pucci, G.N. (2010). Characterization of three nitrogen fixing and hydrocarbon degrading strains isolated from Patagonian soil. *Ecosistema,* Vol. 19, No 2 (May, 2010), pp. 125-136, ISSN 1697-2473

Aeckersberg, F.; Rainey, F.A. & Widdel, F. (1998). Growth, natural relationships, cellular fatty acids and metabolic adaptation of sulfate-reducing bacteria that utilize long-chain alkanes under anoxic conditions. *Archives of Microbiology,* Vol. 170, No 5 (July, 1998), pp. 361–369, ISSN 0302-8933

Amann, R.I.; Ludwig, W. & Schleifer, K.H. (1995). Phylogenetic identification and in situ detection of individual microbial cells without cultivation. *Microbiological Reviews,* Vol. 59, No 1 (March, 1995), pp. 143–169, ISSN 0893-8512

Atlas, R.M. (1991). Microbial hydrocarbon degradation - bioremediation of oil spills. *Journal Chemical Technology and Biotechnology,* Vol. 52, No 1 (January, 1991), pp. 149–156, ISSN 1097-4660

Atlas, R.M. (1981). Microbial degradation of petroleum hydrocarbons: an environmental perspective. *Microbiological Reviews,* Vol. 45, No 1 (March, 1981), pp. 180–209, ISSN 0066-4227

Atlas, R.M. & Bartha, R. (1972). Degradation and mineralization of petroleum in seawater: limitation by nitrogen and phosphorus. *Biotechnology Bioengineering,* Vol. 14, No 3 (May, 1972), pp. 309–318, ISSN 006-3592

Bacosa, H.P.; Suto, K. & Inoue, C. (2010). Preferential utilization of petroleum oil bydrocarbon components by microbial consortia reflects degradation pattern in aliphatic–aromatic hydrocarbon binary mixtures. *World Journal of Microbiology and Biotechnology.* Vol.27, No 5 (September, 2010), pp 1109-1117, ISSB 0959-3993

Bartha, R. (1986). Biotechnology of petroleum pollutant biodegradation. *Microbial Ecology,* Vol. 12, No 1 (January, 1986), pp. 155–172, ISSN 0095-3628

Bartha, R. (1979). Effects of environments Parameters on the biodegradation of oil sludge. *Applied Environmental Microbiology,* Vol. 37, No 4 (April. 1979), pp. 729-739, ISSN 0099-2240

Beckman, E.M.; Tomohiro Kawaguchi, G.; Chandler, T. & Decho, A.W. (2008). Development of a microplate-based fluorescence immunoassay using quantum dot streptavidin conjugates for enumeration of putative marine bacteria, *Alteromonas* sp., associated with a benthic harparticoid copepod. *Journal of Microbiological Methods,* Vol. 75, No 3 (December, 2008), pp. 441-444, ISSN 0167-7012

Bordenave, S.; Goni-urriza, M.S.; Caumette, P. & Duran, R. (2007). Effects of heavy fuel oil on the bacterial community structure of a pristine microbial mat. *Applied and Environmental Microbiology,* Vol. 73, No 19 (October, 2007), pp. 6089–6097, ISSN 0099.2240

Bragg, J.R.; Prince, R.C.; Harner, E.J. & Atlas, R.M. (1994). Effectiveness of bioremediation for the Exxon Valdez oil spill. *Nature*, Vol. 368, No 3 (March, 1994), pp. 413–418, ISSN 0028-0836

Brakstad, O.G. & Lødeng, A.G.G. (2004). Microbial diversity during biodegradation of crude oil in seawater from the North Sea. *Microbiol Ecology,* Vol. 49, No 1 (January, 2004), pp. 94–103, ISSN 0095-3628

Britton, L.N. (1984). Microbial degradation of aliphatic hydrocarbons. *In Microbial Degradation of Organic Compounds. D.T. Gibson,. (Ed.)*,pp 89-129, Marcel Dekker, ISBN 0-824-77333-0 *New York*

Brock, T.D. (1970). Biology of microorganisms. Prentice-Hall. ISBN 1-55581-142-6. Washington new Jersey

Cagnon, C.; Stauffert, M.; Huang, L.; Cravo-Laureau, C.; Goñi Urriza, M.S.; Bordenave, S.; Païssé, S.; Caumette, P. & Duran, R. (2011). Impact of Hydrocarbons on Marine Microbial Communities. *Global Change: Mankind Marine Environment Interactions,* Part. 6, pp. 335-339, DOI: 10.1007/978-90-481-8630-3-59

Harayama , S.; Kasai, Y. & Hara, A. (2004). Microbial communities in oil-contaminated seawater. *Current opinion in biotechnology,* Vol. 15, No 3 (June, 2004), pp. 205-214, ISSN 0958-1669

Head, I.M.; Jones, D.M. & Röling, W.F.M. (2006). Marine microorganisms make a meal of oil. *Nature*.. Vol. 4, No 3 (March, 2006) , 173-182, ISSN 0028-0836.

Hill, T.C.J.; Walsh, K.A.; Harris, J.A. & Moffett, B.F. (2003) Using ecological diversity measures with bacterial communities. *FEMS Microbiology Ecology,* Vol. 43, No 1 (February 2003), pp. 1–11, ISSN 0168-6496

Hoffmann, M.; Fischer, M. & Whittaker, P. (2010). Evaluating the use of fatty acid profiles to identify deep-sea Vibrio isolates. *Food Chemistry,* Vol. 122, No 4 (October 2010), pp. 943–950, ISSN 0308-8146

Holt, H.M.; Gahrn-Hansen, B. & Bruun, B. (2005). Shewanella algae and Shewanella putrefaciens: clinical and microbiological characteristics. *Clincal Microbiology and Infectology,* Vol. 11, No 5 (May 2005), pp. 347–352, ISSN 1198-743X

Hughes, J.B.; Hellmann, J.J.; Ricketts, T.H. & Bohannan, B.J. (2001). Counting the uncountable: statistical approaches to estimating microbial diversity. *Appled Environment Microbiololy,* Vol. 67, No 10 (October 2001), pp. 4399–4406. Erratum in: *Appl Environ Microbiol.* Vol. 68, N° 1 (Enero, 2002), pp. 448, ISSN 0099-2240

Iantanos, N.; Pucci, G.N.; Acuña, A.J. & Pucci, O.H. (2009). Derrame de hidrocarburos en la playa de Caleta Córdova, ciudad de Comodoro Rivadavia, Argentina: su evolución. *AIDIS ARGENTINA - Ingeniería Sanitaria y Ambiental,* Vol. 100, No 4 (Septiembre-Diciembre 2004), pp. 113-117, ISSN 0328-2937

Ivanova, E.; Flavier, S. & Christen, R. (2004). Phylogenetic relationships among marine *Alteromonas*-like proteobacteria: emended description of the family *Alteromonadaceae* and proposal of Pseudoalteromonadaceae fam. nov.,

Colwelliaceae fam. nov., Shewanellaceae fam. nov., Moritellaceae fam. nov., Ferrimonadaceae fam. nov., Idiomarinaceae fam. nov. and Psychromonadaceae fam. nov. *International Journal of Systematic Evolutionary Microbiology,* Vol. 54, pp. 1773-1788, ISSN 1466-5026

Jeffrey, L.M. (1980). Petroleum residues in the marine environment. In: Geyer RA, editor. Marine environmental pollution: 1. Hydrocarbon. Amsterdam. *Elsevier,* Vol. 27, No 1 (May, 1980), pp. 163–179, ISSN 0020-7225

Jiménez, N.; Viñas, M.; Bayona, J.M.; Albaiges, J. & Solanas, A.N. (2007). The Prestige oil spill: bacterial community dynamicsduring a field biostimulation assay. *Applied Microbiology and Biotechnology,* Vol. 77, No 4 (October, 2007), pp. 935–945, ISSN 0175-7598

Jiménez, N.; Viñas, M.; Sabaté, J.; Díez, S.; Bayona, J.M.; Solanas, A.M. & Albaiges, J. (2006). The Prestige oil spill. II. Enhanced biodegradation of a heavy fuel oil by the use of an oleophilic fertilizer under field conditions. *Environmetal Science and Technology,* Vol. 40, No 8 (March, 2006), pp. 2578–2585, ISSN 1064-3389

Kaplan, C.W. & Kitts, C.L. (2004) Bacterial succession in a petroleum land treatment unit. *Appled Environmental Microbiology,* Vol. 70, No 3 (December, 2004), pp. 1777–1786, ISSN 0175-7598

Kasai, Y.; Kishira, H. & Harayama, S. (2002). Bacteria Belonging to the Genus *Cycloclasticus* Play a Primary Role in the Degradation of Aromatic Hydrocarbons Released in a Marine Environment. *Applied Environmental Microbiology,* Vol. 68, No 11 (November, 2002), pp. 5625-5633, ISSN 0175-7598

Leahy, J.G. & Colwell, R.R. (1990), Microbial degradation of hydrocarbons in the environment. *Microbiology Review,* Vol. 54, No 3 (September, 1990), pp. 305-315, ISSN 0066-4227

Lennon Jay, T. (2007). Diversity and metabolism of marine bacteria cultivated on dissolved DNA. *Applied Environmental Microbiology,* Vol. 73, No 9 (May, 2007), pp. 2799-2805, ISSN 0175-7598

MacNaughton, S.J.; Stephen, J.R.; Venosa, A.D.; Davis, G.A.; Chang, Y.J. & White, D.C. (1999). Microbial population changes during bioremediation of an experimental oil spill. *Applied Environmental Microbiology,* Vol. 65, No 8 (August, 1999), pp. 3566–3574, ISSN 0175-7598

Margesin, R. & Schinner, F. (1997). Bioremediation of diesel-oil-contaminated alpine soils at low temperatures. *Applied Microbiology and Biotechnology,* Vol. 47, No 4 (December, 1996), pp. 462–468, ISSN 0175-7598

Margesin, R. & Schinner, F. (1999). Biological decontamination of oil spills in cold environments. *Journal Chemical Technology and Biotechnology,* Vol. 74, No 5 (May, 1999), pp. 381–389, ISSN 0268-7525

Margesin, R. & Schinner, F. (2001). Biodegradation and bioremediation of hydrocarbons in extreme environments. *Applied Environmental Microbiology,* Vol. 56, No 6 (May, 2001), pp. 650–663, ISSN 0175-7598

Maruyama, A.; Ishiwata, H.; Kitamura, K.; Sunamura, M,; Fujita, T.; Matsuo, M. & Higashihara1, T. (2003) Dynamics of Microbial Populations and Strong Selection for

Cycloclasticus pugetii following the Nakhodka oil spill. *Microbial Ecology,* Vol. 46, No 4 (August, 2003), pp. 442–453, ISSN 0095-3628

Minas, W. & kunkel, W. (1995) Oil pollution in the North Sea—a microbiological point of view. *Helgoland Marine Research,* Vol. 49, N° 1-4 (January, 1995), pp. 143–158, ISSN 1438-387X

Mosbech, A. (2002). Potential Environmental Impacts of Oil Spills in Greenland: An Assessment of Information Status and Research Needs. NERI Technical Report No. 415. National Environmental Research Institute, Ministry of the Environment, Denmark.

Oguntoyinbo, F.A. (2006). Monitoring of marine Bacillus diversity among the bacteria community of sea water. *African Journal of Biotechnology,* Vol. 6, No 2 (August, 2006), pp. 163-166, ISSN 1684-5315

Pedrós-Alió, C. (2006). Marine microbial diversity: can it be determined? *Trends in Microbiology.* Vol. 14, No 6 (June, 2006), pp. 257–263, ISSN 0966-842X

Popp, N.; Schlömann, M. & Mau, M. (2006). Bacterial diversity in the active stage of a bioremediation system for mineral oil hydrocarbon-contaminated soils. *Microbiology,* Vol. 152, No 11 (November, 2006), pp. 3291–3304, ISSN 1350-0872

Pucci, G.N.; Acuña, A.J.; Llanes, M.L.; Tiedemann, M.C. & Pucci, O.H. (2009). Diversity Of Bacteria Cultivable From The Coast Of Caleta Olivia, Patagonia, Argentina. *Acta Biológica Colombiana,* Vol. 14, No 3 (Septiembre-Diciembre, 2003), pp. 120-126, ISSN 0120-548X

Pucci, G.N.; Acuña, AJ,; Llanes, M.L.; Tiedemann, M.C. & Pucci, O.H. (2009b). Identification of cultivable marine bacteria from the coastal city of Comodoro Rivadavia, Argentina. *Revista de Biolología Marina Oceanografía,* Vol. 44, No 1 (January, 2009), pp. 49-58, ISSN 0717-3326

Pucci, G.N.; Acuña, A.J.; Tonín, N.L.; Tiedemann, M.C. & Pucci, O.H. (2010). Diversity of culturable bacteria capable of degrading hydrocarbons from the beach of Caleta Córdova, Argentina *Revista Peruana de Biología,* Vol. 17, No 2 (February, 2010), pp. 237-244, ISSN 1561-0837

Pucci, G.N. & Pucci, O.H. (2003). Biodegradabilidad de componentes de mezclas naturales de hidrocarburos previamente sometidas a Landfarming. *Revista Argentina de Microbiología,* Vol. 35, No 2 (Marzo, 2003), pp. 62-68, ISSN 0325-7541

Purkrtova, Z.M.; Jablonska, E. & Kralova, K. (2010). Isolation and characterization of lipid bodies from a marine bacterium Alcanovirax borkumensis. *Journal of Biotechnology,* Vol. 150, No 1 (November, 2010), pp. 224, ISSN 0168-1656

Röling, W.F.M.; Milner, M.G.; Jones, D.M.; Lee, K.; Fabien, D.; Swannell, R.J.P. & Head I-M. (2002). Robust Hydrocarbon Degradation and Dynamics of Bacterial Communities during Nutrient-Enhanced Oil Spill Bioremediation. *Applied and Environmental Microbiology,* Vol. 68, No 11 (November, 2002), pp. 5537–5548, ISSN 0099-2240

Röling, W.F.M.; Milner, M.G.; Jones, D.M.; Fratepietro, F.; Swannell, R.J.; Fabien, D. & Head, I.M. (2004). Bacterial Community Dynamics and Hydrocarbon Degradation during a Field-Scale Evaluation of Bioremediation on a Mudflat Beach Contaminated with Buried Oil. *Applied and Environmental Microbiology,* Vol. 70, No 5 (May, 2004), pp. 2603–2613, ISSN 0099-2240

Rollins, D.M., & Colwell, R.R. (1986). Viable but non-culturable stage of Campylobacter jejuni and its role in survival in the natural aquatic environment. *Applied and Environmental Microbiology,* Vol. 52, No 3 (September, 1986), pp. 531–538, ISSN 0099-2240

Rueter, P.; Rabus, R.; Wilkes, H.; Aeckersberg, F.; Rainey, A.; Jannasch, H.W. et al. (1994). Anaerobic oxidation of hydrocarbons in crude oil by new types of sulfate-reducing bacteria. *Nature,* Vol. 372, No 12 (December, 1994), pp. 455–458, ISSN 0028-0836

Saul, D.J.; Aislabie, J.M.; Brown, C.E.; Harris, L. & Foght, J.M. (2005). Hydrocarbon contamination changes the bacterial diversity of soil from around Scott Base, Antarctica. *FEMS Microbial Ecology,* Vol. 53, No 1 (June, 2005), pp. 141–155, ISSN 0168-6496

Schloss, P.D. & Handelsman, J. (2006). Introducing TreeClimber, a test to compare microbial community structures. *Applied Environmental Microbiology,* Vol. 72, No 4 (April, 2006), pp. 2379–2384, ISSN 0099-2240

Shiaris, M.P. (1989). Seasonal biotransformation of naphthalene, phenanthrene, and benzo[a]pyrene in surficial estuarine sediments. *Applied Environmental Microbiology,* Vol. 55, No 6 (June, 1989), pp. 1391–1399, ISSN 0099-2240

Stabili, L. & Cavallo, R. (2004). Biodiversity of culturable heterotrophic bacteria in the Southern Adriatic Sea Italian coastal waters. *Scientia Marina,* Vol. 68, No (November, 21), pp. 31-41, ISSN 0214-8358

Stephen, J.R.; Chang, Y.I.; Gan, Y.D.; Peacock, A.; Pfiffner, S.M.; Barcelona, M.J.; White, D.C. & Macnaughton, S.J. (1999). Microbial characterization of a JP-4 fuel-contaminated site using a combined lipid biomarker/ polymerase chain reaction-denaturing gradient gel electrophoresis (PCRDGGE)- based approach. *Applied Environmental Microbiology,* Vol. 1, No 3 (June, 1999), pp. 231–241, ISSN 0099-2240

Süss, J.; Engelen, B.; Cypionka, H. & Sass, H. (2004). Quantitative analysis of bacterial communities from Mediterranean sapropels based on cultivation depend methods. *FEMS Microbial Ecology,* Vol. 51, No 1 (December, 2004), pp. 109-121, ISSN 0168-6496

Swannell, R.P.J.; Fabien, D. & Head, I.M. (2004). Bacterial Community Dynamics and Hydrocarbon Degradation during a Field-Scale Evaluation of Bioremediation on a Mudflat Beach Contaminated with Buried Oil. *Applied Environment Microbiology,* Vol. 70, No 5 (May, 2004), pp. 2603–2613, ISSN 0099-2240

Viñas, M.; Sabaté, J. & Solanas, A.M. (2005). Bacterial community dynamics and PAHs degradation during bioremediation of a heavily creosote-contaminated soil. *Appled Environmental. Microbiololy,* Vol. 71, No 11 (November, 2005), pp. 7008–7018, ISSN 0099-2240

Wilkinson, S.G. (1988). Gram negative bacteria. *In: Ratledge, C., Wilkinson, S.G. (Eds.), Microbial Lipids.* Academic Press, London, England, pp. 299–488, ISBN 0-12-582304-5

Yu, K.S.H.; Wong, A.H.Y.; Yau, K.W.Y.; Wong, Y.S. & Tam, N.F.Y. (2005a). Natural attenuation, biostimulation and bioaugmentation on biodegradation of polycyclic aromatic hydrocarbons (PAHs) in mangrove sediments. *Marine Pollution Bulletin,* Vol. 51, No 8-12 (August-December, 2005), pp. 1071-1077, ISSN 0025-326X

Yu, S.H.; Ke, L.; Wong, Y.S. & Tam, N.F.Y. (2005b). Degradation of polycyclic aromatic hydrocarbons (PAHS) by a bacterial consortium enriched from mangrove sediments. *Environment International,* Vol. 31, No 2 (February, 2005), pp. 149-154, ISSN 0160-4120

Zrafi-Nouira, I.; Guermazi, E.S.; Chouari, R.; Safi, N.M.D.; Pelletier, E.; Bakhrouf, A.; Saidane-Mosbahi, D. & Sghir, A. (2009). Molecular diversity analysis and bacterial population dynamics of an adapted seawater microbiota during the degradation of Tunisian zarzatine oil. *Biodegradation,* Vol. 20, No 4 (December, 2009), pp. 467–486, ISSN 0923.9820

7

Human Impacts on Marine Biodiversity: Macrobenthos in Bahrain, Arabian Gulf

Humood Naser
Department of Biology, College of Science, University of Bahrain
Kingdom of Bahrain

1. Introduction

In recent decades, biological diversity has received increased interest. However, most studies are related to terrestrial systems, and knowledge of marine biodiversity lags behind that of land systems (Ellingsen, 2002). Oceans cover about 70% of the earth, and soft-sediment habitats cover most of the bottom of marine environments (Snelgrove, 1998). These habitats support a diverse array of macrobenthic communities that play important roles in ecosystem processes such as recycling nutrients, detoxifying pollutants, dispersion and burial, and secondary production (Gray, 1997; Snelgrove, 1997). Additionally, these organisms provide food for humans and are considered an important source of food for fishes and birds (Snelgrove, 1999; Thrush & Dayton, 2002). Human activities are the primary cause of recent changes to marine biological diversity in coastal and subtidal areas. It is therefore important to improve our understanding of both biodiversity in marine sediments and anthropogenic impacts affecting these habitats in order to effectively incorporate conservation measures.

The Arabian Gulf is a semi-enclosed sea situated in the subtropical zone and characterized by low precipitation and high aridity. It is a shallow sedimentary basin dominated by soft substrate benthos. The average depth of the Arabian Gulf is 35 m, and whole of its substrate lies within the photic zone. The Kingdom of Bahrain is an archipelago composed of 37 islands located in the Arabian Gulf between latitude 25° 32′ and 26° 20′ north and longitude 50° 20′ and 50° 50′ east (Fig. 1). The total land area of Bahrain is about 762 km^2. The length of the coastlines of Bahrain is approximately 126 km and the marine area is estimated to be around 8000 km^2. Despite the limited land area of Bahrain, waters around its islands support a range of coastal and marine habitats such as seagrass beds, coral reefs, mangrove swamps, offshore islands, and mud and sand flats.

High levels of salinity and surface temperature are the main natural stresses in the Arabian Gulf. Salinities around Bahrain are generally high due to the effects of high temperatures associated with high evaporation rates. Salinities on the west coast are higher than those on the east coast, with average means of 50-57 psu for the west coast and 43-45 psu for the east coast (Price et al., 1985). This variation in the salinity gradients may be attributed to a complex system of water circulation around Bahrain enforced by reduced water exchange, particularly in south of Bahrain, where salinity could reach 70 - 80 psu in areas with restricted water flow such as tidal pools and lagoons (Al-Zayani, 2003).

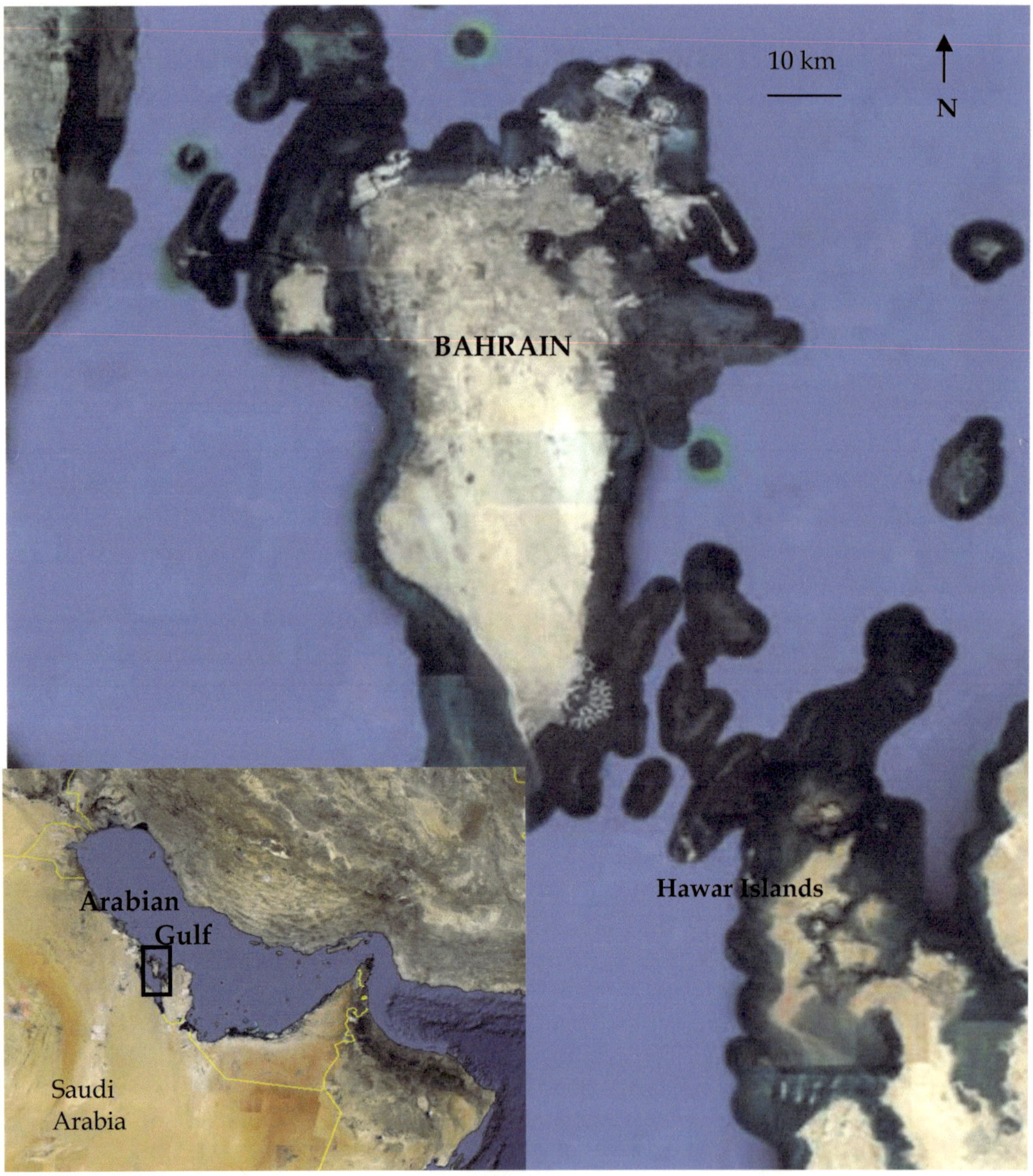

Fig. 1. Maps showing the Arabian Gulf and Bahrain (Google-Earth).

Although macrobenthos in the Arabian Gulf are characterized by high levels of biodiversity, they are distinguished by low species richness due to harsh environmental conditions such as high levels of salinity and temperature (Al-Yamani et al., 2009; Basson et al., 1977; Price, 2002; Sheppard et al., 2010). Additional anthropogenic effects could arguably be critical for biodiversity and abundance of macrobenthos inhabiting the naturally stressed marine environment of the Arabian Gulf, which is considered among the highest anthropogenically impacted regions in the world (Halpern et al., 2008). Such impacts include reclamation and dredging, industrial and sewage effluents, hypersaline water discharge from desalination plants, and oil pollution (Hamza & Munawar, 2009; Sheppard et al., 2010). This chapter explores the biodiversity of macrobenthos and anthropogenic impacts affecting these assemblages. Additionally, measures that may contribute to conservation of macrobenthos are discussed.

2. Macrobenthos biodiversity in the Arabian Gulf

Macrobenthos are a species-rich group of invertebrates that are defined as animals retained on a 300 μm sieve (Snelgrove, 1998). Macrobenthos are mainly composed of polychaetes, crustaceans, molluscs, and many other taxonomic groups. Polychaetes occur in almost all benthic marine sediments, and are typically the dominant component of macrobenthos in terms of number of species and abundance (Hutchings, 1998). Similarly, crustaceans and molluscs constitute diverse taxonomic groups that inhabit all major marine habitats (Feulner & Hornby, 2006; Snelgrove, 1998). Biodiversity and distribution of marine macrobenthos are influenced by sediment type, temperature, salinity, primary productivity, depth and physical disturbance (Basson et al., 1977; Coles & McCain, 1990).

Study	Area	Biodiversity of major groups
Basson et al. (1977)	Eastern coastline of Saudi Arabia	530 species were associated with seagrass, 638 species were recorded in subtidal sand, and 610 species were recorded in subtidal mud
MCain (1984)	Eastern coastline of Saudi Arabia	A total of 624 species of benthic organisms were identified, 452 from sand biotope and 369 from seagrass biotope samples
Coles and McCain (1990)	Eastern coastline of Saudi Arabia	834 species were found associated with seagrasses and sand/silt substrata
Zainal et al. (2007)	Baseline ecological survey for Hawar Islands, Bahrain	119 species were identified (47 Crustacea, 40 Mollusca, 30 Polychaeta)
Al-Yamani et al. (2009)	Kuwait coastal and subtidal areas	270 species were identified (103 Mollusca, 83 Polychaeta, 38 Crustacea)
Naser (2010a)	Subtidal areas influenced by organic and industrial effluents off the eastern coastline of Bahrain	97 species were identified (46 Polychaeta, 25 Mollusca, 18 Crustacea)

Table 1. Examples of studies conducted on macrobenthos in the Arabian Gulf.

Giving that more than 97% of the bottom substrate of the Arabian Gulf is dominated by sand and mud (Al-Ghadban, 2002), macrobenthos form the largest and most diverse marine ecosystem. Indeed, mudflats are among the most productive marine habitats on the southern shore of the Arabian Gulf (Basson et al., 1977; Sheppard et al., 2010). Mudflats and coastal wetlands in Bahrain support high levels of biodiversity and primary productivity, and contribute significantly into fisheries productivity (Abdulqader, 1999) and provide feeding and roosting grounds for important numbers of shorebirds (Mohamed, 1993).

It can be argued that biodiversity of macrobenthos in the Arabian Gulf is underestimated. Generally, studies of soft-sediment macrobenthic assemblages in tropical coastal and marine environment, including the Arabian Gulf, lag behind that of equivalent environments in temperate zones (Mackie et al., 2005). These studies require the identification of organisms, which is typically constrained by time, costs and the shortages of trained taxonomists. Nonetheless, studies investigating macrobenthic communities in coastal and subtidal areas of the Arabian Gulf revealed considerable levels of biodiversity (Table 1). Even though these studies were conducted using different sampling and processing methods, a general trend of reduction in diversity and abundance of macrobenthos could be detected. For instance, higher numbers of species were recorded in the earlier studies (Basson et al., 1977; Coles & McCain, 1990; MCain, 1984) in comparison with the most recent studies (Al-Yamani et al., 2009; Naser, 2010a; Zainal et al., 2007;). This reduction in both species number and abundance could be primarily attributed to the environmental degradation due to escalated anthropogenic activities in the Arabian Gulf (Sheppard et al., 2010).

3. Human impacts on macrobenthos

3.1 Dredging and reclamation

Bahraini coastal and marine environments are the prime target for most of the major housing, recreational, and economic projects, which typically associated with intensive dredging and reclamation activities (Naser et al., 2008). Presently, reclamation activities have resulted in adding 91 km^2 representing an increase of 11% of the total land area. Bahrain National Land Use Strategy 2030 recognizes reclamation as the major option for securing the future needs for land, indicating that coastal environment will continue to be the major focus for developmental projects in the coming future (Naser, 2011). Coastal reclamation is regularly carried out in Bahrain to meet the demand of rapid coastal developments (Fig. 2). Generally, sand and mud characterized by lower levels of biodiversity and abundance are extracted from designated borrow areas within the territorial waters of Bahrain, and then dumped into coastal and subtidal areas characterized by high levels of biodiversity and productivity (Naser, 2010a).

Dredging and reclamation involve the direct removal of macrobenthos and results in physically smothering the coastal and subtidal habitats and deoxygenating the underlying sediments (Allan et al., 2008; Newell et al., 1998). These physical and chemical alterations may reduce biodiversity, richness, abundance and biomass of macrobenthos (Smith & Rule, 2001).

Naser (2011) conducted a microcosm experiment to examine the effects of mud burial on selected macrobenthic species collected from a proposed reclaimed coastal area in the north-east of Bahrain. The study found significant difference in numbers of survived organisms between control and experimental treatments, with a survival percentage of 41.8% for all of the selected species. Dredging and reclamation in Bahrain are major activities causing the direct removal and burial of macrobenthic assemblages in the coastal and marine environments. Therefore, biodiversity and abundance of macrobenthos are severely affected by mortality and smothering associated with dredging and reclamation activities.

Fig. 2. Reclamation of sandy and muddy coasts of Tubli Bay.

3.2 Industrial effluents

Bahrain has witnessed a rapid industrial growth, mainly in the sectors of oil refining, aluminum and petrochemical industries. Several companies and industrial factories are producing effluents that may contain hydrocarbons, ammonia, phenols, phosphorous and heavy metals. In Bahrain, most of the industrial activities are located along the eastern coastal areas. It is estimated that around 1, 793, 294 m^3 day^{-1} of industrial effluents characterized by high inputs of heavy metals and hydrocarbons are discharged to the shallow subtidal areas off the eastern coastline of Bahrain (Naser, 2010a). The eastern coastline of Bahrain is considered a hotspot for high concentrations of hydrocarbons (De Mora et al., 2010) and heavy metals (De Mora et al, 2004). Effects of industrial effluents on macrobenthic assemblages involve changing the composition of the community structure, increasing the numbers of opportunistic species, and reducing the general biodiversity and abundance (Bigot et al., 2006; Frouin, 2000).

Naser (2010a) investigated the community structure of macrobenthos inhabiting a subtidal area subject to effluents containing hydrocarbons and heavy metals from the main oil refinery in Bahrain. The study recorded 44 species of which polychaetes, molluscs, crustaceans and remaining groups accounted for 66.0%, 22.7%, 6.8% and 4.5% respectively. Crustaceans are considered sensitive bioindicators for marine environmental pollution

(Gesteira & Dauvin, 2000, Ugolini et al, 2004). This is reflected in the limited number of species and abundance of crustaceans in the area influenced by the industrial effluents in Bahrain.

3.3 Sewage discharges

Sewage effluents are major sources of coastal pollution in Bahrain. Several sewage treatment plants varying in size and degree of treatment are discharging effluents to the coastal and subtidal areas in Bahrain. The main one is Tubli Water Pollution Control Centre (TWPCC), which discharges around 160,000 m^3 day^{-1} of treated effluents into the shallow water of Tubli Bay (Fig. 3). These effluents are characterized by high-suspended solid reaching up to 290 FTU. Seawaters adjacent the TWPCC are distinguished by high load of nutrients such as ammonia, nitrate and phosphate with concentrations reaching 1.40, 0.90, and 3.60 mg l^{-1} respectively. In particular, phosphate largely exceeded the maximum value of 2.0 mg l^{-1} recommended in the Bahraini effluent standards (Naser, 2010a).

Fig. 3. The outfall of Tubli Water Pollution Control Centre, the main sewage treatment plant in Bahrain.

Changes in macrobenthic community structure associated with moderate organic enrichment are represented by an increase in species richness, abundance and biomass. However, excessive organic enrichment reduces species richness, and increases densities and numbers of few opportunistic species and their associated biomass (Grall & Chauvaud, 2002; Savage et al., 2002). Naser (2010b) investigated the effects of sewage discharge on macrobenthic assemblages inhabiting subtidal areas surrounding the major treatment plant in Bahrain. The study showed a reduction in biodiversity, richness and evenness of macrobenthos due to the increase in organic enrichment mainly ammonia and phosphate. Indeed, some areas adjacent to the outfall that were characterized by high levels of organic content (> 60%) were devoid of macrobenthic assemblages reflecting severe sewage pollution.

3.4 Desalination plants

Bahrain, like most of the Arabian Gulf countries, depends mainly on desalination seawater as a source of potable water. In Bahrain, there are currently four major desalination plants producing fresh water and energy (Zainal et al., 2008). Sitra Power and Water Station (SPWS) is the largest plant in Bahrain (Fig. 4) with a capacity of 125 megawatts of electrical power and around 100 x 10^6 L day^{-1} of desalinated water using multi-stage flash technology (Khalaf & Redha, 2001).

Fig. 4. Sitra Power and Water Station.

SPWS produces around 66,000 m^3 day^{-1} of effluent to the marine environment of which 12,000 m^3 are brine water. This brine water causes an average increase in temperature of 7.5 ^oC above the natural water temperatures of summer and winter (Altayaran & Madany, 1992).

Reduced levels of biodiversity and abundance were recorded in stations adjacent to SPWS outlet reflecting severe impacts on macrobenthic assemblages caused by brine wastewater discharges, which are associated with high temperatures, salinities, and a range of chemical and heavy metal pollutants (Naser, 2010a). Generally, a decline in populations of all macrobenthic species is expected in the mixing zone of an effluent discharge potentially associated with toxic wastes. Survival of species in the adjacent areas of a mixing zone depends on their tolerance and adaptation. This is demonstrated by the dominance of tube anemones (*Cerianthus* sp.) in the areas adjacent to the outlet of SPWS, which suggests that these species might be resistant or adaptable to pollutants and changes in water quality.

3.5 Oil pollution

The Arabian Gulf is considered the largest reserve of oil in the world. Consequently, Bahrain is under a permanent threat from oil related pollution. The sources of oil spills are offshore oil wells, underwater pipelines, oil tanker incidents, oil terminals, loading and handling operations, weathered oil and tar balls, and illegal dumping of ballast water (Literathy et al., 2002). Generally, Bahraini ports and oil terminals are at high risk of major oil spill incidents. Similarly, minor oil spills are frequent in the territorial waters of Bahrain. For instance, six minor oil spills, including leakages in pipelines, over flooding of containers, weathered oil and tar balls, and incidents during loading of tankers in the terminals, were recorded in 2000.

Benthic assemblages are sensitive to oil spills. However, the effects of oil pollution primarily depend on the proportion of hydrocarbon-sensitive species, particularly crustaceans, in the affected community (Gesteira et al., 2003). Suchanek (1993) reviewed in details the impacts of oil on marine invertebrates. This study indicated that macrobenthic invertebrate assemblages respond to severe oil pollution with initial massive mortality and lowered community diversity followed by fluctuations in population of opportunistic species. Studies in the Arabian Gulf reported a reduction in macrobenthos diversity (Jones et al., 1998) and a decline in the abundance of major faunal groups after a major oil spill (Price, 1998).

4. Macrobenthos conservation

4.1 Marine Protected Areas

A marine protected area (MPA) is defined by the World Conservation Union (IUCN) as any area of the marine environment that has been reserved by law or other effective means. Marine protected areas may contribute significantly to both preservation and conservation of genetic characteristics, species, habitats and cultural biodiversity in marine environments (Agardy, 1994; Kelleher & Phillips, 1999; Krupp, 2002).

Coastal and marine protected areas have been established in Bahrain (Table 2). Marine protected areas in Bahrain are considered an affective mean of protecting macrobenthos diversity. For example, Hawar Islands, the largest protected area in Bahrain, support high levels of macrobenthos diversity. Zainal et al. (2007) recorded 119 species in the shallow subtidal habitats of Hawar Islands. Likewise, the sheltered mudflats in Ras-Sand are characterized by high levels of macrobenthos diversity and abundance (Naser, 2010a). It is recognized that mudflats provide several ecosystem services (Zedler & Kercher, 2005), and higher levels of benthos diversity and abundance in these habitats are important to bird populations (Zou et al., 2008). Several nationally and internationally important birds use mudflats in Ras-Sand as feeding and roosting grounds (Evans, 1994). Similarly, Al-Sayed et al. (2008) indicated that the protected mudflat of Duwhat Arad is a productive ecosystem, and macrobenthic assemblages inhabiting this area provide food resources for bird populations.

However, the lack of management plans for the Bahraini protected areas may restrict the achievement of the conservational objectives of these areas. The Convention on Biological

Diversity requires that each of the contracting parties prepares a National Biodiversity Strategy and Action Plan. Toward this, Bahrain is preparing the Bahraini National Biodiversity Strategy and Action Plan (BNBSAP) with support from United Nations Development Programme (UNEDP). The BNBSAP will address the current lack of management plans for the protected areas and suggest measures to protect biodiversity in Bahrain (Naser et al., 2008).

Protected area	Ecological importance
Hawar Islands	Extensive growth of seagrass beds and algal mats that support vulnerable species such as dugongs, turtles and dolphins. These islands host one of the largest breeding colonies of the endemic Socotra Cormorants with a winter population of 200,000 individuals.
Ras-Sanad area in Tubli Bay	A sheltered lagoon hosting the last remaining mangrove ecosystems in Bahrain. Foraging and roosting ground for migratory and breeding birds.
Mashtan Island	Offshore island characterized by extensive growth of seagrass and algal mats for the feeding of dugongs and turtles.
Duwhat Arad	Tidal mudflat used as a feeding and roosting ground for important shorebird populations.
Fasht bulthama	Small reef characterized by relatively high levels of diversity and cover (> 50%)

Table 2. Marine protected areas in Bahrain and their ecological importance.

4.2 Environmental Impact Assessment

Environmental Impact Assessment (EIA) is a systematic process of identifying, predicting, evaluating and mitigating the environmental consequences of a proposed project on the biological and physical environments (Glasson et al., 1999). EIA aims at integrating environmental considerations in the decision-making system, minimizing or avoiding adverse impacts, protecting natural systems and their ecological processes, and implementing principles of sustainable development (IAIA, 1999).

EIA is considered a standard tool for decision-making in most countries throughout the world (Erickson, 1994). It ensures that authorities are provided with necessary knowledge relating to any likely significant effects of a proposed project on the environment prior to the decision-making process. The integration of environmental considerations may result in a rational and structured decision-making process that maintains a balance of interest between the development action and the environment (Glasson et al., 1999). EIA minimizes or avoids the adverse effects of a proposed development on the environment, by addressing effective designs, alternatives, mitigations, cumulative impacts, and monitoring (Cooper & Sheate, 2002). EIA contributes to protect biodiversity, ecosystems and natural resources (Treweek, 1999). Since the early stages of incorporating EIA in The National Environmental Policy Act in 1969 in the USA, ecological considerations have been an integrated part of the EIA process (Gontier et al., 2006). EIA addresses environmental consequences along with economic and social considerations in the planning system, which is widely recognized as a principle of sustainable development (Lee & George, 2000). EIA was formally adopted in

Bahraini environmental system in 1998 to protect the environment and to reduce environmental degradation associated with major developmental projects.

EIA is widely recognized as an important tool for integrating biodiversity considerations into planning and decision-making processes (Slootweg et al., 2003). The Convention on Biological Diversity (CBD) requires the use of EIA to ensure that impacts on biodiversity are taken into account and adverse effects are avoided or minimized. However, there is a recognized need for further enforcement and integration of biodiversity issues into EIA processes (Wegner et al., 2005).

Most of developmental projects in Bahrain are related to the coastal and marine environments. Therefore, macrobenthic assemblages are subject to direct or indirect impacts by associated dredging and reclamation activities. Naser et al. (2008) evaluated the quality of ecological input in 15 Bahraini EIA reports concerning coastal and marine developments produced between 1996 and 2004. The study showed major shortcomings in these reports, including limited new ecological surveys, inadequate evaluation of impacts, neglecting cumulative and long-term impacts, and failing to address adequately mitigation and monitoring measures. Nonetheless, several major developmental projects have been undertaken at a rapid rate in Bahrain in the recent years, and quality of the EIA reports are notably improving.

Macrobenthic assemblages are considered a useful tool for ecological assessment within the framework of EIA. These organisms can facilitate the characterization of existing impacts as well as the prediction of likely impacts due to proposed projects (Naser 2010a). EIA can be used to ensure that necessary measures needed to protect biodiversity and its sustainable use are addressed in the process of development planning (Khera & Kumar, 2010). Therefore, effective EIA can contribute to the protection of macrobenthos biodiversity and the sustainable use of coastal and marine environments in Bahrain.

4.3 Legal instruments and higher environmental policies

The Environmental Protection Committee (EPC), established by Decree No. 7 of 1980 under the Ministry of Health, was the first governmental authority concerned with the protection of the Bahraini environment. The EPC was upgraded by the Legislative Decree No. 21 of 1996, establishing 'Environmental Affairs' under the Ministry of Housing, Municipalities, and Environment, which was reorganized under the Ministry of State for the Municipalities Affairs and Environmental Affairs in 2000. For the purpose of integrating efforts and resources to achieve effective protection of the environment, the three main governmental bodies concerned with the environment, namely Environmental Affairs, National Commission for Wildlife Protection, and Directorate of Fisheries were joined under the umbrella of the 'Public Commission for the Protection of Marine Resources, Environment and Wildlife' (PCPMREW) in 2002. The PCPMREW was established by the Legislative Decree No. 50 of 2002 and reorganized further by Decrees No. 10 and No. 43 of 2005. The PCPMREW is considered as a positive initiative towards the institutionalization of the authorities concerned with the environment in Bahrain.

Recognizing the escalation of environmental degradation, Bahrain has undertaken several initiatives to prevent the ongoing deterioration of the environmental quality. Towards this, Bahrain launched the National Environmental Strategy (NES), which was approved by the Council of Ministers in 2006. NES indentifies mechanisms by which principles of sustainable development can be implemented, including enforcing the role of EIA during planning, implementation and after commissioning phases of major projects, adopting principles of integrated environmental management for coastal and marine environments, applying valuation systems to estimate the costs of environmental degradation and rehabilitation, strengthening institutional and legal frameworks, and increasing public awareness and participation (GDPEW, 2006).

Environmental legislations related to biodiversity in Bahrain are based on a range of national laws as well as regional and international agreements. Nationally, there are several laws with respect to environment, regulation of fishing and exploitation of marine resources, protection of wildlife, environmental quality standards for wastewater effluents, declarations of protected areas, and banning of catching endangered species. Similarly, Bahrain ratified several regional and international agreements and conventions that could contribute to the protection of biodiversity, including Convention on Biological Diversity, Ramsar convention, UN Framework Convention on Climate Change, and Kuwait Regional Convention for Cooperation on the Protection of the Marine Environment from Pollution (PCPMREW, 2010). Such legislations and agreements, which protect the biodiversity and promote the sustainable use of environmental resources, can directly or indirectly contribute to the protection of macrobenthos habitats and associated ecosystems.

4.4 Conservation of associated ecosystems

Conducting ecological studies that investigate other ecosystems such as seagrass, mangroves and coral reefs are essential parts of any effort to conserve macrobenthos in Bahrain. As a result of the shallowness of the Arabian Gulf, a wide range of its bottom is covered with seagrass beds (Phillips, 2003). Seagrass beds are highly productive ecosystems that provide important ecological and economical functions (Duffy, 2006). They contribute to the productivity of local fisheries (Price et al., 1993), and provide food sources and nursery grounds for turtles, dugongs, shrimps and a variety of economically important marine organisms (Preen, 2004). It is widely recognized that seagrass habitats support greater macrofauna species diversity, abundance and biomass than adjacent unvegetated habitats (Al-Khayat, 2007; Ansari et al., 1991; Coles & McCain, 1990). Indeed, about 9% of the Arabian Gulf's faunal taxa have been estimated to occur in its seagrass meadows, about half of which are molluscs (Sheppard et al., 2010).

Coral reefs are characterized by both biological diversity and high levels of productivity. They provide a variety of ecological services such as renewable sources of seafood, maintenance of genetic, biological and habitat diversity, and recreational values (Moberg & Folke, 1999). Coral reefs in the Arabian Gulf have traditionally been important habitats for fisheries (Sheppard et al., 1992). Due to the higher levels of productivity in coral reef ecosystems, macrobenthos associated with these habitats are distinguished by higher levels of biodiversity (Moberg & Ronnback, 2003).

Mangrove swamps are ecologically important coastal ecosystems that provide food, shelter and nursery areas for a variety of terrestrial and marine fauna (Saenger, 2002). Mangrove ecosystems maintain species diversity, including macrobenthos (Ellison, 2008; Lee, 2008). Further, many macrobenthos species utilize mangrove habitats as nursery or feeding grounds during some life stage (Moberg & Ronnback, 2003).

4.5 Environmental monitoring and scientific research

Monitoring is an integral part of any effort to reduce the loss of biodiversity (Dallmeier, 1996). Monitoring could provide decision makers with information on the state of biodiversity, and consequently, assist in identifying management goals and assessing priorities for conservation (Niemela, 2000). Monitoring can be described as systemic observations and measurements of ecosystems to detect changes over time (Treweek, 1999). Soft-sediment macrobenthic assemblages are useful and sensitive indicators of the quality of the environment. They are widely used in biodiversity assessment and monitoring programmes in coastal and marine environments (Gray et al., 1992). This is mainly attributed to their characteristics that make them a useful target for assessment and monitoring purposes. These organisms consist of different species that show different levels of tolerance to stresses and pollution. They can exhibit detectable changes in their community structure in response to stresses and pollution such as changes in biodiversity, abundance, biomass, and numerical dominance of opportunistic species (Gray, 1989). Additionally, macrobenthic assemblages may be utilized to characterize the 'health' of coastal and marine ecosystems represented by their structures (the species and populations involved) and functions (the flow of energy, growth and productivity) (Boesch & Paul, 2001). Indeed, responses of macrobenthic assemblages to environmental variations have been adopted as a tool for evaluating the success of conservation efforts (Winberg et al., 2007), and managing marine environments (Desroy & Retiere, 2004).

The role of taxonomic research in biological research in general and biodiversity conservation in particular is widely recognized (Ziegler & Krupp, 1996). Effective conservation can only be achieved if the state of the environment is fully documented and understood. Taxonomy has a crucial role in preserving the biological diversity (Mace, 2004). Indeed, identifying and describing species are critically required to assess the biodiversity and to understand the structure and function of macrobenthic assemblages in Bahrain. However, taxonomic research in the Arabian Gulf is constrained by the limitation of taxonomic keys and guidelines. For instance, there are no comprehensive up-to-date taxonomic guides to the polychaetes in the Indian Ocean region (Mackie et al., 2005), including the Arabian Gulf. Identification to the species-level is critically required to assess the biodiversity and to understand the structure and function of macrobenthos in the Arabian Gulf. Consequently, producing identification guides for macrobenthic organisms in the Arabian Gulf may contribute to solve the problem of taxonomy. However, it is widely recognized that the process of identifying and describing species and subsequently producing such taxonomic guides is constrained by time, costs and shortages of trained taxonomists (Kendall et al., 2000). Therefore, logistical and financial cooperation between local and regional institutions and organizations concerned with the marine environment in the Arabian Gulf is required.

5. Conclusions

Globally, coastal and marine environments support a diverse array of macrobenthic communities that play important roles in ecosystem processes and provide several ecological and economic services. The Arabian Gulf is a shallow sedimentary basin dominated by sand and mud substrates. Macrobenthic assemblages form the largest and most diverse marine ecosystem in the Arabian Gulf. However, these assemblages inhabit one of the harshest marine environments due to marked fluctuations in sea temperatures and high salinities. Additional anthropogenic effects could arguably be critical for biodiversity and abundance of macrobenthos inhabiting the naturally stressed marine environment. Studies that investigated macrobenthos in the Arabian Gulf indicated that anthropogenic pollution and habitat degradation and destruction reduced biodiversity, richness, abundance and altered biomass of macrobenthic assemblages.

Preserving and conserving genetic, species, and habitat biodiversity in the marine environments are immediate priorities. Several measures could be applied to protect the biodiversity of macrobenthos in Bahrain. Marine protected areas are widely recognized as an effective mean of protecting biodiversity. Several coastal and marine protected areas have been established in Bahrain. However, the lack of management plans for the Bahraini protected areas may restrict the achievement of the conservational objectives of these areas.

Further integration of biodiversity into environmental impact assessment system in Bahrain is needed. This is of crucial importance as coastal and marine environments are the prime target for most developmental projects in Bahrain. Considering effects of dredging and reclamation on macrobenthos in EIA studies and suggesting measures to avoid or reduce adverse impacts could contribute to conserve the sensitive and productive mudflat habitats in Bahrain.

Legal instruments and higher level environmental policies in Bahrain are contributing to preventing environmental degradation and conserving biodiversity. Incorporating biodiversity assessment into the legal system and implementing the Bahraini National Biodiversity Strategy and Action Plan may contribute significantly into the enforcement of biodiversity conservation.

Seagrass, mangrove, and coral reef ecosystems support higher levels of marine biodiversity, including macrobenthic assemblages. Therefore, conserving and maintaining these ecosystems in a holistic approach would contribute to preserve the whole marine environment of the Arabian Gulf.

Monitoring and scientific research are integral part of any effort to reduce the loss of biodiversity. Macrobenthic assemblages are widely used in biodiversity assessment and monitoring programmes in coastal and marine environments. Developing necessary plans and mechanisms for population and habitat conservation require adequate knowledge and description of species. This could be achieved by promoting taxonomic research in the Arabian Gulf.

6. Acknowledgment

Thanks are due to the staff at the Department of Biology, College of Science, University of Bahrain. Technical assistance provided by F. Al-Ammari is highly appreciated.

7. References

Abdulqader, E. (1999). The role of shallow waters in the life cycle of the Bahrain penaeid shrimps. *Estuarine, Coastal and Shelf Science*, Vol. 49, pp. 115-121.

Agardy, T. (1994). Advances in marine conservation: The role of marine protected areas. *Trends in Ecology and Evolution*, Vol. 9, pp. 267-270.

Al-Ghadban, A. (2002). Geological oceanography of the Arabian Gulf. In: *The Gulf ecosystem: Health and Sustainability*, N. Khan, M. Munawar & A. Price (Eds.), pp. 23-39, Backhuys Publishers, Leiden.

Al-Khayat, J. (2007). Macrofauna abundance in seagrass of Qatari waters, Arabian Gulf. *Egyptian Journal of Aquatic Research*, Vol. 33, pp. 257-276.

Allan, S. Ramirez, C. & Vasquez, J. (2008). Effects of dredging on subtidal macrobenthic community structure in Mejillones Bay, Chile. *International journal of Environment and Health,* Vol. 2, pp. 64-81.

Al-Sayed, H. Naser, H. & Al-Wedaei, K. (2008). Observations on macrobenthic invertebrates and wader bird assemblages in a tropical marine mudflat in Bahrain. *Aquatic Ecosystem Health & Management,* Vol. 11, pp. 450-456.

Altayaran, A. & Madany, I. (1992). Impact of desalination plant on the physical and chemical properties of seawater, Bahrain. *Water Research,* Vol. 26, pp. 435-441.

Al-Yamani, F. Boltachova, N. Revkov, N. Makarov, M. Grintsov, V. Kolesnikove, E. & Murina, V. (2009). Winter species composition diversity and abundance of macrozoobenthos in Kuwait's waters, Arabian Gulf. *ZooKeys,* Vol. 31, pp. 17-38.

Al-Zayani, A. (2003). *The selection of marine protected areas: a model for the Kingdom of Bahrain.* PhD thesis, Centre for Environmental Science, University of Southampton, UK.

Ansari, Z. Rivonker, C. Ramani P. & Parulekar, A. (1991). Seagrass habitat complexity and macroinvertebrate abundance in Lakshadweep coral reef lagoons, Arabian Sea. *Coral Reefs,* Vol. 10, pp. 127–131.

Basson, P. Burchard. J. Hardy, J. & Price, A. (1977). *Biotopes of the western Arabian Gulf: Marine life and environments of Saudi Arabia.* ARAMCO, Dhahran, Saudi Arabia.

Bigot, L. Conand, C. Amouroux, J. Frouin, P. Bruggemann, H. & Gremare, A. (2006). Effects of industrial outfalls on tropical macrobenthic sediment communities in Reunion Island (Southwest Indian Ocean). *Marine Pollution Bulletin,* Vol. 52, pp. 865-880.

Boesch, D. & Paul, J. (2001). An overview of coastal environment health indicators. *Human and Ecological Risk Assessment,* Vol. 7, pp. 1409-1417.

Coles, S. & McCain, J. (1990). Environmental factors affecting benthic communities of the western Arabian Gulf. *Marine Environmental Research,* Vol. 29, pp. 289-315.

Cooper, L. & Sheate, W. (2002). Cumulative effect assessment: a review of UK environmental impact statements. *Environmental Impact Assessment Review,* Vol. 22, pp. 415-439.

Dallmeier, F. (1996). Biodiversity inventories and monitoring: essential elements for integrating conservation principles with resource development projects. In: *Biodiversity in managed landscapes: theory and practice,* R. Szaro, & D. Johnston, (Eds.), pp. 221-236, Oxford University Press, Oxford.

De Mora, S. Fowler, S. Wyse, E. & Azemard, S. (2004). Distribution of heavy metals in marine bivalves, fish and coastal sediments in the Gulf and Gulf of Oman. *Marine Pollution Bulletin,* Vol. 49, pp. 410-424.

De Mora, S. Tolosa, I. Fowler, S. Villeneuve, J. Cassi, R. & Cattini, C. (2010). Distribution of petroleum hydrocarbons and organochlorinated contaminants in marine biota and coastal sediments from the ROPME Sea Area during 2005. *Marine Pollution Bulletin,* Vol. 60, pp. 2323-2349.

Duffy, J. (2006). Biodiversity and the functioning of seagrass ecosystems. *Marine Ecology Progress Series,* Vol. 311, pp. 233-250.

Ellingsen, K. (2002). Soft-sediment benthic biodiversity on the continental shelf in relation to environmental variability. *Marine Ecology Progress Series,* Vol. 232, pp. 15-27.

Ellison, A. (2008). Managing mangroves with benthic biodiversity in mind: moving beyond roving banditry. *Journal of Sea Research,* Vol. 59, pp. 2-15.

Erickson, P. (1994). *A practical guide to environmental impact assessment*. Academic Press, London.

Evans, M. (1994). *Important bird areas in the Middle East*. Birdlife Convention Series No._2, Birdlife International, UK.

Feulner, G. & Hornby, R. (2006). Intertidal molluscs in UAE lagoons. *Tribulus*, Vol. 16, pp. 17-22.

Frouin, P. (2000). Effects of anthropogenic disturbances of tropical soft-sediment benthic communities. *Marine Ecology Progress Series*, Vol. 194, pp. 39-53.

GDPEW, General Directorate for the Protection of Environment and Wildlife (2006). *National Environmental Strategy*. GDPEW in Bahrain & United Nations Development Programme (UNDP), Bahrain.

Gesteira, J & Dauvin, L. (2000). Amphipods are good bioindicators of the impact of oil spills on soft-bottom macrobenthic communities. *Marine Pollution Bulletin*, Vol. 40, pp. 1017-1027.

Gesteira, J. Dauvin, J. & Frega, M. (2003). Taxonomic level for assessing oil spill effects on soft-bottom sublittoral benthic communities. *Marine Pollution Bulletin*, Vol. 46, pp. 562-572.

Glasson, J. Therivel, R. & Chadwick, A. (1999). *Introduction to environmental impact assessment*. UCL Press, London.

Gontier, M. Balfors, B. & Mortberg, U. (2006). Biodiversity in environmental assessment-current practice and tools for prediction. *Environmental Impact Assessment Review*, Vol. 26, pp. 268-286.

Grall, J. & Chauvaud, L. (2002). Marine eutrophication and benthos: the need for new approaches and concepts. *Global Change Biology*, Vol. 8, pp. 813-830.

Gray, J. McIntyre, A. & Stirn, J. (1992). *Manual of methods in aquatic environment research: Biological assessment of marine pollution with particular reference to benthos*. FAO fisheries technical paper 234, Rome.

Gray, S. (1989). Effects of environmental stress species rich assemblages. *Biological Journal of the Linnaean Society*, Vol. 37, pp. 19-32.

Gray, S. (1997). Marine biodiversity: patterns, threats and conservation needs. *Biodiversity and Conservations*, Vol. 6, pp. 153-175.

Halpern, B. Walbridge, S. Selkoe, K. Kappel, C. Micheli, F. D'Agrosa, C. Bruno, J. Casey, K. Ebert, C. Fox, H. Fujita, R. Heinemann, D. Lenihan, H. Madin, E. Perry, M. Selig, E. Spalding, M. Steneck, R. & Watson, R. (2008). A global map of human impact on marine ecosystems. *Science*, Vol. 319, pp. 948-952.

Hamza, W. & Munawar, M. (2009). Protecting and managing the Arabian Gulf: past, present and future. *Aquatic Ecosystems Health & Management*, Vol. 12, pp. 429-439.

Hutchings, P. (1998). Biodiversity and functioning of polychaetes in benthic sediments. *Biodiversity and Conservation*, Vol. 7, pp. 1133-1145.

IAIA, International Association for Impact Assessment (1999). *Principles of environmental impact assessment best practice*. IAIA & Institute for Environmental Assessment, UK.

Jones, D. Plaza, J. & Al-Sanei, M. (1998) Long-term (1991-1995) monitoring of the intertidal biota of Saudi Arabia after the 1991 Gulf War oil spill. *Marine Pollution Bulletin*, Vol. 6, pp. 472-489.

Kelleher, G. & Phillips, A. (1999). *Guidelines for marine protected areas*. World Commission on Protected Areas of IUCN. The World Conservation Union (IUCN).

Kendall. M. Paterson, G., & Aryuthaka, C. (2000). On-line exchange of polychaete taxonomic information. *Bulletin of Marine Science*, Vol. 67, pp. 411-420.

Khalaf, A. & Redha, M. (2001). Rehabilitation of water production facilities of the Ministry of Electricity and Water, State of Bahrain. *Desalination,* Vol. 138, pp. 319-328.

Khera, N. & Kumar, A. (2010). Inclusion of biodiversity in environmental impact assessment (EIA): a case study of selected EIA reports in India. *Impact Assessment and Project Appraisal*, Vol. 28, pp. 189-200.

Krupp, F. (2002). Marine protected areas. In: *The Gulf ecosystem health and sustainability*, N. Khan, M. Munawar & A. Price, (Eds.), pp. 447-473, Backhuys Publishers, Leiden.

Lee, N. & George, C. (2000). *Environmental assessment in developing and transitional countries: Principles, methods, and practice*, Wiley, Chichester.

Lee, S. (2008). Mangrove macrobenthos: assemblages, services, and linkages. *Journal of Sea Research*, Vol. 59, pp. 16-29.

Literathy, P. Khan, N. & Linden, O. (2002). Oil and petroleum industry. In: *The Gulf ecosystem: Health and Sustainability*, N. Khan, M. Munawar & A. Price (Eds.), pp. 127-156, Backhuys Publishers, Leiden.

Mace, G. (2004). The role of taxonomy in species conservation. *Philosophical Transactions of the Royal Society B*, Vol. 359, pp. 711-719.

Mackie, A., Oliver, P., Darbyshire, T. & Mortimer, K. (2005). Shallow marine benthic invertebrates of the Seychelles Plateau: high diversity in a tropical oligotrophic environment. *Philosophical Transactions of the Royal Society A*, Vol. 363, pp. 203-228.

MCain, J. (1984). Marine ecology of Saudi Arabia: the nearshore soft-bottom benthic communities of the northern area, Arabian Gulf, Saudi Arabia. Fauna of Saudi Arabia, Vol. 6, pp. 79-101.

Moberg, F. & Folke, C. (1999). Ecological goods and services of coral reef ecosystems. *Ecological Economics*, Vol. 29, pp. 215-233.

Moberg, F. & Ronnback, P. (2003). Ecosystem services of the tropical seascape: interactions, substitutions and restoration. *Ocean & Coastal Management*, Vol. 46, pp. 27-46.

Mohamed, S. (1993). *Birds of Bahrain and the Arabian Gulf*. BCSR, Bahrain.

Naser, H. Bythell, J. & Thomason, J. (2008). Ecological assessment: an initial evaluation of the ecological input in environmental impact assessment reports in Bahrain. *Impact Assessment and Project Appraisal*, Vol. 26, pp. 201-208.

Naser, H. (2010a). *Using macrobenthos as a tool in ecological impact assessment: applications in Environmental Impact Assessment (EIA)*. Lambert Academic Publishing, Saarbrucken.

Naser, H. (2010b). Testing taxonomic resolution levels for detecting environmental impacts using macrobenthic assemblages in tropical waters. *Environmental Monitoring and Assessment*, Vol. 170, pp. 435-444.

Naser, H. (2011). Effects of reclamation on macrobenthic assemblages in the coastline of the Arabian Gulf: A microcosm experimental approach. *Marine Pollution Bulletin*, Vol. 62, pp. 520-524.

Newell, R. Seiderer, L. & Hitchcock, D. (1998). The impact of dredging works in coastal waters: a review of the sensitivity to disturbance and subsequent recovery of biological resources on the sea bed. *Oceanography and Marine Biology: an Annual Review*, Vol. 36, pp. 127-178.

Niemela, J. (2000). Biodiversity monitoring for decision-making. *Annales Zoologici Fennici*, Vol. 37, pp. 307-317.

PCPMREW, Public Commission for the Protection of Marine Resources, Environment and Wildlife (2010). *Bahrain fourth national report to the Convention on Biological Diversity*. PCPMREW, Bahrain.

Phillips, R. (2003). The seagrasses of the Arabian Gulf and Arabian Region. In: *World Atlas of seagrasses*, E. Green & F. Short. (Eds.), pp. 74-81, UNEP-WCMC.

Preen, A. (2004). Distribution, abundance and conservation status of dugongs and dolphins in the southern and western Arabian Gulf. *Biological Conservation*, Vol. 118, pp. 205-218.

Price, A. (1998). Impact of the 1991 Gulf War on the coastal environment and ecosystems: current status and future prospects. *Environment International*, Vol. 24, pp. 91-96.

Price, A. (2002). Simultaneous hotspots and coldspots of marine biodiversity and implications for global conservation. *Marine Ecology Progress Series*, Vol. 241, pp. 23-27.

Price, A. Sheppard C. & Roberts, C. (1993). The Gulf: Its biological setting. *Marine Pollution Bulletin*, Vol. 27, pp. 9-15.

Price, A. Vousden, D. & Ormond, R. (1985). *An ecological study of sites on the coast of Bahrain*. UNEP Regional Seas Reports and studies No. 72, Geneva.

Saenger, P. (2002). *Mangrove ecology, silviculture and conservation*. Kluwer, Dordrecht.

Savage, C. Elmgren, R. Larsson, U. (2002). Effects of sewage-derived nutrients on an estuarine macrobenthic community. *Marine Ecology Progress Series*, Vol. 243, pp. 67-82.

Sheppard, C., Price, A. & Roberts, C. (1992). *Marine ecology of the Arabian region: Patterns and processes in extreme tropical environments*. Academic Press, London.

Sheppard, C. Al-Husiani, M. Al-Jamali, F. Al-Yamani, F. Baldwin, R. Bishop, J. Benzoni, F. Dutrieux, E. Dulvy, N. Durvasula, S. Jones, D. Loughland, R. Medio, D. Nithyanandan, M. Pilling, G. Polikarpov, I. Price, A. Purkis, S. Riegl, B. Saburova, M. Namin, K. Taylor, O. Wilson, S. & Zainal, K. (2010). The Gulf: A young sea in decline. *Marine Pollution Bulletin*, Vol. 60, pp. 3-38.

Slootweg, R. Kolhoff, A. Verheem, R. & Hoft, R. (2003). A generic approach to integrate biodiversity considerations in screening and scoping for EIA. *Environmental Impact Assessment Review*, Vol. 23, pp. 657-681.

Smith, S. & Rule, M. (2001). The effects of dredge-spoil dumping on a shallow water soft-sediment community in the Solitary Island Marine Park, NSW, Australia. *Marine Pollution Bulletin*, Vol. 42, pp. 1040-1048.

Snelgrove, P. (1997). The importance of marine sediment biodiversity in ecosystem processes. *Ambio*, Vol. 26, pp. 578-583.

Snelgrove, P. (1998). The biodiversity of macrofaunal organism in marine sediments. *Biodiversity and Conservation*, Vol. 7, pp. 1123-1132.

Snelgrove, P. (1999). Getting to the bottom of marine biodiversity: Sedimentary habitats. *BioScience*, Vol. 49, pp. 129-138.

Suchanek, T. (1993). Oil impacts on marine invertebrate populations and communities. *American Zoologist*, Vol. 33, pp. 510-523.

Thrush S. and Dayton, P. (2002). Disturbance to marine benthic habitats by trawling and dredging: Implications for marine biodiversity. *Annual Review of Ecological and Systematics*, Vol. 33, pp. 449-473.

Treweek, J. (1999). *Ecological impact assessment*. Blackwell, Oxford.

Ugolini, A. Borghini, F. Calosi, P. Bazzicalupo, M. Chelazzi, G. & Focardi, S. (2004). Mediterranean Talitrus saltator (Crustacea, Amphipoda) as a bioindicator of heavy metals contamination. *Marine Pollution Bulletin*, Vol. 48, pp. 526-532.

Winberg, P. Lynch, T. Murray, A. Jones, A. & Davis, A. (2007). The importance of spatial scale for conservation of tidal flats macrobenthos: An example from New South Wales, Australia. *Biological Conservation*, Vol. 134, pp. 310-320.

Zainal, K. Al-Sayed, H. Ghanem, E. Butti, E. & Nasser, H. (2007). Baseline ecological survey of Huwar Islands, The Kingdom of Bahrain. *Aquatic Ecosystem Health & Management*, Vol. 10, pp. 290-300.

Zainal, K. Al-Sayed, H. & Madany, I. (2008). Coastal pollution in Bahrain and its management. In: *Protecting the Gulf's marine ecosystems from pollution*, A. Abuzinada, H. Barth, F. Krupp, B. Boer, & T. Al-abdessalaam (Eds.), pp. 147-162, Birkhauser Verlag, Basel.

Zedler, J. & Kercher, S. (2005). Wetland resources: status, trends, ecosystem services and restorability. *Annual Review of Environment and Resources*, Vol. 30, pp. 39-74.

Ziegler, W. & Krupp, F. (1996). The role of taxonomic research in biodiversity conservation. In: *A marine wildlife sanctuary for the Arabian Gulf. Environmental research and conservation following the 1991 Gulf War oil spill*, F. Krupp, A. H. Abuzinada and I. Nader (Eds.), NCWCD, Riyadh and Senckenberg Research Institute, Frankfurt.

Zou, F. Zhang, H. Dahner, T. Yang, Q. Cai, J. Zhang, W. & Liang, C. (2008). The effects of benthos and wetland area on shorebird abundance and species richness in coastal mangrove wetlands of Leizhou Peninsula, China. *Forest Ecology and Management*, Vol. 255, pp. 3813-3818.

8

People, Plants, and Pollinators: The Conservation of Beargrass Ecosystem Diversity in the Western United States

Susan Charnley and Susan Hummel
USDA Forest Service, PNW Research Station
USA

1. Introduction

Biodiversity conservation often focuses on strategies that aim to protect a species from extinction and to preserve its functional role within an ecosystem. In this chapter we adopt a broader view of conserving biodiversity that calls for conservation of the ecological and social roles of a species within an ecosystem, which we understand to include humans. Viewed as such, biodiversity conservation entails sustaining ecosystem diversity to support both a species and the web of interdependent social and ecological relations in which it is embedded. Hence, if one component of the ecosystem diversity associated with a species is threatened, conservation interventions may be warranted, even if the species itself is not (yet) threatened or endangered. Thus, biodiversity conservation is not only about preventing the extinction of a species, but also about preserving the diversity of its functional roles - both ecological and social - to sustain biocultural diversity.

Conservation strategies based on knowledge about how people affect and interact with the natural disturbance processes that influence ecosystem diversity are more likely to be successful than strategies that focus on only one or the other (e.g., anthropogenic or natural disturbance). Because the niche (both social and ecological) of a species may vary across its range depending on local disturbance regimes and local sociocultural practices, conservation needs and strategies are also likely to vary across its range. For this reason, traditional and local ecological knowledge can make an important contribution to the conservation of ecosystem diversity.

We selected beargrass (*Xerophyllum tenax* (Pursh) Nutt) to illustrate these points. Beargrass is a perennial monocot with distinctive slim, evergreen leaves and a tall flower spike (Fig. 1). Its range lies in the western United States and southwestern Canada, with two disjunct distributions: (1) from the Coast and Sierra Nevada mountain ranges in California north through Oregon's Coast and Cascade mountain range, to the Olympic Peninsula and Cascade Mountains in northwestern Washington; and (2) from the Rocky Mountains in Idaho, Montana, and northwestern Wyoming north to southeastern British Columbia and southwestern Alberta Provinces in Canada (Crane, 1990; Maule, 1959; Vance et al., 2001) (Fig. 2). The coastal portion of this range is influenced by a maritime or mediterranean climate, while the interior portion is continental. Throughout its entire range, beargrass provides food and habitat for several animals and pollinating insects. Beargrass also has cultural and economic value. For centuries the plant has been used by Native Americans, who harvest the leaves for basketry and other uses. In recent decades, beargrass has become an important part of the international floral industry because its leaves are useful for different decorative products.

Fig. 1. Beargrass (*Xerophyllum tenax*). Courtesy Nevada Native Plant Society

Across the range of beargrass many changes are occurring in the frequency, intensity, and severity of natural and anthropogenic disturbances, including fire regimes, timber management, and leaf harvest practices. These changes impact the plant, its pollinators, and the people who use it in an interconnected web of ecosystem relations that are incompletely understood. We suspect that these impacts are - or could be - negative for multiple species and some of the people that are part of the ecosystems where beargrass grows; indeed, they have already been detrimental for some cultural uses of beargrass. Thus, beargrass offers one example of how terrestrial ecosystem conservation strategies might be developed and implemented, taking into consideration both the ecological and social importance of a species within an ecosystem.

Fig. 2. Distribution of beargrass in western North America. Source: Flora of North America

In the next section we introduce beargrass and review some basics of its natural history. We then discuss the ecological and sociocultural roles of beargrass within the ecosystem, with an emphasis on pollination ecology and the traditional cultural and current commercial importance of the plant. We go on to address how natural and anthropogenic disturbances affect beargrass and its ecosystem functions. These sections provide a foundation for the final two sections of the chapter in which we synthesize the information presented to identify and evaluate potential management strategies that could advance the knowledge and practice of ecosystem diversity conservation within the range of beargrass. Our synthesis and analysis are based on a review of the published literature about beargrass.

2. Natural history of beargrass

Beargrass is adapted to a wide range of ecological conditions and can reproduce either sexually or asexually. Habitat conditions affect plant characteristics, which in turn influence its social and ecological roles.

2.1 Botanical overview

Beargrass is not a true grass, but is instead classified as a lily (Liliaceae) (USDA, 2011). The *Xerophyllum* genus has just two members and beargrass is the only one found in the western United States. The congeneric *X. asphodeloides* (L) Nutt, or eastern turkeybeard, is similar to *X. tenax* in form, but is smaller in stature. Eastern turkeybeard grows only in the southeastern United States, where in some places it is classified as threatened or rare. In contrast, beargrass is not listed (USDA, 2011).

The scientific name of beargrass is derived from Greek: *xeros* (dry), and *phyllon* (leaf); and Latin: *tenax* (clinging or tenacious). The prevalent common name is derived from its anecdotal use as a food plant for bears (Crane, 1990). Several alternate common names reflect other roles of beagrass, namely elk grass, soap grass, Indian basket grass, Quip-Quip, and fire lily. As its scientific name suggests, beargrass has morphological adaptations that are typical of drought-tolerant plants, namely a thickened cuticle and stomatal restrictions to the leaf surface that minimize water loss (Rentz, 2003).

Beargrass is an herbaceous, rhizomatous plant with a perennial mass of narrow, long, basally clustered leaves (Hitchcock & Cronquist, 1973). The rhizome is a woody rootstock (Hitchcock & Cronquist, 1973; Maule, 1959). The basal leaves are fibrous and numerous, and grow in clumps (Hitchcock & Cronquist, 1973; Pojar & MacKinnon, 1994). The leaves are 15-100 cm long and 2-10 mm wide at the base, decreasing in width to a thin, stiff, wiry tip (Maule, 1959; Rentz, 2003; Vance et al., 2001).

The plant can reproduce both vegetatively (by sprouting of offshoots from the rhizome) and sexually (by flowering) (Vance et al., 2001). Plants may live for several years, producing vegetative growth and offshoots (Rentz, 2003; Vance et al., 2004). Each shoot arises from a meristematic region located on the upper surface of the rhizome at the leaf base (Crane, 1990). Vegetative reproduction may occur throughout the life span of an individual. An individual plant often dies after flowering, but since offshoots sprout before flowering occurs, a plant is persistant and long-lived (Crane, 1990; Hitchcock & Cronquist, 1973; Laursen, 1984).

In the event of flowering, beargrass produces a single, tall (~1.5 m), and unbranched stalk that bears a terminal inflorescence with 50 to 400 flowers (Munger, 2003; Vance et al., 2001). The onset and length of flowering appear to vary with differences in soil temperatures, aspect, canopy cover, and elevation (Maule, 1959; Rentz, 2003; Vance et al., 2004). The lily-like flowers are small (ca. 1.3 cm) and are whitish (Munger, 2003). Pollen morphology of beargrass and eastern turkeybeard is similar (Takahashi & Kawano, 1989). Floral nectar is not present and the floral scent of beargrass varies. These flowering properties are important, because different pollinating insects are attracted to different flower qualities.

The relative importance of vegetative vs. sexual reproduction in beargrass appears to be associated with environmental factors, but is not well understood. In general, flowering in this species is most often observed in plants growing in open conditions; it becomes less frequent or disappears entirely as forest canopies close and light to the understory is reduced (Crane, 1990; Maule, 1959; Vance et al., 2004).

2.2 Environment

Beargrass grows in a variety of environmental conditions: in open areas such as clearings, meadows, and bogs; on slopes and ridges; and in coniferous forests (Vance et al., 2001). It is a significant component of subalpine meadows, and also frequently occurs as a dominant understory plant in dry, mixed-coniferous forests (Higgins et al., 2004). Beargrass is adapted to harsh environmental conditions; it grows in a variety of soils and forest types. It is associated with soils of low fertility and productivity (Peter & Shebitz, 2006; Vance et al., 2001). The plant is often found on steep sites where the soils are saturated in the spring and well-drained later in the season (Crane, 1990; Higgins et al., 2004; Maule, 1959). The annual precipitation within the range of beargrass is from 48 to 175 cm (19 to 69 inches) (USDA, 2011). The direction of slope (its aspect) may be important for beargrass distribution, due to the influence of aspect on soil temperature and length of the growing season. Topography, hummus content of the soil, and ground cover characteristics may influence soil temperature and thus impact flowering patterns of beargrass.

Beargrass is moderately shade tolerant and can grow in forests with little or no direct sunlight, and on open slopes (Maule, 1959). The amount of overstory shading appears to impact the reproductive strategy of this plant. Flowering is associated with open or filtered light, whereas the plant reproduces by rhizomes in closed overstory conditions (Higgins et al., 2004; Schlosser & Blatner 1997). Overall, beargrass is reported to achieve the highest densities and reproductive success under canopy openings where it grows vigorously and blooms profusely (Crane, 1990).

3. Beargrass ecosystem dynamics

The ecological and social roles of beargrass range from being a food resource for insects and mammals, to being important in the culture of Native Americans, to being at the heart of a multimillion dollar floral greens industry. We describe some of these diverse roles below, recognizing that many of them are not fully understood.

3.1 Ecological roles

3.1.1 Food resource

Beargrass provides food for animals large and small. Bees consume its pollen and, in spring, bears eat the fleshy part of the leaf base (Pojar & MacKinnon, 1994). Likewise, mice and pocket gophers feed on the fleshy leaf bases and rhizomes (Vance et al., 2001). The flowering stalks of beargrass are eaten by elk and deer in summer, and the more tender leaves are eaten by these animals year-round (Crane, 1990; Vance et al., 2001). Since the leaves remain over the winter, they also provide food for mountain goats in cold conditions, when limited resources are available (Vance et al., 2001). Beargrass pollen provides food for a diversity of insects, including at least 29 species of flies, beetles, and bees from at least 14 different families (Vance et al., 2004). Beargrass leaves and flowers also provide habitat, nesting material, and foraging territory for animals, from mice to grizzly bears.

In this chapter our main interest is in the functional role of beargrass as a source of food for the insects that visit its flowers. The process of pollination is essential for the survival of many plants, and therefore, pollen is part of an extensive food web. Pollinating insects include flies (Diptera), beetles (Coleoptera), bees and wasps (Hymenoptera), and butterflies and moths (Lepidoptera). Of these, we focus on the first three because they have been associated with beargrass in published studies. Butterflies are neither known nor expected to pollinate beargrass due to a lack of nectar. Some moths are attracted to nectarless flowers; moth pollination of beargrass is possible, but unknown.

The appearance and presentation of beargrass flowers suggest a generalist rather than a specialist mode of pollination, which means that several insects probably pollinate the plant. Flies, beetles, and bees may not contribute equally to the cross-pollination of beargrass, however (Vance et al., 2004). In a singular study in the Oregon Cascades, flies were the most efficient and effective vectors of beargrass pollen, rapidly moving between flowers and carrying pollen mostly from beargrass (Vance et al., 2004). In contrast, beetles were less efficient, but still effective, cross-pollinators. They spent more time on an inflorescence than did flies, but did move from plant to plant (Vance et al., 2004). Bees were infrequently observed, but carried the most beargrass pollen grains of all insect visitors over two seasons of observation (Vance et al., 2004).

Cheilosia hoodiana, a hoverfly, is one beargrass pollinator. While many fly pollinators consume nectar as their primary reward, flies such as this one feed chiefly on pollen which they require for reproductive success. Many such flies gather pollen and nectar from a wide variety of flower types and shapes (reviewed in Campbell et al., 2007). This implies that beargrass is a non-exclusive host for this hoverfly, which would also pollinate other plants in the beargrass community.

Cosmosalia chrysocoma is a longhorn beetle that pollinates beargrass. The adults fly from June to at least August in the Rocky Mountains and Pacific Coast Range (Craighead, 1923; MacRae & Rice, 2007), actively seeking out flowers and feeding primarily on pollen (reviewed in O'Neill et al., 2008). This species is found on a great diversity of flowers. The life span of most species in this family ranges from one to three years, most of which are spent in the larval stage, feeding on decaying wood. Adult beetles usually emerge, disperse, reproduce, and die within a few days to a few months. The distinctive antennae of this beetle help it locate host trees needed for reproduction.

All bee species are dependent on flowers for meeting larval and adult nutritional needs. Foraging bees may be categorized as specialists (e.g., forage on a single plant species or species in one or several closely related genera), or generalists (foraging across multiple genera and families) (Roulston et al., 2000). Four different families of bees (Andrenidae, Apidae, Halictidae, and Megachilidae) have been reported foraging on beargrass (Vance et al., 2004).

Increased consumption of pollen, protein, and other nutrients may increase the size, survival, longevity, and fecundity of bees, flies, and other pollinating insects (reviewed in Roulston et al., 2000; Vance et al., 2004).

While findings presented in Vance et al. (2004) clearly demonstrate that beargrass is an important food resource to a diverse assemblage of insects, the relative importance of beargrass compared with other plants is difficult to assess. There is currently insufficient evidence to determine whether some insects require beargrass as part of their life history, or whether some pollinators are essential to beargrass reproductive success.

3.1.2 Ecosystem structure

The structural role of beargrass in forested ecosystems is not well documented. However, given its perennial status and growth form, it is probable that beargrass contributes both above- and below-ground structure during different seasons. For example, roots may provide erosion control, particularly in steep habitats (Vance et al., 2001), and also aid in soil stability. The structure around the basal leaves provides habitat for birds, small mammals, and insects. Because the plant is adapted to survive in snow, the spaces among leaves and at the leaf base also likely provide sub-niveal spaces. It may be that some of the pollinators of beargrass - such as bees - use stucture provided by it for nesting and reproducing.

3.1.3 Ecosystem processes

In addition to pollination, an important ecosystem process to which beargrass likely, but indirectly, contributes is decomposition. Some pollinators of beargrass, such as the flower longhorn beetle *(Cosmosalia chrysocoma),* are part of this fundamental process, which breaks down organic matter and makes nutrients available for new life. Although it is not threatened, beetles like this one may be an indicator of forest conditions and processes, since the species assemblages in primary forest habitats differ from those found in second growth.

3.2 Sociocultural roles

Beargrass is a plant that has been valued by people since prehistoric times. It was used by Native Americans in the Northwest primarily for basketry, decorations on regalia, and jewelry, and it continues to be used today for these purposes. More recently, it has gained commercial importance in the floral industry. It is also valued for its aesthetic properties in the wild and in yards and gardens. Finally, beargrass may be useful in ecological restoration, a use that is only now being investigated. Most of the literature documenting its uses references the northwestern rather than the Rocky Mountain portion of its range.

3.2.1 Native American uses

Since prehistoric times, Native American tribes from northern California, Oregon, Washington, Idaho, and British Columbia have harvested beargrass for making baskets of different types (Anderson, 2005; Lobb, 1990; Shebitz, 2005). Plain baskets were used in gathering wild foods such as berries and clams, for storing food and other goods, for carrying water, and for cooking (Lobb, 1990; O'Neale, 1928; Shebitz, 2005). More decorative baskets were used in ceremonies and dances, given as gifts or traded, and used to hold objects of cultural importance. Beargrass was also used to decorate other baskets. Today, Native American basketweavers in the Northwest weave beargrass baskets for sale, gifts, artistic purposes, as a means of carrying on cultural traditions, and as a medium for recording tribal history and expressing tribal identity.

Specific qualities of beargrass are desirable for these traditional uses. Basket weavers prefer long, straight, pliable leaves that are flexible enough to work with, but fibrous enough to withstand the rigor of weaving (Rentz, 2003; Shebitz et al., 2009). The leaves must also be able to lie flat when woven into a design. Leaves having less pigment are favored. Some California basketweavers prefer leaves that have turned white at the tip (Anderson 2005).

The properties of beargrass vary across its range. Certain environmental conditions promote the plant qualities sought for traditional Indian basketry. Plants occurring at high elevations have longer, stronger, more flexible leaves (Hunter, 1988; Rentz, 2003). Plants that grow in partial shade have leaves that remain pliable for longer periods of time, perhaps because they are protected from the sun (Hunter, 1988; Shebitz, 2005). Leaves harvested from plants in areas that recently burned contain less pigment, are easier to pick, and are also relatively strong, thin, and pliable (Anderson, 2005; Hunter, 1988; Rentz, 2003). Because fire improves leaf quality for basketry, harvest often occurs in areas that have recently burned, although the best harvesting is believed to occur three to seven years after a burn (Rentz, 2003; Shebitz et al., 2009). Traditional harvesters seek sites that have environmental conditions conducive to producing high quality beargrass leaves, but accessibility is also important because many harvesters are elderly. Sites near maintained roads are preferred (Hunter, 1988). Although harvesting occurs on public, private, and tribal lands, habitat fragmentation and land conversion have made public lands the most common gathering place today (Lynch & McLain, 2003).

Apart from basketry, beargrass leaves were also traditionally woven into garments and decorations, and were used to make a variety of everyday items (Anderson, 2005; Lobb, 1990; Rentz, 2003). Today they are also used for jewelry, such as necklaces and earrings (Anderson, 2005). The roots were used for medicinal purposes (Vance et al., 2001). Beargrass also played a ceremonial role (for example, at burials) (Peter & Shebitz, 2006). Beargrass often grew as a dominant plant in places where Native Americans went to seek spiritual refuge. In northern California, tribes traditionally ate the tuberous rhizomes of the plant (Anderson, 2005). In addition, it provided forage for game species, such as elk and deer, that were hunted by Native Americans (Shebitz, 2005).

Several researchers have pointed out the links between biodiversity and cultural diversity (see Maffi, 2005 for a review). Biodiversity supports a broad range of cultural practices and adaptations that in turn create demand for, and forest management to support, a broad range of species. In the case of beargrass, forest management to maintain the cultural uses of the plant not only helps preserve associated cultural traditions among Native Americans today; it calls for management to restore beargrass habitat to protect it and associated species assemblages from decline.

3.2.2 Commercial uses

In the 1980s, commercial harvest of beargrass became prominent in the Pacific Northwest and British Columbia (Higgins et al., 2004; Lynch & McLain, 2003). It is used in three forms: fresh, preserved, and dried (Schlosser & Blatner, 1997). Beargrass flowers are used in fresh flower arrangements (Vance et al., 2001). Leaves are used, fresh or dried in floral arrangements, in which they serve as filler (Hansis, 1998). The rich green leaves of beargrass are attractive in floral arrangements but can also be easily dyed, and are long-lasting (Thomas & Schumann, 1993). Although there is a domestic market for beargrass leaves in the United States (especially in the east), most of them are exported for sale on European and Asian markets (Schlosser & Blatner, 1997; Thomas & Schumann, 1993). Since the 1980s, beargrass has become one of the leading nontimber forest product species harvested commercially in the Pacific Northwest. It was believed to be the most widely-harvested floral green species in the Pacific Northwest in the 1990s (Schlosser & Blatner, 1997), and may still be today. Because harvest and trade of nontimber forest products is not well documented or monitored, harvest volumes can only be estimated. Researchers estimated that 200,909 kilograms were harvested from Pacific Northwest forests between 1999 and 2001 (Kramer, 2001), though other estimates are much higher. According to one study, buyers were purchasing and shipping 68,182 to 90,909 kilograms of beargrass leaves per week during the harvest season (Thomas & Schumann, 1993).

Today, commercial beargrass production has become a multi-million dollar industry in the northwestern part of its range. Who participates? Most harvesters are Southeast Asian and Latino immigrants to the Northwest (Hansis, 1998). Despite physical hardships, there are many advantages to working in the floral greens industry. English language skills are not required, advanced skills are not needed, it is unnecessary to provide legal documentation in order to work, and payment is in cash. Many harvesters prefer working in the forest to low-paying jobs in cities, which may be the only alternative, and for some, harvesting may be the only job they can find. Harvesting can also be a way of maintaining family bonds, as it often takes place as a family activity that includes children and elders. The latter might require expensive care if left at home while adults work. Moreover, beargrass harvesting can be done during lapses in employment, serving as a bridge to fill employment gaps until other jobs, such as agricultural and forestry services work, become available. Thus, it provides an important source of supplemental income for many. Beargrass harvesting may also be a primary source of employment; with the decline in jobs in the agriculture and timber sectors in the Northwest since the late 1980s, harvesting beargrass and other nontimber forest product species has provided an alternative employment opportunity in the natural resources sector for workers. Finally, because many immigrants harvest nontimber forest products in their home countries, doing so in the United States provides cultural continuity with the past, and an opportunity to pass on the tradition of gathering to children (Hansis, 1998). Other participants in the market chain include contractors or buyers who purchase beargrass directly from harvesters at buying stations and transport it to processing sheds, wholesalers, and retailers who operate within regional and global market systems (Hansis, 1998; Higgins et al., 2004).

As with traditional cultural uses, commercial-grade beargrass has specific properties that make it valuable. Commercial harvesters seek leaves that are deep green with no yellowing, that retain their color following harvest, that have wide, firm blades, and that are at least 71 centimeters long (Schlosser & Blatner, 1997). Leaf length is the main factor limiting its commercial value. Older, larger beargrass plants are typically those having leaves long enough to harvest, which are generally located in the center of a tussock. Most commercial harvest occurs in spring, summer, and fall, though it is possible to harvest year-round if snow levels permit (Hansis, 1998).

Beargrass leaves that meet commercial standards are usually found in places having partial shade cover from an elevated forest canopy that provides diffused sunlight, that allow for snowfall to the ground, that have low soil compaction, and where the moisture regime is favorable, especially north and east-facing slopes (Schlosser & Blatner, 1997). On the basis of experiments conducted on the west side of the Cascade Range in southern Washington, Higgins et al. (2004) found that as a general guideline, 60 percent forest overstory cover is necessary to produce leaves having the desired color, regardless of forest type. The highest commercial harvest yields are likely to occur from forests having 60 to 90 percent canopy density. Leaf quality is optimal in climax plant communities and during the later stages of forest succession. Beargrass of commercial quality is not present in open canopy areas where recent burns or clearcuts have taken place (Higgins et al., 2004). Therefore, commercial harvest occurs primarily in cool, moist forests at higher elevations (762 to 1,524 meters), where beargrass is the dominant understory species, is abundant enough to make harvesting profitable, and where leaf quality is optimal (Hansis, 1998; Schlosser & Blatner, 1997). On the eastern slope of the Cascade Range and in eastern Oregon and Washington, northern Idaho, and western Montana, these are typically grand fir (*Abies grandis*), subalpine fir (*Abies lasiocarpa*), and mountain hemlock (*Tsuga mertensiana*) forests. Douglas-fir (*Pseudotsuga menziesii*) forests tend not to have commercial grade plants; and beargrass does not usually occur in high enough quantities in western redcedar (*Thuja plicata*) and western hemlock (*Tsuga heterophylla*) forests to be commercially profitable (Schlosser & Blatner, 1997). On the western slope of the Cascade Range and to the west, the forest zones having the highest potential for commercial beargrass production are the Pacific silver fir (*Abies amabilis*) and mountain hemlock forest types; again, Douglas-fir and western hemlock forests are much less favorable (Schlosser et al., 1991). In contrast to traditional uses, we are unaware of specific habitat management interventions undertaken to promote beargrass populations and properties desirable for commercial harvest.

3.2.3 Spiritual/aesthetic values

Beargrass is an aesthetically-pleasing plant - its colorful leaves and prominent, flowering stalk stand out and delight recreationists and nature lovers who visit the natural areas where it occurs. Its appearance also makes it desirable as a cultivated plant in gardens, though it is difficult to propagate beargrass successfully. Seeds gathered in the wild can be cultivated in greenhouses, and seedlings transplanted; but mature plants do not transplant well, and cuttings taken in the wild tend not to establish themselves (Vance et al., 2001; Wick et al., 2008). Some Native American tribal members may also appreciate beargrass for its aesthetic value, and as one component of the larger ecosystem to which they feel a spiritual attachment (Peter & Shebitz, 2006). As noted earlier, open habitat conditions promote flowering. Crane (1990) finds that beargrass achieves its highest densities and blooms profusely in forest openings having a filtered light environment. Flowering diminishes as the forest canopy closes.

3.3 Summary

Our purpose in this section is to demonstrate that beargrass lies at the nexus of a web of ecological and sociocultural relations (Fig. 3). On the one hand, beargrass serves as an important food source for many animals, pollinators, and decomposers, and plays a role in ecosystem structure and processes. On the other hand, it supports a number of cultural uses among tribes throughout its range, and is one of the most economically-valuable plants in the commercial floral industry in the western United States. Its aesthetic and spiritual values are also important. The goods, values, and ecosystem services provided by beargrass all contribute to ecosystem diversity - a diversity that is influenced by both natural and anthropogenic disturbances throughout its range. A description of the key disturbance agents that influence beargrass follows.

4. Disturbance ecology

Disturbances affect beargrass habitat, and in turn, plant structure and ecosystem function. Some dominant disturbance factors that influence beargrass are fire (both naturally occurring and anthropogenic), silvicultural practices, and leaf harvest. Due to limitations of space, we also consider biotic disturbances (insects, pathogens) here, but not abiotic ones (landslides, wind) because we think the former affect more extensive areas, and cause greater alterations across the range of the plant.

Disturbance regimes in the forested plant communities where beargrass grows are changing throughout its range, influencing the ecosystem services the plant provides. Conservation of the ecosystem diversity that beargrass supports will largely depend on managing the natural and anthropogenic disturbances that impact it.

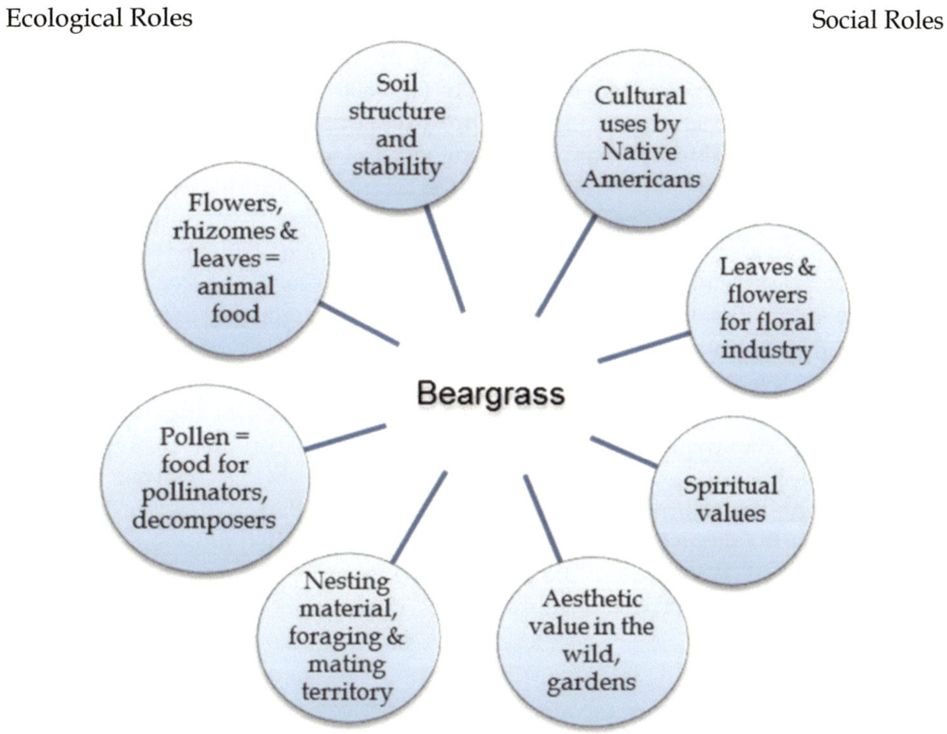

Fig. 3. Ecological and social roles of beargrass

4.1 Fire

4.1.1 Natural ignitions

Wildfire is an important ecological process in shaping plant community structure and function and has been a major natural disturbance in western ecosystems where beargrass grows (Agee, 1993). In forested landscapes of the Pacific Northwest, Rocky Mountains, and Sierra Nevada Range, lightning is the primary natural source of fire ignition (Agee, 1993; Barrett & Arno, 1999; Keeley & Stephenson, 2000). Wildfires generally occur in summer, due to an association with summer lightning storms and the build-up of dry fuel (Bartuszvige & Kennedyre, 2009; Boyd, 1999). Late-season fires (summer to fall) also occur in some areas, such as the eastern Cascades (Wright & Agee, 2004). In northern California, the potential for lightning fires is highest during the period of hot, dry conditions between August and October, especially in inland parts of the region (Agee, 1993).

Within the range of beargrass the extent of wildfire varies, with less than 10 ha (Morrison & Swanson, 1990) to greater than 4000 ha reported in the Oregon Cascades, and from 1 to 800 ha in the Sierra Nevadas. Fire extent in high elevation, subalpine habitat depends largely on the distribution and abundance of forest vegetation, since rock or snow fields may prevent the spread of fires across patchy forests. At lower elevations, wet forests composed of less-flammable tree species can limit the spread of forest fires. Fire extent is also strongly linked to weather patterns, with the tendency for large fires to coincide with periods of annual and seasonal drought.

Natural fire regimes are tied to climate, elevation, forest community structure, and other environmental factors. Thus, the frequency and severity of wildfires varies across the range of beargrass habitat. In general, naturally occuring fires west of the Cascade Range in the maritime, wetter forest types occur infrequently (100-800 years), but burn extensive areas with high severity fire when they do ignite. Forest types at the upper end of the fire return interval include Pacific silver fir, mountain hemlock, and subalpine fir. Forests of Douglas-fir and western hemlock are at the lower end of the range. In contrast, natural fire regimes in the interior, drier forest types east of the Cascade Range occur frequently (1-50 years) and with low severity. These are forests in which ponderosa pine (*Pinus ponderosa*), grand fir, white fir (*Abies concolor*), and Douglas-fir grow. Interior forests of the Rocky Mountains are influenced by a continental climate. Ponderosa pine, Douglas-fir, western larch (*Larix occidentalis*), and lodgepole pine are common forest species in the Rocky Mountains. In dry, lower elevation forests dominated by ponderosa pine, the natural fire regime is characterized by high-frequency, low intensity fires with fire-return intervals from 5 to 40 years. In contrast, low-frequency, high-intensity (stand-replacing) fires are typical at mesic, higher elevation sites; for example, moist western larch forests are reported to experience comparatively long fire return intervals of ~35 to 200 years (Barrett & Arno, 1999). Natural fire regimes across the range of beargrass thus historically included infrequent, high severity fires in moist and cold regions and more frequent, low severity fires in dry regions. Ecotones between the regions were, perhaps not surprisingly, characterized by mixed-severity fires.

4.1.2 Anthropogenic fire

In historic and prehistoric times Native American tribes in the western United States commonly used fire as an environmental management tool. Evidence for anthropogenic fire in the Pacific Northwest dates to about 3500 years ago (Peter & Shebitz, 2006; Wray & Anderson, 2003); in northern California, the evidence goes back 8,000 years (Rentz, 2003); and in the Sierra Nevada of California, Native American burning is believed to have begun at least 9,000 years ago (Klinger et al., 2008). Burning took place to create habitat conditions favored by desired game species such as elk and deer, enhancing hunting opportunities (Boyd, 1999). It was also used to promote the growth of plants favored as food, medicine, for clothing, and for basketry. In places having high precipitation and long natural fire return intervals, anthropogenic fire made it possible to maintain culturally-important plants in the quantity, and of a quality, needed (Shebitz et al., 2009). Tribes from northern California, Washington's Olympic Peninsula, and others are known to have used fire to maintain beargrass populations for use in basketry (Rentz, 2003; Shebitz et al., 2008, 2009). Following a burn, beargrass re-establishes itself, and new leaves having the leaf qualities desirable for basketry are harvested from the burned clumps. In many parts of its range, beargrass occurs with food plants valued by Native Americans – such as camas (*Camassia quamash*), (a bulb), and berries (*Vaccinium spp.*). Thus, burning promoted an assemblage of desired species (Peter & Shebitz, 2006).

The frequency and severity of burns varied, depending on location and purpose. Fires in beargrass habitat in the Pacific Northwest and California were typically slow moving surface fires that burned with low severity (Rentz, 2003; Shebitz et al., 2009). On the Olympic Peninsula, the Skokomish tribe burned beargrass praries at two to three-year intervals (Shebitz et al., 2009). These fires burned most of the old leaves off of the beargrass plants, and up to 95 percent of their live foliage. New leaves from burned tussocks were harvested one to three years later. In northern California, fire was also used frequently (Rentz, 2003). Elsewhere, anthropogenic burning was less frequent (averaging once every nine years in the northern Rocky Mountains, and every 5 to 15 or 20 years in the Sierra Nevada). In most places, burns were conducted in late summer and fall, though early spring burns took place in the Rockies (Barrett & Arno, 1998).

Whereas Native Americans commonly used fire as a forest management tool, nonnative settlers took the opposite approach: in the mid to late 1800s, they began suppressing both natural and anthropogenic fires (Shebitz et al., 2009). Since the early 1900s, Indian burning in the American West has been severely curtailed and natural fires have been actively suppressed in forest and rangeland ecosystems. Consequently, many praries and savannas in the Pacific Northwest and Rocky Mountains that were not settled by people and instead remained undeveloped have undergone succession to woodlands and forest. One result has been a scarcity of beargrass leaves at traditional gathering sites suitable for use in Native American basketry, and a decline in beargrass in some parts of its range resulting from increases in canopy cover (Peter & Shebitz, 2006; Shebitz, 2005; Shebitz et al., 2008). A century of fire suppression has also meant that today, when natural ignitions do occur, they can cause uncharacteristically severe wildland fires that are difficult to suppress. High intensity fire can kill beargrass by burning the meristematic portion of the rhizome, located near the soil surface, from which growth takes place (Vance et al., 2001).

4.1.3 Effects of fire on beargrass and pollinators

The primary adaptation of beargrass to fire is its ability to sprout from underground rhizomes (Crane, 1990). Shebitz et al. (2009) found that beargrass leaves began resprouting from rhizomes within five months of high-severity fire, and Rentz (2003) reported that plants burned to the ground were once again covered with leaves by the following summer. Fire reduces competition for growing space by shrubs and trees, reduces forest canopy to increase light and soil temperatures, and releases nutrients into the soil, all of which favor beargrass. Increases in beargrass flowering have been observed within one or two years following fire (Maule, 1959; Rentz, 2003). Depending on the season, weather, fuels conditions, forest type, and location of a fire, the severity and extent of a burned area could range from small (a few square meters to a few square hectares) to large (several hundred or thousand hectares).

Anthropogenic fire contrasted to naturally-occurring fire in a number of ways. The effects of interactions between natural fire regimes and the intervals created by human ignitions on ecosystem diversity are unknown, but could be expected to vary directly with the magnitude of alteration from historic frequency and severity. For example, in maritime, or coastal-influenced moist forest ecosystems, frequent anthropogenic burning to favor beargrass and associated species like camas shortened the natural fire return interval and probably reduced the area affected by any one fire. Thus instead of hundreds of hectares burning once every several hundred years, there would have been fewer hectares burned every decade. This would have created and maintained a mosaic of early seral conditions in otherwise mid- to-late-seral forests, which would have increased species diversity and altered fuelbed structure. In contrast, in interior, dry forest types frequent anthropogenic burns would have been more consistent with the natural fire regime. The gradient of potential interaction between natural and anthropogenic fire throughout the range of beargrass implies a variety of effects on genetic, species, and ecosystem diversity, rather than a uniform response to fire.

Fire also affects pollinators. The potential impacts of fire on arthropod communities are variable and can be beneficial or detrimental (reviewed in Swengel, 2001). Beneficial effects include more food for survivors and migrants. Pollinating insects generally rely on sunlight to raise their body temperature enough to fly; thus, the reduced live tree density and canopy cover in burned forests may attract these species (Campbell et al., 2007). Moreover, canopy openings provide an environment in which flowering plants can flourish. These changes improve nectar and pollen rewards associated with a shift from annuals to perennials (Potts et al., 2003). For example, pollen production has been found to be highest in freshly burned sites and to decrease with time (Potts et al., 2003).

The degree of fire-related impact and the potential for insects to rebound are related to a number of characteristics, especially exposure to lethal temperature, the stress experienced in the post-fire environment, the suitability of post-fire vegetation as habitat, and the ability of survivors or colonizers to rebuild their numbers at the site (Swengel 2001). Mobility is important in both fire avoidance (e.g., the ability to reach suitable unburned habitat) and in post-fire recolonization (e.g., the ability to reach burned sites from unburned sites) (Hartely et al., 2007). Thus, ground-dwelling arthropods are slower to return after wildfire than groups adept at flight (reviewed in Swengal, 2001). This suggests that any negative effect of fire on the insect pollinators of beargrass would be lower for mobile species like flies and bees and higher for the larval and adult stages of beetles.

4.2 Timber harvest

Timber harvest is an anthropogenic disturbance across the range of beargrass in the Pacific Northwest and Rocky Mountains (Halpern & Spies, 1995). Specific harvest practices (e.g., harvest method and extent, degree of soil disturbance, management of slash, reforestation efforts, rotation length) have varied over time, depending on land ownership, site conditions, and forest type (Halpern & Spies, 1995), and thus affect forest structure and ecosystem processes differently. In general, harvest activities alter natural successional processes and influence the diversity, abundance, and composition of understory vegetation over both the short and long term (Battles et al., 2001; Halpern & Spies, 1995). The canopy openings, increased light levels, and elevated soil temperatures created through logging could potentially benefit beargrass growth and reproduction (Maule, 1959; Vance et al., 2001). Mechanized harvest could, however, result in compacted or poorly-drained soils that would have a negative impact on beargrass.

The available literature focuses on clearcut harvests, which may adversely affect beargrass. One short-term study of plant cover and composition was conducted in the Oregon Cascades prior to clear-cut logging, after logging but before broadcast burning the slash, and during each of five growing seasons following burns. The study found that beargrass disappeared immediately after clearcut logging, remained absent after burning, and reappeared in trace amounts four years later (Dyrness, 1965). Other studies from Oregon (Halpern & Spies, 1995) and Idaho (Crane, 1990) report an absence of beargrass for decades after clearcutting, though it is not clear what effect post-harvest broadcast burning might have had on the plant. The likely reason for this absence is competition from understory shrub and forb species, which proliferate following a clearcut, creating conditions that make it difficult for beargrass to recover for at least 20 years (Shebitz et al., 2009). Commercial harvest of beargrass from stands that have been clearcut does not usually resume until a closed forest canopy has redeveloped and been in place for a long time (Schlosser & Blatner, 1997).

Studies are scarce on the effects of less-intensive silvicultural treatments, such as shelterwood and group selection, on beargrass. Schlosser & Blatner (1997) report that floral greens harvest, including commercial beargrass harvest, can begin about three to seven years after commercial thinning and other intermediate-level stand treatments, though this may vary some, according to local circumstances. One study has implications for the beetle pollinators of beargrass. O'Neill et al. (2008) investigated the effects of shelterwood logging on wood-boring beetle pollinators living in lodgepole pine forests in Montana and found that logged sites had more adult cerambycids (including *Cosmosalia chrysocoma*, the most abundant beetle pollinator of beargrass [Vance et al., 2004]) than unlogged plots and meadows. Logging-related increases in cerambycid abundance may be attributed to the abundance of decaying wood (larval food) and flowers (adult food) in recently-logged areas (O'Neill et al., 2008). Syrphid flies may increase post-logging for the same reasons (Reemer, 2005).

4.3 Harvesting beargrass

Traditional harvest of beargrass for basketry entails removing the leaves; commercial harvest entails removing the leaves and sometimes the flowers. Done correctly, harvesting leaves and flowers does not kill the parent plant (Thomas & Schumann, 1993). If the rhizome is left intact, it regenerates well after harvest. Beargrass is assumed to recover from harvest within four or five years, based on appearance, although no research confirms this observation (Higgins et al., 2004). It is unclear whether harvesting stimulates the growth of new shoots (Shebitz et al., 2009).

Traditional harvest methods used by Native Americans entail removing the longest plant leaves from the center of a clump of beargrass by gently pulling them, or cutting them at the base (Anderson, 2005; Rentz, 2003; Shebitz, 2005; Shebitz et al., 2009). Because most of the plant remains intact following harvest, traditional harvesting techniques are assumed to have a negligible impact on beargrass.

Commercial harvesters also remove leaves from the center of the beargrass plant, as these are the ones that meet commercial quality standards (Thomas & Schumann, 1993). In general, roughly a dozen long leaves are removed per plant. Harvest takes place by pulling the leaf blade from the sheath, or cutting it with a knife as close to the base as possible to maximize length. Poor training and a tendency to work quickly in order to maximize harvests can cause careless harvesting that damages the plant, or the removal of plant material that does not meet commercial standards, causing waste. If the rhizome is cut or torn out during leaf removal, beargrass can take three years to grow back (Kramer, 2001; Thomas & Schuman, 1993; Vance et al., 2001; Vance et al., 2004). Flowering stalks may also be destroyed in the process of removing leaves. And, some commercial harvesters remove entire plants instead of selectively harvesting the leaves, making regeneration more difficult (Shebitz, 2005; Shebitz et al., 2008). The tendency for commercial harvesters to concentrate in specific places increases harvest impacts.

The effects of commercial harvesting appear to vary by who is doing it. Harvesters who are more place-based and who participate regularly in the nontimber forest products trade are likely to be more concerned about the long-term sustainability of harvesting (Brown, 2001; Hansis, 1998). In contrast, those who are migratory, or who move in and out of the commercial beargrass sector, are more likely to be primarily interested in maximizing short-term gain, and to have less concern for long-term, sustainable harvest practices.

The largely unregulated commercial harvest of beargrass poses a potential threat to the species in parts of its range, and to other uses and values of the plant. For example, some Native Americans in the Pacific Northwest are concerned about declines in the quality and quantity of beargrass, which they attribute in part to the commercial floral industry and to commercial harvesters who do not selectively harvest leaves (Shebitz, 2005; Shebitz et al., 2009). Over the past 20 years, federal and state agencies and private landowners have made attempts to regulate access to beargrass and harvest levels. These efforts have met a number of barriers. The cost of harvest permits varies considerably across federal, state, and private lands, and year to year, and is often too expensive to enable harvesters to make a significant profit (Lynch & McLain, 2003). Permitting rules also change over time and ownership, causing confusion among harvesters. In addition, there is concern that too many people are obtaining permits to harvest in the same locations, creating competition. This situation leads to high levels of illegal harvest, exacerbated by the challenges of enforcing management regulations. Because illegal harvesting is difficult to regulate and monitor, it tends to have a more negative impact on beargrass populations than legal harvesting does.

We are not aware of any studies that directly address the impacts of beargrass harvesting on pollinators. For the most part, harvesting targets leaves, not flowers, so unless flowers are damaged in the harvest process (which sometimes occurs), the most likely impact to pollinators would come from reduced photosynthesis (Vance et al., 2004). The removal of plant leaves compromises the photosynthetic capacity of a plant, reducing its ability to grow and reproduce. If the quality and number of beargrass pollen grains is consequently lowered as a result of leaf harvest, then pollinators would have less food available to them, and would expend more energy obtaining food. It is also possible that insects could be killed by trampling or vehicles used to transport harvesters. Any negative impacts on pollinators could, in turn, affect beargrass by reducing pollination.

4.4 Biotic disturbances

Biotic disturbances - including insects and pathogens - affect light conditions and soil properties within the range of beargrass. Defoliating insects, such as western spruce budworm (*Choristoneura occidentalis*), are native to forest ecosystems throughout the northwest and interior west (Hummel & Agee 2003). Budworm populations are regulated by factors including predators (both vertebrate and invertebrate) and weather. Outbreaks occur when the density of host tree species is favorable and regulating factors are weakened. Repeated budworm defoliation over years can weaken or kill individual trees or extensive groups of trees. Non-native insects, such as the balsam woolly adelgid (*Adelges piceae*), introduced from Europe around 1900, can kill or weaken true fir trees throughout the range of beargrass.

Fungal root diseases in forested habitat cause patches of dead and dying trees, which, upon falling or dying back, create openings in the forest canopy. Trees of all sizes and ages may be killed by these diseases, although susceptibility varies greatly among tree species.

Important pathogens in the Pacific Northwest and Rocky Mountains include laminated root rot (*Phellinus weirii*), armillaria root disease (*Armillaria ostoyae*), and annosus root disease (*Heterobasidion annosum*). In the grand fir zone, they contibute to patchy forest structures associated with small scale mortality (individual trees or groups of trees). These three diseases exist in different combinations and severity throughout the range of the plant. They could impact beargrass pollinators because they increase the availability of dying and dead trees that create habitat for guilds such as wood-boring beetles, hover flies, and wood-nesting bees. Furthermore, the canopy openings caused by root disease may further attract pollinators by providing warmer foraging conditions, and by promoting the flowering of beargrass and other species.

4.5 Summary

This review of beargrass disturbance ecology finds that both anthropogenic and natural disturbances influence beargrass ecosystems. Exactly how these disturbances affect beargrass - either alone or in combination - is understudied. Given the long history of anthropogenic fire within the range of beargrass, and the fact that natural fire return intervals are relatively long in many parts of its range, anthropogenic fire was undoubtedly the dominant disturbance agent affecting beargrass ecosystems prior to around 1900. Indicators are that anthropogenic fire favored beargrass, its habitat, its cultural uses, its flowers, and presumably, associated pollinator communities as well as other species that use it for food, habitat, and nesting material. The effects of other natural disturbances on beargrass - such as landslides, insects, and diseases - were relatively small and localized.

Since 1900, anthropogenic disturbance has continued to play a dominant role in influencing beargrass ecosystems, though the nature of the disturbance has shifted from fire (largely because of suppression policies) to commercial timber harvesting, and in recent decades, to the commercial harvest of beargrass itself. These changes have been less beneficial to beargrass and some of its ecosystem services. Fire suppression has brought about forest succession and the gradual disappearance of open beargrass habitat in parts of its range. Timber harvesting could potentially alter this trend, but the impacts of timber harvest on beargrass are mixed, depending on harvest method. Clearcut timber removal, the most common treatment employed during much of the 20th century, appears to have negative impacts on beargrass. However, partial harvest techniques may enhance beargrass and its associates by creating the forest canopy openings and filtered light conditions it responds to. Long-term studies of the relationship between habitat characteristics, beargrass productivity, and beargrass population dynamics are required before the sustainability of commercial beargrass harvesting can be adequately assessed (Higgins et al., 2004).

5. Management implications

5.1 Overview

We selected beargrass to illustrate how one species can have multiple ecological and sociocultural roles, and to assert the importance of considering the social and ecological relations associated with a species in ecosystem conservation. In doing so, we highlighted how some - but not all - natural and anthropogenic disturbances may affect beargrass and its ecosystem functions and relations over time. Now we consider management implications for ecosystem diversity conservation.

Beargrass still grows across much of the range mapped in the 1950s (Maule, 1959), and is not listed as federally-threatened or endangered in the United States. Nonetheless, the plant has declined in abundance in parts of its range (e.g., on the prairies and wetlands of Washington's Olympic Peninsula), and some of its ecosystem services have been affected by changing disturbance regimes. This trend has negative implications for some of the biocultural diversity associated with beargrass ecosystems. The main values at risk to date are traditional Native American uses and their associated cultural roles, and most likely, commercial harvesting and its socioeconomic importance to harvesters in some locations. Ecosystem processes associated with pollination may also be a concern. It is important to emphasize that it is not merely the presence or absence of the plant in a particular location that puts such values at risk; the reproductive strategy of the plant (flowering or vegetative), plant properties (leaf quality, flower quality, pollen nutrient status), and plant distribution are also vital elements of its functional role.

In order to develop and implement effective management strategies to protect the ecosystem values of a species in places where they are a concern, it is important to understand how natural and anthropogenic disturbances interact to affect them. These interactions may be complex, and will likely vary throughout the range of a species. Thus, best management practices to conserve ecosystem diversity will also vary according to local disturbance regimes, within a regional context. What does this mean for ecosystem management to achieve conservation? One implication is that management needs and approaches will vary throughout the range of a species, because ecosystem values at risk in one location may be less of a concern elsewhere. For example, documented concerns over insufficient beargrass leaves for Indian basketry come from the Northwest, but not the Rocky Mountain portion of its range.

Table 1 summarizes the management considerations associated with beargrass. It indicates the social and ecological values of the plant, the plant properties that support these values, and the values that are currently at risk. It also identifies the environmental conditions that favor plant properties associated with each value that could be managed for in order to protect it. Finally, it lists the natural and anthropogenic disturbances that currently threaten each value. It is apparent from Table 1 that beargrass management, and indeed management to protect the ecosystem diversity associated with many species, is a complex proposition, especially when actions to protect one value may not be optimal for another.

Because management needs and approaches are likely to vary locally within the range of beargrass and other species, there is a role for traditional and local ecological knowledge in addition to western scientific knowledge in contributing to effective ecosystem conservation, especially when a threatened value is sociocultural in nature. We define traditional ecological knowledge as a cumulative body of knowledge about the relationships between people, other living things, and the environment, that is handed down across generations through cultural transmission (Berkes, 1999). It includes knowledge, practices, beliefs, and the range of skills and strategies that people use to respond to the environmental circumstances they find themselves in, and is place-based (Berkes, 1999; Ingold, 2004). Similarly, local ecological knowledge (which is more recent) includes knowledge, practices, beliefs, skills, and strategies that people develop as a result of extensive interactions with, and personal observation of, local ecosystems (Charnley et al., 2007). There are several examples of how traditional and local ecological knowledge can be integrated into biodiversity conservation efforts (see Charnley et al., 2007, 2008 for a review). The most desirable way is to engage the knowledge holders directly, as active participants in conservation efforts, using participatory approaches.

Again, beargrass provides an example to illustrate these points. Based on our earlier discussion of disturbance impacts on beargrass, management interventions to conserve its threatened ecosystem values would likely focus on fire management, silvicultural treatments, and/or regulation of commercial beargrass harvest. We give brief examples of each of these below, including actual or potential roles for traditional and local ecological knowledge.

Ecosystem role	**Value at risk?**	**Associated plant part/properties**	**Environmental conditions that favor desirable plant properties**	**Disturbances that may negatively impact value**
Social				
Native American basketry	Yes	Leaves: long, thin, pliable, strong, less pigment	- partial canopy/partial shade - recently burned areas	Anthropogenic: - fire suppression - commercial beargrass harvesting - clearcutting and slash burning Natural: - succession to late-seral forest
Commercial floral greens industry	In some parts of range	Leaves: deep green, long, wide, firm, >71 cm in length	- 60-90 percent canopy cover - higher elevation conifer forest in later stages of succession	Anthropogenic: - overharvesting beargrass for commercial purposes - silvicultural practices that create large canopy openings, reducing needed shade (e.g., clearcutting) -prescribed fire Natural: - wildland fire

Ecosystem role	Value at risk?	Associated plant part/properties	Environmental conditions that favor desirable plant properties	Disturbances that may negatively impact value
Aesthetic/spiritual	?	Flowers	- best flowering occurs in open conditions	Anthropogenic: - trampling, commercial harvesting Natural: - processes that favor vegetative state and suppress flowering state
Ecological				
Food	No	Flowers, leaf base and leaves, pollen	- partial canopy, open or diffuse light	Anthropogenic: - fire suppression - overharvesting beargrass Natural: - processes that favor vegetative state (e.g., closed forest canopy) and suppress flowering state
Habitat and soil structure	No	Basal leaves, leaves, and rhizomes	- diffuse light or shade	Anthropogenic: -timber harvest practices that result in soil compaction and plant death
Pollination, decomposition	In some parts of range	Aggregated flowers with nutrient-rich pollen	- partial canopy, open or diffuse light, dead or dying trees that provide substrate for invertebrate pollinators and decomposers like longhorn beetles	Anthropogenic: - fire suppression Natural: - succession to late-seral forest

Table 1. Beargrass management considerations

5.2 Fire management

Beargrass restoration on the Olympic Peninsula of Washington State, where declines in beargrass populations have been observed since the 1980s at least, provides an example in which Native Americans, public land managers, and western scientists are working together to actively integrate traditional ecological knowledge into efforts to restore traditional cultural values associated with beargrass. In 1995 the Olympic National Forest began a restoration project in an area that was historically Skokomish territory to restore beargrass and other shade-intolerant species having cultural importance (Shebitz, 2005). American Indians, forest managers, and University of Washington scientists collaborated to design the project and implement treatments. Land management practices based on traditional ecological knowledge about historical landscape structure and burning techniques (e.g., the season, frequency, and intensity of the burn) were reintroduced. Plots have been set up in several places to monitor the effects of different fire and thinning treatments on beargrass. In the early 2000s, additional management experiments were initiatied on the Olympic National Forest and on the Quinault Indian Reservation to determine if prescribed burns would help reverse the observed declines in beargrass populations occurring on the Olympic Peninsula (Shebitz, 2005; Shebitz et al., 2009).

The short-term results of these experiments indicate that both high- and low-severity fire decrease the percent cover of beargrass by burning its leaves and damaging some meristems. Nevertheless, high-severity fire created conditions favorable for seed germination and seedling establishment, and increased beargrass vegetative reproduction. Shoot production also increased two years later. In the short term, low-severity fire did not affect shoot production or seedling establishment, but the scientists hypothesize that, if done repeatedly, it would likely bring about the greatest increase in beargrass abundance. Frequent fire controls shrub and tree encroachment, limiting competition, and helping beargrass flourish. It also favors leaf properties desirable for basketry (Shebitz et al., 2009). Burning does not promote beargrass properties desired by commercial harvesters, however.

Elsewhere, fuel specialists, timber planners, and cultural resource managers have collaborated with California Indian basketweavers to design prescribed burns that enhance beargrass and other important basketry plants on national forests in northern California (Anderson, 2005; Ortiz, 1993). These projects have been motivated by a desire to restore species having cultural value to tribes, and in the process restore habitat types and associated species that have declined in the absence of fires.

5.3 Silvicultural treatments

A silvicultural system is a planned series of treatments for a forest stand that implies a process for creating target conditions over time. The timing and intensity of treatments in any system that is designed to manage forest ecosystem diversity within the range of beargrass will depend on site-specific conditions as well as key management objectives. Clearcutting is an even-aged system that removes almost all trees, creating a fully exposed microclimate for a new age class of trees to develop under. In the Pacific Northwest, clearcutting predominated for a century (Curtis et al., 1998; Tesch, 1994). The system might provide sufficient light to stimulate flowering in beargrass, but could adversely affect the structural properties of soil and the quantities of dead wood associated with adequate drainage and with pollinator habitat. Clearcutting has also been found to reduce beargrass cover because it leads to increased competition with woody shrubs.

As an alternative, a two-aged, shelterwood system could create the dappled light environment that promotes flowering in beargrass, leaf properties suitable for traditional harvesters, and standing dead and down wood for decomposers. A shelterwood is one in which most trees are harvested, but some are left to shade the new trees establishing underneath. It involves the intentional use of shade, which can give desired species a growth advantage over competing vegetation during the establishment phase of regeneration. Trees retained in a shelterwood system are generally harvested after a new age class is established. In contrast to even-aged or two-aged systems, an uneven-aged system regenerates a forest stand with three or more age classes. This is typically accomplished with some form of selection system. In these systems, mature and immature trees are felled to create or maintain uneven-aged stands. Single tree selection fells individual trees and generally tends to increase the proportion of shade-tolerant species in mixed-species stands. Group selection cuts trees in units and therefore maintains a higher proportion of shade-intolerant species in mixed species stands than individual tree selection. Uneven-aged systems would likely create a shadier environment that inhibited beargrass flowering but produced leaf properties desired by commercial harvesters. They would also tend to promote dead and down wood for decomposers.

We did not find published studies on the effects of different silvicultural systems or intensities of harvest specifically on beargrass. We did, however, discover that long-term data sets exist that could help provide insight. The only study we are aware of that used a participatory approach included Skokomish and Quinault tribal members in management experiments designed to identify management practices effective for restoring cultural uses of beargrass on the Olympic Peninsula (Shebitz et al., 2009). In this study, vegetation and coarse woody debris were manually removed from plots in lowland beargrass habitat as an alternative to burning. Manual clearing caused beargrass cover to decrease, and shoot production and flowering to increase. It did not stimulate seedling establishment, however, needed to sustain beargrass populations.

5.4 Commercial beargrass harvesting

Earlier we noted the difficulty in regulating commercial beargrass harvest. If forest managers were to engage harvesters in the management of local beargrass populations, they might be able to more effectively address the barriers that cause illegal harvesting. Harvesters could also contribute to beargrass management by participating in biological inventory and monitoring of beargrass populations. Many are frequent visitors to forests where they observe plants, ecosystem conditions and processes, and are comfortable navigating the terrain (Charnley et al., 2007). Guidance for involving harvesters in participatory inventory and monitoring of nontimber forest products exists (Lynch et al., 2004). Harvesters could also participate in management experiments or research about the impacts of harvesting on beargrass. One excellent model for engaging harvesters in this type of research again comes from the Olympic Peninsula (Ballard, 2004; Ballard & Huntsinger, 2006), and focuses on another important commercial floral green in the Pacific Northwest - salal (*Gaultheria shallon*). There, harvesters participated in identifying study sites, developing research methods, and gathering and interpreting data about the effects of different harvest intensities on salal regrowth and sustainability, leading to management recommendations. For harvesters to participate in such activities, they must see a benefit. Their interest in sustainable management of plants that contribute to their livelihoods, and in having secure and sustained access to harvest locations that would help them steward the resource, could provide an incentive to participate (Charnley et al., 2007).

6. Conclusions

In this chapter we used beargrass to illustrate a view of biodiversity conservation that focuses on relations among species within an ecosystem. We think single-species conservation approaches have obvious limitations, and that successful conservation strategies must be concerned with the multitude of social and ecological roles a species has within an ecosystem. Such a view implies the need to identify the multiple values and ecosystem services associated with a species (that may or may not itself be threatened or endangered) throughout its range, and to determine which may be at risk and whether conservation interventions are called for. Associated social and ecological values may be at risk because a species has become extirpated locally, changes in species distribution have occurred, or there have been changes in specific properties of the species needed to support the value. It follows that conservation needs, and the nature of management interventions to promote conservation, are likely to vary locally.

We have also contended that conservation interventions will be most effective if they are based on an understanding of the interacting natural and anthropogenic disturbance factors that put a species or its ecosystem values at risk. Again, these are likely to vary locally. Knowledge about how disturbance regimes are changing, and the localized effects of these changes, is important for helping to prioritize management responses.

Where information about how to conserve or restore values at risk is limited, traditional and local ecological knowledge can contribute. Conducting management experiments and monitoring results can be one way for natural resource managers to participate with local knowledge holders to learn what strategies are effective. In this regard, local metrics - preferably those that are important from both an ecological and sociocultural standpoint - are desirable as indicators of conservation success. One example would be abundance of beargrass flowers, which are desirable for pollinators, decomposers, game animals that people hunt, and nature lovers. If different values are at risk in the same locale (such as traditional and commercial uses of beargrass), but different ecological conditions and management interventions are needed to protect these values, it will be necessary to prioritize what values to favor in a particular place. Conservation strategies that address both the sociocultural and ecological values of a species that are important to people will be more likely to receive support, and to succeed, than those that do not.

7. Acknowledgments

We gratefully acknowledge the research assistance of Sarah Foltz Jordan and Sophia Polasky who conducted a thorough review of the literature on beargrass that provided a foundation of information for this chapter. For a comprehensive review and synthesis of existing literature about beargrass, see Foltz Jordan et al. (In press).

8. References

Agee, J.K. (1993). *Fire Ecology of Pacific Northwest Forests,* Island Press, ISBN 1-55963-229-1, Washington D.C.

Anderson, M.K. (2005). *Tending the Wild: Native American Knowledge and the Management of California's Natural Resources*, University of California Press, Berkeley

Ballard, H.L. (2004). *Impacts of Harvesting Salal (Gaultheria shallon) on the Olympic Peninsula, Washington: Harvester Knowledge, Science, and Participation*, Doctoral dissertation, Environmental Science, Policy, and Management. University of California, Berkeley, California

Ballard, H.L. & Huntsinger, L. (2006). Salal harvester local ecological knowledge, harvest practices and understory management on the Olympic Peninsula, Washington. *Human Ecology,* Vol. 34, pp. 529-547

Barrett, S.W. & Arno, S.F. (1999). Indian fires in the northern Rockies: ethnohistory and ecology. In: *Indian, Fire and the Land in the Pacific Northwest,* R. Boyd (Ed.), pp. 50-64, Oregon State University Press, ISBN: 0-87071-459-7, Corvallis, OR

Bartuszvige, A.M. & Kennedy, P.L. (2009). *Synthesis of Knowledge on the Effects of Fire and Thinning Treatments on Understory Vegetation in U.S. Dry Forests,* Speical Report 1095, Oregon State University, Extension and Experiment Station, Corvallis, Oregon

Battles, J.J.; Shlisky, A.J.; Barrett, R.H.; Heald, R.C. & Allen-Diaz, B.H. (2001). The effects of forest management on plant species diversity in a Sierran conifer forest. *Forest Ecology and Management,* Vol. 146, No. 11, pp. 211-222

Berkes, F., 1999. *Sacred Ecology: Traditional Ecological Knowledge and Resource Management,* Taylor and Francis, Philadelphia, Pennsylvania

Boyd, R. (Ed.). 1999. *Indians, Fire and the Land in the Pacific Northwest,* Oregon State University Press, Corvallis, Oregon

Brown, B.A. (2001). *Challenges Facing Community Forestry: the Role of Low Income Forest Workers,* Jefferson Center for Education and Research, Wolf Creek, Oregon

Campbell, J.W.; Hanula, J.L. & Waldrop, T.A. (2007). Effects of prescribed fire and fire surrogates on floral visiting insects of the Blue Ridge province in North Carolina. *Biological Conservation,* Vol. 134, pp. 393-404

Charnley, S.; Fischer, A.P. & Jones, E.T. (2007). Integrating traditional and local ecological knowledge into forest biodiversity conservation in the Pacific Northwest. *Forest Ecology and Management,* Vol. 246, pp. 14-28

Charnley, S.; Fischer, A.P. & Jones, E.T. (2008). *Traditional and Local Ecological Knowledge about Forest Biodiversity in the Pacific Northwest,* U.S. Department of Agriculture, Forest Service, Pacific Northwest Research Station, General Technical Report PNW-GTR-751, Portland, Oregon

Craighead, F.C. (1923). North American Cerambycid Larvae. A Classification and the Biology of North American Cerambycid Larvae. *Canada Department of Agriculture Entomological Bulletin,* Vol. 23, pp. 1-239. Available from: http://www.archive.org/stream/northamericancer00craiuoft#page/94/mode/2up

Crane, M.F. (1990) Xerophyllum tenax, In: *Fire Effects Information System,* 5 March, 2011. Available from <http://www.fs.fed.us/database/feis/>U.S. Department of Agriculture, Forest Service, Rocky Mountain Research Station, Fire Sciences Laboratory. (Producer)

Curtis, R.O.; DeBell, D.S.; Harrington, C.A.; Lavender, D.P.; St. Clair, J.B.; Tappeiner, J.C. & Walstad, J.D. (1998). *Silviculture for Multiple Objectives in the Douglas-fir Region,* U.S. Department of Agriculture, Forest Service, Pacific Northwest Research Station, General Technical Report PNW-GTR-435, Portland, Oregon

Dyrness, C.T. (1965). *The Effect of Logging and Slash Burning on Understory Vegetation in the H. J. Andrews Experimental Forest*, U.S. Department of Agriculture, Forest Service, Pacific Northwest Forest and Range Experiment Station, PNW-RN-31, Portland, Oregon

Flora of North America. May 2011, Available from: <http://www.efloras.org/florataxon.aspx?flora_id=1&taxon_id=242102053>

Foltz Jordan, S., Polasky, S., Hummel, S. and Charnley, S. (In press). Natural and cultural history of beargrass (X. tenax). USDA Forest Service, PNW Research Station General Technical Report (PNW-GTR-xxx). Portland, OR.

Halpern, C.B. & Spies, T.A. (1995). Plant species diversity in natural and managed forests of the Pacific Northwest. *Ecological Applications,* Vol. 5, pp. 913-934

Hansis, R. (1998). A political ecology of picking: Non-timber forest products in the Pacific Northwest. *Human Ecology,* Vol. 26, No. 1, pp. 67-86

Hartley, M.K.; Rogers, W.E.; Siemann, E. & Grace, J. (2007). Responses of prairie arthropod communities to fire and fertilizer: balancing plant and arthropod conservation. *American Midland Naturalist,* Vol. 157, pp. 92-105

Higgins, S.; Blatner, K.; Kerns. B.K. & Worthington, A. (2004). Relationship between Xerophyllum tenax and canopy density in the southern Cascades of Washington. *Western Journal of Applied Forestry,* Vol. 19, pp. 82-87

Hitchcock, G.L. & Cronquist, A. (1973). *Flora of the Pacific Northwest: An Illustrated Manual,* University of Washington Press, ISBN: 9780295952734, Seattle, Washington

Hummel, S. & Agee, J.K. (2003). Western spruce budworm defoliation effects on forest structure and potential fire behavior. *Northwest Science,* Vol. 77, No. 2, pp. 159-169

Hunter, J.E. (1988). Prescribed burning for cultural resources. *Fire Management Notes* Vol. 49, No. 2, pp. 8-9

Ingold, T. (2004). Two reflections on ecological knowledge, In: *Nature Knowledge: Ethnoscience, Cognition, and Utility,* Sanga, G., & Ortalli, G., pp. 301-311, Berghahn Books, New York

Keeley, J.E. & Stephenson, H.L. (2000). Restoring natural fire regimes to the Sierra Nevada in an era of global climate change, In: *Proceedings of the Wilderness Science in a Time of Change Conference,* Cole, D.N.; McCool, S.F.; Borrie, W.T. & McLoughlin, J., pp. 255-265, U.S. Department of Agriculture, Forest Service, Rocky Mountain Research Station, RMRS-P-15, Vol. 5, Ogden, Utah

Klinger, R.; Wills, R. & Brooks, M.L. (2008). Fire and nonnative invasive plants in the Southwest Coastal Bioregion, In: *Wildland Fire in Ecosystems: Fire and Nonnative Invasive Plants,* Zouhar, K.; Smith, J.K.; Sutherland, S. & Brooks, M.L., Chapter 9, U.S. Department of Agriculture, Forest Service, Rocky Mountain Research Station, RMRS-GTR-42 Vol. 6, Ogden, Utah

Kramer, B. (2001). Dangerous harvests? Natural Resources: Forest Service studying whether popularity of beargrass is harming ecosystem. *The Spokesman Review* (June 5, 2001)

Kremen, C.; Williams, N.M; Aizen, M.A.; Gemmill-Herren, B.; LeBuhn, G.; Minckley, R.; Packer, L.; Potts, S.G.; Roulston, T.; Steffan-Dewenter, I.; Vázquez, D.P.; Winfree, R.; Adams, L.; Crone, E.E.; Greenleaf, S.S.; Keitt, T.H.; Klein, A.-M.; Regetz, J. & Ricketts, T.H. (2007). Pollination and other ecosystem services produced by mobile organisms: a conceptual framework for the effects of land-use change. *Ecology Letters,* Vol. 10, pp. 229-314

Larson, B.M.H.; Kevan, P.G. & Inoyue, D.W. (2001). Files and flowers: taxonomic diversity of anthophiles and pollinators. *Canadian Entomologist,* Vol. 133, pp. 439-465

Laursen, S.B. (1984). *Predicting Shrub Community Composition and Structure Following Management Disturbance in Forest Ecosystems of the Intermountain West,* University of Idaho Doctoral Dissertation, Moscow, Idaho

Lobb, A. (1990). *Indian Baskets of the Pacific Northwest and Alaska,* Graphic Arts Center Publishing Co., Portland, Oregon

Lynch, K.A.; Jones, E.T. & McLain, R.J. (2004). *Nontimber Forest Product Inventorying and Monitoring in the United States: Rationale and Recommendations for a Participatory Approach,* National Commission on Science for Sustainable Forestry, Washington, D.C.

Lynch, K.A. & McLain, R.J. (2003). *Access, Labor, and Wild Floral Greens Management in Western Washington's Forests,* U.S. Department of Agriculture, Forest Service, Pacific Northwest Research Station, PNW-GTR-585, Portland, Oregon

MacRae, T.C. & Rice, M.E. (2007). Biological and distributional observations on North American Cerambycidae (Coleoptera). *The Coleopterists Bulletin,* Vol. 61, No. 2, pp. 227-263

Maffi, L. (2005). Linguistic, cultural, and biological diversity. *Annual Review of Anthropology,* Vol. 29, pp. 599-617

Maule, S.M. (1959). Xerophyllum tenax, squawgrass, its geographic distribution and its behaviour on Mount Rainier, Washington. *Madroño,* Vol. 15, pp. 39-48

Morrison, P.H. & Swanson, F.J. (1990). *Fire History and Pattern in a Cascade Range Landscape,* U.S. Department of Agriculture, Forest Service, Pacific Northwest Research Station, PNW-GTR-254, Portland, Oregon

Munger, S.H. (2003). *Common to this Country: Botanical Discoveries of Lewis and Clark,* Artisan, New York

O'Neale, L.M. (1928) [reprinted 1995]. *Yurok-Karok Basket Weavers,* Phoebe Hearst Museum of Anthropology, Berkeley, California

O'Neill, K.M.; Fultz, J.E. & Ivie, M.A. (2008). Distribution of adult Cerambycidae and Buprestidae (Coleoptera) in a subalpine forest under shelterwood management. *The Coleopterists Bulletin,* Vol. 62, No. 1, pp. 27-36

Ortiz, B. (1993). Contemporary California Indian basketweavers and the environment, In: *Before the Wilderness: Environmental Management by Native Californians,* Blackburn, T.C. & Anderson, K., pp. 195-211, Ballena Press, Menlo Park, California

Peter, D. & Shebitz, D.J. (2006). Historic anthropogenically maintained bear grass savannas of the southeastern Olympic Peninsula. *Restoration Ecology,* Vol. 14, No. 4, pp. 605-615

Pojar, J. & MacKinnon, A. (1994). *Plants of the Pacific Northwest Coast: Washington, Oregon, British Columbia, and Alaska,* Lone Pine Publishing, Vancouver, B.C.

Potts, S.G.; Vulliamy, B.; Dafni, A.; Ne'eman, G.; O'Toole, C.; Roberts, S. & Willmer, P. (2003). Response of plant–pollinator communities to fire: changes in diversity, abundance and floral reward structure. *Oikos,* Vol. 101, pp. 103-112

Reemer, M. (2005). Saproxylic hoverflies benefit by modern forest management (Diptera: Syrphidae). *Journal of Insect Conservation,* Vol. 9, pp. 49-59

Rentz, E. (2003). *Effects of Fire on Plant Anatomical Structure in Native Californian Basketry Materials,* Masters Thesis, San Francisco State University, San Francisco, California

Roulston, T.H.; Cane, J.H. & Buchmann, S.L. (2000). What governs protein content of pollen: pollinator preferences, pollen-pistil interactions, or phylogeny? *Ecological Monographs,* Vol. 70, No. 4, pp. 617-643

Schlosser, W.E.; Blatner, K.A. & Chapman, R.C. (1991). Economic and marketing implications of special forest products harvest in the coastal Pacific Northwest. *Western Journal of Applied Forestry,* Vol. 6, No. 3, pp. 67-72

Schlosser, W.E. & Blatner, K.A. (1997). *Special Forest Products: an East-side Perspective,* U.S. Department of Agriculture, Forest Service, Pacific Northwest Research Station, PNW-GTR-380, Portland, OR

Schneider, F. (1969). Bionomics and physiology of aphidophagous Syrphidae. *Annual Review of Entomology,* Vol. 14, pp. 103-124

Shebitz, D.J. (2005). Weaving traditional ecological knowledge into the restoration of basketry plants. *Journal of Ecological Anthropology,* Vol. 9, pp. 51-68

Shebitz, D.J.; Reichard, S.H. & Woubneh, W. (2008). Beargrass (Xerophyllum tenax) on the Olympic Peninsula, Washington: Autecology and population status. *Northwest Science,* Vol. 82, No. 2, pp. 128-140

Shebitz, D.J.; Reichard, S.H. & Dunwiddie, P.W. (2009). Ecological and cultural significance of burning beargrass habitat on the Olympic Peninsula, Washington. *Ecological Restoration,* Vol. 27, No. 3, pp. 306-319

Swengel, A.B. (2001). A literature review of insect responses to fire, compared to other conservation managements of open habitat. *Biodiversity and Conservation,* Vol. 10, pp. 1141-1169

Takahashi, M. & Kawano, S. (1989). Pollen morphology of the Melanthiaceae and its systematic implications. *Annals of the Missouri Botanical Garden,* Vol. 76, No. 3, pp. 863-876

Tesch, S.D. (1994). The Pacific Northwest Region, In: *Regional Silviculture of the United States, Third Edition,* Barrett, J.W., Wiley & Sons, Inc., New York

Thomas, M.G. & Schumann, D.R. (1993). *Income Opportunities in Special Forest Products: Self-help Suggestions for Rural Entrepreneurs,* U.S. Department of Agriculture, Forest Service, Agriculture Information Bulleting AIB-666. State and Private Forestry, Washington, D.C.

United States Department of Agriculture (USDA) (2011). Plants database. May, 2011, Available from:
< http://plants.usda.gov/java/profile?symbol=XETE>

Vance, N.; Bernhardt, P. & Edens, R.M. (2004). Pollination and seed production in Xerophyllum tenax (Melanthiaceae) in the Cascade Range of Central Oregon. *American Journal of Botany,* Vol. 91, pp. 2060-2068

Vance, N.C.; Borsting, M.; Pilz, D. & Freed, J. (2001). *Special Forest Products: Species Information Guide for the Pacific Northwest,* U.S. Department of Agriculture, Forest Service, Pacific Northwest Research Station, PNW-GTR-513, Portland, Oregon

Wick, D.; Evans, J. & Luna, T. (2008). *Propagation Protocol for Production of Container Xerophyllum tenax (Pursh) Nutt. Plants (160 ml containers).* U.S. Department of the Interior, National Park Service, Glacier National Park, West Glacier, Montana, Retrieved from <http://www.nativeplantnetwork.org>

Wray, J. & Anderson, M.K. (2003). Restoring Indian-set fires to prairie ecosystems on the Olympic Peninsula. *Ecological Restoration* Vol. 21, No. 4, pp. 296-301

Wright, C.S. & Agee, J.K. (2004). Fire and vegetation history in the eastern Cascade Mountains, Washington. *Ecological Applications*, Vol. 14, No. 2, pp. 443-459

9

Sown Wildflower Strips – A Strategy to Enhance Biodiversity and Amenity in Intensively Used Agricultural Areas

Christine Haaland and Mats Gyllin
Swedish University of Agricultural Sciences, Alnarp
Sweden

1. Introduction

1.1 Agricultural changes and effects on biodiversity

Biodiversity in agricultural areas in Europe has been in constant decline (European Environment Agency [EEA], 2010). Agricultural land is a habitat for many species, both plants and animals, which have adapted to the special conditions of these human-influenced ecosystems. The decline in diversity is due to the intensification of agricultural production and the abandonment of marginalised agricultural areas that are no longer profitable. Intensive agricultural production involves increased application of herbicides, pesticides and fertiliser, which has a profound effect on many plant and animal communities (Stoate et al., 2001, 2009). Often emphasised examples are fertilisation or aerial deposition of nitrogen on semi-natural grassland, which leads to a decline in many herbal plants adapted to less nitrogenous conditions (e.g. Dupre et al., 2010). The application of insecticides has been shown to affect whole food webs, for example a decline in bird species due to the lack of insect prey (Vickery et al., 2009).

The intensification of agricultural production has not only changed the quality of habitats, but also their quantity, with areas of low production, for example semi-natural grasslands, being in dramatic decline. The removal of certain biotopes to enlarge arable field size and its consequences for biodiversity have been discussed for several decades, for example the removal of hedges and field margins (Robinson & Sutherland, 2002). These measures have caused profound changes to the landscape pattern of agricultural landscapes, resulting in fragmentation of semi-natural areas and homogenisation (Jongman, 2002).

The decline in many species groups with habitats on farmland has been widely acknowledged. Birds have been one of the most studied groups and loss of bird species is well documented (e.g. Chamberlain et al., 2000; Donald et al., 2001; Newton, 2004; Wretenberg et al., 2010). The decline in birds is related to agricultural intensification practices, such as increased in herbicide use, change in sowing regimes (from spring to autumn sowing), land drainage and changes in landscape pattern. Kragten et al. (2011) emphasise the importance of farming system (organic versus conventional) for the abundance of invertebrate prey for birds. Dramatic declines in species have also been recorded for insects, for example butterflies (Nilsson et al., 2008; van Swaay et al., 2006), bumblebees (Goulson et al., 2008) and bees (Biesmeijer et al., 2006). The picture is similar for beetles (Kotze & O'Hara, 2003) and bugs (Frank & Künzle, 2006). Again, the reasons cited for the decline are habitat loss and fragmentation, loss of foraging opportunities and the general homogenisation of the agricultural landscape (Benton et al., 2003; Diekotter et al., 2008; Tscharntke et al., 2005).

1.2 Sown wildflower strips

In order to decrease the negative effects of intensive farming and support extensive farming practices, agri-environmental schemes were introduced within the EU Common Agricultural Policy (CAP) in the late 1980s. Today, agri-environmental schemes are part of the Rural Development Programme (RDP), which is compulsory for all EU member states. However, the schemes can be designed differently by the member states within the given framework. Sown wildflower strips have been introduced in several European countries as a measure to enhance and support biodiversity especially in intensively farmed areas (Haaland et al., 2011, Noordijk et al., 2010). The overall goal with the establishment of wildflower strips is to enhance biodiversity and the abundance of certain species groups such as insects, birds and plants. This is especially important in areas with intensive agriculture. Such strips are often particularly intended to enhance the abundance of particular functional insect groups, such as pollinators and predators of pest species, through the provision of adult food resources, for example nectar-rich flowers. Such pollinators and pest predators are important in the context of agricultural production. Birds can also benefit from high insect numbers or seeds in the wildflower strips (Vickery et al., 2009). Some of the schemes are designed to support (rare) plant species that used to be common and typical for field margins prior to the introduction and wider application of fertilisers and herbicides (Marshall & Moonen, 2002)

Sown wildflower strips are usually established on arable land by sowing a seed mixture of wild flowers. Establishment and management are carried out by the farmer, who is compensated for the cost of seed mixtures, any management operations required and loss of income according to the regulations of the particular agri-environmental scheme. Sown wildflower strips vary regarding the seed mixture applied, size of strip, duration of the scheme and management. The seed mixture applied varies between countries and also between the different agri-environmental schemes within a country. Thus in some seed mixtures only a few wildflower species are present (1-5), while others can comprise up to 30 species (Nentwig, 2000). Two examples of seed mixtures are given in an appendix to this chapter. The agri-environmental schemes specify which seed mixture the farmers are allowed to use among the seed mixtures offered by commercial suppliers. One distinctive difference is whether seed mixtures contain grass species or not. The importance of using local provenance when establishing wildflower strips has been pointed out (Bischoff et al., 2010). Strip width can vary between from a few metres up to 24 metres, while certain schemes allow for areas instead of strips. Strips can be sown along field edges or within fields but a minimum size of total area sown is often required. Some schemes last only one or two years, while other run for up to five or seven years.

Management of wildflower strips also varies, a common feature being that the use of pesticides, herbicides and fertilisers is prohibited. However, in special cases the use of spot treatments with herbicides may be permitted when certain weed species become a problem. In some countries, for example Switzerland, wildflower strips are often not managed at all after their establishment and are left uncut over a period of up to seven years. Due to succession processes the plant composition of wildflower strips can change greatly during that time, for example due to the invasion of grasses, the passing of annuals and sometimes the increasing dominance of a single plant species in the seed mixture, for example *Dipsacus fullonum* (see also Noordijk et al, 2011, for an example from the Netherlands).

On the other hand uncut strips offer an excellent overwintering habitat for many insect species (Frank & Reichhart, 2004; Pfiffner & Luka, 2000). In other countries it is more common to cut wildflower strips once a year in autumn.

The potential of sown wildflower strips for amenity purposes in the countryside has only recently been discussed as a strategic option (see examples from Germany where wildflower strips are promoted as '*Blühende Landschaften*', flowering landscapes). Wildflower strips can be very attractive landscape features when flowering and thus appreciated by people visiting the agricultural landscape for recreational purposes (Junge et al., 2009; Lindemann-Matthies et al., 2010).

The combination of biodiversity goals and recreational goals makes wildflower strips a particularly interesting element in the context of intensively used agricultural landscapes, where available space for recreation and wildlife is equally scarce. This is particularly the case in peri-urban areas, where the need for recreation opportunities is high and the pressure on land is increased due to housing developments. In the following sections we illustrate some biodiversity benefits from wildflower strips for insects and then suggest some ways in which biodiversity goals and recreation interests can be combined in the same greenway system. By the term greenway, we refer to linear features that are established on arable land, covered by vegetation and suitable for walking and possibly other forms of recreation. Paved paths, gravel paths or walking paths are not included in the definition, although the large body of existing research on greenways (e.g. Ahern, 1995; Fabos & Ryan, 2006), also includes much broader definitions. We conclude by suggesting possible approaches for implementation.

2. Wildflower strips for biodiversity conservation

2.1 Abundances and species diversity in wildflower strips

Quite a number of studies have investigated the abundance and diversity of insects (Haaland et al., 2011) and spiders in sown wildflower strips. These studies have identified several factors that can affect abundance and species diversity in the strips: flower abundance (e.g. Pywell et al., 2006), plant diversity (e.g. Aviron et al., 2007) seed mixture (e.g. Marshall, 2007), vegetation structure (e.g. Woodcock et al., 2005), management (e.g. Woodcock et al., 2008), age (e.g. Noordijk et al., 2010) and landscape factors (e.g. Aviron et al., 2007). Species groups that have been studied include bees and bumblebees, butterflies, beetles, bugs, grasshoppers. Sown wildflower strips have proven capable of containing high numbers of bumblebees (Carvell et al., 2006, 2007; Haaland & Gyllin, 2010; Pywell et al., 2005, 2006), with strips sown with few plant species particularly rich in pollen and nectar attracting most bumblebees (e.g. Pywell et al., 2005). Nevertheless a greater variety of wildflower plant species has the advantage of providing food resources during a longer period of the year and might be able to support particular species (Carvell et al., 2007). Sown wildflower strips can attract more butterflies compared with other margin types or other habitats typical for the open agricultural landscape (Aviron et al., 2007; Feber et al., 1996; Haaland & Gyllin, 2010). Interestingly, Jacot et al. (2007) found higher butterfly diversity in strips sown with both grasses and wildflowers compared with strips sown with wildflower seeds only. This might be explained by the fact that several butterfly species have grasses as the food plant for their larvae. In sown wildflower strips often the more common butterfly species are found, but a relatively large percentage of a region's species pool can be observed within them (Haaland & Bersier, 2011).

Beetles are a well-studied insect taxa in sown wildflower strips. Studies have shown that wildflower strips have a greater diversity of beetles than other types of field margins or agricultural habitats (Aviron et al., 2007). However, beetle abundances are rather dependent on factors such as management and vegetation structure, and thus higher beetle abundances are not necessarily found in sown wildflower strips compared with other field margins (Woodcock et al., 2008). For grasshoppers, high species richness and abundances have been recorded in margins sown with both grass seeds and a wildflower mixture (Jacot et al., 2007, Marshall, 2007). Bugs can reach similar high abundances and diversity in wildflower strips as in meadows (Zurbrügg & Frank, 2006).

Spider densities can be significantly increased in fields adjacent to sown wildflower strips compared with other types of field margins (Schmidt-Entling & Döbeli, 2009). In addition, since wildflower strips can contain a larger number of insect species, they can act as a good food resource for birds (Vickery et al., 2009).

2.2 An example from Sweden

In the most southerly province of Sweden, Scania, we have studied butterfly and bumblebee diversities and abundances in two different types of green structures: sown wildflower strips and greenways established primarily for recreation purposes (Haaland & Gyllin, 2010). Until recently, there were no agri-environmental schemes for wildflower strips in Sweden, but the Swedish University of Agricultural at Alnarp (near Malmö) established wildflower strips in the vicinity of the university campus for various purposes such as demonstration, amenity, research projects (Figures 1 and 2). These wildflower strips were mostly sown with a seed mixture of grasses (5 species) and wildflowers (15-25 species). In two cases, wildflower strips were established using cut hay from a nearby meadow. In addition, three greenway systems were studied. These were established either on the initiative of a farmer (n=1, Tottarp, Figure 3) or the municipality of Lund on private land after negotiations with the concerned land-owners (n=2, Arendala and Lund, Figure 4). In all three cases the landowner is being paid compensation by the municipality.

Butterflies and bumblebees were recorded by visual observance along transects in both wildflower strips and greenways during one summer (2007). Each transect was divided into sections of varying length, with each section having more or less the same characteristics. The recorded transect length was 2.9 km in the wildflower strips at Alnarp and 6.8 km in the greenways at Tottar, Arendala and Lund (for more details see Haaland & Gyllin, 2010).

Fig. 1. Sown wildflower strip at Alnarp, Scania, Sweden. Photo: Mats Gyllin.

Fig. 2. Sown wildflower strip at Alnarp, Scania, Sweden. Photo: Christine Haaland.

Fig. 3. Greenway established by farmer primarily for recreation purposes (walking, horse riding) on arable land. Tottarp, Staffanstorp, Scania, Sweden. Photo: Christine Haaland.

Fig. 4. Greenway established primarily for recreation purposes by the municipality of Lund on arable land. Bushes and tree species were also used in order to meet biodiversity goals. Lund, Scania, Sweden. Photo: Christine Haaland.

1769 butterflies of 18 species (including one day flying moth) and 1216 foraging bumblebees (8 species) were recorded during the study. Although the greenways investigated were more than twice as long as the wildflower strips, nearly all butterflies (86%) and bumblebees (83%) were observed in the wildflower strips. The mean number of butterflies was about 20 times higher in the wildflower strips than in the greenways. Bumblebees were virtually absent from the grass sown greenways, but occurred in slightly higher numbers in the greenways with plantings and some flowering edge vegetation (greenways in Lund, Figure 4). Most butterfly and bumblebee species recorded are common species in southern Sweden. Species numbers were higher for both butterflies and bumblebees in the sown wildflower strips (Figure a, b). In terms of the flowers visited by butterflies and bumblebees, it was noted that a few plant species were visited very often (Figure 6). Thus two-thirds of all butterflies observed on a flower were seen on brown knapweed (*Centaurea jacea*) and greater knapweed (*C. scabiosa*) and field scabious (*Kanutia arvensis*) (Figure 6a). Three-quarters of all foraging bumblebees were observed on *Centaurea* ssp. and 14% on *Knautia arvensis* (Figure 6b). This shows that few plant species attract most individuals of these two species groups and that preferences differ between species groups. However, flower visit patterns change during the season as the availability of flowering plants changes with their flowering period.

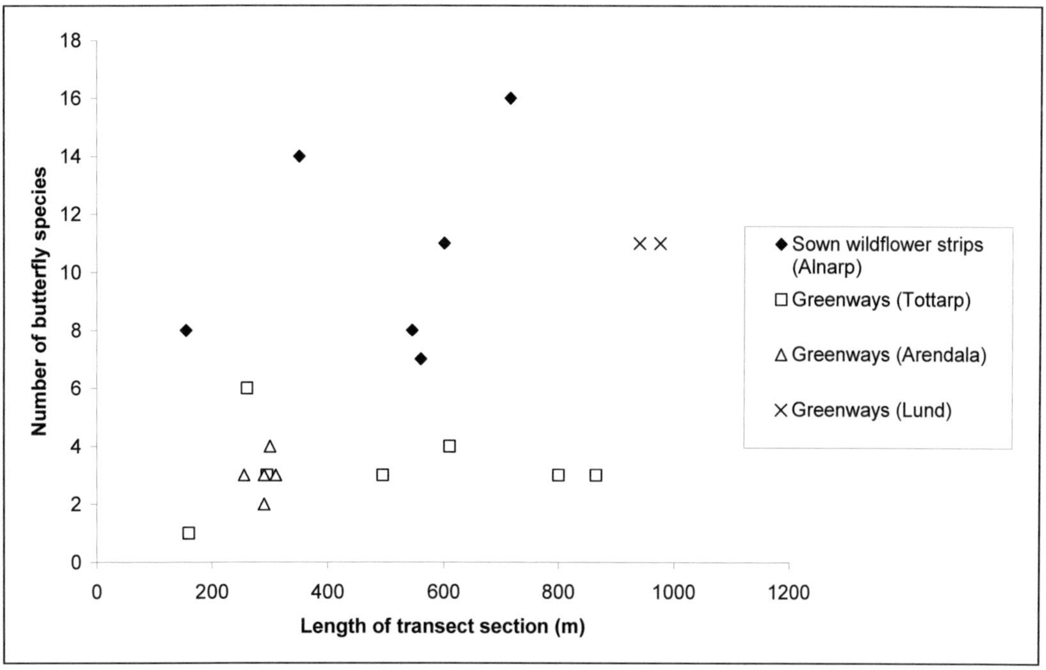

Fig. 5a. Butterfly species numbers in sown margins and greenways at the different study sites in relation to length of transect section, Scania, South Sweden.

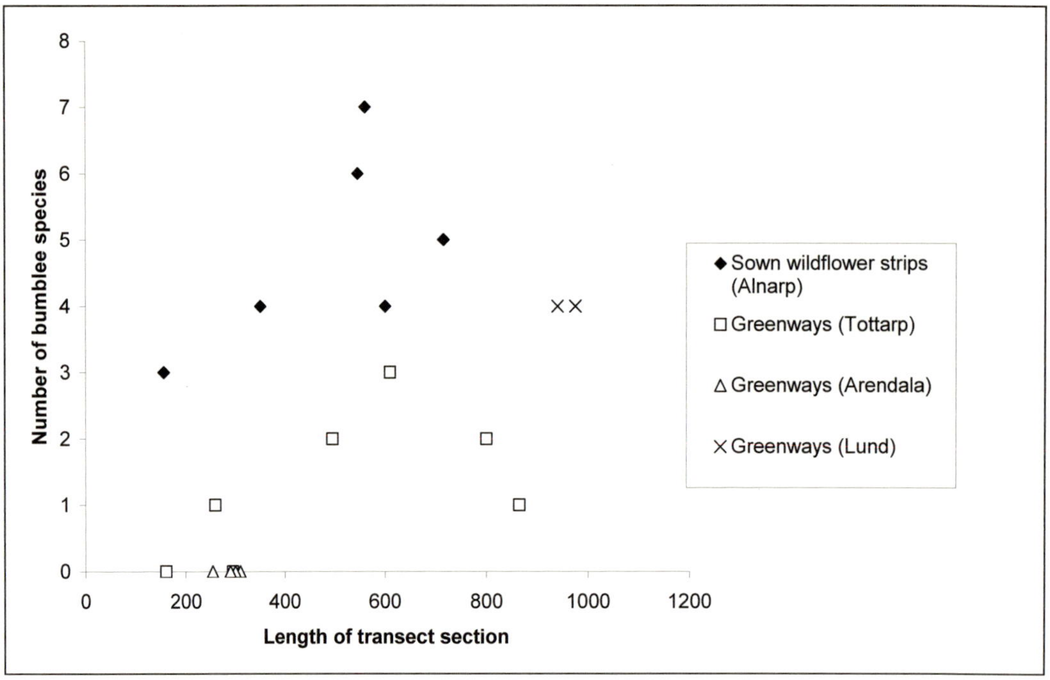

Fig. 5b. Bumblebee species numbers in sown margins and greenways at the different study sites in relation to length of transect section, Scania, South Sweden.

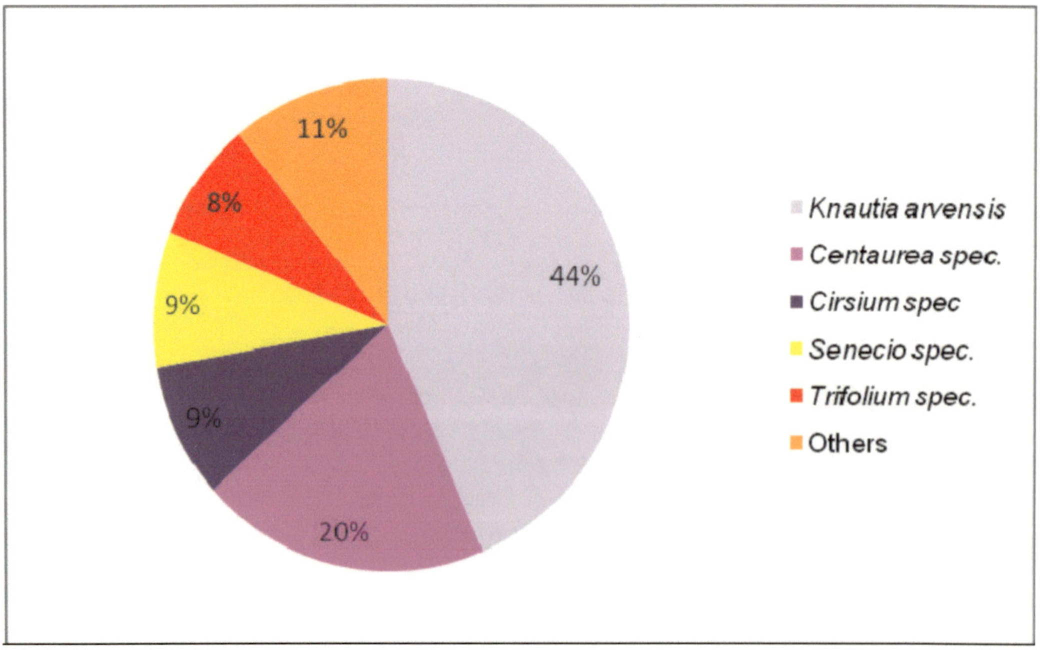

Fig. 6a. Flower visits by butterflies in wildflower strips and greenways, n=347 (Scania, Sweden).

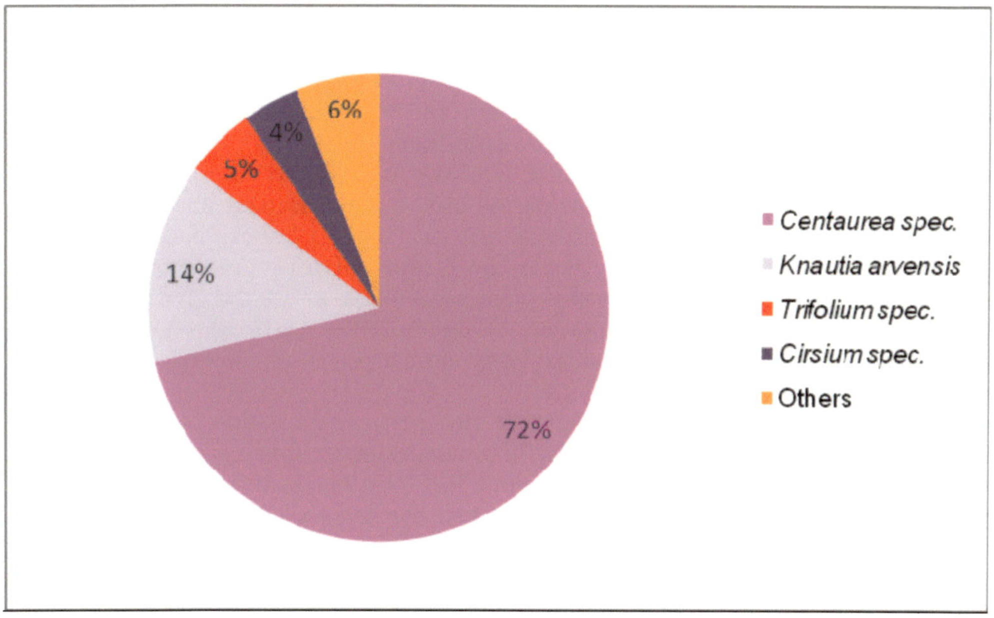

Fig. 6b. Flower visits by bumblebees in wildflower strips and greenways, *n*=1216 (Scania, Sweden).

3. Combining biodiversity and recreation goals

Sown wildflower strips can provide food and other resources for insects and thus also for other species groups such as birds. Potentially they are also suited to increase the amenity of intensively farmed agricultural landscapes. The types of greenways that have been typically established on arable land in the most Southern part of Sweden fulfil recreation goals in that they provide possibilities for walking and horse riding in a landscape that is otherwise rather inaccessible due to the removal of farm tracks and field boundaries to enlarge field size. However, these greenways do not contribute to biodiversity conservation or biodiversity enhancement, despite this being a goal that is often stated in policy documents in connection with their establishment. Even though the primary aim of greenway establishment is recreation, there is still the ambition that these green structures should positively contribute to biodiversity goals. In landscapes with intensive agriculture which are simultaneously under pressure from urban development, arable land for developing recreation opportunities or biodiversity goals is scarce. Thus it would be desirable to create green structures that optimally serve both needs. In our project 'Multifunctional Greenways' we are trying to find design solutions for greenways where recreation and biodiversity goals are better integrated than is currently common practice in southern Sweden. We suggest a combination of approaches partly originating in agri-environmental schemes and partly in the (European) greenway tradition (Fumagalli & Toccolini 2007).
There are several challenges in the design of such local networks. Within most agri-environmental schemes regarding sown wildflower strips, the farmer is relatively free where to establish the strips: on arable land at the edge or within the field. No network approach or connectivity of strips is needed. In addition, the strips are established on arable land and can be ploughed up again after the scheme has expired. Furthermore, the strips are only intended

to meet biodiversity goals and thus farmers do not have to deal with an increased number of people moving across their farmland. Thus when creating multifunctional greenways that serve recreation and enhance species richness, a network approach is needed, continuity is desirable, but not necessary (as long as no bush and tree plantings are involved) and the farmer must be willing to allow public access on his farmland.

Regarding visitors, there are also different interests to take into account (Peterson et al., 2010). In southern Sweden, most greenways are used by both walkers and for horse riding. This can work well as long as the number of visitors is limited, but can become problematic when walkers are afraid of meeting horses on a rather narrow greenway or horses are disturbed by walkers. In wet weather it is often not suitable to use the greenways for horse riding because of damage to the vegetation. Mountain bikers are another group that might increasingly wish to use the greenways, and here similar problems can occur in wet weather conditions in terms of damage to vegetation.

Multifunctional greenways can be designed to suit different recreation purposes and at the same time be more beneficial to wildlife than simple grass sown greenways (Figure 7). Depending on available width, local conditions and local needs, different solutions can be chosen. When riders and walkers frequently use the same network, plantings or sown wildflower strips can be established in the middle of the greenway to separate these two visitor groups. The width of a mown path can also be adjusted to the kind of recreation experience intended or to the number or type of visitors. In this type of green structure, both recreation values and biodiversity can be enhanced. Plantings of bushes or trees are only sensible in networks that are intended to be continuous. Sown wildflower strips within agri-environmental schemes are regularly ploughed up, so an approach combining grass strips with wildflower strips can also function in a non-continuous network.

A. Greenways without tree and bush plantings

a) 'Classic' greenway, sown with grasses only, cut short (several times per year)

b) Greenway where grass areas are combined with sown wildflower strips at both edges

c) Greenway with sown wildflower strip along one edge

d) Greenway with sown wildflower strip in the middle

e) Sown wildflower strip with mown path in the middle

f) Sown wildflower strip with two mown paths

B. Greenways with bush plantings

a) Greenway with planted bushes and sown wildflower strips along one edge

b) Greenway with bush planting and two sown wildflower strips in the middle

c) Greenway with bush plantings along both edges, sown wildflower strips and mown path

Fig. 7. Possible design for multifunctional greenways. A) without bushes and B) with bush plantings. Illustration: Christine Haaland

4. Possible approaches for implementation

There are several ways to implement multifunctional greenways in which measures for biodiversity and recreation are combined on agricultural land. In Sweden we have idnetified three possible approaches for implementation (Haaland et al., 2010). Greenways have so far been established on the farmer's initiative, the initiative of the local authority (municipality) and private initiative, for example sporting organisations. In addition, agri-environmental schemes can be viewed as a framework for financing the establishment and management of greenways.

The first greenway network in Sweden on arable land was initiated by a farmer, but subsidised by the municipality (Figure 3). (However, farmers' reluctance to allow public access to farmland is an acknowledged problem, Ryan & Walker, 2004). There can be a pressing need to channel recreation on farmland, especially in peri-urban areas and particularly when horse riding is involved. Swedish law provides the right of public access to the countryside, but this right is difficult to exercise in areas of intensive agriculture with large fields, few farm tracks and field boundaries. Thus there may be a tendency for walkers and riders to access areas of the farmland where the farmer does not want them (small verges between fields, farm tracks near the farm) or along road verges of trafficked roads, which can be very unsafe. Farmers themselves can therefore have an interest in channelling visitors to parts of the farm where they disturb farm operations least. There are also farmers who welcome visitors, like to offer access for recreation on their farmland and are willing to prepare these - at least when they are compensated. An open question is how interested these farmers would be in combining access issues and biodiversity measures within the same green structure network.

Several greenway networks have been established by municipal authorities to provide recreation possibilities for the urban population in peri-urban areas. Here land is often leased by the municipality and farmers may be additional compensated for management. This approach allows for well-designed greenways for both recreation and biodiversity means, but is not fully exploited yet. There are examples where both aims are considered, for example choosing certain bush and tree species, which are intended to enhance biodiversity (in this case birds, Figure 4). Here, relatively easily sown wildflower strips or patches could be added.

In other cases private initiatives have been taken to establish greenways to improve accessibility, for example by equestrian organisations (Larsson et al., 2011). In these cases, where recreation and access issues have priority for the initiative takers, biodiversity issues are easily neglected (Larsson et al., 2011). This might especially be the case where organisations pay compensation to farmers. When municipalities are involved in paying compensation, it is easier to ask for biodiversity measures in addition.

Agri-environmental payments could potentially be used to finance the establishment and management of multifunctional greenways (Von Haaren & Reich, 2006). In Sweden, several greenways projects are financed at least partly through payments regulated in the Rural Development Programme, for example Leader projects. Another possibility would be to enhance the function of buffer strips along water courses to include recreation access. In some municipalities in Scania this is already common practice, but it may occasionally contravene the rules of the agri-environmental schemes. These buffer strips would also be suitable for part sowing with wildflowers (e.g. along the field edge) instead of grass mixtures only.

It can be assumed that the use of agri-environmental payments to increase access to intensively used agricultural areas can enhance the acceptance of such payments by the public - or at least among those interested in access to these areas.

5. Conclusions

Sown wildflower strips can support and enhance species richness in intensively farmed areas. The number of species in wildflower strips is dependent on the type of seed mixture sown, which affects other factors such as flower abundance, plant species diversity and vegetation structure. These factors are in turn influenced by the type of management. The age of the strips is relevant for species numbers, how is often depending on the type of management, which steers if and how fast successional processes take place. Where suitable, we advocate combining sown wildflower strips with recreation possibilities by creating multifunctional greenways. Multifunctional greenways enhance species richness and also increase access for visitors in areas with intensive agriculture. This can be particularly relevant in peri-urban areas. The most suitable design and management of greenways is dependent on the species groups they are intended to support and the visitor groups that will use them. The question of how long the greenways are supposed to exist is crucial for issues such as bush and tree plantings. Thus greenway design can be adapted to local conditions and needs. There are several possible approaches to implement multifunctional greenways, where farmers, organisations and municipalities play key roles. An important stimulant for the establishment of multifunctional greenways would be the possibility to finance them with the help of agri-environmental payments (agri-environmental schemes, Leader).

6. Acknowledgements

This study was financed by the Swedish research council FORMAS.

7. Appendix

Example 1 of plant species sown in wildflower strips (applied at Alnarp, Sweden)

Herbs
Achillea millefolium
Anthyllis vulneraria
Campanula rotundifolia
Centaurea jacea
Filipendula vulgaris
Galium verum
Hypericum perforatum
Hypochoeris maculata
Hypochoeris radicata
Knautia arvensis
Leucanthemum vulgare
Lotus corniculatus
Plantago lanceolata
Plantago media
Primula veris
Prunella vulgaris
Ranunculus acris
Ranunculus bulbosus
Rhinanthus serotinus
Rumex acetosa
Scabiosa columbaria
Senecio jacobaea
Succisa pratensis
Trifolium pratense
Vicia cracca
Grasses
Anthoxanthum odoratum
Avenula pratensis
Cynosurus cristatus
Festuca ovina
Festuca rubra
Phleum pratense ssp. bertolonii

Example 2 of plant species sown in wildflower strips (applied in wildflower strips in Switzerland)

Achillea millefolium
Agrostemma githago
Anthemis tinctoria
Centaurea cyanus
Centaurea jacea

Cichorium intybus
Daucus carota
Dipsacus fullonum
Echium vulgare
Hypericum perforatum
Leucanthemum vulgare
Malva moschata
Malva sylvestris
Origanum vulgare
Papaver rhoeas
Pastinaca sativa
Silena pratensis
Tanacetum vulgare
Verbascum lychnitis
Verbascum thapsus ssp

8. References

Ahern, J. (1995). Greenways as a planning strategy. *Landscape and Urban Planning*, Vol.33, No.1-3 (October 1995), pp. 131-155, ISSN 0169-2046

Aviron, S., Herzog, F., Klaus, I., Luka, H., Pfiffner, L., Schupbach, B. & Jeanneret, P. (2007). Effects of Swiss agri-environmental measures on arthropod biodiversity in arable landscapes. *Aspects of Applied Biology*, Vol.81, pp. 101-109, ISSN 0265-1491

Benton, T.G., Vickery, J.A. & Wilson, J.D. (2003). Farmland biodiversity: is habitat heterogeneity the key? *Trends in Ecology & Evolution*, Vol.18, No.2 (April 2002), pp. 182-188, ISSN 0169-5347

Biesmeijer, J.C., Roberts, S.P.M., Reemer, M., Ohlemuller, R., Edwards, M., Peeters, T., Schaffers, A.P., Potts, S.G., Kleukers, R., Thomas, C.D., Settele, J. & Kunin, W.E. (2006). Parallel declines in pollinators and insect-pollinated plants in Britain and the Netherlands. *Science*, Vol.313, No.5785 (July 2006), pp. 351-354, ISSN 0036-8075

Bischoff, A., Steinger, T. & Muller-Scharer, H. (2010). The Importance of Plant Provenance and Genotypic Diversity of Seed Material Used for Ecological Restoration. *Restoration Ecology*, Vol.18, No.3 (May 2010), pp. 338-348, ISSN 1061-2971

Carvell, C., Westrich, P., Meek, W.R., Pywell, R.F. & Nowakowski, M. (2006). Assessing the value of annual and perennial forage mixtures for bumblebees by direct observation and pollen analysis. *Apidologie*, Vol.37, No.3 (May-June 2006), pp. 326-340, ISSN 0044-8435

Carvell, C., Meek, W.R., Pywell, R.F., Goulson, D. & Nowakowski, M. (2007). Comparing the efficacy of agri-environment schemes to enhance bumble bee abundance and diversity on arable field margins. *Journal of Applied Ecology*, Vol. 44, No.1 (February 2007), pp. 29-40, ISSN 0021-8901

Chamberlain, D.E., Fuller, R.J., Bunce, R.G.H., Duckworth, J.C. & Shrubb, M. (2000). Changes in the abundance of farmland birds in relation to the timing of agricultural intensification in England and Wales. *Journal of Applied Ecology*, Vol.37, No.5 (October 2000), pp. 771-788, ISSN 0021-8901

Diekotter, T., Billeter, R. & Crist, T.O. (2008). Effects of landscape connectivity on the spatial distribution of insect diversity in agricultural mosaic landscapes. *Basic and Applied Ecology*, Vol.9, No.3 (May 2008), pp. 298-307, ISSN 1439-1791

Donald, P.F., Green, R.E. & Heath, M.F. (2001). Agricultural intensification and the collapse of Europe's farmland bird populations. *Proceedings of the Royal Society B*, Vol.268, No.1462 (January 2001), pp. 25-29, ISSN 0962-8452

Dupre, C., Stevens, C.J., Ranke, T., Bleeker, A., Peppler-Lisbach, C., Gowing, D.J.G., Dise, N.B., Dorland, E., Bobbink, R. & Diekmann, M. (2010). Changes in species richness and composition in European acidic grasslands over the past 70 years: the contribution of cumulative atmospheric nitrogen deposition. *Global Change Biology*, Vol.16, No.1 (January 20101), pp. 344-357, ISSN 1354-1013

European Environment Agency [EEA] (2010). The European environment — state and outlook 2010: synthesis. European Environment Agency, ISBN 978-92-9213-114-2, Copenhagen, Denmark

Fabos, J.G. & Ryan, R.L. (2006). An introduction to greenway planning around the world. *Landscape and Urban Planning*, Vol.76, , No.1-4 (April 2006), pp. 1–6, ISSN 0169-2046

Feber, R.E., Smith, H. & Macdonald, D.W. (1996). The effects on butterfly abundance of the management of uncropped edges of arable fields. *Journal of Applied Ecology*, Vol.33, No.5 (October 1996), pp. 1191-1205, ISSN 0021-8901

Frank, T. & Künzle, I. (2006). Effect of early succession in wildflower areas on bug assemblages (Insecta : Heteroptera). *European Journal of Entomology*, Vol.103, No.1 (January 2006), pp. 61-70, ISSN 1210-5759

Frank, T. & Reichhart, B. (2004). Staphylinidae and Carabidae overwintering in wheat and sown wildflower areas of different age. *Bulletin of Entomological Research*, Vol.94, No.3 (June 2004), pp. 209-217, ISSN 0007-4853

Fumagalli, N. & Toccolini, A. (2007). Greenways and ecological networks: synergisms and conflicts. Estimo e Territorio, Vol.70, No.12, pp. 49-62, ISSN 1824-8918

Goulson, D., Lye, G.C. & Darvill, B. (2008). Decline and conservation of bumble bees. *Annual Review of Entomology*, Vol.53, pp. 191-208, ISSN 0066-4170

Haaland, C. & Gyllin, M. (2010). Butterflies and bumblebees in greenways and sown wildflower strips in southern Sweden, *Journal of Insect Conservation*, Vol.14, No.2 (April 2010), pp. 125-132, ISSN 1366-638X

Haaland, C., Larsson, L., Peterson, A. & Gyllin, M. (2010). Implementing multifunctional greenways in Sweden - challenges and opportunities. *Proceedings of Fábos Conference on Landscape and Greenway Planning*, ISBN 978-963-503-409-3, July 2010, Budapest, Hungary

Haaland, C. & Bersier, L.-F. (2011). What can sown wildflower strips contribute to butterfly conservation?: an example from a Swiss lowland agricultural landscape. *Journal of Insect Conservation*, Vol.15, No.1-2 (April 2011), pp. 301-309, ISSN 1366-638X

Haaland, C., Naisbit, R.E. & Bersier, L.-F. (2011). Sown wildflower strips for insect conservation - a review. *Insect Conservation and Diversity*, Vol.4, No.1 (February 2011), pp. 60-80, ISSN 1752-458X

Jacot, K., Eggenschwiler, L., Junge, X., Luka, H. & Bosshard, A. (2007). Improved field margins for a higher biodiversity in agricultural landscapes. *Aspects of Applied Biology*, Vol. 81, pp. 277-283, ISSN 0265-1491

Jongman, R.H.G. (2002). Homogenisation and fragmentation of the European landscape: ecological consequences and solutions. *Landscape and Urban Planning*, Vol. 58, No.2-4 (February 2004), pp. 211-221, ISSN 0169-2046

Junge, X., Jacot, K.A., Bosshard, A. & Lindemann-Matthies, P. (2009). Swiss people's attitudes towards field margins for biodiversity conservation. *Journal for Nature Conservation*, Vol.17, No.3 (August 2009), pp. 150-159, ISSN 1617-1381

Kotze, D.J. & O'Hara, R. (2003). Species decline - but why? Explanations of carabid beetle (Coleoptera, Carabidae) declines in Europe. *Oecologia*, Vol.135, No.1 (March 2003), pp. 138-148, ISSN 0029-8549

Kragten, S., Tamis, W.L.M., Gertenaar, E., Ramiro, S.M.M., Van der Poll, R.J., Wang, J. & De Snoo, G.R. (2011). Abundance of invertebrate prey for birds on organic and conventional arable farms in the Netherlands. *Bird Conservation International*, Vol.21, No.1 (March 2011), pp. 1-11, ISSN 0959-2709

Larsson, A, Peterson, A., Bjärnborg, E., Haaland, C. & Gyllin, M. (2011). Regional Landscape Strategies - a pilot study concerning the implementation of the European Landscape Convention and public participation in Sweden. In: *The European Landscape Convention – Challenges of Participation*, Jones, M., Stenseke, M. (eds.), pp. 261-274, Landscape Series Vol.13, ISBN 978-90-481-9931-0, Springer

Lindemann-Matthies, P., Junge, X. & Matthies, D. (2010). The influence of plant diversity on people's perception and aesthetic appreciation of grassland vegetation. *Biological Conservation*, Vol.143, No.1 (January 2010), pp. 195-202, ISSN 0006-3207

Marshall, E.J.R. & Moonen, A.C. (2002). Field margins in northern Europe: their functions and interactions with agriculture. *Agriculture Ecosystems & Environment*, Vol.89, No.1-2 (April 2002), pp. 5-21, ISSN 0167-8809

Marshall, E.J.P. (2007). The effect of arable field margin structure and composition on Orthoptera assemblages. *Aspects of Applied Biology*, Vol.81, pp. 231-238, ISSN 0265-1491

Nentwig, W. (Ed.) (2000). Streifenförmige ökologische Ausgleichsflächen in der Kulturlandschaft: Ackerkrautstreifen, Buntbrachen, Feldränder. 293 pp., Verlag Agrarökologie, ISBN 3-919293-14-9, Bern, Switzerland, Hannover, Germany

Newton, I. (2004). The recent declines of farmland bird populations in Britain: an appraisal of causal factors and conservation actions. *Ibis*, Vol.146, No.4 (October 2004), pp. 579-600, ISSN 0019-1019

Nilsson, S.G., Franzén, M. & Jönsson, E. (2008). Long-term land-use changes and extinction of specialised butterflies. *Insect Conservation and Diversity*, Vol.1, No.4 (November 2008), pp. 197-207, ISSN 1752-458X

Noordijk, J., Musters, C.J.M., van Dijk, J. & de Snoo, G.R. (2010). Invertebrates in field margins: taxonomic group diversity and functional group abundance in relation to age. *Biodiversity & Conservation*, Vol.19, No.11 (October 2010), pp. 3255–3268, ISSN 0960-3115

Noordijk, J., Musters, C.J.M., van Dijk, J. & de Snoo, G.R. (2011). Vegetation development in sown field margins and on adjacent ditch banks. *Plant Ecology*, Vol.212, No.1 (January 2011), pp. 157–167, ISSN 1385-0237

Pfiffner, L. & Luka, H. (2000). Overwintering of arthropods in soils of arable fields and adjacent semi-natural habitats. *Agriculture Ecosystems & Environment*, Vol.78, No.3 (May 2000), pp. 215-222, ISSN 0167-8809

Peterson, A., Gyllin, M., Haaland, C. & Larsson, A. (2010). Recreation in Swedish agricultural areas - public attitudes to multifunctional greenway design. *Proceedings of Fábos Conference on Landscape and Greenway Planning*, ISBN 978-963-503-409-3, July 2010, Budapest, Hungary

Pywell, R.F., Warman, E.A., Carvell, C., Sparks, T.H., Dicks, L.V., Bennett, D., Wright, A., Critchley, C.N.R. & Sherwood, A. (2005). Providing foraging resources for bumblebees in intensively farmed landscapes. *Biological Conservation*, Vol.121, No.4 (February 2005), pp. 479-494, ISSN 0006-3207

Pywell, R.F., Warman, E.A., Hulmes, L., Hulmes, S., Nuttall, P., Sparks, T.H., Critchley, C.N.R. & Sherwood, A. (2006) Effectiveness of new agri-environment schemes in providing foraging resources for bumblebees in intensively farmed landscapes. *Biological Conservation*, Vol.129, pp. 192-206, ISSN 0006-3207

Robinson, R.A. & Sutherland, W.J. (2002). Post-war changes in arable farming and biodiversity in Great Britain. *Journal of Applied Ecology*, Vol.39, No.1 (February 2002), pp. 157–176, ISSN 0021-8901

Ryan, R.L. & Walker, J.T.H. (2004). Protecting and managing private farmland and public greenways in the urban fringe. *Landscape and Urban Planning*, Vol.68, No.2-3 (May 2004), pp. 183-198, ISSN 0169-2046

Schmidt-Entling, M. & Döbeli, J. (2009). Sown wildflower areas to enhance spiders in arable fields. *Agriculture, Ecosystems & Environment*, Vol.133, No.1-2 (March 2009), pp. 19-22, ISSN 0167-8809

Stoate, C., Boatman, N.D., Borralho, R.J, Rio Carvalho, C., De Snoo, G.R. & Eden, P. (2001). Ecological impacts of arable intensification in Europe. *Journal of Environmental Management*, Vol.63, No.4 (December 2001), pp. 337-365, ISSN 0301-4797

Stoate, C., Báldi, A., Beja, P., Boatman, N.D., Herzon, I., Van Doorn, A., De Snoo, G.R., Rakosy, L. & Ramwell, C. (2009). Ecological impacts of early 21st century agricultural change in Europe--a review. *Journal of Environmental Management*, Vol.91, No.1 (October 2009), pp. 22-46, ISSN 0301-4797

Tscharntke, T., Klein, A.M., Kruess, A., Steffan-Dewenter, I. & Thies, C. (2005). Landscape perspectives on agricultural intensification and biodiversity - ecosystem service management. *Ecology Letters*, Vol.8, No.8 (August 2005), pp. 857-874, ISSN 1461-023X

Van Swaay, C., Warren, M. & Lois, G. (2006). Biotope use and trends of European butterflies. *Journal of Insect Conservation*, Vol.10, No.2 (June 2006), pp. 189-209, ISSN 1366-638X

Vickery, J.A., Bradbury, R.B., Henderson, I.G., Eaton, M.A. & Grice, P.V. (2004). The role of agri-environment schemes and farm management practices in reversing the decline of farmland birds in England. *Biological Conservation*, Vol.119, No.1 (September 2004), pp. 19-39, ISSN 0006-3207

Vickery, J.A., Feber, R.E. & Fuller, R.J. (2009). Arable field margins managed for biodiversity conservation: A review of food resource provision for farmland birds. *Agriculture, Ecosystems and Environment*, Vol.13, No.1-2 (September 2009), pp. 1-13, ISSN 0167-8809

Von Haaren, C. & Reich, M. (2006). The German way to greenways and habitat networks. *Landscape and Urban Planning*, Vol.76, No.1-4 (April 2006), pp. 7-22, ISSN 0169-2046

Woodcock, B.A., Westbury, D.B., Potts, S.G., Harris, S.J. & Brown, V.K. (2005). Establishing field margins to promote beetle conservation in arable farms. *Agriculture Ecosystems & Environment*, Vol.107, No.2-3 (May 2005), pp. 255-266, ISSN 0167-8809

Woodcock, B.A., Westbury, D.B., Tscheulin, T., Harrison-Cripps, J., Harris, S.J., Ramsey, A.J., Brown, V.K. & Potts, S.G. (2008), Effects of seed mixture and management on beetle assemblages of arable field margins. *Agriculture Ecosystems & Environment*, Vol.125, No1-4 (May 2008), pp. 246-254, ISSN 0167-8809

Wretenberg, J., Pärt, T. & Berg, A. (2010). Changes in local species richness of farmland birds in relation to land-use changes and landscape structure. *Biological Conservation*, Vol.143, No.2 (February 2010), pp. 375-381, ISSN 0006-3207

Zurbrügg, C. & Frank, T. (2006). Factors influencing bug diversity (Insecta: Heteroptera) in semi-natural habitats. *Biodiversity and Conservation*, Vol.15, No.1 (January 2006), pp. 275-294, ISSN 0960-3115

10

Combining Historical and Ecological Knowledge to Optimise Biodiversity Conservation in Semi-Natural Grasslands

Eva Gustavsson[1], Anna Dahlström[2], Marie Emanuelsson[3], Jörgen Wissman[2] and Tommy Lennartsson[2]
[1]*Lake Vänermuseum of Natural and Cultural Sciences, Lidköping*
[2]*The Swedish Biodiversity Centre, Uppsala*
[3]*Research and Innovation Service, University of Gothenburg, Gothenburg*
Sweden

1. Introduction

The species-rich semi-natural grasslands, i.e. unfertilised, uncultivated pastures and hay meadows, are among the most threatened habitats in the European agricultural landscape. Most European countries have lost more than 90% of their semi-natural grasslands due to abandonment or productivity intensification during the last century (Bernes, 1993; Piessens & Hermy, 2006; Stanners & Bourdeau, 1995). This drastic loss of habitat has caused population declines and extinctions of large numbers of grassland species (Baillie, 2004; Cheffings & Farrel, 2005; Gärdenfors, 2010). To stop further declines of grassland species, habitats and cultural values, national and EU subsidiary programmes pay farmers to uphold and reintroduce management of semi-natural grassland (European Union, 2011; Kleijn & Sutherland, 2003; Myrdal, 2001). To merit the payment, certain management criteria have to be fulfilled, e.g. in Sweden mainly aiming at minimising litter accumulation and shrub encroachment in the grassland (Swedish Board of Agriculture, 2007).

A growing number of studies express concern that these conservation efforts, and the current grassland management in general, are not sufficiently reaching the conservation goals, i.e. to rescue threatened grassland organisms and ecological functionality of the habitats. Invertebrates, birds and management-dependent vascular plants in particular have been reported to decline in spite of ongoing grassland management for conservation purposes (e.g. Boschi & Baur, 2007; Bühler & Schmid, 2001; Kruess & Tscharntke, 2002; Öckinger et al., 2006; Söderström et al., 2001; Willcox et al., 2010). Ecologically insufficient management quality is a potential threat to semi-natural grasslands that has attracted considerably less attention than has the cessation of management.

The species-richness in semi-natural grassland is to a large extent a legacy of historical agricultural practices (Bruun et al., 2001; Pykälä, 2000; Reitalu et al., 2010). The agricultural system preceding the modernisation processes in Sweden and other countries in north

Europe, i.e. mixed farming (Grigg, 1974), formed a wide range of grassland habitats and disturbance regimes through a variety of land-use practices, differing in terms of type, timing, intensity and dynamics of disturbance. Common main types of land use were grazing, mowing, burning and irrigation. Grassland ecosystems were central in this system as a base for livestock husbandry and for nutrients, which were reallocated from pastures and hay meadows to the permanent arable fields through livestock manure (Emanuelsson, 1988). In Scandinavia this system was established around AD 800-1000 (Welinder et al., 1998). It remained until a new agricultural system was introduced, between c. 1860 and 1950, which was based on artificial fertilisers as the nutrient base (Gadd, 2000).

The main types of historical grassland use in Europe was grazing for summer forage and mowing for winter fodder, both constituting a, more or less, yearly biomass removal. This causes nutrient depletion or reallocation, reduction of vegetation height and damage to plants. Competition patterns between plant species thus generally favour low-competitive species because the disturbance counteracts the succession towards taller and more species-poor vegetation (Elisseou et al., 1995; Plantureux, 1996). Lower vegetation increases ground sun exposure and reduces litter thickness, favouring seed germination and ground arthropods. Damage to plants reduces flowering and seed production of plants, in turn negatively affecting e.g. insects dependent on seeds, nectar and pollen. The net effects on grassland vegetation largely depend on timing, intensity and variability of management. Thus, different anthropogenic land-use regimes may be considered as analogous to different natural disturbance regimes (Pykälä, 2000), creating a set of ecological conditions which are the basic prerequisites for grassland biodiversity.

This chapter discusses management quality in an ecological-historical context. It relies on the basic assumptions that current grassland biodiversity is largely a legacy of past land use, that species still present in a grassland have the same ecological needs today as historically (Lennartsson & Linkowski, 2011), and that biodiversity therefore is favoured by management practices which are ecologically similar to the historical practices, and threatened by management that ecologically differ from the historical conditions. For example, Gustavsson et al. (2007) found that the switch of management from mowing to grazing around one century ago had been equally detrimental to the number of grassland plant species (cf. Fischer & Wipf, 2002) as had around four decades of abandonment. In order to design conservation management for habitats that are formed by historical land use we thus need to know which land-use components that were present historically, and which of the components that are ecologically significant and therefore need to be reintroduced or imitated in today's management.

We identify ecological variables related to type, intensity, timing, and dynamics of disturbance and discuss how these variables are related to the historical use of grasslands compared to the current management practices. We address the following questions: 1) Which necessary ecological variables in semi-natural grasslands can be identified, based mainly on plant life cycles? 2) How may different components of the pre-industrial grassland use have corresponded to the identified ecological variables? 3) What are the qualitative and quantitative differences between the current and the historical landscape in terms of the landscape's content of ecologically critical management components? 4) How can differences between past and current grassland management be expected to affect grassland biodiversity, and which are the implications for future management and restoration of species-rich grasslands?

2. Study regions and methods

2.1 Study regions

The study concerns 66 villages in five regions in southern Sweden, covering in all 18,620 hectares (Table 1, Fig. 1). The climate is rather similar between the regions: annual mean temperature ranges from 5 to 6 °C, yearly precipitation is around 600 mm, and the vegetation period is 180-190 days (National Atlas of Sweden, 1995).

Name of region	Lat. Long.	Area (ha)	Upland or lowland	Soil properties	Main historical agricultural production
Källstorp/ Söne	58°50′N 12°91′E	1100	Mosaic	Clay and thin till	Livestock[1]
Selaön	59°25′N 17°12′E	4084	Lowland	Clay and till	Grain[2]
Kristberg	58°34′N 15°13′E	3912	Upland	Clay and till	Livestock[2]
Fornåsa	58°29′N 15°14′E	1604	Lowland	Limey clay and till	Grain[2]
Alseda	57°25′N 15°15′E	7920	Upland	Till with fine soil fractions	Livestock[2]

[1] Jansson, 1998;
[2] Dahlström, 2006a

Table 1. Basic information about the five study regions

Fig. 1. Study regions. Kristberg and Fornåsa share the same position mark. Östuna study region was used only for estimating current stocking density (see section 2.3).

2.2 Definitions

For a systematic use of grassland management vocabulary, we have decided to use the following: Land use is reserved for the historical conditions, when grasslands were "used" as a necessary part of the livelihood. **Management** is used for the current conditions, when grasslands are artificially "kept" for conservation purposes. Land-use/management **type** is used for the main division between grazing and haymaking, which historically were the dominating types of grassland use in the study regions. Land-use/management **regime** is the systematic application of different variants of grazing or haymaking. Historically the land-use regimes were governed by the agricultural system and in the study regions they occurred in hay meadows (including pasture fenced with hay meadow), permanent pastures and grassland (mainly pasture) enclosed with arable fields. A range of variants of these main regimes existed historically, consisting of either a certain type being applied in the same way every year, or varying in type, timing and intensity more or less systematically between years. The regimes can then be subdivided into land-use/management **components**, thus comprising the smallest unit in the regimes. A component may be unique to one of the two types, such as handling of hay, or may be applicable in both, such as aspects of management timing. The land-use/management components create, or otherwise affect, the conditions for plants and animals in the grassland habitats. Different habitat conditions are denoted ecological factors, for example microclimate, soil properties and habitat processes such as disturbance of the vegetation by grazing or mowing. The term **traditional** is only used in the cases where local, current practices can be reasonably expected to be the same as the historical practices.

	18th century	Current	Example
Type	Land-use	Management	Grazing, mowing
Regime	Land-use	Management	Full-season grazing, alternating mowing and grazing
Component	Land-use	Management	Late onset of grazing, handling of hay, weak grazing intensity

Table 2. Terms used in the chapter, as defined in the text above

2.3 Methods

First, literature on grassland management, historical land-use regimes, grassland plants, and grassland vegetation was examined in order to identify ecological factors necessary for grassland plants and components of historical grassland use that created these factors. "Historical" in this case refers to the 18th century, unless differently stated. Ecologically, grassland management can be viewed as a disturbance to the vegetation and single plant individuals, and identification of necessary ecological factors and land-use components was based mainly on how disturbance affects the main steps in the life cycle of grassland plants. The components comprise different aspects of timing, intensity and dynamics of management.

Secondly, the likely abundances of the identified land-use/management components in the five regions were analysed, qualitatively and quantitatively, for both current and historical conditions. For some components this could be done by quantifying the area of grassland managed by the main land-use/management regimes, which provide different combinations of land-use/management components related to timing, intensity and dynamics of management. Historical abundances were estimated using large-scaled historical maps (typically 1:4000 and 1:5000) that depict the agricultural system in detail (Kain & Baigent, 1992; Tollin, 1991). All 66 villages were thus analysed (Archival references: Aurell, 1787; Dahlström, 2006a, p. 277; Hierpe, 1712). Current abundances of these management regimes were estimated by measuring the areas based on information from TUVA, a national database on Swedish semi-natural grasslands, where management type and status of each grassland is recorded (Swedish Board of Agriculture, 2011).

More thorough quantifications of some land-use components could be done by using historical data available for one or a few of the study regions. For Selaön the 18th century grazing intensity was estimated by relating pasture size to the data on livestock type and numbers provided by the land surveyor for each pasture. Livestock data was converted to grazing equivalents according to Dahlström (2006a, pp. 87-89) to allow for comparisons of stocking densities irrespective of livestock type. To compare with current stocking densities, a dataset from Östuna parish, Uppland County, was used. In 2003 the individual areas of 28 pastures were measured by means of GIS and the livestock owners provided data on livestock numbers in each pasture. Current and 18th century size distribution of permanent pastures in the five regions was compared. Current data was extracted from TUVA, whereas historical sizes were measured on 18th century maps.

2.4 Interpretation of historical maps

Due to their characteristics, accessibility and extensive geographic cover, historical cadastral maps is the most widely used historical source in Sweden addressing landscape research related to pre-industrial times. There are large scale cadastral maps showing land use from early 17th century onwards (Tollin, 2004). The interpretation of cadastral maps relies on a number of historical source critical aspects. Desired information may not be available and the image accuracy may be seductive, leading to over-interpretations. Furthermore the intentions behind the production of maps changed through time, affecting what type of information was included. Eighteenth century maps were generally created for cadastral information or as tools to modernise the agricultural system. Ephemeral features, e.g. temporary arable fields or pastures, were therefore mostly omitted. Hence, these maps can be viewed as representations of the main spatial arrangement of land use during a much longer time period than the specific years they were made. The maps in this study represent a period spanning the time of the first enclosure act in mid-18th century (in Swedish: *storskifte*, see Gadd, 2000) and back several centuries, due to the relative stability of the village organisation (Riddersporre, 1995; cf. Sporrong, 1971). Land-use components at an ecologically relevant degree of detail are rarely directly shown by the maps, but need to be derived through historical-ecological interpretation. This study uses information about main land use, spatial relation to other land-use types (as shown by fencing) and areas in order to interpret land-use regimes and ecologically relevant land-use components. The land-use regimes depicted in the maps can be expected to have been present before and after the studied century. Data from other historical sources have therefore been used, mainly from the 17th and 19th centuries to support our interpretations.

3. Results

In the five study regions, an area of 15,563 hectares of semi-natural grassland was in use in the 18th century. Of this, only 685 hectares remain today, thus a decrease by almost 96%. In addition, the different land-use regimes have decreased disproportionately (Fig. 2). Seven different variants of the main land-use regimes were identified (Fig. 2) in the historical landscape of the study regions, today almost all remaining grassland consists of permanent pasture. Pasture enclosed with arable fields or hay meadows has disappeared completely and mowing with aftermath grazing has decreased by 99.7%.

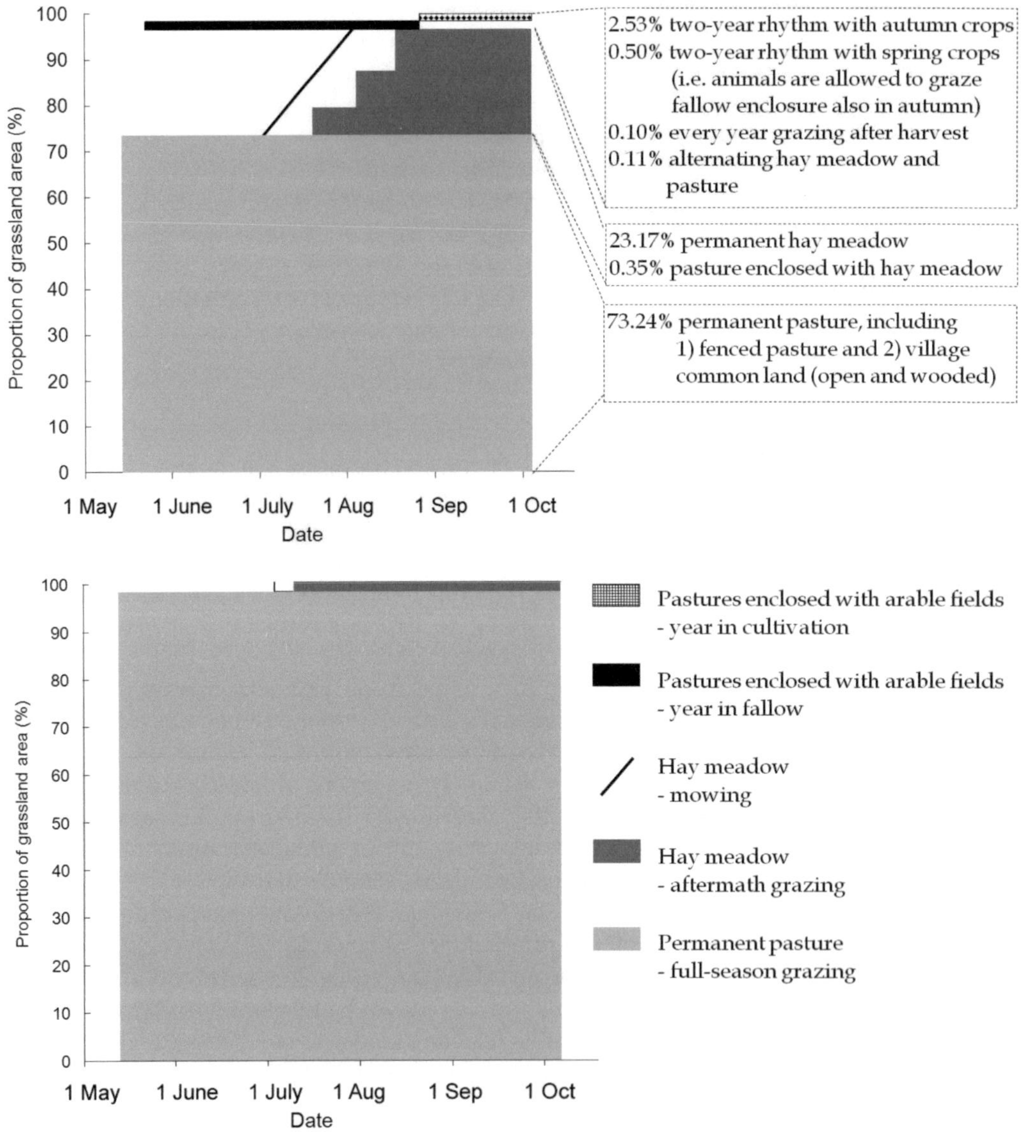

Fig. 2. The proportion of the total area of managed grassland subject to different land-use regimes in the 18th century (upper panel) and management regimes today (lower), and the temporal extension of the regimes over the growing season. Total grassland area in the 18th century, 15,6 hectares; today, 685 hectares. Aftermath grazing is depicted with three steps, symbolising that one hay meadow at a time is opened for grazing

3.1 Timing of management

Management timing refers to when the disturbance of the vegetation occurs, thereby dictating e.g. how much the vegetation is able to grow before the disturbance and which stage of the life cycle of a species that is affected by the disturbance.

3.1.1 Ecological significance of timing of disturbance

Timing of disturbance affects seed production, because it is the main factor deciding the proportion of plants that will have time to flower and set seeds before disturbance. An early onset of grazing (April, May) and which continues throughout the season restricts flowering and seed production of vascular plants (Brys et al., 2004; Lennartsson & Oostermeijer, 2001). The seeds of the majority of grassland plant species ripen between late July to mid August (Dahlström et al., 2008; Wissman, 2006). Even small adjustments of the timing of disturbance within this time window will have large effects on seed production (cf. Jantunen et al., 2007; Oostermeijer et al., 2002; Wissman, 2006), especially in cool summers, when the ripening of seeds is slow (Svensson & Carlsson, 2005).

Late disturbance on the other hand, reduces the viability of small plant species, seedlings and juvenile rosettes by forcing them to grow in tall vegetation for much of the summer (Brys et al., 2005; Lanta et al., 2009; Lennartsson & Oostermeijer, 2001). Late onset of grazing may also reduce the proportion of biomass removed during a growth season, because old vegetation to some extent is rejected by the grazers (Bailey et al., 1998).

Grassland plants cope with damage by tolerance mechanisms decreasing the negative effects of damage or avoid damage by resistance (or defence) mechanisms aiming at decreasing the degree of damage (Rosenthal & Kotanen, 1994). A common tolerance mechanism is regrowth of new flowers and branches after damage (Lennartsson et al., 1997) and is related to timing of damage (Lennartsson et al., 1998). Timing of mowing has likely caused the development of early and late flowering ecotypes. Whereas early flowering ecotypes are sensitive to damage before the seeds mature, late flowering ecotypes need early enough damage to regrow and set seeds afterwards (Simán & Lennartsson, 1998).

3.1.2 Management components affecting timing of disturbance

All treatments included in haymaking - spring raking, mowing and grazing of aftermath - occurred historically in a specific sequence from April to September, each interspaced by a shorter or longer resting period (e.g. Arrhenius & Lindquist, 1904; Ohlsson, 2006). Hay meadows were historically mown during a certain mowing period, often c. one month of length and starting in late June to mid July (Dahlström et al., 2008; Vestbö-Franzén, 2004, pp. 171-172). The process of mowing is swift and non-selective; hence the timing of disturbance within each meadow unit is roughly equal to the mowing onset of that meadow. However, the onset of management differed with about one month from the first to the last meadow. The drying of the grass has been shown to allow some immature seeds to ripen, thereby increasing the seed output similar to a later onset of mowing (Lennartsson, 2003).

Historically, hay meadow enclosures could include unproductive areas not subject to mowing, such as stony and dry areas. These areas were probably not grazed until the meadows were opened for aftermath grazing after the harvest of hay. Pastures in meadow enclosures thus experienced a later disturbance than the meadows themselves, in particular considering that it takes some time after onset of grazing until the vegetation is grazed off, Even later onset of grazing occurred in pastures enclosed with arable fields, in which grazing could not start until after the crop harvest (Fig. 2).

In permanent pastures, the grazing season in late 19th century lasted from late May/early June to late September/early October, according to ethnological sources (Dahlström et al., 2008; Israelsson, 2005, pp. 192-193). The timing of disturbance to the vegetation in such pastures is largely governed by size of enclosure and grazing intensity. In small enclosures, timing of disturbance on the patch level will be close to the timing of grazing onset, i.e. the vegetation is grazed off within a few days after the animals are let out. Larger enclosures take longer to graze off, which creates a variation in timing of disturbance between patches within the pasture (Brunsell, 2002). Also manual direction of grazing through herding and fencing affected timing of disturbance and created variation both within the pastures during the season (Kardell, 2006) and between pastures (Dahlström, 2010).

3.1.3 Historical and current timing of management in the study regions

Historically, late disturbance mainly occurred in hay meadows, but likely also in pastures fenced with arable fields. In the 18th century, about 23% of the grassland area consisted of hay meadows, i.e. late disturbance (Fig. 2). Another 3% was pastures fenced together with hay meadow or arable fields and thus likely being grazed late. This was e.g. to be found in Alseda, where every-year cultivation was employed, indicating that pastures enclosed with the arable fields were grazed very late every year. Hay meadow area has decreased by 99.7% and today only c. 2% of the currently managed grassland area is subject to late onset of management, all of which by mowing (Fig. 2).

The historical tradition of drying the hay is not generally used in today's conservation mowing (Overud & Lennartsson, 2004), but is since 2008 liable for extra subsidies and is now applied in c. 10% of the Swedish hay-meadows with special botanical values (Swedish Board of Agriculture, pers. comm.).

Seeing as timing of disturbance at the patch level in permanent pastures is largely a function of pasture size, a comparison between current and historical size distribution was made. The 600 pastures existing in the 18th century study areas ranged between 0.07 and 710 ha (mean 18.1, median 6.0 ha), whereas the current 210 pastures ranged between 0.1 and 24 ha (mean 3.2, median 1.9). Although the smallest sized pasture was found in the 18th century landscape, the higher historical mean, median and maximum values in the size distribution imply a later mean timing of disturbance in the pastures historically (Fig. 3). One source critical aspect of the current pasture size is that the database provides data on semi-natural grassland area, not size of enclosure. This dataset should thus be viewed as an indication of enclosure size.

Manual herding has been common in Sweden since at least medieval times (Kardell, 2006; Myrdal, 1999, pp. 132-135) and indications of it also exist in the study regions historically (Dahlström, 2010; Mats Bunner, herdsman's aid in the 1950's, pers. comm.). Today, full-season grazing is virtually the only management regime left (Fig. 2).

3.2 Management intensity

Disturbance intensity relates to the degree of biomass removal, which is the key factor behind a number of habitat conditions in grasslands, e.g. vegetation height, litter accumulation, gap creation and nutrient removal or reallocation. These habitat conditions in turn largely determine the competition relationships among species (Brys et al., 2005; Elisseou et al., 1995).

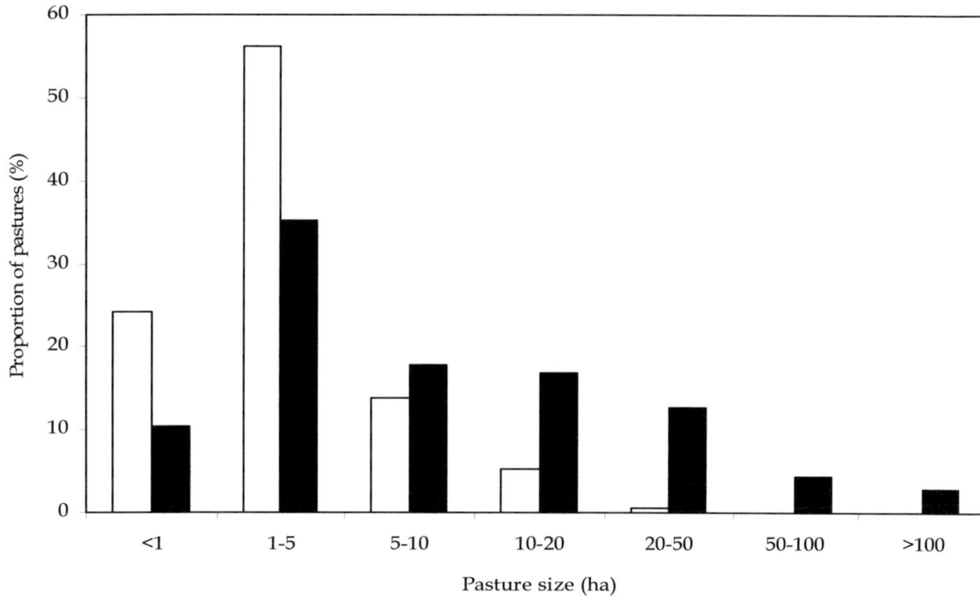

Fig. 3. Proportional size distribution of current (open bars) and 18th century (filled bars) permanent pastures in the study areas

3.2.1 Ecological significance of disturbance intensity

In general, disturbance intensity determines several aspects of damage to plant individuals, in turn essential for plant population viability, e.g. risk of being damaged, proportion of the plant removed when damaged, risk of repeated damage and risk of being damaged before completed reproduction (Lennartsson, 2002; Lennartsson & Oostermeijer, 2001). Seed production of plants decrease with increasing grazing intensity, both because of increased herbivory of reproductive organs (Bühler & Schmid, 2001; Oostermeijer et al., 2002; Pihlgren, 2007), and because of adverse effects on resistance and tolerance mechanisms (Gatehouse, 2002; Huhta et al., 2000; Huhta et al., 2003). Resistance is affected because intense grazing reduces the selectiveness of the herbivores, thereby reducing benefits of plant traits reducing palatability (Olff & Ritchie, 1998). Tolerance is affected because mechanisms such as regrowth has limited functionality under repeated damage (Del Val & Crawley, 2005; Moser & Schutz, 2006), in particular at sites where plants are stressed, e.g. by drought (Lennartsson, 2000; Oconnor, 1991).
On the other hand, the probability of successful germination and seedling establishment increases with increasing disturbance intensity creating a low vegetation, a thin litter layer and, by means of trampling and grazing, gaps in the litter and vegetation cover (Ehrlén et al., 2005; Fischer & Matthies, 1998; Kaligaric et al., 2006). Trampling can also "plant" seeds, thereby protecting them from granivory and favouring the germination in some species (Eichberg et al., 2005). Although too high trampling intensity impairs establishment by increasing seedling mortality (Oconnor, 1991; Owens & Norton, 1992), there is from a disturbance intensity point of view in general a trade-off between seed production and establishment (Pihlgren, 2007; cf. Watt & Gibson, 1988; Wissman, 2006).
Growth and survival of juvenile and adult plants are also affected by disturbance intensity as a trade-off function. High intensities decrease growth and survival because of increased

damage frequency (Bühler & Schmid, 2001), but increases growth and survival of rosettes and low plant stages because of reduced light competition (Lennartsson & Oostermeijer, 2001; Wallin, 2007).

Grazing, in opposite to mowing, forms a mosaic of shortly grazed and tall ungrazed vegetation patches (Adler et al., 2001; Bakker, 1989; Pratt et al., 1986) due to preferences of the animals towards certain plant species and communities (Huntly, 1991; Jerling & Andersson, 1982; Putman et al., 1991), avoidance of grazing close to the faeces of conspecifics (Loucougaray et al., 2004; Ritchie & Olff, 1999) and random grazing behaviour (Brunsell, 2002; Pihlgren, 2007). The proportion of ungrazed patches is important for seed production in the grassland (Pihlgren 2007) and for several groups of invertebrates (e.g. Oostermeijer et al., 2002). High grazing intensity decreases the possibility for grazers to reject patches of vegetation, thereby reducing the described mosaic.

3.2.2 Management components affecting intensity of disturbance

In general, estimates of disturbance intensity need to be related to a specific response variable. For example, mowing can be viewed as a more intense disturbance than grazing, because mowing is non-selective, usually non-mosaic and because all harvested biomass is removed, as opposed to grazing, which is selective, causes a mosaic vegetation structure and reallocates nutrients through manure and urine. On the other hand, grazing may be considered more intense than mowing if the risks of early and repeated damage to individual plants are the response variables in focus.

In hay meadows disturbance intensity is not only the result of mowing *per se*, but is affected also by the two traditional land-use components of raking and burning of debris in spring and aftermath grazing in autumn. Both reduce the litter layer and disturb the ground surface (Carlsson, 1991; Svensson & Carlsson, 2005; Wallin, 2007; Wissman, 2006) and aftermath grazing in addition implies a repeated, albeit late, damage to plants.

In historical pastures the grazing intensity in a village is difficult to estimate: data on stocking density (grazing area and number of animals) are often available but not data on the pastures' productivity. There are indications of increased grazing intensity over time, e.g. Dahlström (2006a, pp. 132-149) found stocking densities to increase during the period 1620-1850. Moreover, the proportion of individually fenced pastures increased at the cost of common outland pastures, especially in regions directed at crop production and during periods of intense land use. By the 18th century almost all outland was subdivided by fences (Kardell, 2006). This process can be seen as a sign of intensified grassland use, but is also a result of the privatisation of common land (Dahlström, 2006a, pp. 181-203). Some areas reported a shortage of grazing resources until the enclosures of meadows and arable fields in cultivation were opened for grazing in August-September (e.g. some in Kristberg, late 19th century, Dahlström, 2006a, pp. 142-144). In those villages grazing may have been intense in permanent pastures in early but less so in late summer.

Historical stocking densities have been found to differ considerably between villages, thus potentially creating a patchwork of different grazing intensities across the regions (Dahlström, 2006b).

Burning has been used historically in pastures to remove unwanted vegetation, mainly shrubs and dwarf-shrubs but to some extent also old grass (Ekstam & Forshed, 2000; Moreno & Villafuerte, 1995; Webb, 1998). Burning can be regarded as increasing the disturbance intensity in pastures, as it contributes to the removal of shrubs and litter.

3.2.3 Historical and current management intensity in the study regions

Stocking densities in 18th century Selaön and present day Östuna are not directly comparable (Fig. 4). However, in both cases stocking densities are generally higher in very small enclosures compared to the larger ones (Fig. 4). Seeing as the median pasture size has decreased by c. 68%, there is an indication of increased stocking densities since the 18th century. If pasture productivity has remained fairly constant, this would imply a higher mean grazing intensity in today's pastures compared with the 18th century.

Based on between-year variation in livestock number, it has been estimated that the average grazing intensity has been equal to consumption of maximum 60 per cent of the biomass in the early 17th century (Dahlström 2006a, pp. 179-180). This is probably less than in current pastures, because subsidiaries for semi-natural pastures are accompanied by a demand to remove most of the biomass annually (Overud & Lennartsson, 2004).

Although there is no direct evidence of the historical use of spring raking and aftermath grazing in the study regions, both are likely to have occurred, having been commonly applied components of the historical mowing regimes in Sweden (e.g. Lithberg, 1934; Sjörs, 1954, pp. 16-18; von Linné, 1741, August 12th). It is not known if those components occur in the current hay-meadows, but neither is common in meadows in Sweden as a whole (Wallin, 2007). In the regulations for subsidies for hay-meadow management in Sweden, aftermath grazing is not compulsory, but merits extra payment, and spring raking "should be done when needed" (Swedish Board of Agriculture, 2007).

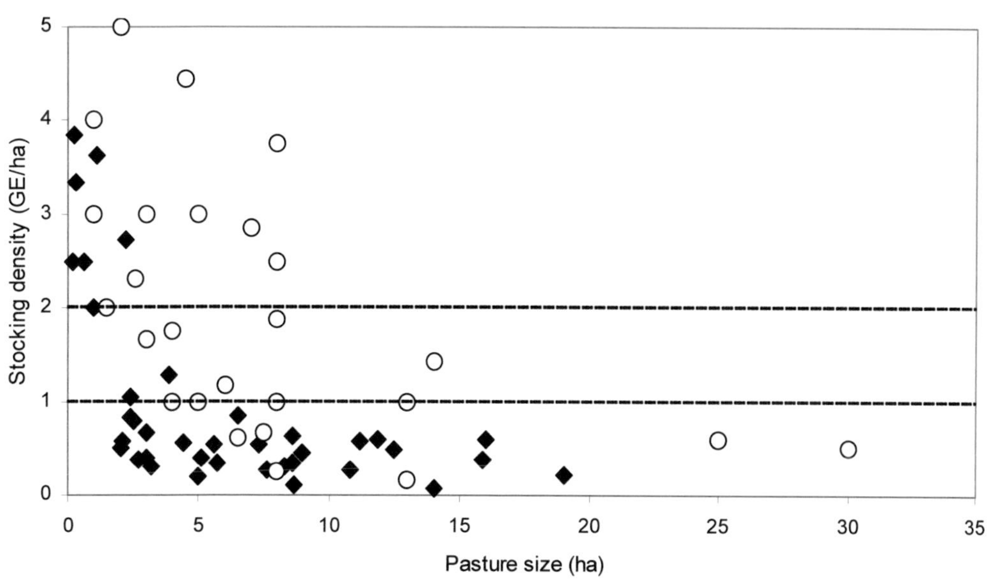

Fig. 4. Eighteenth century stocking densities (grazing equivalents, GE, per hectare) at Selaön (◆), and current stocking density in Östuna (○). Each data point represents one pasture. The dashed lines indicate current stocking recommendations of non-lactating cows, when pastures are grazed for conservation; mesic pastures above and dry below (Höök-Patriksson, 1998)

3.3 Dynamics of management

Dynamics of management is here defined as the between-season variation in timing, intensity and/or type of land-use/management.

3.3.1 Ecological significance of dynamics of disturbance

By varying the two components timing and intensity of disturbance the balance in the previously discussed trade-offs can be shifted, namely those between seed production vs. germination (Lennartsson & Oostermeijer, 2001) and between growth of small, shade-sensitive plants and life-stages vs. their survival in very intensively grazed pastures (Helldin & Lennartsson, 2006). Specifically, in years of late, weak or more selective disturbance the survival of plants as well as the seed production would be favoured, whereas in years of intense disturbance seedling establishment and growth of shade-sensitive plants would instead be favoured. The expected net effect on plant populations should be that variable disturbance intensity gives higher population growth rates than either intense or weak disturbance (cf. Lennartsson & Oostermeijer, 2001). By changing every-year grazing to every-second-year grazing, seed production increased by 7-14 times and seedling recruitment by 3-4 times (Wissman, 2006). This partly confirms the expected trade-off shift, but indicates also that grasslands are more limited in seed availability than in suitable germination sites (Eriksson & Ehrlén, 1992; Turnbull et al., 2000). Similarly, a variable grazing intensity may affect the vegetation differently than a constant grazing intensity, even if the mean intensity is equal. For example, in years with intense grazing also unpalatable species, which in the long-run may become dominants, may be grazed and reduced in abundance (Linkowski & Lennartsson, 2006). Conversely, also palatable species may reproduce in years with weak grazing pressure.

Alternation between grazing and haymaking in two-field systems between years can be expected to create special conditions for grassland plants. The meadow year would favour seed production and growth of good competitors, whereas the pasture year would favour germination, seedling establishment and growth of weak competitors and grazing-resistant species.

Management dynamics also include periods without management. Shorter periods of abandonment, i.e. a few years, affect plants mainly by favouring seed production as well as species and life stages that do not suffer from tall vegetation or increasing litter thickness. During longer periods without management, however, succession processes start, mainly leading to increased abundances of competitive species at the cost of low-competitive ones (Baur et al., 2006; Helldin & Lennartsson, 2006; Van Andel et al., 1993).

3.3.2 Management components affecting dynamics of disturbance

A considerable amount of randomness in the dynamics of disturbance is due to natural variations in a number of grassland habitat parameters. For example, weather conditions affect the timing of mowing between years and cause variations in grassland productivity. In pastures this variation is today counteracted by adjusting the number of livestock (Höök Patriksson, 1998, p. 71). More or less random variations in grazing intensity has been shown for early 17th century, due to considerable variation in local livestock numbers, caused by varying availability of summer fodder during the previous season and winter fodder during the current and also depending on natural demographic fluctuations (Dahlström, 2006a, pp. 96-102; Dahlström, 2006b).

Historically, large variations appear to have existed due to alternating land-use types, including periods without management. For example, Vestbö-Franzén (2004, chapter 7) compared 17th century maps of the same villages, but drawn with different levels of detail. Some hay meadows that in one map were drawn as continuous meadow, were in a contemporary map drawn as a patchwork of hay meadows, grazed areas, temporary arable fields and areas of slash-and-burn. This indicates a more flexible use of the meadows than is generally depicted in historical map material, even indicating that parts of the hay meadows were systematically left to succession and then reopened after a period of abandonment (Vestbö-Franzén, 2004). This is consistent with Sjöbecks (1933) description of the early 20th century use of wooded hay meadows in south Sweden.

More regular dynamics, indicated in many maps, is created for example by the rhythm of grazing in the widespread two-field system (Ekstam & Forshed, 1996). The majority of a property's arable fields were divided into two areas, each enclosed by a fence and each comprising one year's worth of crop production. Each enclosure was used for crop production every second year; the one in fallow was instead used for grazing (Gadd, 2000, p. 115). A varying amount of pasture could also be enclosed with the arable fields and were thus in a two-year land-use rhythm. Hay meadows included in the two-field rhythm were common in the region around Källstorp/Söne (Jansson, 1998, p. 121). They were treated as hay meadow during the year in cultivation (spring raking, mowing and aftermath grazing) and as pasture (full-season grazing) during the year in fallow. Herding within the arable field enclosures was common in the lowlands (Kardell, 2006), thus in some cases counteracting the described two-year rhythms of haymaking and grazing.

3.3.3 Historical and current management dynamics in the study regions

Historically, about 3% of the grassland area was located in two-field enclosures, indicating that grazing and, to a small extent, mowing were subject to a two-year rhythm (Fig. 2). Three variants of such land-use regimes were identified for the study regions historically (Fig. 2). Today, there is no systematically dynamic management regime in the study area.

Pastures being smaller today than in the 18th century (Fig. 3) imply that the variation at the patch level between years is smaller, especially because it is likely that grazing intensity is higher today. The variation in grassland productivity today is, to a larger extent than historically, counteracted by varying the stocking densities or by mechanical trimming of ungrazed patches of vegetation (Overud & Lennartsson, 2004). The trimming ensures that the farmer fulfils the management goal of the vegetation being "well grazed", thereby avoiding sanctions (Swedish Board of Agriculture, 2007).

Today there are many abandoned grasslands that are in succession at a stage corresponding to the historical abandonment periods (Sjöbeck, 1933; Vestbö-Franzén, 2004). Today the process of succession is rarely broken by restoration (Swedish University of Agricultural Sciences, 2010). Recently the Swedish regulations for environmental subsidies introduced the possibility of one grazing free year out of five (Swedish Board of Agriculture, 2007), which opens up for the possibility of introducing a certain amount of short-term variation. There is as yet no national statistics on how widely it is applied (Swedish Board of Agriculture, pers. comm.). Longer periods of abandonment or shorter intervals between grazing free years are not liable for environmental subsidies.

4. Discussion

This study confirms the loss of area of semi-natural grassland described from other parts of Scandinavia and Europe. It furthermore indicates a loss of habitat quality. The pre-industrial agricultural systems contained a range of land-use regimes and components many of which are absent or rare today: the large overall abundance and distribution of grassland during hundreds of years (e.g. Eriksson et al., 2002), the presence of many different grassland types in every agricultural landscape, the presence of land-use types which form disturbance regimes that are particularly beneficial for biodiversity (this study), the long continuity of specific land-use types and combinations of types within single grasslands (Fischer & Wipf, 2002; Gustavsson et al., 2007), and the dynamic land-use regimes (this study; Dahlström, 2006a,b; Sjöbeck, 1933; Vestbö-Franzén, 2004). Several of these components can be expected to be ecologically necessary, as they provided ecological conditions necessary for grassland biodiversity. Such change of habitat conditions may well explain the reported cases of decline of grassland biodiversity in semi-natural grassland subject to conservation management.

Timing of management has changed, in particular in terms of loss of late disturbance. Today, most of the remaining grasslands are managed with grazing during the entire season irrespective of historical land-use regime. The loss is caused by loss of mowing and of late grazing in pastures fenced with meadow or arable field; both losses imply a loss of late disturbance at the enclosure level. In addition, drastically decreased median pasture size and indications of increased stocking densities imply an earlier mean disturbance at the patch level because the vegetation is grazed faster.

As indicated by the literature review performed in this study, as well as by several empirical studies, the loss of late managed grassland can be expected to have large effects on grassland biodiversity (Linusson et al., 1998; Matìjková et al., 2003; Mitlacher et al., 2002). Plants are affected mainly through reduced flowering and seed production, but increased probability of a seed establishing a new plant, especially in small and low-competitive species (Jutila, 2003; Jutila & Grace, 2002; Lennartsson & Oostermeijer, 2001). Empirical studies have suggested that the recruitment of grassland plants is limited rather by seed production than by germination sites (Eriksson & Ehrlén, 1992; Turnbull et al., 2000; Wissman, 2006). Empirical studies have also demonstrated increasing densities of plant individuals in the sward when late disturbance was introduced (Wissman, 2006), and an increased content of forbs in the vegetation (Pavlů et al., 2006).

Early-flowering plant species, which have historically been favoured by late management, can be expected to be particularly negatively affected by the loss of late disturbance. This is the case also for early-flowering ecotypes which may have been evolved in man-made habitats along with the use of mowing and other late grassland use (Karlsson, 1984; Lennartsson, 1997 and references therein).

Subsequently, the effects of management timing on plants strongly affect a number of grassland invertebrate groups, in particular phytophagous, nectar- and pollen feeding insects (Kruess & Tscharntke, 2002; Söderström et al., 2001). Reduced pollen- and nectar resources have been suggested to be an important threat to, for example, wild bees, of which several species are red listed (Linkowski et al., 2004; Pekkarinen, 1998; Pekkarinen, 1999). For phytophagous insects, timing of disturbance to their host plant is crucial, especially for sedentary larvae that are killed when the plant is eaten or cut (Johst et al., 2006; Valtonen & Saarinen, 2005). Moreover, different groups of ground-dwelling, predatory arthropods are favoured by either early or late onset of grazing due to the differences in vegetation structure (Lenoir & Lennartsson, 2010).

Intensity of management appears to have increased from the 18th century until today due to earlier management in pastures and hay-meadows, smaller pastures, and probably increased stocking densities. In the remaining hay-meadows, however, some changes of management, namely reduced use of spring raking and aftermath grazing, may however have reduced the intensity. Although current grazing intensity probably is higher than the historical, it is sometimes argued that management intensity at the landscape level has decreased because of large areas of abandoned grasslands (cf. Helldin & Lennartsson, 2006). This may be true during a succession phase, but in the long run too weak management (or none) in some grasslands cannot compensate for too intense management in others.

Some ecological effects of intensification of management are similar to the effects of managing grasslands earlier, such as reduced probability of finishing reproduction before disturbance, reduced resources of pollen, nectar, and host-plants for invertebrates, better conditions for shade-sensitive species of plants and invertebrates which occur close to the ground. Intensification of grazing also increases mortality of plants induced by trampling and repeated grazing, reduces the selectivity and patchiness of grazing, and reduces the litter layer. All these effects can be expected to change the species composition of grassland plants and invertebrates, and to threaten certain species.

Dynamics of management has decreased considerably in terms of between-year variation of timing, intensity, and type of management. Some of the historical long-term grassland dynamics, long enough for succession processes to start, can be said to be temporarily imitated by the currently abandoned successional habitats. The loss of dynamics can be expected to change the habitat conditions in single grasslands because sequences of different land use are replaced by more uniform management, and to decrease the variation between grasslands because the uniformity applies to most grasslands in a landscape.

Pre-industrial grassland use provided both predictable dynamics in the form of sequences of different land-use regimes and unpredictable variation in the form of, for example, varying number of livestock combined with between-year variations in productivity. One obvious ecological effect of this is that historical land use formed a mosaic landscape; although conditions could be less favourable for a grassland species in a certain place or at a certain time, more benign conditions were seldom far away, in space or in time. For mobile organisms like insects, dynamic management implies that suitable vegetation structures are always available (e.g. Brown, 1984; Cameron & Bryant, 1999). The review in this study indicates that between-year variation in grassland management can also favour plant diversity in a single grassland, through favouring of some life-stages one year and other stages the next year.

4.1 Implications for conservation of grassland biodiversity

Acknowledging the past and ongoing loss of semi-natural grassland ecosystems, the European Common Agricultural Policy (CAP) aims at increasing the area of semi-natural grassland by awarding environmental subsidies for management and restoration (Swedish Board of Agriculture, 2007). Although the area of semi-natural grassland is thus given high priority in the CAP, the subsidiary systems also contain regulations in order to obtain sufficient ecological quality in the remaining grasslands. For example, an awareness of the differences between grassland types in terms of threats, values and management requirements has motivated different national schemes for the application of the CAP.

An important question is, however, whether the quality aspect of grassland management is enough acknowledged and implemented in grassland conservation within the CAP and in other conservation efforts, such as national protection of nature. For example, the Swedish application of the CAP has been criticized for intensifying and homogenising grassland management through being concerned mainly with regulations for keeping the litter layer thin, producing a short grass sward, and minimising shrub cover (Höök-Patriksson, 1998; Overud & Lennartsson, 2004), and through excluding several ecologically important and historically widespread grassland types and management components (Andersson & Paltto, 2008).

Based on this study, we make some suggestions for future conservation of grassland biodiversity:

Always question whether management for conservation in a particular grassland is ecologically similar enough to the historical land use to be considered "continued" or "resumed" management and thus whether the management can be expected to preserve the grassland's biodiversity. Some of the habitats discussed in this study have experienced such a profound shift of disturbance regime that they should probably be considered new habitats; examples are hay-meadows and late grazed pastures, now managed with full-season grazing every summer. Such grasslands should in many cases probably be disregarded in area estimates of semi-natural grassland that aim at estimating functional, or effective, area of grassland, analogous to, for example, effective population size.

1. Apply and develop further the interdisciplinary approach employed in this study, in which detailed ecological and historical knowledge are combined in order to identify which ecological variables that are most crucial for grassland species, both in general and for the current species composition in specific grasslands, and what historical land-use components that created these factors. Grassland ecology and land use can preferably be interpreted in terms of type, timing, intensity, and dynamics of disturbance, as in this study, but also other aspects of the grassland ecology may be important in certain grassland habitats, such as temporary cultivation and the use of trees and shrubs. In this study vascular plants were used as target organism group, but in order to preserve the entire grassland biodiversity it is important to perform similar studies also for other groups.
2. Apply knowledge-based, adaptive grassland management from the identified links between species, habitat and components of the historical disturbance regimes. Several studies of plant population viability and plant diversity have indicated the benefits of local, traditional management compared to new or modified methods of grassland management (Fischer & Wipf, 2002; Hansson & Fogelfors, 2000; Köhler et al., 2005; Maurer et al., 2006). Restoring habitat quality by reintroducing the place-specific historical regime is however not always possible. For example, manual mowing is often too labour intensive, and all the land-use variations that existed in the pre-industrial landscape cannot be introduced into the small habitat fragments managed today. Instead, some missing ecological variables must be created by applying novel management methods, i.e. ecological engineering, thus functionally mimicking the historical land-use components (Lennartsson, 2003; Lennartsson & Linkowski, 2011). Non-historical management has in some cases been shown to function equally well as historical from a biodiversity point of view (e.g. Ruzickova et al., 2001), such as historical mowing being replaced by late onset of grazing (Wissman 2006).
3. Identify and prioritise in the CAP such regions in which traditional land use is still practiced and in which the grassland ecosystems can thus be expected to be ecologically functional. One of the most striking examples is the Romanian Carpathians which contains the largest areas of traditionally mown hay-meadows in Europe (Lennartsson & Helldin, 2007). It is essential that the CAP regulations are not resulting in

deterioration of the land-use methods in such regions, but instead are designed based on the knowledge of grassland ecology that can be derived from these land-use systems.

5. Conclusions

In conclusion, in the pre-industrial agricultural landscape, suboptimal management in some grasslands and during some time periods was probably not a problem because the vast grassland areas created a variety of management regimes and could support viable metapopulations of grassland species. In the current landscape, few species exhibit functioning metapopulations, hence restoration and management measures need to ensure sufficient habitat quality more or less in each single grassland. We therefore urge for knowledge-based management for conservation and regulation of environmental subsidies. This study indicates the need to focus on restoration of grassland quality. It also shows that more multi-disciplinary research is needed that combines historical, ecological and agricultural knowledge. Although restoration of habitat quality should thus be based on ecological-historical analyses of the grassland habitats and their species, the restoration methods may well be based on novel tools as long as the new methods provide all necessary ecological variables. Seeing as grasslands owe their species richness to past land-use practices, studying the components of historical land use in greater detail and in an ecological context may reveal not only new details of grassland ecology as such. It may also contribute to significant knowledge also of the ecology of the agricultural landscape in general, thereby providing keys as to the development of new conservation tools and strategies.

6. Acknowledgement

This study was funded by The Bank of Sweden Tercentenary Foundation and The Royal Academy of Letters, History and Antiquities through Nordiska Museet Research School, The Foundation for Strategic Environmental Research (HagmarksMISTRA), the Swedish Council for Forestry and Agricultural Research (award 34.0297/98 to T. Lennartsson) and the Swedish National Heritage Board (Biocultural Heritage).

7. References

Adler, P.B., Raff, D.A. & Lauenroth, W.K. (2001). The effect of grazing on the spatial heterogeneity of vegetation. *Oecologia*, Vol. 128, pp. 465-479

Andersson, L. & Paltto, H. (2008), *Hur har natur- och kulturvärden påverkats av röjning/avverkning i betesmarker?*, Pro Natura. http://www.pro-natura.net/Rapport%20-%20uppf%F6ljning%20av%2050-tr%E4dsregeln%20-%20slutversion.pdf

Arrhenius, J. & Lindquist, C. A. (1904). *Lantbrukspraktika* Beijers Bokförlags-Aktiebolag, Stockholm.

Aurell, J. J. Söne socken, storskifte. [16-SÖN-1g]. 1787. Mariestad, Cadastral map, Swedish Land Survey Agency.

Bailey, D. W., Dumont, B., & WallisDeVries, M. F. (1998). Utilization of heterogeneous grasslands by domestic herbivores: Theory to management. *Annales de Zootechnie,* Vol 47, pp. 321-333.

Baillie, J. E. M. (2004). *2004 IUCN Red List of Threatened Species: A Global Species Assessment* International Union for the Conservation of Nature, 2831708265.

Bakker, J. P. (1989). *Nature management by grazing and cutting. On the ecological significance of grazing and cutting regimes applied to restore former species-rich grassland communities in the Netherlands.* Kluwer Academic Press, 0792300688, Dordrecht, Netherlands.

Baur, B., Cremene, C., Groza, G., Rakosy, L., Schileyko, A. A., Baur, A., Stoll, P., & Erhardt, A. (2006). Effects of abandonment of subalpine hay meadows on plant and invertebrate diversity in Transylvania, Romania. *Biological Conservation,* Vol. 132, pp. 261-273.

Bernes, C. (1993). *The Nordic Environment - Present State, Trends and Threats.* The Nordic Council of Ministers, Stockholm.

Boschi, C. & Baur, B. (2007). Effects of management intensity on land snails in Swiss nutrient-poor pastures. *Agriculture Ecosystems & Environment,* Vol. 120, pp. 243-249.

Brown, V. K. (1984). Secondary succession: Insect-plant relationships. *BioScience* 34, 710-716.

Brunsell, M. (2002), *Betesrator och kärlväxters reproduktion - En studie av naturbetesmarkers heterogenitet i tid och rum. (Grazing rejects and the reproduction of vascular plants - A study on the spatial and temporal heterogeneity in semi-natural grasslands, in Swedish),* Department of Conservation Biology, Swedish University of Agricultural Sciences, Uppsala, 79.

Bruun, H. H., Fritzbøger, B., Rindel, P. O., & Hansen, U. L. (2001). Plant species richness in grasslands: the relative importance of contemporary environment and land-use history since the Iron Age. *Ecography,* Vol. 24, pp. 569-578.

Brys, R., Jacquemyn, H., Endels, P., De Blust, G., & Hermy, M. (2004). The effects of grassland management on plant performance and demography in the perennial herb Primula veris. *Journal of Applied Ecology,* Vol. 41, pp. 1080-1091.

Brys, R., Jacquemyn, H., Endels, P., De Blust, G., & Hermy, M. (2005). Effect of habitat deterioration on population dynamics and extinction risks in a previously common perennial. *Conservation Biology,* Vol. 19, pp. 1633-1643.

Bühler, C. & Schmid, B. (2001). The influence of management regime and altitude on the population structure of Succisa pratensis: implications for vegetation monitoring. *Journal of Applied Ecology,* Vol. 38, pp. 689-698.

Cameron, G. N. & Bryant, E. H. (1999). Species composition and trophic structure of insect communities in Texas prairies. *Prairie Naturalist,* Vol. 31, pp. 221-242.

Carlsson, Å. (1991). Kransrams - västsvensk fagningsblomma. *Svensk botanisk tidskrift,* Vol. 85, p. 81.

Cheffings, C. M. & Farrel, L. (2005). *The Vascular Plant Red Data List for Great Britain - Species Status No 7* Joint Nature Conservation Committee, 2831708265.

Dahlström, A. (2006a). *Betesmarker, djurantal och betestryck 1620-1850. Naturvårdsaspekter på historisk beteshävd i Syd- och Mellansverige. (Pastures, livestock number and grazing pressure 1620-1850. Ecological aspects of grazing history in south-central Sweden. In Swedish with English summary)* Swedish Biodiversity Centre, Uppsala.

Dahlström, A. (2006b). Grazing dynamics at different spatial and temporal scales: examples from the Swedish historical record a.d. 1620–1850. *Vegetation History and Archaeobotany,* Online First, DOI 10.1007/s00334-006-0087-1.

Dahlström, A. (2010). Markanvändningsdynamik - rekonstruktion med hjälp av bondedagböcker och historiska kartor. In *Nycklar till kunskap. Om människans bruk av naturen.* eds. H. Tunón & A. Dahlström, Centrum för biologisk mångfald, Uppsala & Kungl. Skogs- och lantbruksakademien, Stockholm.

Dahlström, A., Lennartsson, T., Wissman, J., & Frycklund, I. (2008). Biodiversity and traditional land use in south-central Sweden - the significance of timing of management. *Environment and History*, Vol. 14, pp. 385-403.

Del Val, E. K. & Crawley, M. J. (2005). Are grazing increaser species better tolerators than decreasers? An experimental assessment of defoliation tolerance in eight British grassland species. *Journal of Ecology*, Vol. 93, pp. 1005-1016.

Ehrlén, J., Syrjänen, K., Leimu, R., Garcia, M. B., & Lehtilä, K. (2005). Land use and population growth of Primula veris: an experimental demographic approach. *Journal of Applied Ecology*, Vol. 42, pp. 317-326.

Eichberg, C., Storm, C., & Schwabe, A. (2005). Epizoochorous and post-dispersal processes in a rare plant species: Jurinea cyanoides (L.) Rchb. (Asteraceae). *Flora*, Vol. 200, pp. 477-489.

Ekstam, U. & Forshed, N. (1996). *Äldre fodermarker. Betydelsen av hävdregimen i det förgångna. Målstyrning. Mätning och uppföljning.* Naturvårdsverkets förlag, 91-620-1147-2, Stockholm.

Ekstam, U. & Forshed, N. (2000). *Svenska Naturbetesmarker, historia och ekologi.* Naturvårdsverkets förlag, 91-620-1202-9, Stockholm.

Elisseou, G. C., Veresoglou, D. S., & Mamolos, A. P. (1995). Vegetation productivity and diversity of acid grasslands in Northern Greece as influenced by winter rainfall and limiting nutrients. *Acta Oecologica-International Journal of Ecology*, Vol. 16, pp. 687-702.

Emanuelsson, U. (1988). A model describing the development of the agricultural landscape. In *The Cultural Landscape - Past, Present and Future*, eds. H. H. Birks et al., pp. 111-122. Cambridge University Press, Bergen.

Eriksson, O., Cousins, S. A. O., & Bruun, H. H. (2002). Land-use history and fragmentation of traditionally managed grasslands in Scandinavia. *Journal of Vegetation Science*, Vol. 13, pp. 743-748.

Eriksson, O. & Ehrlén, J. (1992). Seed and Microsite Limitation of Recruitment in Plant-Populations. *Oecologia*, Vol. 91, pp. 360-364.

European Union. (2011). *Common Agricultural Policy (CAP) in the European Union*, 3-18-2011, Available from http://ec.europa.eu/agriculture/capreform/index_en.htm

Fischer, M. & Matthies, D. (1998). Experimental demography of the rare Gentianella germanica: seed bank formation and microsite effects on seedling establishment. *Ecography*, Vol. 21, pp. 269-278.

Fischer, M. & Wipf, S. (2002). Effect of low-intensity grazing on the species-rich vegetation of traditionally mown subalpine meadows. *Biological Conservation*, Vol. 104, pp. 1-11.

Gadd, C.-J. (2000). *Den agrara revolutionen. 1700-1870.* Natur och Kultur/LTs förlag, 91-27-35222-6.

Gärdenfors, U. (2010). *Rödlistade arter i Sverige 2010 - The 2010 redlist of Swedish species* ArtDatabanken, 978-91-88506-35-1, Uppsala.

Gatehouse, J. A. (2002). Plant resistance towards insect herbivores: a dynamic interaction. *New Phytologist*, Vol. 156, pp. 145-169.

Grigg, D. B. (1974). *The agricultural systems of the world: an evolutionary approach* Cambridge U.P., London.

Gustavsson, E., Lennartsson, T., & Emanuelsson, M. (2007). Land use more than 200 years ago explains current grassland plant diversity in a Swedish agricultural landscape. *Biological Conservation*, Vol. 138, pp. 47-59.

Hansson, M. & Fogelfors, H. (2000). Management of a semi-natural grassland; results from a 15-year-old experiment in southern Sweden. *Journal of Vegetation Science,* Vol. 11, pp. 31-38.

Helldin, J.-O. & Lennartsson, T. (2006). Lagom är bäst - ett måttligt betestryck ger den största mångfalden (A moderate grazing pressure favours the greatest range of biodiversity, in Swedish). *Biodiverse*, Vol. 4, pp. 17-19.

Hierpe, C. Källstorps säteri, geometrisk avmätning. [P266-6:1]. 1712. Gävle, Swedish Land Survey Agency. Ref Type: Map

Höök Patriksson, K. (1998). *Skötselhandbok för gårdens natur- och kulturvärden.* Swedish Board of Agriculture, 9188264-20-3, Jönköping.

Huhta, A. P., Hellström, K., Rautio, P., & Tuomi, J. (2003). Grazing tolerance of Gentianella amarella and other monocarpic herbs: why is tolerance highest at low damage levels? *Plant Ecology*, Vol. 166, pp. 49-61.

Huhta, A. P., Lennartsson, T., Tuomi, J., Rautio, P., & Laine, K. (2000). Tolerance of Gentianella campestris in relation to damage intensity: an interplay between apical dominance and herbivory. *Evolutionary Ecology*, Vol. 14, pp. 373-392.

Huntly, N. (1991). Herbivores and the dynamics of communities and ecosystems. *Annual Review of Ecology and Systematics*, Vol. 22, pp. 477-503.

Israelsson, C. (2005). *Kor och människor. Nötkreatursskötsel och besättningsstorlekar på torp och herrgårdar 1850-1914,* Doctoral thesis, Department of Economics, Section of Agricultural History, Swedish University of Agricultural Sciences, Uppsala.

Jansson, U. (1998). *Odlingssystem i Väneromrädet. En studie av tidigmodernt jordbruk i Västsverige.*, Meddelanden Nr 103, Kulturgeografiska institutionen, Stockholms universitet.

Jantunen, J., Saarinen, K., Valtonen, A., & Saarnio, S. (2007). Flowering and seed production success along roads with different mowing regimes. *Applied Vegetation Science*, Vol. 10, pp. 285-292.

Jerling, L. & Andersson, M. (1982). Effects of selective grazing by cattle on the reproduction of Plantago maritima. *Holarctic Ecology*, Vol. 5, pp. 405-411.

Johst, K., Drechsler, M., Thomas, J., & Settele, J. (2006). Influence of mowing on the persistence of two endangered large blue butterfly species. *Journal of Applied Ecology*, Vol. 43, pp. 333-342.

Jutila, H. M. (2003). Germination in Baltic coastal wetland meadows: similarities and differences between vegetation and seed bank. *Plant Ecology*, Vol. 166, pp. 275-293.

Jutila, H. M. & Grace, J. B. (2002). Effects of disturbance on germination and seedling establishment in a coastal prairie grassland: a test of the competitive release hypothesis. *Journal of Ecology*, Vol. 90, pp. 291-302.

Kain, R. J. P. & Baigent, E. (1992). *The cadastral map in the service of the state: a history of property mapping* The University of Chicago Press, London.

Kaligaric, M., Skornik, S., Ivancic, A., Rebeusek, F., Sternberg, M., Kramberger, B., & Sencic, L. (2006). Germination and survival of endangered Pulsatilla grandis (ranunculaceae) after artificial seeding, as affected by various disturbances. *Israel Journal of Plant Sciences*, Vol. 54, pp. 9-17.

Kardell, Ö. (2006). Vallning, bete, mjölkning och hägnader kring sekelskiftet 1900: en kontextuell skiss. *Svenska Landsmål och Svenskt Folkliv B* 129, 49-77.

Karlsson, T. (1984). Early-flowering taxa of Euphrasia (Scrophulariaceae) on Gotland, Sweden. *Nordic Journal of Botany*, Vol. 4, pp. 303-326.

Kleijn, D. & Sutherland, W. J. (2003). How effective are European agri-environment schemes in conserving and promoting biodiversity? *Journal of Applied Ecology*, Vol. 40, pp. 947-969.

Köhler, B., Gigon, A., Edwards, P. J., Krüsi, B., Langenauer, R., Lüscher, A., & Ryser, P. (2005). Changes in the species composition and conservation value of limestone grasslands in Northern Switzerland after 22 years of contrasting managements. *Perspectives in Plant Ecology Evolution and Systematics*, Vol. 7, pp. 51-67.

Kruess, A. & Tscharntke, T. (2002). Grazing intensity and the diversity of grasshoppers, butterflies, and trap-nesting bees and wasps. *Conservation Biology*, Vol. 16, pp. 1570-1580.

Lanta, V., Dolezal, J., Lantova, P., Kelisek, J., & Mudrak, O. (2009). Effects of pasture management and fertilizer regimes on botanical changes in species-rich mountain calcareous grassland in Central Europe. *Grass and Forage Science*, Vol. 64, pp. 443-453.

Lennartsson, T. (1997). *Demography, reproductive biology and adaptive traits in Gentianella campestris and G. amarella. - Evaluating grassland management for conservation by using indicator plant species*, Doctoral thesis, Acta Universitatis Agriculturae Suecicae - Agraria 46.

Lennartsson, T. (2000). Management and population viability of the pasture plant Gentianella campestris: the role of interactions between habitat factors. *Ecological Bulletines*, Vol. 48, pp. 111-121.

Lennartsson, T. (2002). Extinction thresholds and disrupted plant-pollinator interactions in fragmented plant populations. *Ecology*, Vol. 83, pp. 3060-3072.

Lennartsson, T. (2003). Vilken traditionell kunskap är relevant för naturvården? - några svar från fältgentianans horisont. (Which traditional knowledge is relevant in conservation? - some answers from the horizon of the field gentian, in Swedish). *Biodiverse* 8, 7.

Lennartsson, T. & Helldin, J.-O. (2007). Agricultural landscapes in Eastern Europe as reference areas for Swedish land management. *Kungl.Skogs- och Lantbruksakademiens tidskrift*, Vol. 5, pp. 26-30.

Lennartsson, T. & Linkowski, W. Markanvändningshistoria och biologisk mångfald. Almstedt, M. and Ebenhard, T. *Naturvårdskedjans avnämarbok* . 2011. Uppsala, Swedish Biodiversity Centre. Ref Type: In Press

Lennartsson, T., Nilsson, P., & Tuomi, J. (1998). Induction of overcompensation in the field gentian, Gentianella campestris. *Ecology*, Vol. 79, pp. 1061-1072.

Lennartsson, T. & Oostermeijer, J. G. B. (2001). Demographic variation and population viability in Gentianella campestris: effects of grassland management and environmental stochasticity. *Journal of Ecology*, Vol. 89, pp. 451-463.

Lennartsson, T., Tuomi, J., & Nilsson, P. (1997). Evidence for an evolutionary history of overcompensation in the grassland biennial Gentianella campestris (Gentianaceae). *American Naturalist*, Vol. 149, pp. 1147-1155.

Lenoir, L. l. l. s. s. & Lennartsson, T. (2010). Effects of timing of grazing on arthropod communities in semi-natural grasslands. *Journal of Insect Science (Tucson)* , Vol. 10.

Linkowski, W., Cederberg, B., & Nilsson, L. A. (2004). *Vildbin och fragmentering* Swedish Board of Agriculture, Jönköping.

Linkowski, W. & Lennartsson, T. (2006), *Renbete och biologisk mångfald. Kunskapssammanställning. (Biodiversity in subalpine birch forest. Literature review. In Swedish.)*, Länsstyrelsen i Norrbottens län, 18/2006.

Linusson, A.-C., Berlin, G. A. I., & Olsson, E. G. A. (1998). Reduced community diversity in semi-natural meadows in southern Sweden, 1965-1990. *Plant Ecology*, Vol. 136, pp. 77-94.

Lithberg, N. (1934). Gotlandsängen. *Sveriges Naturskyddsförenings årsskrift,* pp. 92-102.

Loucougaray, G., Bonis, A., & Bouzillé, J.-B. (2004). Effects of grazing by horses and/or cattle on the diversity of coastal grasslands in western France. *Biological Conservation*, Vol. 116, pp. 59-71.

Matìjková, I., van Diggelen, R., & Prach, K. (2003). An attempt to restore a central European species-rich mountain grassland through grazing. *Applied Vegetation Science*, Vol. 6, pp. 161-168.

Maurer, K., Weyand, A., Fischer, M., & Stocklin, J. (2006). Old cultural traditions, in addition to land use and topography, are shaping plant diversity of grasslands in the Alps. *Biological Conservation*, Vol. 130, pp. 438-446.

Mitlacher, K., Poschlod, P., Rosén, E., & Bakker, J. P. (2002). Restoration of wooded meadows - a comparative analysis along a chronosequence on Öland (Sweden). *Applied Vegetation Science*, Vol. 5, pp. 63-73.

Moreno, S. & Villafuerte, R. (1995). Traditional Management of Scrubland for the Conservation of Rabbits Oryctolagus-Cuniculus and Their Predators in Donana-National-Park, Spain. *Biological Conservation*, Vol. 73, pp. 81-85.

Moser, B. & Schutz, M. (2006). Tolerance of understory plants subject to herbivory by roe deer. *Oikos*, Vol. 114, pp. 311-321.

Myrdal, J. (1999). *Jordbruket under feodalismen. 1000-1700* Natur och Kultur/LTs Förlag, 91-27-35221-8, Stockholm.

Myrdal, J. (2001), *Den nya produktionen - det nya uppdraget: jordbrukets framtid i ett historiskt perspektiv*, Jordbruksdepartementet, Regeringskansliet, Stockholm, 2001:68.

National Atlas of Sweden (1995). *Climate, lakes and rivers.* SNA Publ., Stockholm.

Öckinger, E., Eriksson, A. K., & Smith, H. G. (2006). Effects of grassland abandonment, restoration and management on butterflies and vascular plants. *Biological Conservation*, Vol. 133, pp. 291-300.

Oconnor, T. G. (1991). Local Extinction in Perennial Grasslands - A Life-History Approach. *American Naturalist*, Vol. 137, pp. 753-773.

Ohlsson, E. W. (2006). *Det gotländska änget* Ödin, Visby.

Olff, H. & Ritchie, M. E. (1998). Effects of herbivores on grassland plant diversity. *Trends in Ecology & Evolution*, Vol. 13, pp. 261-265.

Oostermeijer, J. G. B., Luijten, S. H., Ellis-Adam, A. C., & den Nijs, J. C. M. (2002). Future prospects for the rare, late-flowering Gentianella germanica and Gentianopsis ciliata in Dutch nutrient-poor calcareous grasslands. *Biological Conservation*, Vol. 104, pp. 339-350.

Overud, S. & Lennartsson, T. (2004), *Skötsel och restaurering av betesmarker och slåtterängar. En sammanställning av den regionala naturvårdens kunskaper och erfarenheter. (Management and restoration of seminatural grasslands. A review of practical conservation knowledge. In Swedish).*, Swedish Board of Agriculture, Swedish Environmental Protection Agency, The National Heritage Board, Stockholm, No. 2004:11.

Owens, M. K. & Norton, B. E. (1992). Interactions of Grazing and Plant-Protection on Basin Big Sagebrush (Artemisia-Tridentata Ssp Tridentata) Seedling Survival. *Journal of Range Management*, Vol. 45, pp. 257-262.

Pavlů, V., Hejcman, M., Pavlů, L., Gaisler, J., Nežerková, P., & Meneses, L. (2006). Changes in plant densities in a mesic species-rich grassland after imposing different grazing management treatments. *Grass and Forage Science*, Vol. 61, pp. 42-51.

Pekkarinen, A. (1998). Oligolectic bee species in Northern Europe (Hymenoptera, Apoidea). *Entomologica Fennica*, Vol. 8, pp. 205-214.

Pekkarinen, A. Oligolectic bee species and their decline in Finland (Hymenoptera: Apoidea). Proceedings of the XXIV Nordic Congress of Entomology, 151-156. 1999. Conference Proceeding

Piessens, K. & Hermy, M. (2006). Does the heathland flora in north-western Belgium show an extinction debt? *Biological Conservation*, Vol. 132, pp. 382-394.

Pihlgren, A. (2007). *Small-Scale Structures and Grazing Intensity in Semi-Natural Pastures*, Doctoral thesis, Acta Universitatis Agriculturae Sueciae, No. 2007:13.

Plantureux, S. (1996). Biodiversity, type of soil and management intensity of permanent pastures of Plateau Lorrain. *Acta Botanica Gallica*, Vol. 143, pp. 339-348.

Pratt, R. M., Putman, R. J., Ekins, J. R., & Edwards, P. J. (1986). Use of habitat by free-ranging cattle and ponies in the New Forest, Southern England. *Journal of Applied Ecology*, Vol. 23, pp. 539-557.

Putman, R. J., Fowler, A. D., & Tout, S. (1991). Patterns of use of ancient grassland by cattle and horses and effects on vegetational composition and structure. *Biological Conservation*, Vol. 56, pp. 329-347.

Pykälä, J. (2000). Mitigating human effects on European biodiversity through traditional animal husbandry. *Conservation Biology*, Vol. 14, pp. 705-712.

Reitalu, T., Johansson, L. J., Sykes, M. T., Hall, K., & Prentice, H. C. (2010). History matters: village distances, grazing and grassland species diversity. *Journal of Applied Ecology*, Vol. 47, pp. 1216-1224.

Riddersporre, M. (1995). *Village Lands in the Rear-view Mirror: the Cultivated Landscape Before the Age of the Maps*, Doctoral thesis, Department of Geography, Lund University.

Ritchie, M. E. & Olff, H. (1999). Herbivore diversity and plant dynamics: compensatory and additive effects. In *Herbivores: Between Plants and Predators. The 38th Symposium of the British Ecological Society*, eds. H. Olff, V. K. Brown, & R. H. Drent, pp. 175-204. Blackwell Science, Oxford, UK.

Rosenthal, J. P. & Kotanen, P. (1994). Terrestrial plant tolerance to herbivory. *Trends in Ecology & Evolution*, Vol. 9, pp. 145-148.

Ruzickova, H., Halada, L., David, S., & Gerhatova, K. (2001). Management of meadows in the Biosphere Reserve East Carpathians. II. Results after 7 years. *Ekologia-Bratislava*, Vol. 20, pp. 76-87.

Simán, S. & Lennartsson, T. (1998). Slåtter eller bete i naturliga fodermarker? - ett skötselförsök med slåtteranpassade växter. *Svensk botanisk tidskrift*, Vol. 92, pp. 199-210.

Sjöbeck, M. (1933). Lövängskulturen i Sydsverige. *Ymer*, Vol. 53, pp. 33-66.

Sjörs, H. (1954). *Slåtterängar i Grangärde Finnmark. (Summ.: Meadows in Grangärde Finnmark, SW Dalarna, Sweden.)* Svenska Växtgeografiska Sällskapet, ISBN 91-7210-034-6, Uppsala.

Söderström, B., Svensson, B., Vessby, K., & Glimskär, A. (2001). Plants, insects and birds in semi-natural pastures in relation to local habitat and landscape factors. *Biodiversity and Conservation*, Vol. 10, pp. 1839-1863.

Sporrong, U. (1971). *Kolonisation, Bebyggelseutveckling och Administration : Studier i Agrar Kulturlandskapsutveckling Under Vikingatid och Tidig Medeltid med Exempel från Uppland och Närke*, Doctoral thesis, Department of Human Geography, Stockholm University.

Stanners, D. & Bourdeau, P. (1995). *Europe's Environment. The Dobris Assessment.* European Environment Agency, Copenhagen.

Svensson, B. M. & Carlsson, B. Å. (2005). How can we protect rare hemiparasitic plants? Early-flowering taxa of Euphrasia and Rhinanthus on the Baltic island of Gotland. *Folia Geobotanica*, Vol. 40, pp. 261-272.

Swedish Board of Agriculture (2007). *Environmental subsidiaries.* http://www2.sjv.se/webdav/files/SJV/trycksaker/Jordbruksstod/JS61.pdf

Swedish Board of Agriculture. (2005). *TUVA - Ängs- och betesmarksinventeringen (survey of semi-natural grasslands)*, 2-10-2011, Available from https://eidservice.sjv.se/tuva2/site/index.htm

Swedish University of Agricultural Sciences (2010), *Redovisning av uppdrag om halvtidsutvärdering av Landsbygdsprogram för Sverige 2007-2013. Fourth National Report to the Convention on Biological Diversity, Sweden.*, Landsbygdsdepartementet Jo 10.013, Regeringsbeslut 2009-04-20, M2009/385/Na. http://www.sweden.gov.se/sb/d/13746/a/155799

Tollin, C. (1991). *Ättebackar och ödegärden. De äldre lantmäterikartorna i kulturmiljövården.* Riksantikvarieämbetet, 91-7192-819-7, Stockholm.

Tollin, C. (2004). When Sweden was put on the map. In *European Rural Landscapes: Persistance and Change in a Globalising Environment.*, ed. H. Palang, pp. 191-208. Kluwer Academic Publisher.

Turnbull, L. A., Crawley, M. J., & Rees, M. (2000). Are plant populations seed-limited? A review of seed sowing experiments. *Oikos*, Vol. 88, pp. 225-238.

Valtonen, A. & Saarinen, K. (2005). A highway intersection as an alternative habitat for a meadow butterfly: effect of mowing, habitat geometry and roads on the ringlet (Aphantopus hyperantus). *Annales Zoologici Fennici*, Vol. 42, pp. 545-556.

Van Andel, J., Bakker, J. P., & Grootjans, A. P. (1993). Mechanisms of vegetation succession: a review of concepts and perspectives. *Acta Botanica Neerlandica*, Vol. 42, pp. 413-433.

Vestbö-Franzén, A. (2004). *Råg och Rön. Om mat, människor och landskapsförändringar i norra Småland, ca 1550-1700.* Jönköpings läns museum, Jönköping.

von Linné, C. (1741). *Carl von Linnés öländska och gotländska resa 1741.*

Wallin, L. (2007). *Plant population dynamics and conservation in wooded hay-meadows*, Department of Ecology and Evolution, Plant Ecology, Uppsala University.

Watt, T. A. & Gibson, C. W. D. (1988). The Effects of Sheep Grazing on Seedling Establishment and Survival in Grassland. *Vegetatio*, Vol. 78, pp. 91-98.

Webb, N. R. (1998). The traditional management of European heathlands. *Journal of Applied Ecology*, Vol. 35, pp. 987-990.

Welinder, S., Pedersen, E. A., & Widgren, M. (1998). *Jordbrukets första femtusen år. 4000 f.Kr-1000 e.Kr* Natur och Kultur/LTs förlag, 91-27-35542-X, Uppsala.

Willcox, E. V., Tanner, G. W., Giuliano, W. M., & McSorley, R. (2010). Avian Community Response to Grazing Intensity on Monoculture and Mixed Florida Pastures. *Rangeland Ecology & Management*, Vol. 63, pp. 203-222.

Wissman, J. (2006). *Grazing regimes and plant reproduction in semi-natural grasslands*, Doctoral thesis, Acta Universitatis Agriculturae Sueciae, No. 2006:40.

11

Landowners' Participation Behavior on the Payment for Environmental Service (PES)

Wan-Yu Liu
Department of Applied Natural Resources, Aletheia University, Tainan
Taiwan

1. Introduction

Since private landowners have their rights to determine whether to go afforestation, the government intends to achieve the purposes on ecological conservation of land, soil and water conservation, as well as environmental greening. Because it takes a long time to harvest the forests, such long-term investments make private landowners difficult to raise funds. Therefore, it is common to see that each country in the world agrees on the loan and subsidy schemes for afforestation, and related incentive policies include the free supply of seeds, preferential loans on afforestation, preferential taxes, afforestation subsidies, and so on[1].

Taiwan has introduced many afforestation incentive policies to raise people's intention to go afforestation from 1951 to 2002. Since 1951, in order to reward the afforestation in the ecological conservation land that falls into disuse, the Taiwan Provincial Government has announced the "Detailed Rules and Regulations of Rewarding Afforestation of Ecological Conservation Land in Taiwan Province" to reward the people to invest in the afforestation in ecological conservation land, with the offerings of no payment for rents and the adoption of major/side products without price. At that time, in order to promote local people to go afforestation, the government agreed on "Regulations of Private Forests", "Rules and Regulations of Rewarding Afforestation in Private Forest Land", and so on. In 1974, the government agreed on "Regulations of Revenue and Expenditure, Safekeeping and Manipulation of Afforestation Loan Funds of Taiwan Province". In 1983, in order to foster forest resources and enhance the guidance and assistance of private afforestation, the government agreed on "Regulations of Rewarding Private Afforestation in Taiwan Province", in which the subsidy for afforestation was NT$ 1,200 per hectare.

---

[1] According to Nagubadi et al. (1996), the related incentive policies for afforestation policy in the U.S. are divided into two categories in general. The first is the direct payment policy, including tax free, direct subsidy and free technical assistance, e.g., Indiana State belongs to this policy; the second is the cost-sharing policy, in which the partial cost is subsidized by the government for landowners to go afforestation. In other words, the cost occurred from afforestation programs include new afforestation land, afforestation and forestry management expenses, in which 50-75% of the expenses were covered by the government. Many forestry policies in the U.S. all belong to this category of policy, including forestry incentive program (FIP), Stewardship incentive program (SIP), agricultural conservation program (ACP) and conservation reserve program (CRP).

In 1991, the government revised "Regulations of Rewarding Private Afforestation in Taiwan Province", and agreed on "Regulations of Rewarding Afforestation in Cropland" to guild and assist the afforestation in marginal cropland and make use of land resources in a reasonable way, in which the subsidy was increased to NT$ 32,000 per hectare, and was revised again in 1994 to NT$ 150,000 at most per hectare. In 1996, Typhoon Herb attacked Taiwan and resulted in severe damage. In order to recover the function of soil and water conservation, the government had carried out the "General Afforestation Campaign" to achieve the objective of greening afforestation and water conservation. Under this program, the afforestation subsidy was increased up to NT$ 530,000 in total per hectare for 20 years. In August 2008, in order to respond to the impact on the trade liberalization of international farm products due to Taiwan's joining WTO, the production of the cropland resources was reduced accordingly. For the released cropland due to the reduction of production, the government has approved the Plain Landscape Afforestation Program (PLAP)[2] since 31 August 2001, and implemented it on 1 January 2002, to guide and assist farmers and Taiwan Sugar Corporation to leave their land fallow in a long-term period and further afforest it. The government provided afforestation subsidies and direct payments, which amounted to NT$ 1,610,000 per hectare for 20 years, and the main objectives were the marginal cropland in ordinary farm areas.

Reward for the afforestation in cropland has become one of the major trends for the forestry policies in the world, which plays an essential role in the development of forestry policies for the time being. As far as Taiwan is concerned, due to the rising ecological consciousness of forest conservation and the rising living standard of the people, human beings' demand for forests is getting more and more. However, under the policy of emphasizing agricultural production previously, most plain areas in Taiwan are used for agricultural production, the forests in plain areas are insufficient, and the cropland acreages planned by Taiwan government are too huge, so that the ecology is threatened. Therefore, if the marginal cropland can be applied for the PLAP, the cropland use will be raised effectively.

Despite the positive comments on the good intention of PLAP, there are still many difficulties during the process of implementation; as shown in Table 1, the total area that implemented the PLAP is 8,829.18 hectares while the area for the afforestation in private cropland is 869.18 hectares (only occupying 9.84% of all), which is against government's original good intention on drafting the policy earlier on. According to the previous work, the possible reasons for the poor implementation of the afforestation in private cropland include insufficient professional afforestation technology, private landowners' low interest in participating in PLAP, insufficient subsidies, and so on.

From the literature, since private landowners normally have less efficiency in land use and do not attain the land production potential, hence the interference of public policies is required. The main goal of the policy tools related to private landowners is to make the personal objective consistent with the objective of maximizing social welfare. As shown in Table 2, many previous results indicated that the subsidy policies and the assistance of

---

[2] The idea of the afforestation policy in plain areas has been fermenting for a long time, mainly because of insufficient greenery resources of plain areas in Taiwan, entry of WTO, and the bulky croplands that continue to go fallow and reduce production. Hence, the government is actively promoting the afforestation policy in plain areas, which not only attains the fundamental mission for the reduction of fallow area, but also achieves landscaping to increase carbon sequestration and enhance the purpose of enhancing the quality of the environment.

related technologies from the government have positive effects on landowners' decision for afforestation, while there will be negative effects on landowners' decision for afforestation if the government imposes more related regulations on afforestation.

Year	2002	2003	2004	2005	2006	2007	Total
Private Cropland (NIPF)	229.82	132.08	174.83	145.10	124.40	65.17	869.18
Taiwan Sugar Corp. (IPF)	1,361.40	3,940.09	1,150.00	877.00	263.63	367.88	7,960.00
Total	1,591.22	4,072.17	1,324.83	1,022.10	388.03	433.05	8,829.18

Unit: hectare

Note: NIPF denotes nonindustrial private forest, while IPF denotes industrial private forest

Table 1. Afforestation areas for the PLAP in Taiwan (Years from 2002 to 2007).

Government's policy	Effects on decision-making		Related literature
	Effects on afforestation decision	Effects on logging decision	
Knowledge of cost-share programs	positive effect	literature indicated that there was a positive impact, but no concrete conclusion	(Hyberg and Holthausen, 1989) (Royer, 1987) (Zhang and Flick, 2001) (Megalos, 2000) (Hardie and Parks, 1991)
Knowledge of public technical assistance	positive effect	positive effect	(Hyberg and Holthausen, 1989) (Royer, 1987) (Zhang and Flick, 2001) (Hardie and Parks, 1991)
Regulation	negative effect	no related literature	(Zhang and Flick, 2001) (Boyd and Hyde, 1989)

Table 2. Effects of government's related policies on private landowners' afforestation and logging decisions. Source: Cubbage et al. (2003)

Hardic and Parks (1991) indicated that the offerings of cost-sharing programs and public technical assistance give obvious and positive effects on private landowners' decision for afforestation policies. They also conducted an analysis on the cross effects of the public technical assistance and cost-sharing programs, and their result indicated that the cost-sharing programs perform better than the public technical assistance. Cubbage (2003) proposed that the public policy or technical assistance provided by the government would boost landowners' afforestation revenues and the quality of forest management. In addition, a lot of literature analyzes the behavior of participation in afforestation programs, in which the mainly investigated objects are cost-sharing programs. For example, English et al. (1997) found that higher income and lower cost would lead to higher probability of afforestation;

Nagubadi et al. (1996) proposed that older age and larger land area would enhance the probability of participating in afforestation; Stevens et al. (1999) proposed that older age would reduce the probability of participating in afforestation, but higher income would raise it; Megalos (2000) and Lorenzo and Beard (1996) proposed that the people who have larger land area and are not farmers would have higher probability to participate in afforestation. Esseks and Moulton (2000) conducted an examination on private landowners' (NIPF) participation in Forest Stewardship Program (FSP) and Stewardship Incentives Program (SIP), both of which have been implemented since 1990 in the USA and supported by Farm Bill since 2002. Listed as the policies in Forest Land Enhancement Program, these include afforestation, reforestation, forest improvement, forest stewardship plans, agroforestry policy, soil and water quality and wetlands maintenance, and so on. Among them, the subsidy scheme provided by the government has remarkable effects on attracting landowners to participate in afforestation.

In Taiwan, the government has begun carrying on the PLAP since 2002, but the outcome is under expectation. Therefore, the Taiwanese government needs to desperately study the following important issues: What are the key factors that influence private landowners to participate in the PLAP? What are the factors that affect private landowners' afforestation behavior? Are these factors the same as those listed in other literature? In light of the above, the primary purpose of this paper is to analyze the decision behavior of private landowners' participation in the PLAP in Taiwan, and analyze the factors that influence private cropland owners' participation in afforestation. In addition, we compare the similarities and dissimilarities between Taiwanese private landowners' decision and the decision factors listed in the literature. The empirical result of this paper expects to provide as a reference to those policy institutors of forest department in Taiwan for related policy institution.

This paper is organized as follows. The theoretical model of landowners' decision on afforestation is given in section 2, the empirical model and the analysis on our empirical result are given in section 3, whereas Conclusions and suggestions are given in section 4.

2. Theoretical model of landowners' decision

2.1 Theoretical model of landowners' decision on afforestation

As for the landowners' behavior on land use, Rahm and Huffman (1984) were the first to propose a general equilibrium model based on farmers' pursuing the utility maximization. Let $t = 1$ (resp., $t = 0$) represent that the private landowner does (resp., does not) participate in the PLAP. The utility function of private landowner i is defined as $U(H_{ti}, M_{ti})$, in which H_{ti} represents the landowner characteristics factor, including age, education, income and occupation; M_{ti} represents the management characteristics factor, including information source, attitude, current land use, etc. The utility function $U(H_{ti}, M_{ti})$ is expressed as follows:

$$U_{ti} = \alpha_t F_{ti}(H_{ti}, M_{ti}) + e_{ti}, \quad \text{for } t = 1,0 \text{ and } i = 1,2,...,n \tag{1}$$

The above equation is not limited as a linear function, in which U_{ti} is a random function. Therefore, if $U_{1i} > U_{0i}$, then landowner i will participate in the PLAP (t =1); otherwise, landowner i will not participate in the PLAP (t = 0). Let y_i^* (i.e., $U_{1i} - U_{0i}$) be an unobservable afforestation tendency variable. Therefore, the observable variable for afforestation decision (y_i) can be expressed as follows:

$$y_i = \begin{cases} 1, \text{ if } y_i^* > 0; \\ 0, \text{ otherwise.} \end{cases} \tag{2}$$

in which $y_i = 1$ represents that landowner i chooses to participate in the PLAP while $y_i = 0$ represents no. Therefore, the probability P_i of landowner i's participation in PLAP is given as follows:

$$\begin{aligned} P_i &= P_r(y_i = 1) = P_r(U_{1i} > U_{0i}) \\ &= P_r[\alpha_1 F_{1i}(H_{1i}, M_{1i}) + e_{1i} > \alpha_0 F_{0i}(H_{0i}, M_{0i}) + e_{0i}] \\ &= P_r[e_{1i} - e_{0i} > F_{ti}(H_{ti}, M_{ti})(\alpha_0 - \alpha_1)] \\ &= P_r[\mu_i > F_{ti}(H_{ti}, M_{ti})\beta] \\ &= F(X_i^{'}\beta) \end{aligned} \tag{3}$$

in which P_i is the probability of landowner i 's participation in PLAP; $\mu_i = e_{1i} - e_{0i}$ is a random disturbance term; $\beta = \alpha_0 - \alpha_1$ is a predicate parameter vector, X_i' is an explanatory variable vector; $F(X_i', \beta)$ is a cumulative distribution function. If there is no function form of $F(X_i', \beta)$ in the above equation, then it is not allowed to predict it directly. The form of $F(X_i', \beta)$ is determined according to the distribution of $\mu_t = e_{1t} - e_{0t}$. If μ_t is a normal distribution, then F is a cumulative normal distribution; if μ_t is a uniform distribution, then F is a triangular distribution. The landowners of private cropland depend upon the above decision model to determine whether to participate in the PLAP. Since the dependent variables include participation and nonparticipation, the Probit model and Logit model can be applied to the analysis on this type of problems.

2.2 Binary choice model

A so-called binary choice model[3] is to suppose that a representative individual needs to choose one out of two items. The regression model of normal linear probability mode is stated as follows:'"

$$\begin{gathered} Y_i = \alpha + \beta_1 X_{1i} + \beta_2 X_{2i} + \ldots + \beta_j X_{ji} + \varepsilon_i \\ Y_i = \begin{cases} 1, & \text{if "Yes";} \\ 0, & \text{if "No".} \end{cases} \end{gathered} \tag{4}$$

in which Y_i is a binary choice variable, X_{ji} is an independent variable, and ε_i is a deviation item. Since Y_i represents only two numbers, i.e., 1 and 0. Therefore, we can let $P_i = \text{Prob}(Y_i = 1)$ and $1 - P_i = \text{Prob}(Y_i = 0)$ to explain the distribution of Y probability, whose expected value is:

$$E(Y_i) = \alpha + \beta_1 X_{1i} + \beta_2 X_{2i} + \ldots + \beta_j X_{ji} = P_i \tag{5}$$

To transform Equation (4) as an estimate equation, we obtain:

---

[3] Binary choice issue is a issue with two possibilities, such as whether to pass the admission exam, whether to come down with disease, whether to participate in afforestation program and so on. All these belong to Binary or Dichotomous, while the Probit model and Logit model can be used to analyze this kind of binary choice issues.

$$Y_i = \alpha + \beta_1 X_{1i} + \beta_2 X_{2i} + ... + \beta_j X_{ji} \quad (6)$$

That is, the linear probability model in Equation (6) takes the estimated value of Y as its probability, and therefore, if the estimated value Y exceeds the range of [0, 1], then there will be a problem on estimation. In order to solve this problem, we may re-estimate the parameters a and β under the limitation of $0 \le Y_i \le 1$, and determine minimum square parameter estimated value according to the limitation of inequality, but this is a nonlinear way to estimate the value. In the linear probability model, there exist some variances which cannot be explained through the model. In order to solve this problem, the previous research used a conversion probability from 0 to 1, i.e. to use a cumulative probability function to convert the variable. The probability distribution is listed as follows:

$$P_i = F(\alpha + \beta_1 X_{1i} + \beta_2 X_{2i} + ... + \beta_j X_{ji}) = F(Z_i) \quad (7)$$

It will lead to a probability model with restricted conditions if the above equation is applied to the variable conversion. There will be a variety of probability models if we apply different cumulative probability functions for variable conversion. A probability model via the normal cumulative random function for variable conversion is the Probit model, while the probability model via the variable conversion of the cumulative logarithmic probability is the Logit model. The results of these two models are roughly the same, but the differences of explanatory variables will be affected drastically if some explanatory variables are too large numbers and have huge variances. But the Logit model can make up for this drawback. In this paper, we apply the Logit model to conducting analysis, since the numbers of explanatory variables varies a lot in comparison to the numbers of dependent variables.
A probability model via the variable conversion of the cumulative logarithmic probability function is the Logit model, which is shown as follows:

$$P_i = F(Z_i) = \frac{1}{1 + e^{-Z_i}}, \qquad \log\left(\frac{P_i}{1 - P_i}\right) = Z_i, \; Z_i = \alpha + \beta_1 X_{1i} + \beta_2 X_{2i} + ... + \beta_j X_{ji} \quad (8)$$

in which P_i represents the probability of landowner number i for the participation of afforestation; a is the intercept; β_j is the coefficient of independent X_{ji}. From Equation (8), it is estimated that once the independent variable is changed by one unit, the value of dependent variable is changed to:

$$\frac{\partial P}{\partial X} = \frac{\partial F(Z_i)}{\partial X} = f(Z_i)\beta_j X_{ji} \quad (9)$$

This paper will conduct an empirical analysis on the above theoretical model.

3. Empirical model and empirical result analysis

3.1 Variable selection and data source

3.1.1 Variable selection

From the literature, there exist many results regarding the analysis on nonindustrial private forest landowners' behavior on afforestation (e.g., Royer, 1987; Alig et al. 1990; Hardie and

Parks 1991; Nagubadi et al., 1996; Zhang and Flick, 2001; Sills and Abt, 2003). This paper considers the possible factors proposed by Sills and Abt (2003) which might affect private landowners' behavior on afforestation–including landowner characteristics (*OC*), cropland characteristics (*CC*), and forestry programs (*FP*). The landowner characteristics include gender, household income, occupation, age, number of family members, etc.; the cropland characteristics include cropland price, cropland rental, distance away from downtown, area, amount of cropland area, etc.; the forestry programs include afforestation subsidy amount, duration of subsidy, rules on adjoining and adjacent areas, etc.
Hence, this paper bases upon the above variables to establish the following regression model to examine the choice of landowners' decision on afforestation:

$$PP = f\,(OC, CC, FP) \tag{10}$$

In other words,

$$PP = a + \Sigma\, \beta_i \cdot OC_i + \Sigma\, \gamma_j \cdot CC_j + \Sigma\, \delta_k \cdot FP_k + \varepsilon$$

in which *PP* represents whether to participate in the PLAP (*PP* is 1 if participation, while *PP* is 0 if no); *OC* represents private landowners characteristics; *CC* represents the cropland characteristics owned by private landowners; *FP* represents forestry programs[4]; OC_i, CC_j, and FP_k represent each characteristic of *OC*, *CC*, and *FP*, respectively, and β_i, γ_j and δ_k are their corresponding weights in the formula.

3.1.2 Data source

From Table 1, in spite of the limitation of the afforestation area, Pintung County reported the most significant result in terms of afforestation area among the 16 cities and counties that implemented the PLAP in Taiwan, and there were private landowners in Pintung County participating in the PLAP from 2002 to 2005, which was the only region that participated in PLAP for the four years consecutively. Therefore, this paper has conducted a survey in Pintung County during August to November 2004, to analyze farmers' participation and nonparticipation in PLAP.

---

[4] According to a study on landowner's effective theory (Binkley, 1981), the empirical research of landowners' forestry management pointed that the landowners' characteristics including landowners' income, occupation (agriculture and non-agriculture), residence type (local residents or external residents), education level and age based on the research (Cubbage, 2003) said that, the most important about landowner characteristics was "income"; it showed that the higher the income, then the higher participation on the afforestation policy. According to Alig et al. (1990), the impact on logging decision is far more than the impact on afforestation among the landowner characteristics. This probability implies that the landowner characteristics will have a more direct impact on the market (logging revenues). Also, according to the efficiency model for the decision-making behavior of the private landowners (Binkley, 1981; Dennis, 1989), "career factor" is normally a more indirect factor and according to Romm et al. (1987), for those local landowners who live here for a long time have higher probability in participating the afforestation program while those temporary residents have longer probability in participating the afforestation program. Moreover, Romm et al. (1987) also found that older elderly tends to reduce investment behavior, while Zhang & Flick (2001) pointed that the age does not have significant relevance to the choice of afforestation. As far as cropland characteristics are concerned, cropland area has significant relevance to the participation probability. Likewise, the higher percentage the forest land is, the higher participation probability it is. The residents who live here for a long time also reported higher probability in participating in the afforestation program.

The respondents of the survey included the farmers that do or do not participate in the PLAP. There are 33 villages and towns in total at Pintung County, deducted from those mountain aboriginal towns with less or none normal agriculture areas (eight towns/villages in total) and Luoqiu Village. Hence, we considered the 24 remaining villages/towns. Then we conducted a random sampling out of these 24 villages/towns, and selected out 11 villages/towns, including Pintung City, Lin-lo, Qui-zu, Kao-shu, Wang-luan, Ne-pu, Hsin-bei, Fang-liao, Hsin-yuan, Ken-ding, and Ling-beng Villages/Towns.

For these 11 villages/towns, the sampling number of each village/town was determined according to the size of its normal agricultural area, such that the sampling data falls into the cropland in the normal agricultural area. What we chose included 38 samples from Pintung City, 15 samples from Lin-lo, 21 samples from Qui-zu, 115 samples from Kao-shu, 74 samples from Wang-luan, 57 samples from Ne-pu, 43 samples from Hsin-bei, 71 samples from Fang-liao, 16 samples from Hsin-yuan, 15 samples from Ken-ding, and 16 samples from Lin-beng. There were 481 samples in total and 304 effective samples. Since there were not many farmers participating in the PLAP, thus we applied census to investigating the farmers who have participated in the PLAP. There were 39 samples in total.

3.1.3 Variable description

As mentioned above, according to the previous literature, Sills and Abt (2003) pointed that the possible factors of affecting the behavior of private landowners' participation in afforestation includes landowner characteristics–gender, annual household income, occupation, age, number of family members, cropland characteristics–cropland price, cropland rental, distance from the downtown, cropland area, number of cropland parcels, forestry programs–afforestation subsidy, duration of afforestation, rules on adjoining and adjacent areas, etc. This paper uses the survey data of the farmers in Pintung County to conduct the estimation on the decision behavior of the participation in afforestation.

Based upon Equation (10), the definition of variables in the model is listed in Table 3. According to this equation, the explanatory variables of afforestation programs can be divided into three types: landowner characteristics (*OC*), cropland characteristics (*CC*) and forestry programs (*FP*). The landowner characteristics include six variables – gender, age, education level, occupation, household income, and number of family members. Cropland characteristics include six variables – total cropland area, cropland, cropland price, distance from Township Office, distance from county government, and number of cropland parcels. Finally, the forestry program variables include three variables as follows – whether landowners' cropland area meets the regulation of adjoining and adjacent areas, whether landowners agree that the subsidy for PLAP is reasonable, and whether landowners agree that the duration for the subsidy of PLAP is reasonable (e.g., the participation in the PLAP might make the future cropland use lack for flexibility; the participation in the PLAP might reduce the opportunity for the change of land use; not clear about the value of afforestation after 20 years).

3.2 Analysis of whether landowners participate in PLAP

From Table 4, the reason for the landowners' non-participation in the PLAP is that they are not aware of the PLAP, representing up to 60%; those who are aware of the PLAP and choose not to participate in the PLAP still occupies 68.9%, among which "landowners' croplands fail to meet the requirement of afforestation policy's adjoining and adjacent areas, and it is not easy for a joint application with other people" occupies 25%, followed by "the participating in the PLAP might lead to inflexibility of land use" occupying 20.0%, and "the participation in the PLAP might reduce the opportunity for the change of land use", which occupies 18.9%. Other reasons such as "higher revenues if used in other purposes", or "participation in the PLAP will lose flexibility of land use in future" occupies less proportion. Other related parameters that determine whether to participate in afforestation program are shown in Table 4.

Variable type	Variable name	Variable description	Prediction direction
The decision dependent variable (*PP*)	whether to participate in PLAP	yes=1, no=0	
Landowner characteristics (*OC*)	sex (SEX)	male=1, female=0	+
	age (AGE)	age of landowner (year)	+
	education level (EDU)	junior high school=1, above senior high school/vocation school=0	?
	occupation (OCC)	agriculture=1, non-agriculture=0	+
	annual household income (INC)	landowner's INC (NT$ 10,000/year)	+
	number of family members (POP)	number of landowner's family members (person)	?
Cropland characteristics (*CC*)	cropland area (ALAND)	total operation areas of the cropland	+
	cropland rental (RLAND)	landowners' cropland rental (NT$ 10,000/hectare/year)	–
	cropland price (PLAND)	landowners' cropland price (NT$ 10,000/hectare)	–
	distance from the county government (LOCA1)	the distance of landowners' cropland from county government (km)	+
	distance from the township office (LOCA2)	the distance of landowners' cropland from ownership office (km)	+
	number of cropland parcels (NLAND)	landowners' number of cropland parcels	–
Forestry programs (*FP*)	regulation of rewarding afforestation (REG)	landowners' cropland area that meets the regulation of adjoining and adjacent areas = 1, if not=0	+
	duration of afforestation (YSUB)	those who agree that the duration for the subsidy of PLAP is reasonable=1, if not=0 if not	+
	afforestation subsidies (ASUB)	those who agree that the subsidy for PLAP is reasonable=1, if not=0	+

Table 3. Instructions for the parameters used in the model

<table>
<tr><td rowspan="2">Unaware of the PLAP (60%)</td><td>Participating in the PLAP after being aware of it (31.1%)</td><td rowspan="3">Reasons for not participating in the PLAP after being aware of it:
Landowners' cropland fails to meet the regulation on land area (25.0%)
Low afforestatoin subsidies (11.2%)
Lose flexibility of land use for 20 years (20.0%)
Reduce the opportunity for the change of land use(18.9%)
Unfamiliar with afforestation tasks (11.8%)
Unfamiliar with the value of afforestation after 20 years (13.1%)</td></tr>
<tr><td>Not participating in the PLAP after being aware of it (68.9%)</td></tr>
<tr><td colspan="2">Choose not to participate in the PLAP after being aware of it (40%)</td></tr>
</table>

Table 4. Analysis on the landowners who do not participate in PLAP

3.3 Mean and significance of independent variables in afforestation decision-making

The mean and standard deviation of independent variables are listed in Table 5. Generally speaking, no matter whether to participate in PLAP, the male is the majority as always; there is no remarkable difference of ages of the landowners who participate in PLAP or not. This result is consistent with the results from Ervin and Ervin (1982), Korsching et al. (1983),

Independent variable	(participants) $n = 39$		(nonparticipants) $n = 304$		F-statistics
	Mean	(S.D.)	Mean	(S.D.)	
SEX	0.87	0.34	0.80	0.40	1.07
AGE	56.67	10.87	55.77	12.82	0.18
EDU	0.28	0.46	0.57	0.50	11.78***
OCC	0.31	0.47	0.53	0.50	6.91***
INC	55.05	45.15	51.88	46.51	0.16
POP	4.18	1.94	4.69	2.50	1.53
ALAND	3.65	4.70	1.16	1.76	41.30***
RLAND	5.18	3.99	5.18	7.85	0.10
PLAND	592.19	416.41	566.71	690.90	0.05
LOCA1	27.86	19.15	24.88	13.87	1.45
LOCA2	4.35	1.24	4.44	6.50	0.01
NLAND	1.18	0.60	2.51	1.76	22.01***
REG	1.00	0.00	0.51	0.50	32.27***
YSUB	1.00	0.00	0.78	0.42	11.17***
ASUB	1.00	0.00	0.49	0.50	40.87***

Note: a sample size of 343 in total, in which there are 39 participants and 304 nonparticipants. (*), (**) and (***) represent a significance level of 10%, 5% and 1%, respectively

Table 5. Mean and standard deviation of independent variables in the model

McNamara et al. (1991) and Nagubadi (1995), which proposed that older landowners tend to participate in afforestation programs. Moreover, according to the survey conducted by this paper, the education level of those who participate in PLAP is apparently less than that of the landowners who do not participate, which is different from the results from Boyd (1984),

Hammentt (1992) and Nagubadi et al. (1996), which proposed that the landowners with higher education level tend to participate in afforestation programs. As for annual household income, those who have participated in PLAP reported a higher income than those who do not participate in PLAP.

The occupation type for those who do not participate in PLAP tends to be agriculture industry. Those who do not participate in PLAP reported more family members but yet obvious. On the other hand, participants reported larger cropland areas than the landowners who do not participate in PLAP. Such a conclusion is consistent with the results from Napier et al. (1984), Korsching et al. (1983), McNamara et al. (1991) and Nagubadi et al. (1996). As for the cropland rental and cropland price, there is no major difference between participants and nonparticipants while nonparticipants reported higher rental and land cost. The croplands owned by nonparticipants reported far distance from county government and township office. The cropland owned by participants reported more complete cropland while the cropland owned by nonparticipants is more fragmentary. As for the forestry program variables, it is obvious that participants recognize the contemporary policy design more while nonparticipants who fail to meet the rules of adjoining and adjacent areas are less agreeable with the subsidies and duration of PLAP, in which most of the respondents wish that the subsidies can be increased and the rewarding period can be shortened.

3.4 Logit model analysis of decision-making behavior for afforestation

3.4.1 Empirical result analysis of decision-making behavior for afforestation

According to the different estimate results produced by inducing different variables, since there are too many virtual variables in the model, thus we substitute the logarithmic values of AGE, INC, RLAND, PLAND, LOCA1, LOCA2, etc. to the model, in order to obtain more precise estimates. Based on this, we conduct the Logit analysis, and the analysis result is given in Table 6.

From Table 6, we observe that age, education level, as well as annual household income among the landowner characteristics reported remarkable influence on the decision-making of whether to participate in PLAP, under a 10% confidence level. The older a private landowner is, the higher the probability of her/his participation in the PLAP is. For each unit increase on age, a 3.5% participation probability will be increased accordingly. This positive relationship is consistent with the conclusion of Nagubadi et al. (1996), but is contrary to the result of Stevens et al. (1999).

Second, the present study also revealed that the higher the education level of a landowner is, the lower the probability of her/his participation in the PLAP is, with a significance level of up to a = 10%. Our result indicated that when the education level increases from under junior high school to over senior high school level, the probability of landowners' participation in the PLAP will be decreased by 6.4%.

Third, the higher the annual household income of a landowner is, the lower the probability of her/his participation in the PLAP is, with a significance level of a = 10%. When the annual household income is increased by one unit, the participation probability will be increased by 6.9%. This positive relationship is consistent with the conclusion of English et al. (1997). Moreover, Megalos (2000) and Lorenzo and Beard (1996) proposed that non-farmers had reported a higher participation probability. The direction of occupation variables shown in this paper is consistent with the conclusion of the above literature results, but yet at a not much significance level. This has proved the results of Binkley (1981) and Dennis (1989), which found that the occupation was a more indirect factor in the decision utility model of landowners' participation behavior. Generally speaking, landowner characteristics remarkably affect the probability of landowners' participation in the PLAP.

As far as the cropland characteristics is concerned, cropland area, cropland price and number of cropland parcels reported the most significant influence among the factors that affect landowners' participation in the PLAP. The larger the cropland area is, the higher the probability of landowners' participation in the PLAP is. When cropland area is increased by one unit, the participation probability will be increased by 3.3% accordingly. This positive relationship is consistent with the conclusions of Megalos (2000) and Lorenzo and Beard (1996). Besides, from Table 6, the higher the cropland price is, the lower the probability of landowners' participation in the PLAP is. This meets the expectation of the opportunity cost theory.

Our results also indicated that when cropland price is increased by one unit, the participation probability will be decreased by 55.5%. It is evident that the cropland price shows significance on the marginal effect of participation decisions. We also observe from Table 6 that the more the number of cropland parcels is the lower the participation probability is. Hence, whether the cropland is fragmentary will affect the participation probability significantly. This is associated with the regulation that requires 2-hectare adjoining area or over 5-hectare adjacent area. Our results also indicated that when the number of cropland parcels is increased by one unit, the participation probability will be decreased by 74.7% accordingly. It is evident that the fragmentary level reported a significant relationship with the participation probability. Our analyses also showed an insignificant relationship with the distance of cropland from the township office or county government.

The forestry program factors include whether the subsidy, subsidy duration and rewarding policy are reasonable. This paper indicated that these three variables reported a positive relationship with the participation probability though the statistics yet achieve a significance level, i.e., the more the landowners agree on the subsidies, duration and rewarding policy, the higher the participation probability is. This meets the theoretical expectation.

In overall, cropland characteristics reported the most significant impact on the participation probability among the three types of factors, in which the cropland area, cropland price and number of cropland parcels are the utmost important in particular. Besides, many cropland factors achieve over 1% significance level. The empirical result of this paper indicated that apart from the distance of cropland from county government and number of family members are under expectation, the remaining meets the expectation in Table 3 as well. Among the landowner characteristics, age (+), education level (−), and annual household income (+) have achieved a significance level, while gender (+) and occupation (+) have yet achieved a significance level, which are consistent with previous literature results and the expectation. On the other hand, among the cropland characteristics, cropland area (+), cropland price (−), and number of cropland parcels (−) have achieved a significance level, and the factors that fail to achieve a significance level include level-cropland rental (−) and the distance of cropland from villages/ towns (−), which have met the expected direction of the existing theory. Among the forestry programs, the amount of subsidies (+), subsidy duration (+), and rules of adjoining and adjoining areas have yet achieved a significance level while the signs are consistent with the expected direction.

3.4.2 Goodness of fit analysis of afforestation decision model

As far as the goodness of fit in the Logit model is concerned, we analyze the difference level (Hosmer and Lemeshow, 1989) between the designed regression model and the full saturated model[5], under the measurement of log likelihood function[6]. The observed value can be completely estimated by the full saturated model. Based on the likelihood function, we usually introduce -2 multiplied by log of maximum likelihood value between the designed regression model and the full saturated model, to compare the estimated value with observed value. If the value is set larger, then it implies that this regression model has better goodness of fit, and vice versa. The value calculated by this paper is 93.715, which represents a goodness of fit.

Secondly, Hosmer and Lemeshow (1989) proposed a method that examines the goodness of fit of Logit model, called *Hosmer and Lemeshow index*, which is appropriate for the Logit regression with null hypothesis (H_0). The χ^2 value examined by Hosmer-Lemeshow test in this paper is 1.074, whose significance level is 0.998 that is not high, and yet achieves a significance level under 5% of significance level. Hence, we cannot reject H_0. That is, the model and data derived is appropriate, which show that the model established by this paper has well goodness of fit.

As for the precision of predication, based the log likelihood model, this paper introduces an index similar to R^2 index, such as likelihood ratio index (LRI) (Hosmer and Lemeshow 1989), which depicts the percentage for the variance of dependent variable explained by independent variables within the model. Nagelkerke (1991) revised the definition of R^2 proposed by Cox and Snell (1989), which enabled a better prediction power of the index. When R^2 gets close to 1, it implies that this model has a more precise prediction power. Nagelkerke's R^2 value in this paper achieves 0.695, which represents a good prediction power and the regression of this model performs well, with up to 69.5%.

In addition, through the omnibus test of model coefficients, it helps determine whether this regression model is helpful to the predication of the PLAP. The null hypothesis (H_0) in this test is "the Logit regression derived does not help the prediction". The χ^2 value in the omnibus test of model coefficients in this paper is 148.774, with 0.000 significance level, which shows that the model in this paper performs well and achieves a significance level under 5% significance level. It means that the model is helpful to predicting the participation in the PLAP. Moreover, as far as the Logit model established by this paper is concerned, as shown in Table 7, it reported a probability of 98.0% in precisely predicting the nonparticipation in the PLAP and a probability of 71.8% in precisely predicting the participating in the PLAP, with 95% probability for prediction in overall. It represents a good prediction result.

[5] Hair et al. (1998) have suggested to conduct goodness of fit test on Logistic regression model, with the numerous methods above introduced in the meantime, which is more objective to make a consolidated judgment.

[6] Likelihood function means the probability for obtaining the observation result under a certain parameter estimate requirement.

Variable	Coefficients	S.D.	Wald value	*P*-value	Movement value of probability
INTERCEPT	-21.38	31.16	0.16	0.693	-
OC					
SEX	0.77	0.77	1.01	0.316	0.035
ln_AGE	2.23	1.36	2.70	0.100*	0.080
EDU	-1.23	0.69	3.15	0.076*	-0.064
OCC	0.50	0.58	0.75	0.387	0.017
ln_INC	0.58	0.31	3.58	0.058*	0.069
POP	-0.14	0.15	0.95	0.330	-0.063
CC					
ALAND	0.402	0.15	7.36	0.007***	0.033
ln_RLAND	-0.31	0.60	0.28	0.600	-0.056
ln_PLAND	-1.00	0.38	6.97	0.008***	-0.555
ln_LOCA1	0.37	0.39	0.90	0.343	0.051
ln_LOCA2	-0.09	0.49	0.04	0.852	-0.010
NLAND	-1.70	0.47	13.29	0.000***	-0.747
PF					
REG	9.28	42.52	0.05	0.827	0.079
YSUB	9.38	56.13	0.03	0.867	0.080
ASUB	9.66	42.34	0.05	0.820	0.079
-2 Loglikelihood	93.715				
Good/bad of overall model	Cox-Snell R square=0.354, Nagelkerke R square=0.695				
Overall model test	Omnibus test of model coefficient, Chi-square=148.774 with 0.000 significance level***				
Goodness of fit test	Hosmer-Leme show test, Chi-square=1.074 with 0.998 significance level				

Note: a sample size of 343 in total, there are 39 participants (PP = 1) and 304 nonparticipants; (*), (**) and (***) represent a significance level of 10%, 5% and 1% for α, respectively. The "movement value of probability" on the right column is the movement value of probability for the change of unit quantity of independent variables

Table 6. The analysis of Logit model

Prediction groups / Real groups	*PP*=0	*PP*=1	Total
PP=0	296	6	302
PP=1	11	28	39
precision rate(%)	98.0 %	71.8%	95.0 %

Table 7. Statistics for precise values predicted by the Logit model

4. Conclusions and suggestions

The afforestation policy in the plain areas (PLAP) was certified by Taiwan government on 31 August 2001 and has been implementing from 1 January 2002 to 31 December 2007. Despite the positive comments on the good intention of afforestation policy, still, there are many difficulties during the process of implementation, such as insufficient technology for afforestation, private landowners' low interest in the participation in the PLAP, insufficient subsidies, etc., which are all potential threats that hinder the PLAP from moving forward in future. This paper selects Pingtung County as a region for sampling and targets those private landowners with or without intention to participate the PLAP. Then the Logit model for empirical analysis is used in order to analyze the factors that determine whether to join the PLAP respectively in terms of those farmer samples that do or do not participate in the afforestation scheme, and to understand the effect of PLAP on the personal decision of afforestation. This paper indicates that the possible factors that might determine private landowners' participation in PLAP include landowner characteristics, cropland characteristics and forestry programs.

The possible reasons for landowners who do not participate in the PLAP include: not aware of the PLAP with up to 60%, while those who are aware of the PLAP and choose not to participate occupies 68.9%, and among the other possible reasons, "landowners' cropland fails to meet the requirement of afforestation policy's adjoining and adjacent areas, and it is not easy for a joint application with other people" occupies 25.0%, followed by "landowners' participation in the PLAP might lead to inflexibility of land use" occupying 20.0%, and "landowners' participation in the PLAP might reduce the chance for the change of land use", which occupies 18.9%. In addition, according to the empirical analysis of this paper, age, education level and annual household income reported the most significant influence among the factors that determine landowners' participation in PLAP. The older the private landowner is, the higher probability she/he has in participating in the PLAP. The higher the education level of a landowner is, the lower participation probability is. The higher the annual household income of a landowner is, the lower participation probability is. Among the cropland characteristics, cropland area, cropland price and number of cropland parcels reported the most significant influence on participating in the afforestation program. The larger the cropland area is, the higher participation probability is. The higher the cropland price is, the lower participation probability is. The more the number of cropland area is, the lower the participation probability is, which meets the theoretical expectation in this paper.

The total afforestation area implemented in the plain areas between 2002 and 2005 is 8,010.32 hectares (Taiwan Sugar Corporation and private cropland), which comprised only 8.51% of total afforestation area in the plain areas and had a diminishing tendency from year to year. In order to increase the incentives to the private cropland that participates in the PLAP, it is necessary for the government to conduct propaganda and establish related incentive schemes; especially, the first-tier city governments and township office units should be more active on guidance and promotion. According to the survey result in this paper, the reason why the landowners do not participate in the PLAP is their unawareness of the PLAP, which represents up to 60%. The government needs to reinforce the outreach and promotion mechanism. The reasons for the farmers who are aware of the PLAP but choose not to participate include too lengthy contract duration (20 years), uncertainty of cropland value and purpose of wood use, in addition to their primary cropland areas failing to meet the regulation, which occupies 50%. Only 10% is recognized a low subsidy amount. It is evident that the government should make an explanation in detail in terms of 20-year subsidy duration, cropland value, wood use after 20 years so as to solve farmers' confusion to effectively enhance the incentives that attract farmers to participate in the PLAP.

According to the empirical result in this paper, age, education level, and annual household income factors reported a significant influence among the landowner characteristics. The older the private landowner is, the higher probability she/he has in participating in the PLAP. For every year increase on age, a 3.5% probability will be increased accordingly. The higher the education level of a landowner is, the lower participation probability is. When the education level of a landowner is increased from under junior high school to over senior high school level, the participation probability for PLAP will be decreased by 6.4%. The higher the annual household income is, the lower participation probability is. When annual household income is increased by one unit, the participation probability will be increased by 6.9%. Therefore, the government should actively encourage the landowners with three overlapping characteristics of non-farmer occupation, higher education level (junior high school above) and older landowners to establish promotion mechanism and assist with moral persuasion to encourage these landowners to participate in the PLAP.

Moreover, according to the empirical result in this paper, those landowners with larger cropland area, less number of cropland parcels and lower cropland price have higher probability in participating in the PLAP, in which the number of cropland parcels reported the most significant influence. This paper also indicated that when the number of cropland parcels is increased by one unit, the participation probability will be decreased by 74.7% accordingly. When the cropland area is increased by one unit, the participation probability will be increased by 3.3% accordingly. When cropland price is increased by one unit, the participation probability will be decreased by 55.5%. Therefore, the government should conduct promotion to the cropland that meets these characteristics to encourage the farmers to participate in PLAP.

Moreover, it helps understand the situation of goodness of fit of the model established by this paper through many statistics indices, such as Nagelkerke's R^2 value is 0.695 which implies that the regression capability of the model is good. Hosmer-Lemeshow test and χ^2 value of omnibus test also revealed that the Logit model in this paper may provide fine goodness of fit. The Logit model established by this paper has a probability of 98.0% in predicting nonparticipants, and a probability of 71.8% in predicting the participants, with 95% probability for prediction in overall. The empirical result of this paper expects to help implement the PLAP in Taiwan.

5. References

Alig, R.J., K.J. Lee and R.J. Moulton (1990). Likelihood of Timber Management on Nonindustrial Private Forests: Evidence from Research Studies, General Technical Report SE-60, U.S. Department of Agriculture, Forest Service, Southeastern Forest Experiment Station Press, Asheville, North Carolina

Binkley, C.S. (1981). Timber Supply from Nonindustrial Forests, Bulletin No.92., School of Forest and Environmental Studies Press, Yale University, New Haven

Boyd, R. and W. Hyde (1989). "Forestry Sector Intervention: The Impacts of Public Regulation on Social Welfare." Ames, IA: Iowa State University Press. 295p.

Cubbage, F.W. (2003). The Value of Foresters. Forest Landowner, Vol. 62, No. 1, pp. 16-19

Cox, D.R. andSnell, E.J. (1989). Analysis of Binary Data, 2nd Ed, Chapman and Hall Press, London

Dennis, D.F., 1989. "An Economic Analysis of Harvest Behavior: Integrating Forest and Ownership Characteristics." Forest Science 35(4): 1088–1104.

English, B.C., C.D. Bell, G.R. Wells, and R.K. Roberts, 1997. "Stewardship Incentives in Forestry: Participation Factors in Tennessee." The Southern Journal of Applied Forestry 21(1): 5–10.

Ervin, C.A., and D.E. Ervin, 1982. "Factors Affecting the Use of Soil Conservation Practices: Hypotheses, Evidence, and Policy Implications." Land Economics 58: 277–292.

Esseks, J.D. and R.J. Moulton, 2000. "Evaluating the Forest Stewardship Program through a National Survey of Participating Forestland Owners." DeKalb, IL: The Center for Governmental Studies, Social Science Research Institute, Northern Illinois University Press. 113p.

Hammett, A.L., F.W. Cubbage and W.G. Luppold, 1992. "A Logistic Regression Model of Southern Hardwood Lumber Export Participation." Wood and Fiber Science 24(3): 315–329.

Hair, J.R., R.E. Anderson, R.L. Tatham and W.C. Black, 1998. Multivariate Data Analysis. 5th Ed., NJ: Prentice Hall Englewood Cliffs Press.

Hardie, I.W. and P.J. Parks, 1991. "Individual Choice and Regional Acreage Response to Cost-Sharing in the South, 1971–1981." Forest Science 37(1): 175–190.

Hosmer, D.W., and S. Lemeshow, 1989. Applied Logistic Regression. New York: John Wiley and Sons, Inc. Press.

Hyberg, B. and D. Holthausen, 1989. "The Behavior of Nonindustrial Private Forest landowners." Canadian Journal of Forest Research 19: 1014–1023.

Korsching, P.F., C.W. Stofferahn, P.J. Nowak, and D. Wagener, 1983. "Adoption Charateristics and Adoption Patterns of Minimum Tillage: Implication for Soil Conservation Program." Journal of Soil and Water Conservation 38: 428–430.

Lorenzo, A.B., and P. Beard, 1996. "Factors Affecting the Decisions of NIPF Owners to Use Assistance Programs." pp. 264-275 in Proc. of Symp. on Non-Industrial Private Forests: Learning from the Past, Prospects for the Future, University of Minnesota, St Paul, MN.

McNamara, K.T., M.E. Wetzstein, and G. K. Douce, 1991. "Factors Affecting Peanut Producer Adoption of Integrated Pest Management." Review of Agricultural Economics 13(1): 129–139.

Megalos, M.A., 2000. "North Carolina Landowner Responsiveness to Forestry Incentives." Raleigh, NC: Ph.D. Dissertation, North Carolina State University Press. 119p.

Nagelkerke, N.J.D., 1992. "Maximum Likelihood Estimation of Functional Relationships." Lecture Notes in Statistics 69, 110p.

Nagubadi, V., 1995. "Private Landowner Forestry Program Participation Behavior." West Lafayette, IN: M.S. thesis, Pudue University Press.

Nagubadi, V., K.T. McNamara, W. L. Hoover, and W. L. Miller, 1996. "Program Participation Behavior of Nonindustrial Forest Landowners: A Probit Analysis." Journal of Agricultural and Applied Economics 28(2): 323–336.

Napier, T.L., C.S. Thraem, A. Gore, and W. R. Goe, 1984. "Factors Affecting Adoption of Conventional and Conservation Tillage Practices in Ohio." Journal of Soil and Water Conservation 38: 205–205.

Rahm, M.R., and W.E. Huffman, 1984. "The Adoption of Reduced Tillage: The Role of Human Capital and Other Variables." American Journal of Agricultural Economics 66: 405–413.

Romm, J., R. Tuazon, and C. Washburn, 1987. "Relating Forestry Investment to the Characteristics of Nonindustrial Private Forestland Owners in Northern California." Forest Science 33(1): 197–209.

Royer, J., 1987. "Determinants of Reforestation Behavior among Southern Landowners." Forest Science 33(3): 654–667.

Sills, E.O. and K.L. Abt, 2003. Forests in a Market Economy. Kluwer Academic Publishers Press.

Stevens, T.H., D. Dennis, D. Kittredge, and M. Rickenbach, 1999. "Attitudes and Preferences toward Co-operative Agreements for Management of Private Forestlands in the Northeastern United States," Journal of Environment Management 55(2): 81–90.

Zhang D. and W. Flick, 2001. "Sticks, Carrots and Reforestation Investment." Land Economics 77(3): 443–456.

12

Impact of Charcoal Production on Biodiversity in Togo (West Africa)

Jérémie Kokou Fontodji[1], Honam Atsri[1], Kossi Adjonou[1], Aboudou Raoufou Radji[1], Adzo Dzifa Kokutse[1], Yaovi Nuto[2] and Kouami Kokou[1]
[1]*Laboratoire de Botanique et Ecologie Végétale, Faculté des Sciences, Université de Lomé, BP 1515, Lomé*
[2]*Département de Zoologie, Faculté des Sciences, Université de Lomé, BP 1515, Lomé*
Togo

1. Introduction

In Togo, firewood and charcoal contribute more than 80% to the total domestic energy requirements (Kokou et al., 2009). This traditional energy represents about an annual average consumption of 419,964 tons per year on national level (Fontodji, 2007). This incessant growing request for traditional energy is ensured by national production. Unfortunately, this important consumption of charcoal had enormous consequences on the ecosystems since the methods used by the producers and the consumption modes of charcoal by households still remain archaic. Indeed, the production yield is about 15 to 20% i.e. 150 to 200 kg of wood produced per ton of woody material (Girard, 2002) and during the cooking with charcoal, the loss in energy is about 8 to 13% (Thiam, 1991). This situation is an important waste of wood causing the felling of a great number of trees. Moreover, the producers claim that the charcoal is of good quality when part of the wood used is fresh. Therefore, the excessive exploitation of the forestry resources for energy purposes puts an important pressure on the ecosystems and leads consequently to serious harmful effects on the environment and the biodiversity preservation. The objectives of this article are: (1) to analyze the impact of charcoal production on the natural ecosystems; (2) to assess the pressure on the key plant species as well as (3) on the soil physico-chemical properties soil and fauna diversity inside the kilns; (4) to assess how national environment policies and laws regulate charcoal production activities.

2. Material and methodology

2.1 Socio-economic and environmental status of Togo

Togo is situated on the coast of the Benin Gulf (Fig. 1). It has an area of 56,600 km^2 and is bordered at the south by the Atlantic Ocean, at the north by Burkina Faso, at the east by Benin Republic and at the west by Ghana. Situated between the 6° and 11° of latitude north and between 0° and 2° of longitude east, the country stretches over 600 km from north to south, and over 50 km from east to west (at the level of the coast). Its landscape is generally flat, except the Atakora range which crosses it diagonally from the south-west to the north-east, with tops reaching sometimes 900 m in the southern part.

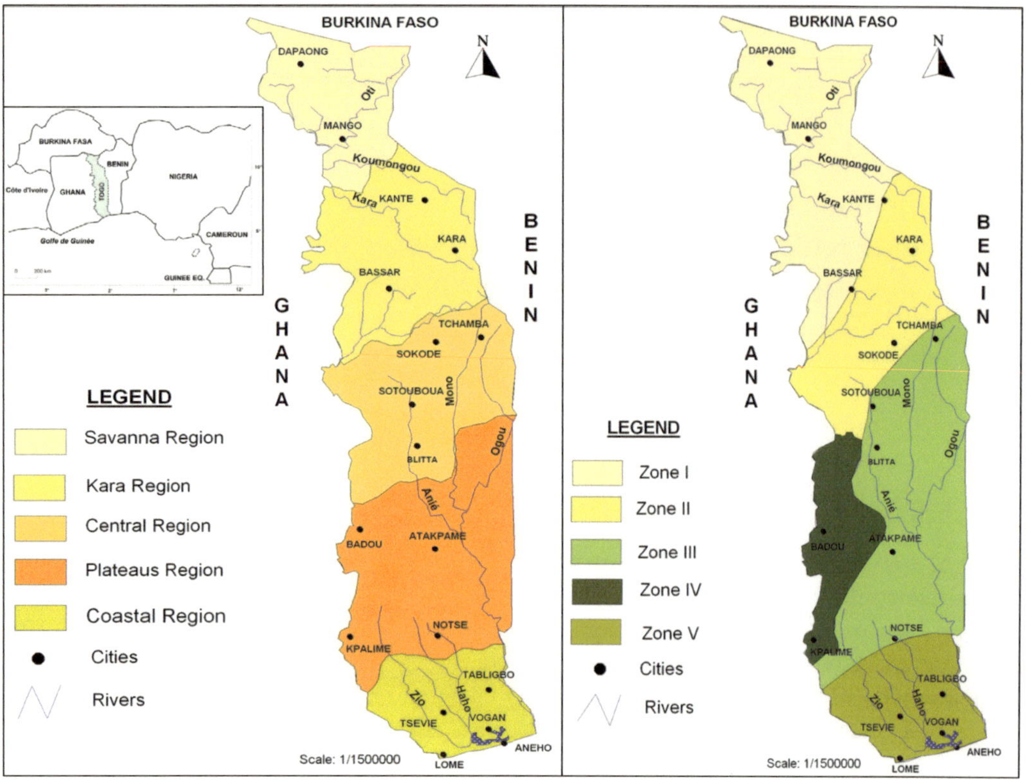

Fig. 1. Administrative regions (left) and phytogeographical zones (right) of Togo. **Legend.** Zone I: Refers to the plains in the north with a sudanian climate characterized by a rainy season running from June to October and a dry season from November to May, with an average of 6–7 dry months. Total rainfall is between 800 and 1000 mm. The predominant vegetation is Sudanian savanna. Zone II: Concerns the hilly dry forest and savannah mosaic zones in the north with a soudano-guinean climate including cool nights at elevated regions, and with a rainy season from April to October and a dry season from October to March, characterized by the harmattan. Yearly rainfall is 1200–1300 mm, but is very irregular within the year and between years. Zone III: Concerns Guinean savanna in the plains in the centre of the country having a tropical climate characterized by a rainy season from May till October and a dry season of at least 4 months. Total rainfall is between 1200 and 1500 mm per year. The savanna of Central Togo is interspersed with islands of semi-deciduous forest in the southern part and with dry forest in the northern parts. Zone IV: Corresponds to the southern part of Togo Mountains with a transitional subequatorial climate characterized by a long rainy season occurring from March to October and disrupted by a slight decrease in rainfall in August or September. The rainfall amounts range from 1300 to 1600 mm per year. Here, semi-deciduous moist forests are the major vegetation type. Zone V: Refers to the coastal plain in the south under a subequatorial climate characterized by a shortage of rainfall in the coastal part (800 mm/year in Lomé, the capital). The landscape offers a mosaic of savanna, agricultural land and preserved forest. Islands of dry semi-deciduous forest are found mainly as sacred forest or as classified forest.

The country is divided into five ecological zones (Ern, 1979). It is also subdivided into five administrative regions (Fig. 1). The impact of poverty is estimated at 61.7% of the population. Poverty is essentially found in rural areas where the impact is of 74.3%. Poverty in rural areas is very acute in the Savanna, Central and Kara regions with respective impacts of 92.5%, 84%, and 80%. There is a rapid tendency of urbanization, caused by a great rural exodus. The Togolese population is extremely young, with a high percentage of women. It is also characterized by an extreme growing poverty (152nd country on 177). The current unemployment rate is estimated between 25% and 33%. That is why charcoal production is one of the main activities in rural areas all over the country.

2.2 Surveys

Investigations have been conducted in four supplying basins of charcoal, SB1, SB2, SB3, and SB4, situated inside the ecological zones I, II, III, and IV, respectively (Fig. 2). First, surveys were conducted with 310 charcoal producers from supply basin. Semi-structured questionnaires, oral interviews and group discussions (focus groups) were held with charcoal producers. Woody species used in making charcoal and the preferred species were identified and their availability was systematically recorded. Data relatives to the quantity of charcoal produced between 2004 and 2008 have been collected in the forest check points and regional directions.

The second step was to carry out relevés in homogeneous geomorphologic and physiognomic units. The number of plots applied lies between 8 and 18 in each basin. A total of 100 relevés were carried out inside the four charcoal production areas. The surface plot covered 900 m^2 (30 × 30 m). In each plot, all woody species were identified as present/absent. The nomenclature adopted was that of Lebrun & Stock (1991-1997). In addition, a forest inventory was conducted to measure the dendrometric parameters (circumference and height) of the woody species. The circumference ≥10 cm was measured at breast height (1.30 m of soil) by means of a ribbon. For multi-stemmed trees, all the stems have been measured. The height of woody plants has been measured by means of a Blum-Leiss. Inside the charcoal production area, 10 plots where randomly selected were five subplots of 25 m^2 (5 × 5 m) were created (in two of the four corners of each plot) to study the regeneration capability of the species. The regeneration capability in exploited plots has been compared with unexploited plots selected inside the charcoal production areas. In total, 200 subplots (100 unexploited plots and 100 in exploited plots) were registered. The woody species <10 cm in circumference are the individuals taken into account in the regeneration. These stems were counted by species and by methods of regeneration (seedling, sprouting or coppicing and suckering).

At the level of the kiln, soil bulk density, total porosity and soil permeability were determined as described by Soltner (1982). The bulk density is a measure of the weight of the soil per unit volume (g/cc), usually given on an oven-dry (105°C) basis, while the total porosity was calculated from values of soil bulk density. For instance, organic carbon was evaluated by wet oxidation using the Walkley method (Walkley, 1947). The organic matter content was computed from the organic carbon content through a conversion factor of 1.72. Soil pH was determined by conventional standard procedures (AFNOR, 1992) in soil water (1:2) medium using digital electronic pH meter.

Fauna assessment was carried out inside the kiln by microbial biomass quantification and fauna species inventory. The microbial biomass was quantified using the technique of soil respiration measurements, i.e. with regard to the amount of CO_2 produced (Schinner et al., 1996). First of all, the soil samples were tested with a hand magnifier and the macro organisms were collected. Then, samples were put through a "Berlèse" device to collect smaller arthropods. The experience consisted in placing the soil samples in a funnel fitted with a mesh sieve from 1 to 5 mm and to expose them to a light source. The insects fleeing the heat, the dryness, and especially the light fall into a becker containing a fixer alcohol 70° under the funnel. Species were identified using a determination key (Coineau, 1974) and a reference of samples of the University of Lomé (Togo).

The last stage of the study consisted in the review of the texts regulating the charcoal production activity, extended to the laws in use with respect to the protection and environment management. The review was also focused on the action plan of the forestry sector. The information collected were completed with the discussions with the Regional Directors of the Environment and the *Office de Développement et d'Exploitation des Forêts (ODEF)*, which is the national institution in charge of charcoal production and trade in Togo.

3. Impact of the charcoal production on the natural ecosystems

In Togo, charcoal production is greatly concentrated in the Plateaus, Central, and Savanna Regions (Table 1). It is in those regions that there is still enough vegetation, mostly in the Plateaus Region because of its forests. On the other hand, natural vegetation is almost inexistent in the Kara and Coastal regions. Kara Region is one of the regions where charcoal production was very high but which today is the region where the quantity of charcoal produced is lowest because of the poor vegetation.

Region	Year				
	2004	2005	2006	2007	2008
Coastal	7,845.6	5,238.3	-	-	-
Plateaus	44,785.2	23,314.2	-	-	23,825.4
Central	19,873.5	12,027	37,441.8	46,042.5	72,312.3
Kara	1,613.4	75	-	-	127.9
Savannas	-	-	8,403.3	10,170.9	12,009.9

Table 1. Quantity of charcoal produced per region (in tons) -: *Data not available*

The extent of charcoal production during these past years in Togo is becoming more and more alarming. For instance, the average production of charcoal on national level, calculated over five years (2004-2008), is of 419,963.8±47,896.5 tons per year (Table 2).

This important consumption of wood is one of the main reasons of deforestation and vegetation degradation in Togo, especially in Plateaus, Central and Savanna regions considered as the main basins of provision in the country (Table 1). The woody plant biomass destroyed by this activity is about 2,799,759 tons per year (Table 2). MEMEPT (2002) indicates that logging for fuel wood is the third reason of the deforested areas, i.e. 5,000 hectares each year. This high rate of deforestation is changing with population growth; for an average population estimated at 5,596,000 inhabitants, the yearly average consumption per inhabitant is then 75 kg.

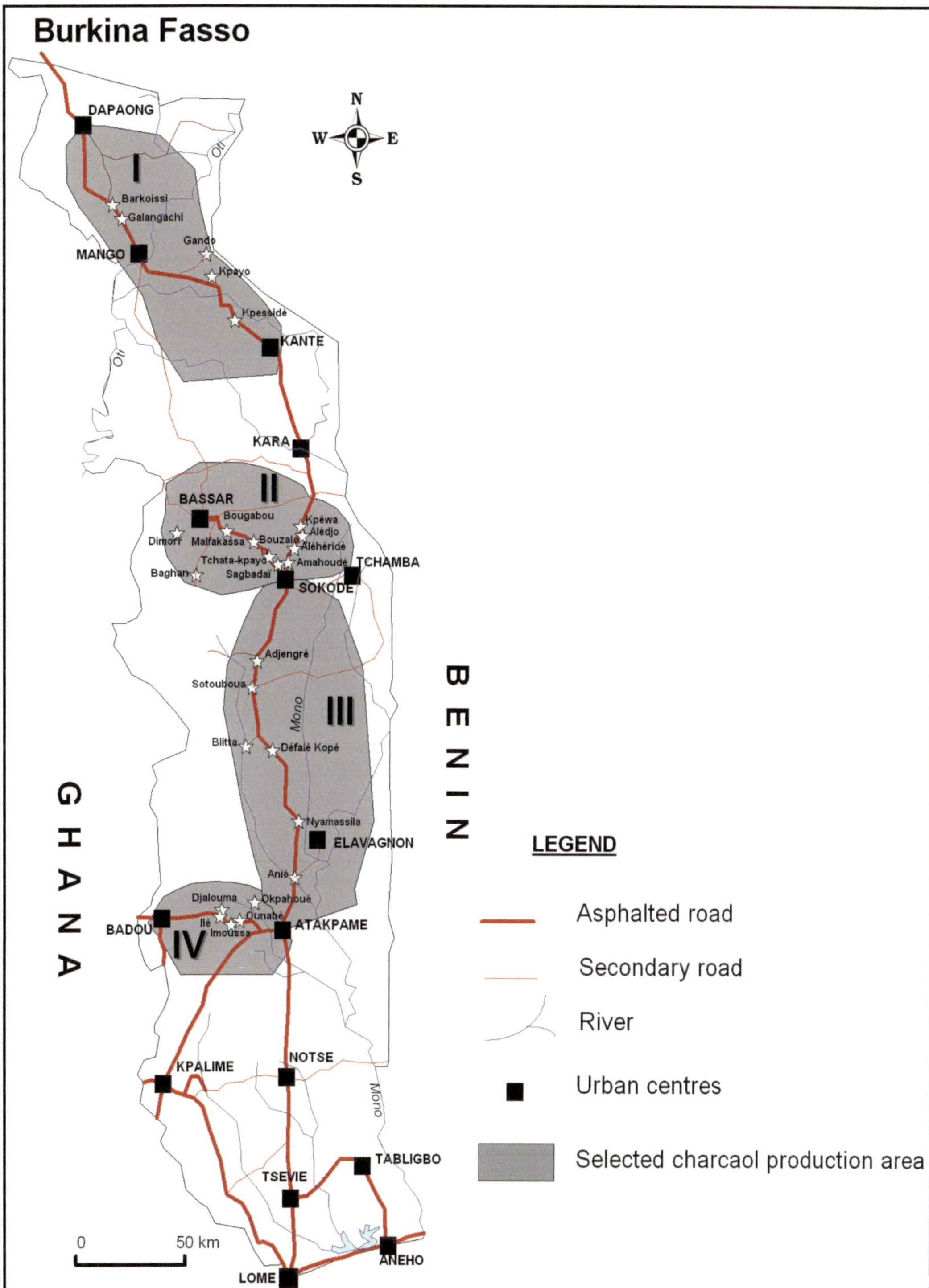

Fig. 2. Location of charcoal production areas in the phytogeographical zones of Togo

Year	Quantity of charcoal controlled toward Lomé (tons)	Quantity of charcoal controlled at the national[1] level (tons)	Real quantity of charcoal produced at the national[2] level (tons)	Quantity of wood destroyed[3] (tons)
2004	62,017.4	151,262	453,785.9	3, 025,239.3
2005	54,056.9	131,846.1	395,538.3	2,636,922.0
2006	51,140	124,731.7	374,195.1	2,494,634.0
2007	53,269.8	129,926.3	389,779	2,598,526.8
2008	66,491.2	162,173.7	486,521	3,243,473.3
Average per year	57,395.06±6,545.9	139,987.95±15,965.5	419,963.8±47,896.5	2,799,759.1±319310

[1]Calculated on the basis of the data from the check points at the entry of Lomé, capital of Togo; it represents 41% of charcoal trade in Togo (MEMEPT, 2002)
[2]Estimation on the basis of the results of surveys; only a third of charcoal toward Lomé is controlled
[3]Calculated on the basis of the minimum output of wood carbonization, i.e. 15% (Thiam, 1991)

Table 2. Assessment of the quantity of charcoal produced/consumed on national level

This situation affects particularly protected areas in Togo. For instance, in the Savanna Region (ecological zone I), the most affected protected area is the Oti-Keran National Park (OKNP). The production of charcoal is the main activity that weighs today on the park while formerly it was just the activity of some few women to cater for their needs. But today, charcoal production is the main activity around the OKNP (Fig. 3), especially produced by fresh wood felled from the park. Charcoal depots in sacs of 50 and 100 kg can be counted by dozens inside and around the park (Fig. 4).

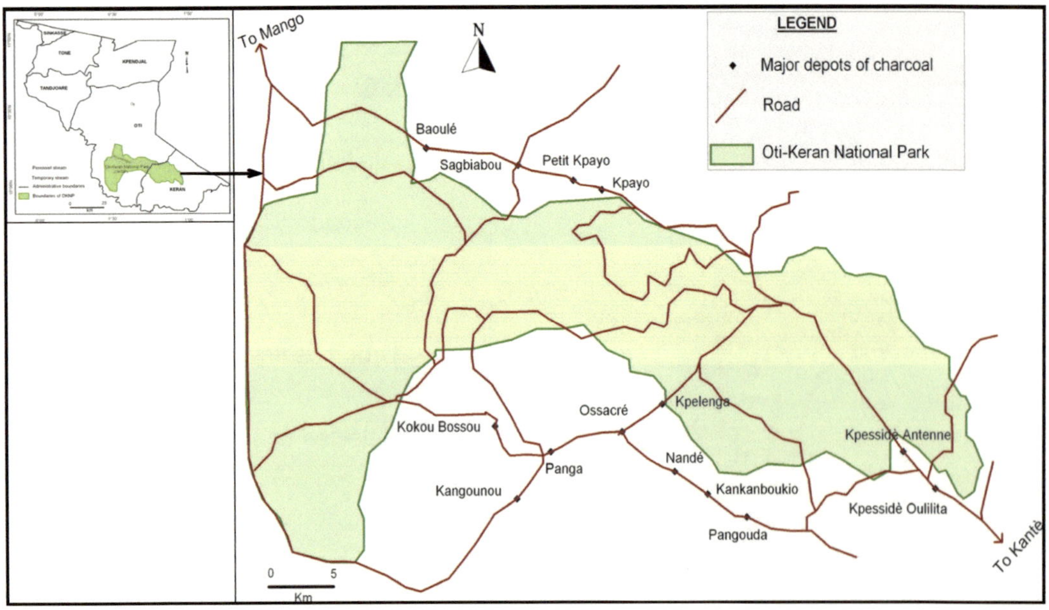

Fig. 3. Main charcoal depots inside and around Oti-Keran National Park

Fig. 4. Depot of a week's production by a village of 200 people, to be loaded to town

4. Impact on plants species used in charcoal production

During a survey in the main charcoal production basin in Togo, Kokou et al. (2009) found that the production is focused on 15 native plant species classified as the species of the first choice or species of category 1 among which *Anogeissus leiocarpa, Erythrophleum suaveolens, Prosopis africana, Burkea africana, Detarium microcarpum, Lophira lanceolata, Vitellaria paradoxa,* etc. (Table 3). In areas of Sudanian savanna and dry forest, *A. leiocarpa, L. lanceolata, B. africana, V. paradoxa* and *P. africana* are by far the most exploited species which provide the best wood for charcoal production. In Guinean savanna and semi-deciduous forest areas, *E. suaveolens* is added to these "key" species. The interest shown in these species is related to the quality of their charcoal, which is highly valued by the consumers in the cities. Due to a lack of adequate policy in reforestation in Togo, these species are overexploited and many of them become rare. The charcoal producers are of the view that highly valued species (*Afzelia africana, L. lanceolata* and *P. africana*), are no more available.

Apart from these first choice species, others are increasingly being exploited because of their availability. These species are mainly used by default because, a decade ago, were not used in charcoal production in Togo. These species exploited by default or species of category 2 or alternative species [sometimes with very small diameter, inferior to 3 cm]. They are 19 of number, among which *Cola gigantea, Gmelina arborea, Parinari curatellifolia, Parkia biglobosa, Tectona grandis, Diospyros mespiliformis,* etc. (Table 4). The use of these species of the second category, mainly those used in reforestation such as teak, is an alternative towards the reduction of the current pressure on natural vegetation. The choice of the resource is not only related to the energy quality but also to the availability. Therefore 34 woody species are popularly used for charcoal production. Two families (Caesalpinaceae and Combretaceae) consist of 60% of these species.

In terms of the regeneration of the resources on the field, two most common types of regeneration of these species are seedlings and coppices (Kokou et al., 2009). The highest densities are observed in both types of regeneration all over the charcoal production areas.

Inside Sudanian savanna and dry forest areas, the densities are very high and can double after exploitation. Inside Guinean savanna and semi-deciduous forest, density differences are not so high. After exploitation, coppicing becomes the preferred method of regeneration of the stumps inside all the charcoal production areas. In the Sudanian savanna and dry forest areas, the exploitation lead to suckering that resulted in doubling or tripling of densities. Inside the unexploited areas, the distribution of the regeneration, all methods inclusive, is inconsistent (Fig. 5). The high penetration of light due to the destruction of part of the woody biomass favours the germination of seed stock of exploited species. These results are consistent with those of Ouédraogo (2006) in Sudanian savannas and those of Neke et al. (2006) in South Africa, which revealed that the high numbers of seedlings may be due to a response to the thinning of the woodland and the opening up of the canopy through harvesting.

Species	Diversity Index of the charcoal plant species							
	SB1		*SB2*		*SB3*		*SB4*	
	N_u	*IVI*	N_u	*IVI*	N_u	*IVI*	N_u	*IVI*
Anogeissus leiocarpus	79^{d}	44.6	_	_	96^{d}	15.1	70^{pd}	50
Afzelia africana	27.4^{pd}	0	20.6^{d}	22.4	_	8.6	30^{pd}	37.5
Burkea africana	72.6^{d}	75.9	67.6^{d}	37.4	50^{d}	39.1	_	14.2
Canthium schimperianum	_	_	_	_	_	_	16.3^{d}	0
Combretum spp.	8.8$^{td;d}$	89.2	_	30.2	21^{td}	14.5	37.5^{td}	28.5
Crossopteryx febrifuga	_	58	_	65.5	_	14.5	25^{d}	70.1
Detarium microcarpum	50.1^{td}	58.8	75^{td}	94.2	_	6.3	_	0
Dialium guineense	_	_	_	_	_	_	13.8^{d}	0
Erythrophleum suaveolens	_	_	_	_	88^{d}	0	68.8^{pd}	64.9
Lophira lanceolata	53.2^{d}	0	_	63.7	37^{d}	28.8	42.5^{d}	43.2
Prosopis africana	88.7^{d}	6.3	82.4^{d}	15.2	64^{pd}	12.5	26.3^{d}	0
Pterocarpus erinaceus	48.4^{d}	53	8.8^{d}	32.9	60^{pd}	78.6	50^{d}	29.1
Terminalia glaucescens	_	53.9	73.5^{td}	71.7	_	62.5	36.3^{d}	18.9
Terminalia laxiflora	_	_	_	_	33^{d}	16.1	_	0
Vitellaria paradoxa	56.1^{td}	46.4	20.6^{d}	59	87^{d}	52.3	37.5^{d}	0

Table 3. Preferred species for charcoal production in Togo (Category 1)
SB: Supply Basin; _: species non quoted by the producers; td: very available; d: available; pd: little available; $N_u = \frac{N_{ui}}{N_{t1}} \times 100$ (N_{ui} =Level of use of the species, N_{t1} number of people interviewed in favour of the use of the species, and N_{t1}: total number of respondents in each supply Basin in Togo) [Equation 1]; Importance Value Index (IVI) per species is obtained by the relationship IVI relative density + relative frequency + relative dominance [Equation 2], (relative density $D_r = 100x\frac{N_i}{N_{t2}}$ [Equation 3], where N_i = number of individuals of *i* species; relative frequency $F_r = 100x\frac{N_r}{N_{t3}}$ [Equation 4], where N_r = number of plots in which the I species is present and N_{t3} = total number of relevés; relative dominance $D_{or} = 100x\frac{g_i}{G_T}$ [Equation 5], where gi = total basal area of a species and GT = total basal area of all species

Species	Diversity Index of the charcoal plant species							
	SB1		*SB2*		*SB3*		*SB4*	
	N_u	*IVI*	N_u	*IVI*	N_u	*IVI*	N_u	*IVI*
Afromorsia laxiflora	24.2[d]	0	–	21	–	–	–	–
Albizia spp.	–	–	–	5.6	–	12.5	15[d]	29.2
Bridelia ferruginea	–	40.9	–	17.5	24[td]	51.8	–	28.3
Cola gigantea	–	6.3	–	5.6	–	25	13.8[d]	25.7
Daniellia oliveri	24.2[td]	9.8	–	76.5	15[d]	13.9	20[td]	23.3
Dichrostachys cinerea	–	50.8	–	55.6	–	100	20[td]	50.7
Diospyros mespiliformis	29[pd]	13.9	–	–	69[pd]	25	31.3[d]	0
Faurea speciosa	–	–	–	–	–	–	60[td]	16.2
Gmelina arborea	–	6.3	–	21	–	37.5	33.8[td]	0
Hexallobus monopetalus	–	–	–	–	–	–	27.5[td]	0
Hymenocardia acida	–	13	–	38.3	–	37.5	20[td]	41.9
Isoberlinia spp.	–	14	44.1[td]	48.5	–	51.2	–	25
Khaya spp.	45.2[d]	15	–	9.4	42[d]	12.7	18.8[pd]	25.3
Parinari curatellifolia	–	15.2	–	46.1	–	42.1	27.5[td]	41.8
Parinari glabra	–	–	–	–	–	–	11.3[d]	56.1
Parkia biglobosa	21[td]	12.9	–	5	–	12.8	17.5[d]	13.3
Pseudocedrela kotschyii	–	20.2	–	27.8	–	64.9	25[td]	50
Tectona grandis	–	–	–	2.1	39[td]	–	–	25
Vitex doniana	–	–	–	11.9	–	–	20[d]	2.3

Table 4. Species used by default for charcoal production in Togo (Category 2)

The deterioration of the vegetation in Togo because of the charcoal production activities (Fig. 6) is not a situation unique to Togo but observed generally in the developing countries. Indeed, it has been proved that the extraction of wood from the tropical forests for charcoal production, firewood, and timber constitutes 68% of the causes of deforestation in Africa, 89% in Asia and 51% in Latin America (Geist & Lambin, 2001). A study conducted by the World Bank in six countries of West Africa, reveals that charcoal production is the main source of deforestation in countries where they produce a lot (Ninnin, 1994). In China, the excessive consumption of charcoal which goes with the population growth is the main source of deforestation (Chen et al., 2005). In Uganda, the high request for charcoal has an incessant pressure on the forest, which leads to a deforestation rate of 600 km^2 per year (Banana & Gombia-Ssembajjwe, 1996) in Uganda. Makundi (2001) relates that 75% of the destroyed forests are attributable to the consumption of ligneous energy (charcoal and firewood) in Tanzania.

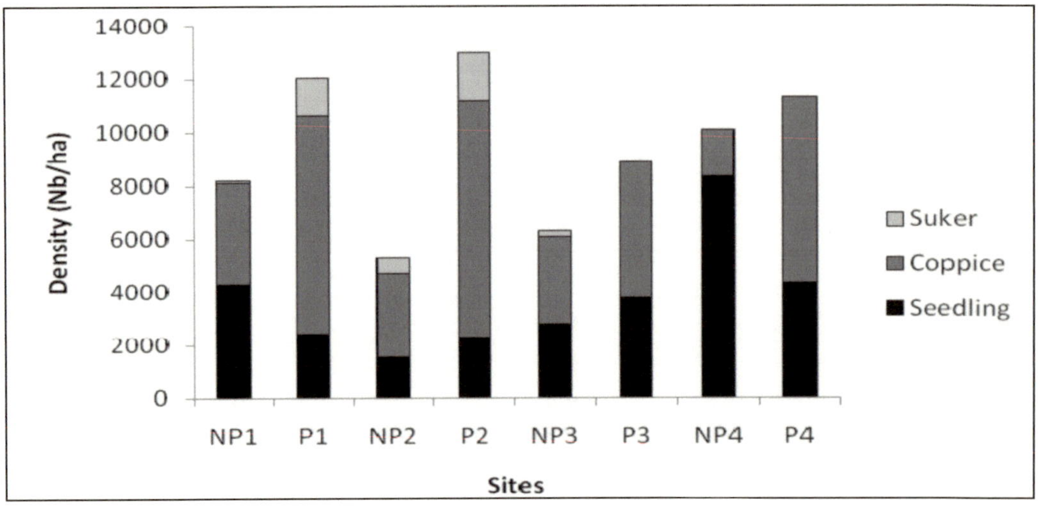

Fig. 5. Mode and density of regeneration in Ps and NPs.

P1: exploited zone in SB1, NP1: unexploited zone in the SB1; P2: exploited zone in SB2, NP2: unexploited zone in SB2; P3: exploited zone in SB3, NP3: unexploited zone in SB3; P4: exploited zone in SB4, NP4: unexploited zone in SB4.

Fig. 6. Exploited area for charcoal production in Sudanian savanna zone

5. Impacts of charcoal production on soil properties and biodiversity: Case study of hypogeal soil fauna

Fontodji et al. (2009) counted and measured the sizes of charcoal kilns in order to know their area and their distribution in the supplying basins in Togo. The results show that the average area of the kilns is of 13.29±4.4 m^2. But the setting up of the kilns needs a total clearing of the vegetation on average area of 11.5±1.29 m^2. The number of kilns per hectare is of 71.5±18.3 in average. The area occupied by the kilns on the carbonization site has been estimated at 950.2±2.73 m^2/ha, that is, approximatively 10% of the farming areas in charcoal production zones.
After burning, the physico-chemical characteristics soils inside the kilns change, notably the pH values ranged between 4.61 and 8.00. The values are more important in Guinean area (Guinean savanna and Semi-deciduous forest) than in Sudanian savanna (Table 5). Bulk density was from 1.12 to 1.57, while the total porosity was between 0.39 and 0.56 and permeability between 0.015 and 0.711. Bulk density of soil under charcoal kilns decreased as compared to witness plots. The fire increased the permeability at the kilns level by raising the total porosity of soil. Comparing different charcoal production areas, soil total porosity is higher in Guinean savanna area than in Semi-deciduous forest. The total porosity of the soil is more important in Sudanian savanna than in Guinean savanna (Table 6).

Charcoal production area	Kiln	Witness
Sudanian savanna	5.58	5.48
Guinean savanna	6.11	5.23
Semi-deciduous forest	8.00	4.61

Table 5. Soil pH variability

Soil characteristic	Kiln			Witness		
Charcoal production area	Permeability (cm/mn)	Bulk density (D′)	Total Porosity (1-D′/D)	Permeability (cm/mn)	Bulk density (D′)	Total Porosity (1-D′/D)
Sudanian savanna	0.022	1.46	0.438	0.015	1.575	0.394
Guinean savanna	0.675	1.257	0.516	0.056	1.486	0.428
Semi-deciduous forest	0.711	1.120	0.569	0.284	1.171	0.549

Table 6. Soil bulk density, total porosity and permeability

Fontodji et al. (2009) dealt also with the activity of the soil fauna at the level of the kiln by measuring the change in microbial mass with regard to the amount of CO_2 produced (soil respiration). As CO_2 evolution in the soil results from the decay of organic matter (soil biological activity), the results of this survey showed that the reduction of CO_2 caused by fire is different according to the ecosystem. It is the most important in Sudanian savanna

(Fig. 7). In savannas (Sudanian or Guinean) the variation in microbial biomass reaches a radius of 15 m around the kilns while this radius is 5 m in Semi-deciduous forest.

In addition, the inventory of soil fauna allowed to record 81 micro-organisms belonging essentially to beetles, Hymenoptera, Heteroptera, Diptera, Orthoptera, myriapods, Homoptera, mites, nematodes, Isoptera, etc. The hypogeal biodiversity indexes (average species richness, Shannon average index and average evenness) were more important in Semi-deciduous forest areas than in Sudanian or Guinean savannas (Table 7). Farther moving away from the kilns, greater the species diversity was. The curve representing specific richness (N_0) and Shannon index tend to stabilise beyond a radius of 10 m from the kiln (Fig. 8). Semi-deciduous forests provided the most microorganisms and Sudanian savannas provided less micro-organism than Guinean savannas. The soil fauna is more diversified in surface than in depth. But in Guinean savanna, the results of this study revealed more diversified fauna in depth than in surface (Fig. 9). This could be explained by the fact that in savannas the surface of soil is completely dried by the effect of the dry season (6 to 8 months of dryness). In the semi-deciduous forests, whatever the season, there is always favorable moisture for the soil fauna because of the abundant waste, which not only keeps this preferential moisture, but serves as nutrients. In savanna, animals of the soil, in search of preferential moisture run off to the depth because of the heat of the dry season. The species are already facing survival difficulties before the event of fire which destroys them completely. In the dry season, the effect of fire only worsens temperature variations that have very serious effects on soil fauna. It is therefore reasonable that the effect of fire be broader in savannas than in forest.

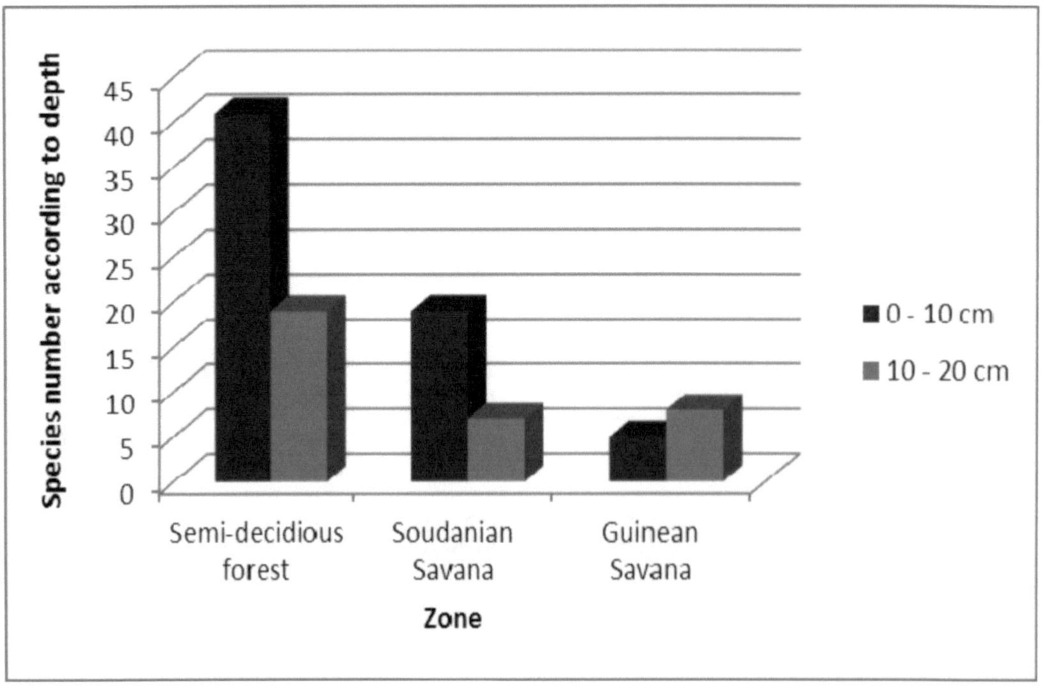

Fig. 7. CO_2 variation per zone

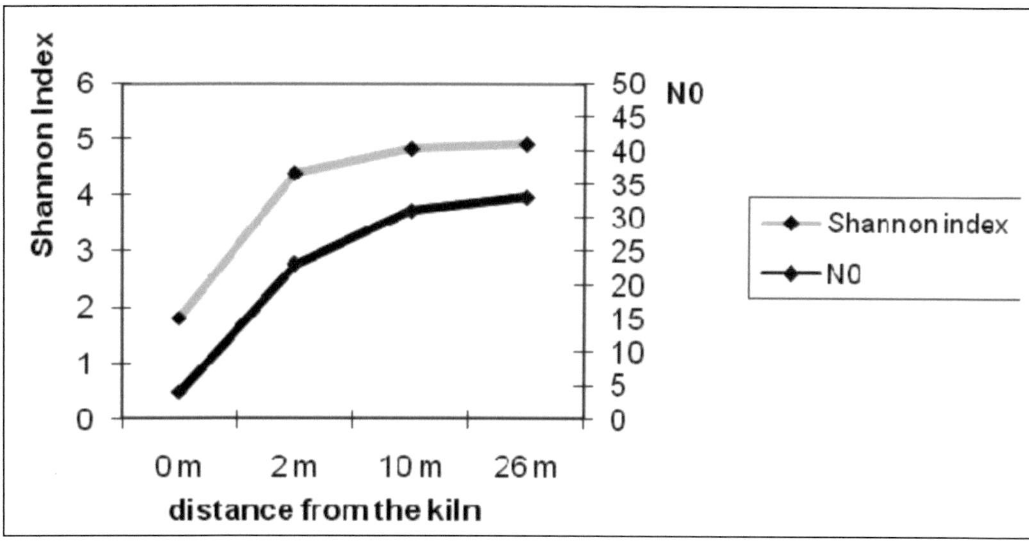

Fig. 8. Variability of diversity index in semi-deciduous forest area

Diversity indices	Semi-deciduous forest	Guinean savanna	Sudanian savanna
Specific richness	46.0	10.0	21.0
Shannon index	5.15	2.64	3.83
Evenness	0.93	0.80	0.87

Table 7. Variability of species diversity

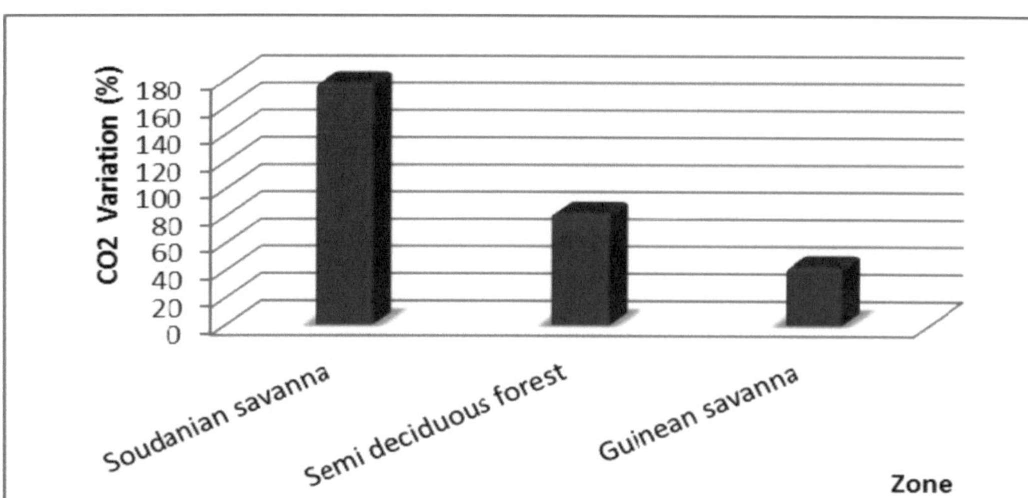

Fig. 9. Species number according to depth

Species richness (N_o) which represents the total number of species; Shannon index: $I_{sh} = -\sum_{i=1}^{n} pi \log(pi)$ [Equation 6], where $pi = \frac{qi}{Q}$, qi representing the population of i species and Q being the total population. Its value increases when the number of species of the collection gets large or reveals slightly different frequencies between the species encountered.

6. Environmental policies impacting charcoal production

The charcoal production has brighter days ahead to damage more the natural ecosystems and contributes more to the loss of the biodiversity because of the failure of the forestry policy (Kokou & Nuto, 2009). This authors show that there is a very significant decoupling between the policies and laws enforcement by the officials in charge of the protection and management of the environment. The direct consequence of this situation is the anarchical trees logging by the farmers for firewood and charcoal supply. Moreover the producers concerned at first place by the protection of the ecosystem are not aware of any laws. They are not adapted to actual realities in terms of protection and regulation of forest products movements. Also, land ownership legislative applied in Togo does not allow the government to conduct checking exercises on exploitation lands. Indeed, the customary land ownership regulations come from the clan organisation of the society which is characterized by the occupation of the autonomous territories by communities made up of all the descendants of a common ancestor. The lands located within the limits of clan's territories are at the disposal of all the members of the same lineage. They can clear the land for farming and produce charcoal when need be. Up to 2007, Togo does not have real forest resources laws. This explains the confusion that comes at the level of the sawyer and woodcutters arrests. There is a lack of modern material means which are needed by public services for the protection and the management of the environment. Though the existence of an Intervention Fund for the Environment, the public services which handle the management of the environment do not have financial means to carry out their daily duties. No coordination between different departments, while the Agricultural Ministry encourages the populations to develop cotton production by practising intensive slash-and-burn farming, the Ministry of Environment banned this technique. Inside the Ministry of Environment, the forest technical directions work independently and without an effective coordination. These administrative structures imply an overlapping of competences, a harmful competition, a weakening of responsibilities, a loss of means and opacity of responsibilities vis-à-vis the population and other public services. An intricate relationship does exist amongst the forest officials on one hand, and between these officials and the rural populations on the other hand. One of the directions is the ODEF created in 1971 to exploit process and sell forest products. In perspective, the Government intended to entrust the production and the marketing of charcoal to this institution. Unfortunately, ODEF gave up the job with time because of the financial and material difficulties that it encountered. In this context, these structural and organisational problems become subsided by the social ones. Some officials work for themselves and do not really bother for their being in the service only at the time of control by their seniors. Moreover, some recruit charcoal producers who produce charcoal for them. Therefore, policies and laws do not have any influence on charcoal production.

The lack of enforcement that followed increased the pressure on plant cover and the direct consequence being uncoordinated trees logging. The government does not play any role to regulate the charcoal market and does not influence the speculations. This situation is not specific to Togo but consistent with most countries in the sub-region (Kersten et al., 1998).

7. Conclusion

The production of charcoal has increased considerably these past years. The current study enabled to know the impact of this activity on the natural ecosystems in Togo. It results that the average annual consumption per inhabitant must be reassessed. Fixed at 59 kg/person/year in 1991 and at 62 kg/person/year in 2007, at present it have reached 75 kg/person/year. The deforestation rate caused by this activity is also rising, 2,799,759 tons/year. There is therefore an uncertain future for the forestry sector of Togo, since the production is not followed by a rational reforestation program. The protected areas in the country are therefore severely affected leading to irreversible degradation. The preferential species are overexploited and some of them are likely to disappear. The soil and its biodiversity (fauna) are highly damaged by fire and carbonization.

But, despite the negative effects that its production causes, charcoal can nevertheless be seen as a product offered gracefully by nature and constitutes a future energy, capable of meeting energy and environmental challenges of the XXIst century. In this situation, the development, the rationalization and the modernisation of the biomass uses, are therefore essential to a sustainable management of resources and to their renewal. It is therefore urgent to review the global policy concerning the setting of prices for energy wood to take into account, on the one hand, the increasing erosion situation of the forestry resources and to well develop the energy wood. It is also urgent to introduce a more adapted modern technique of carbonization such as the improved kiln used in Casamanse in Senegal Republic, called "Casamanse" kiln.

8. Acknowledgements

This study has been sponsored by African Academy for Sciences (AAS) and the African Forest Research Network (AFORNET).

9. References

AFNOR NF X 31-100. (1992). Soil quality. Sampling, Soil sampling method. Paris.

Banana, A.Y. & Gombia-Ssembajjwe, W. (1996). *Successful Forest Management: The importance of Security of Tenure and Rule Enforcement in Uganda Forest*, International Forestry Ressources and Institutions Research Program Series, Bloomington, United States

Chen, L.; Heerink, N.; & Berg, M.V.D. (2005). Energy consumption in rural China: A household model for three villages in Jiangxi Province. *Ecological Economics,* Vol. 58, pp. 407-420

Coineau, Y. (1974). *Introduction à l'étude des microarthropodes du sol et de ses annexes. Document pour l'enseignement pratique de l'écologie*, Doin, Paris

Ern, H. (1979). Die Vegetation Togos. Gliederung, Gefährdung, Erhaltung. *Willdenowia*, Vol. 9, pp. 295-312

Fontodji, K.J. (2007). *Impact de la production du charbon de bois sur les propriétés du sol et la biodiversité au Togo.* DEA Report, Univeriity de Lomé, Togo

Fontodji, K.J. ; Mawussi, G. ; Nuto Y. & Kokou, K. (2009). Effects of charcoal production on soil biodiversity and soil physical and chemical properties in Togo, West Africa. *International Journal of Biolology and Chemistry Sciences* Vol. 3, No. 5, pp. 870-879

Girard, P. (2002). Charcoal production and use in Africa: What future? Unasylva 211: 30-34

Geist, H. J. & Lambin, E.L (2002). Proximate causes and underlying driving forces of tropical deforestation. *BioScience,* 52(2): 143–150.

Kersten, I.; Baumbach, G.; Oluwolé, A., F.; Oboch, I. B. & Ogunsola, O.J. (1998). Urban and rural fuelwood situation in the tropical rain forest area of South-West Nigeria. *Energy* Vol. 23, No. 10, pp. 887–898.

Kokou, K. & Nuto, K. (2009). Assessment of Charcoal Production and Impact of Environmental Policies in Limited Forest Resources Countries: The case of Togo, West Africa. *Discovery & Innovation* Vol. 21 n° ½.

Kokou K.; Nuto, Y. & Atsri, H. (2009). Impact of charcoal production on woody plant species in West Africa: A case study in Togo. *Scientific Research and Essay* Vol.4 (8), pp. 881-893.

Lebrun, J.P. & Stock, A.L. (1991-1997). *Enumération des plantes à fleurs d'Afrique tropicale. 4 volumes.* Conservatoires et Jardin Botanique de la ville de Genève.Makundi, W.R., 2001. Potential and cost of carbon sequestration in the Tanzanian Forest Sector. *Mitigation and Adaptation Strategies for Global Change,* 6 : 335-353.

Ministère des Mines, de l'Energie et des Postes et Télécommunication (MMEPT) (2002). *Analyse, stratégie et programme d'action du sous-secteur des énergies traditionnelles.* Rapport national, 100 p.

Neke, KS; Owen-Smith, N. & Witkowski, T.F. (2006). Comparative resprouting response of Savanna woody plant species following harvesting: the value of persistence. *Forest Ecology and Management* 232: 114–123.

Ninnin, B. (1994). *Eléments d'économie spatiale des énergies traditionnelles : Application au cas de cinq pays sahéliens : Burkina Faso, Gambie, Mali, Niger, Sénégal.* Programme Régional pour le Secteur des Energies Traditionnelles (RPTES), Africa Region, World Bank, Washington, DC.

Ouédraogo, A. (2006). *Diversité et dynamique de la végétation ligneuse de la partie orientale du Burkina Faso.* Thèse de Doctorat, Univ. Ouagadougou, 195 p.

Schinner, F.; Ohlinger, R.; Kandeler, E. & Margesin, R. (1996). *Methods in Soil Biology.* Springer Verlang Berlin Heidelberg; 92-98.

Soltner, D. (1982). *Les Bases de la Production Végétale* (11e édn). Collection Sciences et Techniques Agricoles: Sainte-Gemmes-Sur-Loire.

Thiam, T.A. (1991). *Etude de marché des produits forestiers ligneux au Togo. Projet PNUD/FAO.* Rapport, 225p.

Walkley, A. (1947). A critical examination of a rapid method for determining organic carbon in soils-effect of variations in digestion conditions and of inorganic constituents. *Soil Science,* 63(4): 251-264

13

Infectious Diseases, Biodiversity and Global Changes: How the Biodiversity Sciences May Help

Serge Morand
Institut des Sciences de l'évolution, CNRS-IRD-UM2, CCO65
Université de Montpellier 2
France

1. Introduction

Biodiversity has been reduced on an unprecedented scale over the last decades (MEA, 2005). Concomitantly, new infectious diseases have emerged (or re-emerged) at an increasing rate (Wilcox & Gubler, 2005; Jones et al., 2008) through human impacts on ecosystems, which are supposed to influence this (re-)emergence of diseases (Chivian & Berstein, 2004; Patz et al. 2004). According to the Millenium Ecosystem Assessment (MEA, 2005), biodiversity helps at regulating diseases, with disease regulation as one of the services provided by ecosystems (Walpole et al., 2009). According to this report, high level of species diversity is supposed to reduce or to buffer the risk of disease transmission (MEA 2005).

However, greater overall biodiversity is often shown to be linked to a greater diversity of pathogens (Poulin & Morand, 2004). Thereby, high biodiversity should be linked to high incidence of infectious diseases (Dunn et al., 2010). The reasons of the discrepancy between biodiversity loss and the increasing rate of infectious diseases and the persistence of numerous infectious diseases need a special attention on the likely mechanisms that could link biodiversity and infectious diseases.

Several studies provide evidences that support both of these views. Here, the main questions are how biodiversity is linked to infectious diseases, and how biodiversity sciences may help at improving health.

2. The biodiversity of parasites and pathogens

Although the terms pathogen and parasite are used in different scientific fields (human medicine, veterinary medicine, ecology) and may have different meanings (i.e. pathogens for micro-parasites such as viruses and parasites for macro-parasites such as worms), here we make the choice to use them as synonyms.

Parasites are found in all living organisms. Parasite diversity is estimated by the number of parasite species, but also is reflected in relation to their life cycles (direct, vector-transmitted) or their size (from a few micrometres to metres). More than 10% of the metazoan species that have been so far described are parasites (often referred as macro-parasites) (Poulin & Morand, 2004). When micro-organisms are included (i.e. protozoans and pathogenic bacteriae, viruses, often referred as micro-parasites), parasites may represent more than half of all living species (de Meeûs & Renaud, 2002). Then, no free-living species are free from infection. However, at an interspecific level, a great variability is observed among hosts, with few host species harbour high parasite species diversity and many other harbour less parasite diversity. Several studies have attempted to investigate the reasons of this variability by searching the likely determinants of parasite species richness. The likely explanations for this observed variability among free-living animals can be classified into five major groups of determinants (Table 1):

1. Macro-environmental determinants.
   The biodiversity of plants or animals is highly correlated with some major gradients, such as latitudinal or altitudinal gradients. What about parasites? Do tropical regions host more parasites? The answers to this question highly depend on groups of parasites and/or groups of hosts. For example, the latitudinal gradient in parasite species richness was not observed in many mammalian parasites (such as endoparasitic worms or ectoparasitic arthropod (see Bordes et al., 2010). For humans, Guernier et al. (2004) found a strong positive correlation between proximity to the equator and parasite species richness (Fig. 1a). Moreover, they found that the maximum range of precipitation was the best correlated environmental variable with the species diversity of helminths, protozoa, fungi, and indirectly transmitted viruses. Lindenfors et al. (2007) for helminths and carnivores, and Krasnov et al. (2004) for fleas and rodents (Fig.1b) found opposite results with an increase of parasite species richness towards higher latitudes. Nunn et al. (2005) showed that parasite species richness increases toward lower latitudes only for protozoan parasites in Primates. Interestingly, the recent discovery of new *Plasmodium* species in primates and the potential risks for humans are in favour of non-human primate origins of *Plasmodium falciparum* (Rayner et al., 2011). Recently, Bordes et al. (2010), using a large data base on helminths in mammals, concluded on a lack of correlation between helminth diversity and latitude. All these studies that focused on mammal parasites suggest some special characteristic of human parasite diversity.
2. Epidemiological determinants.
   Epidemiological models are used to estimate the success of invasion or spread, i.e. epidemic, of a parasite in a population of non-infected hosts. Several epidemiologically-related factors may promote and facilitate the spread of parasite in a host population, such as host density. The application of theoretical epidemiological results to the study of parasite diversity has shown that for many animal groups, their species diversity in parasites is well connected to host density (expressed as individual numbers per unit area) (Arneberg et al. 1998; Torres et al., 2006). Epidemiological theory postulates also a positive effect of host longevity on parasite transmission, which however has rarely been found to influence parasite diversity (Linderfors et al., 2007).
3. Ecological determinants.
   Some features of the ecology of host may facilitate transmission. The geographical range of host species has been hypothesized as a potential determinant of parasite species richness. In several groups of mammals, it was indeed shown that species richness of parasites is positively correlated with geographical range of their hosts (Linderfors et al., 2007). Home range, although less investigated, is hypothesized as a potential determinant, with the prediction that large home range should facilitate host-parasite contacts and then the increase of parasite diversity. However, Bordes et al.

(2009) reported the opposite patterns in mammals where an increase in home range area is associated with a decrease in parasite species richness in carnivores, rodents and lagomorphs (Fig. 2). As they also showed that home range size is negatively correlated with host density in mammals, they hypothesized that the pattern observed was an epidemiological determinant. By increasing their home range area, mammals reduce their densities but also the transmission successes of their parasite.

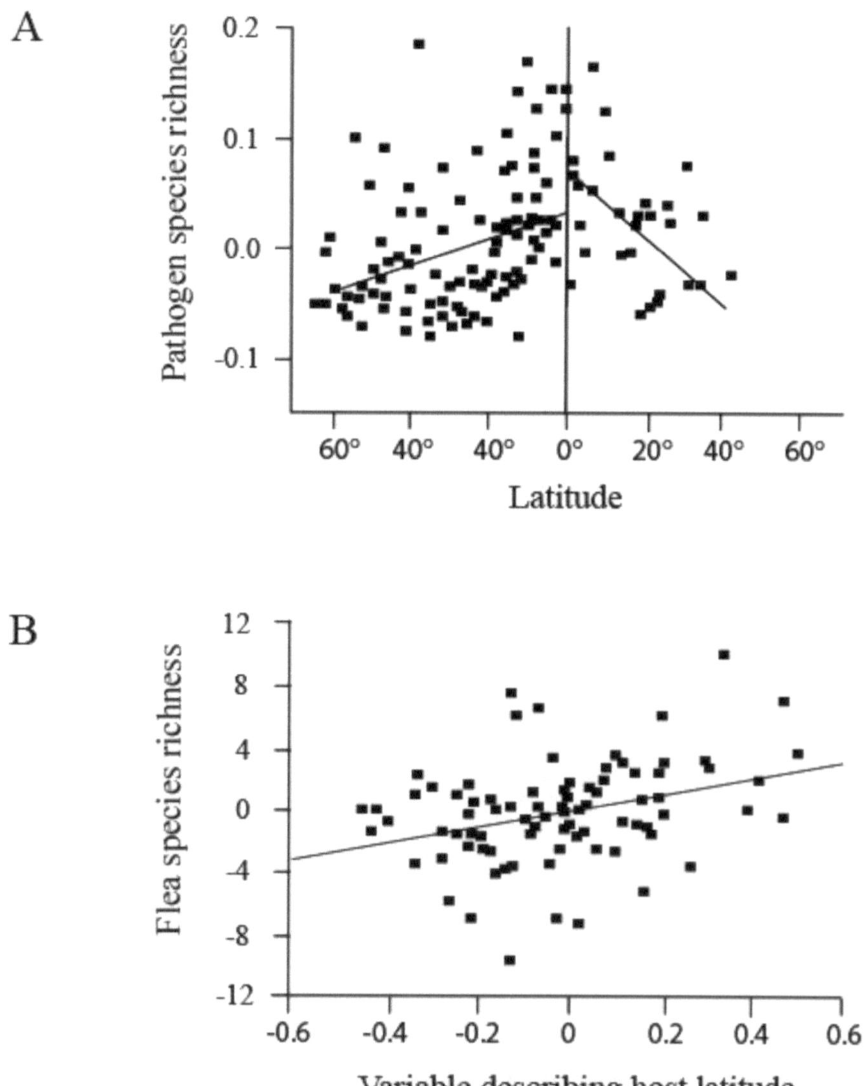

Fig. 1. The latitudinal gradients of parasite and pathogen species richness. (a) Relationship between pathogen species richness of humans and latitude of human populations sampled for pathogen infectious diseases across the two hemispheres (modified after Guernier et al., 2004); (b) relationship between flea species richness in rodents and the latitude of centre of geographic range (modified after Krasnov et al., 2004)

4. Evolutionary determinants.
   Host-parasite interactions occur in ecological time (on a relative small time scale). However, we must not forget that these interactions are also the result of an evolutionary history. Hence, a positive correlation between parasite diversity and diversification of their hosts has been observed in several host-parasite associations. Typically, host groups that harbour highly diverse parasite communities are also highly species diversified. However, there is no clear explanation. Either parasites can be the driving force of host diversification or they just follow their host diversification.

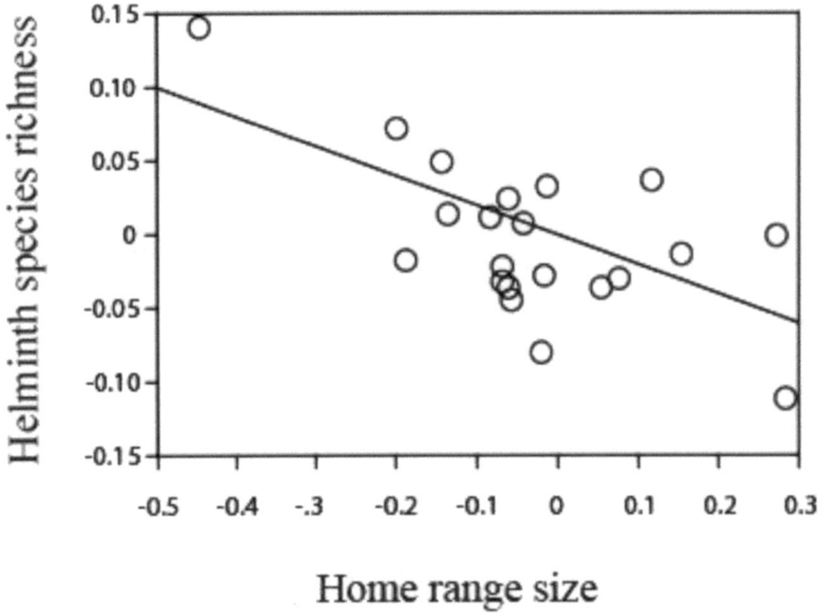

Fig. 2. Relationship between helminth species richness of carnivores and home range (modified after Bordes et al., 2009)

Bordes et al. (2007) found that the increase of social complexity in rodents is negatively correlated with ectoparasite species richness. They suggested that the evolution of sociality in rodents could be associated with behavioural traits that may reduce ectoparasitism load, such as allogrooming, or may increase parasite avoidance or dilute infection risks (but see Alitzer et al., 2003).

Determinant	Organism	Correlation	Reference
Latitudinal gradient	Helminths and mammals	No	Poulin (1995)
	Helminths and mammals	No	Morand (2000)
	Helminths and mammals	No	Bordes et al. (2010)
	Helminths and primates	No	Nunn et al. (2005)
	Helminths and carnivores	Yes (positive)*	Lindefors et al. (2007)
	Fleas and rodents	Yes (positive)*	Krasnov et al. (2004)
	Protozoans and primates	Yes (negative)	Nunn et al. (2005)
	Microparasites and humans	Yes (negative)	Guernier et al. (2004)
	Microparasites and rodents	Yes (negative)	Bordes et al. (2011)
Area size (geographic distribution)	Helminths and rodents	Yes (positive)	Feliu et al. (1997)
	Fleas and rodents	Yes (positive)	Krasnov et al. (2004)
	Helminths and carnivores	Yes (positive)	Torres et al. (2006)
	Macro- , microparasites and carnivores	Yes (positive)	Lindefors et al. (2007)
Host body size	Helminths and mammals	No	Morand & Poulin (1997)
	Helminths and rodents	No	Feliu et al. (1998)
	Macro- , microparasites and primates	No	Nunn et al. (2003)
	Macro- , microparasites and ungulates	Yes (positive)	Ezenwa et al. (2006)
Host density	Helminths and mammals	Yes (positive)	Morand & Poulin (1998)
	Nematodes and mammals	Yes (positive)	Arneberg (2002)
	Fleas and rodents and insectivores	Yes (positive)	Stanko et al. (2002)
	Helminths and primates	Yes (positive)	Nunn et al. (2003)
	Helminths and carnivores	Yes (positive)	Torres et al. (2006)
	Macro- , microparasites and carnivores	Yes (positive)	Lindefors et al. (2007)
Host longevity	Helminths and mammals	Yes (positive)	Morand & Harvey (2000)
	Fleas and rodents and insectivores	No	Stanko et al. (2002)
	Helminths and carnivores	No	Torres et al. (2006)

	Macro- , microparasites and ungulates	Yes (positive)	Ezenwa et al. (2006)
Group size	Macro- , microparasites and primates	No	Nunn et al. (2003)
	Macro- , microparasites and ungulates	Yes (negative)	Ezenwa et al. (2006)
Host sociality	Helminths and rodents	No	Bordes et al. (2007)
	Ectoparasitic arthropods and rodents	Yes	Bordes et al. (2007)
Home range	Helminths and primates	Yes (negative)	Nunn et al. (2003)
	Helminths and ungulates	No	Ezenwa et al. (2006)
	Direct-transmitted parasites and carnivores	Yes (negative)	Lindefors et al. (2007)
	Helminths and ungulates	No	Bordes et al. (2009)
	Helminths and carnivores	Yes (negative)	Bordes et al. (2009)
	Helminths and glires	Yes (negative)	Bordes et al. (2009)
Host genetic diversity (MHC)	Helminths and rodents	Yes	Goüy de Bellocq et al. (2008)
Host genetic diversity (HLA)	Virus and humans	Yes	Prugnolle et al. (2005)
Anthropization gradient	Helminths and rodents	Yes	Chaisiri et al. (2010)

Table 1. Some examples of determinants of parasite diversity (* opposite trend as expected) of mammals (humans included)

3. Global change, biodiversity and pathogens

Global changes are the results of increases in all factors of anthropogenetic origins: atmospheric conditions, land uses, over-exploitation of resources, biotic invasions and pollutants. Global changes affect biodiversity by increasing the rate of extinction, by modifying the functioning of ecosystems and then by affecting the health of plants, animals and humans.

3.1 Climate change

Climate change is altering the ecology of pathogens, and the interactions among hosts and their pathogens (Shope, 1991; Patz et al., 1996, 2000; Sutherst, 1998, 2001, 2004; Rogers & Randolph, 2000; Kovats et al., 2001; Harvell et al., 2002; Mouritsen & Poulin, 2002) (IPPC, Intergovernmental Panel on Climate Change reports at http://www.ipcc.ch/). Temperatures and precipitations will be affected globally, regionally and locally with both increases in average values and more frequent extreme events (storms, warm events…). Among others, such changes will affect the geographical distributions of animals, plants but also their parasites, pathogen reservoirs, and vectors (Wilson, 2000; Stenseth et al., 2002; Walther et al., 2002; Lovejoy & Hannah, 2005).

Temperature is one important abiotic parameter that may affect parasites, but also vectors, at all their life-cycle stages. Temperature impacts on the release of eggs or larvae by parasites and their vectors, embryonic development and hatching rates, longevity of free-living stages, infectivity to intermediate hosts, development of either microparasites (virus, bacteria) or macroparasites (worms) in these hosts, infectivity to definitive hosts, time to maturation, and the longevity and mortality (Marcogliese, 2001; Poulin, 2006). Temperature also plays a key role in host feeding and behaviour, host range size and host resistance to infection.

Kutz et al. (2005) have shown how a host-parasite system may respond to climate change. They presented an empirical and predictive model to elucidate the impact of climate warming on development rates of a parasitic nematode of muskoxen in the Canadian Arctic, a region that is particularly vulnerable to climate change, and showed that warming in the Arctic may have already radically altered the transmission dynamics of this parasite with the number of parasite generation increasing due to the lessened detrimental effect of the outside environment. The infection pressure is then expected to continue to escalate for muskoxen.

A second example can be given by tick-borne diseases for which the tick vector is affected by climate change, such as tick-borne encephalitis (TBE) and Lyme disease. The ticks may live for several years and their survival, re-production rate and activity are affected by seasonal climate, which indirectly influences the risk of disease. Warmer temperatures increase vector and pathogen reproduction and blood feeding activity. Changes in rainfall, humidity but also large-scale meteorological phenomena such as ENSO (El Niño-Southern Oscillation) may affect the number and the quality of vector breeding sites (Liang et al., 2002). These influences may operate in synergy with other changes in the habitat and in biodiversity. Modifications of the habitat and of the occurrence of animals that are carriers of different pathogens may result in changes in the spread of tick-borne pathogens. Several studies have shown that in recent decades the tick *Ixodes ricinus*, transmitting Lyme borreliosis and TBE, has spread into higher latitudes and altitudes, and has become more abundant in many places (Daniel et al., 2003; Materna et al., 2005). Climate change in Europe seems likely to facilitate the spread of Lyme borreliosis and TBE into higher latitudes and altitudes, and to contribute to the extension and transmission of these diseases in some new areas. Modelling suggests that the disease will no longer persists along the southern edge of its present range but may have new foci in northern parts of Europe (Randolph, 2001).

Several other examples show that global warming has a significant impact on host-parasite interactions (McIntyre et al., 2010). However, scenarios of the future of these interactions are not easy to produce without the help of modelling. Moreover, the non-linear interactions between temperature and various aspects of parasite transmission are likely to occur and climate changes can work synergistically with other anthropogenic changes.

3.2 Habitat fragmentation

Escalating human activities, habitat fragmentation and degradation, and increasing the proximity between wildlife and humans have greatly impaired the health of both humans and wildlife (Daszak et al., 2000; Jones et al., 2008). These global changes are occurring at an unprecedented rate and speed, which creates opportunities for parasites to negatively affect their hosts through emergence, outbreak and higher virulence (Cleaveland et al., 2001).

Fragmentation of ecosystems, mostly due to the conversion of forests and wildlife habitats into agriculture, results in human-made island ecosystems. Knowing that true islands are places of high extinction rates, the fragmentation processes are likely to drive many isolated populations and species to extinction.

One potential consequence of habitat fragmentation is related to the increasing contacts among animals, both domestic and wild animals, and humans. The increasing proximity to wildlife and domestic animals results in the increased transmission of zoonotic diseases (Blouin et al., 1984; Wallis & Lee, 1999; Daszak et al., 2000) (Table 2). For example, some

Disease	**Emergence mechanism**	**Anthropogenic drivers**
Malaria	niche invasion, vector expansion	deforestation, water projects
Lymphatic filariasis	habitat alteration	water projects, urbanization
Schistosomiasis	intermediate host expansion	water projects, irrigation
Dengue fever	vector expansion	Urbanization, poor housing conditions
Onchocerciasis	habitat alteration	water projects
Chagas disease	habitat alteration	deforestation, urban sprawl, encroachment
Leishmaniasis	vector expansion, host transfer	habitat alteration
Meningitis	habitat alteration, dust storms	desertification
Hantavirus	variations in population density of natural food sources	climate variability
Rabies	biodiversity loss, altered host selection	Deforestation, mining
Trypanosomiasis	habitat alteration	deforestation
Encephalitis	vector expansion	irrigated rice fields
Rift Valley fever	heavy rains	climate variability, dam building
Lyme disease	depletion of predators	biodiversity loss, habitat fragmentation, reservoir expansion
SARS	host transfer	intensive livestock operations mixing wild and domestic animals
West Nile virus and other encephalitides	niche invasion	international travel, climate variability
Cholera	sea surface temperature rising	climate variability and change
Cryptosporidiosis	contamination by oocysts	poor watershed management where livestock exist
Coccidioidomycosis	disturbing soils	climate variability global
Ebola	forest encroachment, bushmeat hunting	forest encroachment
Leptospirosis	habitat alteration, agricultural development	urban sprawl
Nipah/Hendra viruses	niche invasion	industrial food production
H5N1 avian influenza	expansion of livestock Production	wild bird migration, poultry trade, wild bird trade

Table 2. Infectious diseases, their mechanisms of emergence in relation to anthropogenic drivers (adapted from Patz & Confalonieri, Chapter 14, MEA, 2005) (Li & Li, 1998; Ostfeld & Keesing, 2000; Liu et al., 2001; Fa et al., 2002; Patz et al., 2002; Brook et al., 2003; LoGiudice et al., 2003; Milner-Gulland & Bennett, 2003; Bell et al., 2004; Foley et al., 2005; Wolfe et al., 2005; Ezenwa et al., 2006 Fergus et al., 2006; Fevre et al., 2006; Kilpatrick et al., 2006)

management practices in national parks and natural reserves allowing or favouring multi-usages, in which domestic animals graze the same lands as wild animals, facilitate the spread of diseases among them (Deem et al., 2001). Increasing encroachment of farms on wildlife habitats has also increased the overlap between livestock and wild animals, with the consequence that the vast majority of emerging diseases of livestock have been acquired from wild animals (Cleaveland et al., 2001).

The epidemiological environment is greatly altered by the recent intensification of agriculture, with massive uses of nutrients and pesticides. This may favour rodents as they are important consumers of agricultural crops. Rodents are hosts of many parasites that may cause diseases to humans, including bacteria, viral hemorrhagic fevers, tick-borne encephalites, but also macroparasites such as trematodoses and nematodoses (Meerbur et al., 2009). Agricultural intensification favours rodent outbreaks through the removal of predators and other natural enemies, while supplementing their food supply, with subsequent disease outbreaks in rural human populations (Chaisiri et al., 2010).

In many Western countries the epidemiological environment is significantly altered by reforestation and suburbanization. These changes in land uses may have consequences for humans and their domestic animals by increasing the contact between wild animals, and in particular rodents. A good example of this is given by Lyme disease, which has spread over the last decades due to increases in deer populations, and also rodent and tick populations, human recreational activities and contacts with ticks (see Daily & Ehrlich, 1996).

Disturbance seems to favour generalist hosts, i.e. hosts that are able to use different types of habitats (Marvier et al., 2004). These generalist hosts often have a broad geographical distribution and a high resilient ecology (Smart et al., 2006), two potential determinants for hosting high diversified parasite communities (see above). These species seem also to have higher reservoir or vector competence than species that are not favoured by disturbance (Mills, 2006; Vittor et al., 2006; Molyneux et al., 2008; Chaisiri et al., 2010). Moreover, these species being released for competitors may rich high population density.

3.3 Biological invasion and parasite release

Parasites are linked to biological invasions either as invader, often through the help of carriers, or by facilitating the invasion success of their hosts. The worldwide introduction of parasites, infectious diseases and their vectors and carriers is linked to globalization, international trades and increasing human travels. The introduction of the mosquito *Aedes albopictus* in many parts of the world has facilitated the transmission of the Chikungunya virus (Benedict et al., 2007; Charrel et al., 2007).

Less known are the indirect effects of parasitism and the success of biological invaders. The "parasite release hypothesis" has been proposed as an ecological mechanism to explain the success of introduced or invasive species, as they often lose their parasites when invading new habitats. Introduced species have a competitive advantage over local species because they are released from control by their natural enemies. Mitchell & Power (2003) and Torchin et al. (2003) found that parasitism is significantly reduced in organisms in their introduced range, supporting this hypothesis. The parasite release hypothesis may then explain the success of introduced species and the demographic explosion of some introduced species (Clay, 2003; Keane & Crawley, 2003; Mitchell & Power, 2003; Torchin et al. 2003).

3.4 Extinction

Pathogens are considered to be a serious threat in conservation biology (Daszak & Cunningham, 1999; Daszak et al., 2000; Cleaveland et al., 2001). Pathogens have been implicated in the extinction of numerous species (McCallum & Dobson, 1995; Vitousek et al., 1997), such as several endemic Hawaiian birds (VanRiper et al., 1986, 2002) or the thylacine (an Australian marsupial carnivore). Daszak et al. (2000) mentioned 19 parasites as important dangers for conservation or for human health (major zoonotic agents). Theoretically, several mechanisms can produce extinction due to parasitism (Table 3):

Mechanism	Host species	Pathogen / Impact	References
Small population syndrome	Thylacine (*Thylacinus cynocephalus*)	Virus/probable extinction	Guiler, 1961 (in McCallum & Dobson, 1995)
	Random Golden toad (*Bufo periglenes*)	Virus/probable extinction	Pounds et al. (1997)
	Black-footed ferret (*Mustela nigripes*)	Virus/probable extinction	Thorne & Williams (1988)
	Feral goats (*Capra pyrenaica hispanica*)	Mite/reduction in population size	Leon-Vizcaino et al. (1999)
Reduced genetic variability	Cheetah (Acinonyx jubatus)	Virus/increased susceptibility to diseases	O'Brien et al. (1985)
Independent transmission	Rabbit (*Oryctolagus cuniculus*)	Virus/haemorrhagic disease, potential extinction (model)	White et al. (2003)
Reservoir	White-tailed deer (*Odocoileus virginianus*), elk (*Alces alces*)	Nematode/reduction in population size	Schmitz & Nudds (1994)
	Grey squirrel (*Sciurus carolinensis*), red squirrel (*Sciurus vulgaris*)	Virus/reduction in population size	Rushton et al. (2000)

Table 3. Several examples of a proven or potential role for parasitism in extinction of the host population (modified from Gog et al., 2002)

1. Small population syndrome.
   Small populations are characterized by low densities, which are reduced in the presence of a pathogen. Fluctuations below a persistence threshold can increase and proceed to extinction because of random environmental processes. The impact of parasites can also drive the host population close to the threshold for the Allee effect (the threshold below which the population inexorably declines), leading the host population into a spiral to extinction (Deredec & Courchamp, 2003).
2. Reduced genetic variability.
   Small populations are often characterized by a high degree of relatedness, with a reduced genetic variability, especially for genes involved in resistance against parasites such as major histocompatibility complex (MHC) genes.

3. Density-independent or vector-transmitted parasites.
   Sexually or vector- transmitted parasites show a transmission independent of host population size because of the dilution effect (i.e. transmission is density independent). Then, regulatory effects are maintained or may increase with a reduction in the host population.
4. Reservoir effect and apparent competition.
   Non-specific parasites are maintained on abundant populations but continue regulating all populations, even small ones.

4. How biodiversity is related to the emergence of infectious diseases

Diseases are now considered as important factors in the conservation of biodiversity (Meffe, 1999; Daszak et al., 2000; Deem et al., 2001). Several reviews have emphasized the potential impact of parasites and pathogens on their host population dynamics under global changes. Several studies have also emphasized the need to preserve vertebrate biodiversity and community composition in order to significantly reduce the risk of emergence (LoGiudice et al., 2003; Keesing et al., 2010). In a recent review, Keesing et al. (2010) have given some examples of the impacts of biodiversity changes on diseases (Table 4).

Diseases	**Mechanism**	**References**
Bacteriophage of *Pseudomonas syringae*	host/vector/parasite behaviour	Dennehy et al. (2007)
Coral diseases	host/vector abundance	Raymundo et al. (2009)
Fungal disease of *Daphnia*	host/vector/parasite behaviour	Hall et al. (2009)
Hantavirus disease	host/vector abundance, host/vector/parasite behaviour	Tersago et al. (2008); Clay et al. (2009); Dizney & Ruedas (2009); Suzan et al. (2009)
Helminthic parasite of fish	host/vector abundance	Kelly et al. (2009)
Lyme disease	host/vector abundance, host/vector/parasite behaviour	Brunner & Ostfeld (2008); LoGiudice et al. (2008), Keesing et al. (2009)
Malaria	host/vector abundance	Carlson et al. (2009)
Schistosomiasis	host/vector/parasite behaviour	Johnson et al. (2009)
Trematode diseases of snails and birds	host/vector/parasite behaviour	Kopp & Jokela (2007); Thieltges et al. (2008, 2009)
West Nile fever	host/vector abundance host/vector/parasite behaviour	Ezenwa et al. (2006); Swaddle & Calos (2008); Allan et al. (2009); Koenig et al. (2010)

Table 4. Biodiversity loss and increase disease transmission (adapted from Keesing et al., 2010)

Anthropogenic drivers of global change affect biodiversity (through climate change, habitat fragmentation, land-use changes and bioinvasions). Changes at all levels in biodiversity (genetic, population and community) affect ecosystem functioning and, in particular, host-pathogen interactions, with major consequences in health ecology (emergence and re-emergence, outbreaks) (Fig. 3).

Any change in species richness may affect the dynamics of infectious diseases with potential increase or decrease in parasite transmission:

1. Dilution effect.
   The dilution effect refers to describe a pattern when increased species diversity reduces disease risk. The processes may be linked to a variety of mechanisms (Keesing et al., 2006). This applies to vector-borne and directly transmitted diseases, although the concept of dilution has been developed most with regards to the tick-borne Lyme disease (Allan et al., 2003; LoGiudice et al., 2003, 2008).

The hypothesis underlying the dilution effect is that for many diseases, the competence of reservoirs, i.e. the ability to become infected and retransmit the pathogen, varies according to the host species (Haydon et al., 2002). The composition of the host community thus can influence the transmission dynamic of the disease. Similarly, since vectors have different competence to transmit pathogens, the composition of the vector community may also likely influence transmission dynamics. Different mechanisms are thought to be involved, but they are difficult to differentiate (Keesing et al., 2006; Begon, 2008). One is the modification of the encounter rate (when reduced, this corresponds to a real "dilution effect"). In the presence of species that are poorly or not competent, the infectious agent may be transmitted to a non-competent individual rather than to a susceptible individual. For vector-borne diseases, the increased species diversity of poorly competent hosts on which the vector feeds increases the proportion of vector bites that are wasted for the parasite transmission (Table 3). For direct transmission, the addition of non competent hosts can decrease transmission if these hosts remove infective stages (Begon, 2008).

Another mechanism at work is that high host species diversity regulates the abundance of the competent host population. This regulation can be mediated by interspecific competition for limiting resources or by predation upon competent hosts. It could also mediated by parasite interactions through apparent competition. By decreasing densities of competent hosts, high biodiversity modulates the parasite population dynamics.

The potential link between parasitism and the interaction between predators and the diversity of preys may also be involved. This may happen when predators modify the mortality rate of a host and lower pathogen transmission by feeding on a heavily diseased individual (Packer et al., 2003).

The dilution effect has been demonstrated through theoretical works and empirically observed in several host-parasite systems (Table 3). One example is Lyme disease in the USA that is caused by pathogenic bacteria transmitted by ticks. These ticks feed readily on many species of vertebrates and these species vary in their degrees of reservoir competence. The white-footed mouse (*Peromyscus leucopus*) is thought to be the most competent host and dominates in fragmented forests. In native forests, which harbour a higher diversity of species than fragmented forests, ticks have a higher probability to dilute their bites by feeding on less competent hosts (Allan et al., 2003; LoGiudice et al., 2003, 2008). Another example is the West Nile virus. Increased species diversity of on passerine birds, which are less competent reservoir hosts compared to passerines, was found associated with decreased West Nile virus infection in mosquitoes and humans (Ezenwa et al. 2006, Swaddle & Calos 2008).

There have been few examples of directly transmitted diseases, but studies on hantaviruses and rodents have shown that higher diversity of these small mammals appears to regulate reservoir host populations through competition or predation with the net consequence to decrease hantavirus transmission. High rodent species richness might also inhibit intraspecific aggressive encounters between reservoir hosts that result in hantavirus transmission (Suzán et al., 2009).

2. Amplification effect.
   Amplification effects are typically associated with the consequences of species introduction that radically modifies encounter rates. The introduced species can introduce new pathogens that may infect native hosts (i.e. spillover) (Bruemmer et al., 2010). But the introduced species can amplify the circulation of local pathogens (i.e. spillback) (Kelly et al., 2009), although the introduction of a resistant host may favour dilution (Kopp & Jokela, 2007). The introduction of additional species as providing new hosts for vectors may increase vector numbers or their activities (Saul, 2003). For example, the introduced Siberian chipmunks (*Tamias sibiricus*) in suburban forests increase the risk of Lyme disease because of its higher competence in regard to native hosts (Vourc'h et al., 2007).

5. How biodiversity sciences may contribute to health

5.1 Predicting bioinvasions

Biotic invaders are species that establish in a new area in which they proliferate to the detriment of local species and the environment. They are the biggest ecological outcomes of the global alterations and distributions of the earth's biota due to human transport and commerce (Williamson, 1996). As emphasized by Mack et al. (2000), biotic invasions can be compared with epidemics because many important factors in disease epidemiology are common to invasions. These factors are the minimum population size necessary for successful establishment, population growth and the fate of interacting species in the new range.

The movement of parasites, potential vectors, or disease reservoirs is greatly facilitated by intense modern transport. A recent case of transfer of a dangerous vector was the introduction from Asia to the United States and several European countries of a mosquito (*Aedes albopictus*) capable of transmitting dengue (Scholte & Schaffner, 2007).

Identifying future parasite or vector invaders and vulnerable communities, and taking effective measures to prevent their establishment and their dispersal, are great challenges for conservation biology and ecological health. The features associated with bio-invasions are:

1. Attributes of parasite invaders.
   Parasite invaders are generally those parasites that are found in high local abundances or prevalences, with direct life-cycles and low host specificity. Local abundances of a parasite species are positively correlated with their geographical distribution, with highly abundant parasites found in a larger number of host populations. The distribution of parasites with indirect life-cycles is dependent on those of all hosts in the life cycle. Any modification of host distribution will determine if parasites can persist and where they colonize (Dobson & Carper, 1992). Nevertheless, parasites that have indirect life-cycles such as *Schistosoma mansoni, Fasciola hepatica* or *Angiostrongylus spp.* can be invaders. Low host specificity may not always be a crucial characteristic of parasite invaders, a more important characteristic being the ability of the parasite to switch host.

2. Attributes of invading hosts, vectors and reservoirs.
   Many morphological and ecological characteristics may favour invasion success. The introduced species may have a competitive advantage over local species because in the new area they are released from control by their natural enemies. Mitchell & Power (2003) and Torchin et al. (2003) showed that parasite diversity is significantly reduced in organisms in their introduced range compared to their native range, supporting the "parasite release hypothesis".
   Invasive hosts are also at an advantage during concomitant invasion when they have evolved strong immune defences in their original range. High immune investment confers a better capacity to control parasites that they may acquire or be in contact with in the introduced range (Lee & Klasing, 2004; Møller & Cassey, 2004).
3. Community vulnerability.

Vacant niches within communities, i.e. low species richness, both in hosts and in their parasites is widely assumed to provide opportunities for the settlement and spread of biotic invaders (Elton, 1958). The vacant niches hypothesis suggests that species-poor communities cannot offer biological resistance to invasion and particularly to pathogen or vector invasion.

5.2 Predicting extinctions

Climate changes, but also habitat uses' changes, will extend the geographical ranges of many species along with their parasites, potentially overlapping with endangered species and driving them to extinction (Dobson & Carper, 1992). The seasonal and spatial distributions of parasites and their hosts are often temperature dependent and the synchronicity of their population dynamics is threatened by climatic changes. Many parasites have adapted their life cycles, reproduction, and transmission to overlap with definitive and intermediate hosts (Poulin, 2006; see above Kutz et al., 2005). Any increase in temporal asynchrony between infective parasite stages and hosts has significant effects on the persistence of the interaction. Some parasites will adapt to these new conditions, whereas others may not. We may expect parasite extinctions to occur.

The global extinction of parasites may be more pronounced than that of free-living animals. According to Poulin & Morand (2004), the 11% of threatened mammalian species imply that a total of 409 helminth species are at risk of extinction. This predicted parasite extinction may also have consequences on the epidemiological environment.

5.3 Predicting distribution using modeling environmental niche

Ecological niche modelling is used in biogeography to predict the distributional range of species from existing occurrence data (Anderson et al., 2003). Using appropriate algorithms in a GIS containing layers of environmental information (such as topography, climate, and vegetation), epidemiological and spatial risk stratification can be achieved from data on the location of vectors or pathogens. This approach has been used in the case of Chagas disease and for vectors of leishmaniasis and filovirus infections (Peterson et al., 2002, 2004a,b). Moreover, using scenarios of climate change, it is then possible to project scenarios of pathogen and vector distribution changes.

5.4 The importance of preserving parasite diversity

Parasite extinction may count. Hence, for any host extinction, at least one parasite may go with it (Poulin & Morand, 2004). This may not be seen as a positive consequence as parasites should not only be thought of as harmful, however. Parasites have their own intrinsic value (Sprent, 1992). They contribute to maintain high diversity as ecological engineers (Thomas et al., 1999), they may control biotic invaders, and they stabilize ecosystems (Marcogliese & Cone, 1997; Mouritsen et al., 2005; Arias-González & Morand, 2006; Pedersen & Fenton, 2006). At the individual host level, parasites may help to equilibrate immune responses and prevent the spread of other parasites. There is then a balance between the need to control parasites and the benefits of their presence (Pedersen & Fenton, 2006).

6. Ecosystem services, biodiversity and health

Variability among ecosystems is a key element of biodiversity. Climate changes, land-use changes, the over-use of living resources and bioinvasion all affect biodiversity, not only by increasing species loss but also by potentially damaging the functioning of ecosystems. Theoretical and empirical works have both identified links between global changes, biodiversity and the way in which ecosystems function (Chapin et al., 2000).

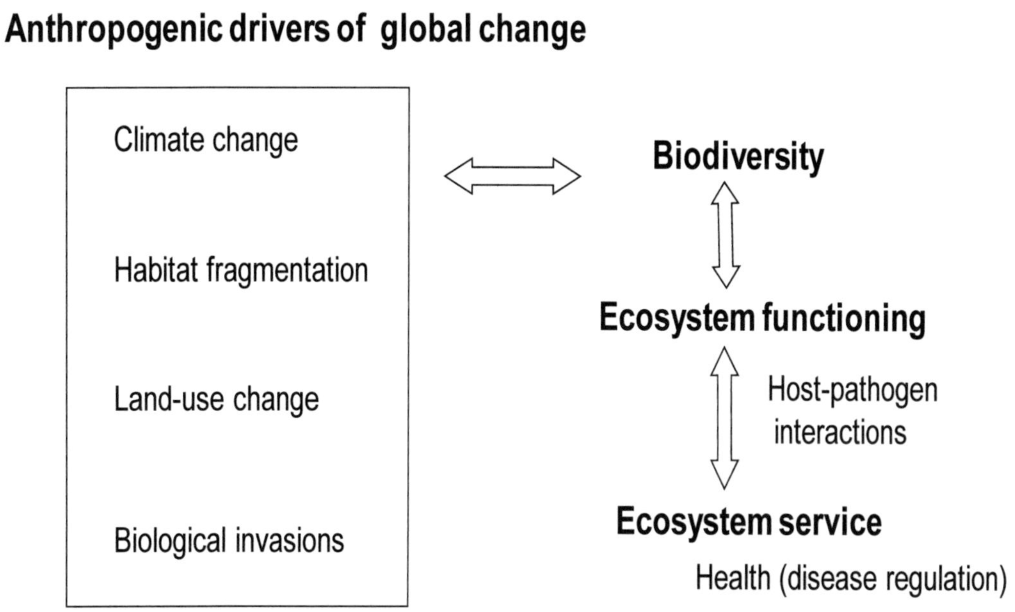

Fig. 3. Anthropogenic drivers of global change may affect biodiversity (climate change, habitat fragmentation, land-use changes, and bioinvasions). Changes at all levels of biodiversity (from genes to ecosystems) affect ecosystem functioning with major consequences in health ecology (emergence and re-emergence, the evolution of virulence and resistance). All theses impact the services providing by ecosystems, such as disease regulation, with a potential effect on human well-being

The Millennium Ecosystem Assessment was initiated in 2001. Its objectives were to assess the consequences of ecosystem change for human well-being and to provide the scientific basis for the actions needed to enhance conservation and sustainable use. More than 1,360 experts throughout the world worked on this project, with the goals of examining the present condition and future trends of the ecosystems of the world, assessing the services that such ecosystems provide, and developing solutions to enhance the sustainable use of these ecosystems.

Among the benefits obtained by humans from the regulation of ecosystem processes are the regulation of human and animal diseases (since any changes in ecosystems can directly affect human pathogens or disease vectors, such as mosquitoes), and biological control (since any changes can affect crop pests and plant diseases). One important concept has emerged from the ecosystem approach, the idea of 'ecosystem health'. This term is often applied in the evaluation of ecosystems, although it is also used to refer to the links between ecosystems and human (and animal) health, by emphasising the regulating role of ecosystems on pathogens.

The next step was to develop biodiversity indicators. Twenty-two biodiversity indicators have been developed by the international organisations which participated in the 2010 Biodiversity Indicators Partnership (www.twentyten.net). However, the development of these indicators has recently been found to be incomplete. Moreover, these indicators do not include any measure of the impact of climate change on biodiversity, and few may be used to estimate the goods and services that people gain from biodiversity and ecosystems (Walpole et al., 2010).

7. Conclusion

Several studies now recognize the need to preserve biodiversity in order to maintain high ecological health (Aguirre et al., 2002). This task needs to be given a higher priority and increase its rate of discovery, in particular by collecting data, and organizing databases on parasites and their hosts (Brooks & Hoberg, 2000).

Finally, as a conclusion, we aim at emphasizing that biodiversity sciences, ecology and evolution, are central to the study of zoonotic diseases in relation to climate changes. We should aware that if climate, and its variability, by itself is important as a key factor in pathogen transmission, this factor works in synergy with other drivers of the global changes such as land-use changes, habitat fragmentation and bioinvasion. Anthropogenic drivers of global change affect biodiversity, including parasite biodiversity, potentially modifying the epidemiological environment (Daily & Ehrlich, 1996).

Increasing collaboration between biodiversity scientists and the health sciences corresponds to the call of scientific expertises conducted by IPCC, MEA, IPBES (« Intergovernmental Science-Policy Platform on Biodiversity and Ecosystem Services»), and other international agencies, such as FAO, OIE, and WHO. The "One Health" approach is one initiative that points out the need to modify disease control strategies. The current approach to disease prevention and control emphasizes transmission disruption (early warning, early detection and early response mechanisms targeting), while the new approach promoted by the "One Health" is to act at the driver level of diseases, i.e. at the human-animal-ecosystems interface (Burgos & Slingenbergh, 2011).

8. Acknowledgment

The author thanks Dr. Jordi López-Pujol, editor of this book, for his kind invitation. The author also thanks the ANR CERoPath (ANR 07 BDIV 012).

9. References

Aguirre, A.A.; Ostfeld, R.S.; House, C.A.; Pearl, M.A. & Tabor, G.M. (2002). Conservation medicine: Ecological health in practice. Oxford University Press, Oxford

Altizer, S.: Nunn, C.L.; Thrall, P.H; Gittleman, J.L.; Antonovics, J. et al. (2003). Social organization and parasite risk in mammals: integrating theory and empirical studies. *Annual Review of Ecology and Systematics*, Vol.34, pp. 517–547

Allan; B.F.; Keesing, F. & Ostfeld, R.S. (2003). Effect of forest fragmentation on Lyme disease risk. *Conservation Biology*, Vol.17, pp. 267-272.

Allan, B.F.; Langerhans, R.B.; Ryberg, W.A.; Landesman, W.J.; Griffin, N.W. et al. (2009). Ecological correlates of risk and incidence of West Nile virus in the United States. *Oecologia*, Vol.155, pp. 699–708.

Anderson, R.P.; Lew, D. & Peterson, A.T. (2003). Evaluating predictive models of species' distributions: criteria for selecting optimal models. *Ecological Modelling*, Vol.162, 211–232.

Arias-González, J.E. & Morand, S. (2006). Trophic functioning with parasites: a new insight for ecosystem analysis. *Marine Ecology Progress Series*, Vol.320, 43-53.

Arneberg, P. (2002). Host population density and body mass as determinants of species richness in parasite communities: comparative analyses of directly transmitted nematodes of mammals. *Ecography*, Vol.25, 88–94.

Arneberg, P.; Skorping, A.; Grenfell, B. & Read, A. F. (1998). Hosts densities as determinants of abundance in parasite communities. *Proceedings of the Royal Society B-Biological Sciences*, Vol.266: 1283–1289.

Begon, M. (2008). Effect of host diversity on disease dynamics, In: *Effects of Ecosystems on Disease and of Disease on Ecosystems*, R.S. Ostfeld, F. Keesing & Eviner V.T. (Eds.), 12-29, Princeton University Press, Princeton.

Bell, D.; Roberton, S. & Hunter, P. R. (2004). Animal origins of SARS coronavirus: possible links with the international trade in small carnivores. *Philosophical Transactions of the Royal Society of London B Biological Sciences*, Vol.359, 1107-1114.

Benedict, M.Q.; Levine, R.S.; Hawley, W.A. & Philipp Lounibos, L. (2007). Spread of the tiger: Global risk of invasion by the mosquito *Aedes albopictus*. *Vector-Borne and Zoonotic Diseases*, Vol.7, 76-85.

Bordes, F.; Blumstein, D.T. & Morand, S. (2007). Rodent sociality and parasite diversity. *Biology Letters*, Vol.3: 692–694.

Bordes, F.; Guégan, J.F.; Morand, S. (2011). Microparasite species richness in rodents is higher at lower latitudes and is associated with reduced litter size. *Oikos* (in press)

Bordes F.; Morand, S.; Kelt, D.A. & Van Vuren, D.H. (2009). Home range and parasite diversity in mammals. *The American Naturalist*, Vol.173, 467-474

Bordes, F.; Morand, S.; Krasnov, B.R. & Poulin, R (2010). Parasite diversity and latitudinal gradients in terrestrial mammals. In: *The Biogeography of Host-Parasite Interactions*, S. Morand & B.R. Krasnov (Eds), 89-98, Oxford University Press.

Brooks, D.R. & Hoberg, E.P. (2000). Triage for the biosphere: The need and rationale for taxonomic inventories and phylogenetic studies of parasites. *Comparative Parasitology*, Vol.67, 1–25.

Bruemmer, C. M., Rushton, S. P., Gurnell, J., Lurz, P. W., Nettleton, P. et al. (2010). Epidemiology of squirrelpox virus in grey squirrels in the UK. *Epidemiology and Infection*, Vol.138, 941-950.

Brunner, J. & Ostfeld, R.S. (2008). Multiple causes of variable tick burdens on small mammal hosts. Ecology, Vol.89, 2259–2272.

Burgos, S. & Slingenbergh, J. (2011). Thoughts on human-animal-ecosystems interface. *Transboundary and Emerging Diseases*, on line version, doi:10.1111/j.1865-1682.2011.01214.x

Chaisiri, K.; Chaeychomsri, W.; Siruntawineti, J.; Bordes, F.; Herbreteau, V. & Morand, S. (2010). Human-dominated habitats and helminth parasitism in Southeast Asian murids. *Parasitology Research*, Vol.107, 931-937.

Carlson, J.C.; Dyer, L.A., Omlin, F.X. & Beier, J.C. (2009). Diversity cascades and malaria vectors. *Journal of Medical Entomology*, Vol.46, 460–464.

Charrel, R.N.; de Lamballerie, X.; Raoult. D. (2007). Chikungunya outbreaks - The globalization of vectorborne diseases. *New England Journal of Medicine*, Vol.356, 769-771.

Chapin, F.S. III; Zaveleta, E.S.; Eviner, V.T.; Naylor, R.L.; Vitousek, P.M.; Lavorel, S.; Reynolds, H.L.; Hooper, D.U. et al. (2000). Consequences of changing biotic diversity. *Nature*, Vol.405, 234-242

Chivian, E. & Bernstein, A.S. (2004). Embedded in nature: human health and biodiversity. *Environmental Health Perspectives*, Vol.112, 12-13.

Clay, C.; Lehmer, E.M.St.; Jeor, S. & Dearing, M.D. (2009). Sin Nombre virus and rodent species diversity: a test of the dilution and amplification hypotheses. *PLoS ONE*, Vol.4, e6467.

Clay, K. (2003). Parasites lost. *Nature*, Vol.421, 585-586.

Cleaveland, S.; Laurenson, M.K. & Taylor, L.H. (2001). Diseases of humans and their domestic mammals: Pathogen characteristics, host range and the risk of emergence. *Philosophical Transactions of the Royal Society London B*, Vol.356, 991–999.

Daily, G.C. & Ehrlich, P.R. (1996). Global change and human susceptibility to disease. *Annual Review of Energy and Environment*, Vol.21, 125–144.

Daniel, M.; Danielova, V.; Kriz, B.; Jirsa, A. & Nozicka, J. (2003). Shift of the tick *Ixodes ricinus* and tick-borne encephalitis to higher altitudes in central Europe. *European Journal of Clinical Microbiology Infectious Diseases*, Vol.22, 327–328.

Daszak, P. & Cunningham, A.A. (1999). Extinction by infection. *Trends in Ecology and Evolution*, Vol.14, 279.

Daszak, P., Cunningham, A.A. & Hyatt A.D. (2000). Emerging infectious diseases of wildlife: threats to biodiversity and human health. *Science*, Vol.287, 433–449.

Deem, S.L., Karesh, W.B. & Weisman, W. (2001). Putting theory into practice: Wildlife health in conservation. *Conservation Biology*, Vol.15, 1224–1233.

Dennehy, J.J., Friedenberg, N.A., Yang, Y.W. & Turner, P.E. (2007). Virus population extinction via ecological traps. *Ecology Letters*, Vol.10, 230–240.

Deredec, A., Courchamp, F. (2003). Extinction thresholds in host-parasite dynamics. *Annales Zoologi Fennici*, Vol. 40: 115–130.

Dizney, L.J. & Ruedas, L.A. (2009). Increased host species diversity and decreased prevalence of Sin Nombre virus. *Emerging Infectious Diseases*, Vol.15, 1012–1018.

Dobson, A.P. & Carper, R. (1992). Global warming and potential changes in host-parasite and disease-vector relationships. In: *Global warming and biological diversity*, R.L. Peters & T.E. Lovejoy (Eds), 201–217, Yale Univ Press, New Haven.

Dunn, R.R.; Davies, T.J.; Harris, N.C., & Gavin, M.C. (2010). Global drivers of human pathogen richness and prevalence. *Proceedings of the Royal Society B-Biological Sciences*, Vol.277, 2587-2595.

Elton, C.S. (1958). *The ecology of invasions by animals and plants*. Methuen, London.

Ezenwa, V., Price, S.A.; Altizer, S.; Vitone, N.D. & Cook, C. (2006). Host traits and parasite species richness in even- and odd-toed hoofed mammals, Artiodactyla and Perissodactyla. *Oikos*, Vol.115, 526–537.

Ezenwa, V.O., Godsey, M.S., King, R.J. & Guptill1, S.C. (2006). Avian diversity and West Nile virus: testing associations between biodiversity and infectious disease risk. *Proceedings of the Royal Society B-Biological Sciences*, Vol.273, 109-117.

Feliu, C.; Renaud, F.; Catzeflis, F.; Durand, P.; Hugot, J.-P. & Morand, S. (1997). A comparative analysis of parasite species richness of Iberian rodents. *Parasitology*, Vol.115, 453-466.

Fergus, R.; Fry, M.; Karesh, W.B.; Marra, P. P.; Newman, S. & Paul, E. (2006). Migratory birds and avian flu. *Science*, Vol.312:845-846.

Fevre, E.M.; Bronsvoort, B.; Hamilton, K.A. & Cleaveland, S. (2006). Animal movements and the spread of infectious diseases. *Trends in Microbiology*, Vol.14, 125-131.

Foley, J. A.; DeFries, R.; Asner, G. P.; Barford, C.; Bonan, G. et al. (2005). Global consequences of land use. *Science*, Vol.309, 570-574.

Gog, J.; Woodroffe, R. & Swinton, J. (2002). Disease in endangered metapopulations: the importance of alternative hosts. *Proceedings of the Royal Society B: Biological Sciences*, Vol.269: 671–676.

Goüy de Bellocq, J.; Charbonnel, N. & Morand, S. (2008). Coevolutionary relationship between helminth diversity and MHC class II polymorphism in rodents. *Journal of Evolutionary Biology*, Vol.21, 1144–1150

Guernier, V.; Hochberg, M.E. & Guégan, J.F. (2004). Ecology drives the worldwide distribution of human infectious diseases. *PloS Biology*, Vol.2: 740–46.

Hall, S.R.; Becker, C.R.; Simonis, J.L.; Duffy, M.A., Tessier, A.J. et al. (2009). Friendly competition: evidence for a dilution effect among competitors in a planktonic host-parasite system. *Ecology*, Vol.90, 791–801.

Harvell, C.D.; Mitchell, C.E.; Ward, J.R.; Altizer, S.; Dobson, A.P.; Ostfeld, R.S. & Samuel M.D. (2002). Climate warming and disease risks for terrestrial and marine biota. *Science*, Vol.296, 2158–2162.

Haydon, D.T.; Cleaveland, S.; Taylor, L.H. & Laurenson, M.K. (2002). Identifying reservoirs of infection: a conceptual and practical challenge. *Emerging Infectious Diseases*, Vol.8, 1468-1473.

Jones, K.E.; Patel, N.G.; Levy, M.A., Storeygard, A.; Balk, D. et al. (2008). Global trends in emerging infectious diseases. *Nature*, Vol.451, 990-993.

Johnson, P. T. J.; Lund, P.; Hartson, R.B .& Yoshino, T. (2009). Community diversity reduces *Schistosoma mansoni* transmission and human infection risk. *Proceedings of the Royal Society B-Biological Sciences*, Vol.276, 1657–1663.

Keane, R.M. & Crawley, M.J. (2002). Exotic plant invasions and the enemy release hypothesis. *Trends in Ecology and Evolution*, Vol.17, 164-179.

Keesing, F.; Brunner, J., Duerr, S.; Killilea, M.; LoGiudice, K. et al. (2009). Hosts as ecological traps for the vector of Lyme disease. *Proceedings of the Royal Society B-Biological Sciences, Vol.*276, 3911–3919.

Keesing, F.; Belden, L.K.; Daszak, P.; Dobson, A.; Harvell, C.D. et al. (2010). Impacts of biodiversity on the emergence and transmission of infectious diseases. *Nature*, Vol.468, 647-652.

Keesing, F.; Holt, R.D. & Ostfeld, R.S. (2006). Effects of species diversity on disease risk. *Ecology Letters*, Vol.9, 485-498.

Kelly, D.W.; Paterson, R.A.; Townsend, C.R.; Poulin, R. & Tompkins, D. M (2009). Parasite spillback: a neglected concept in invasion ecology? *Ecology*, Vol.90, 2047-2056.

Kelly, D.W., Paterson, R.A., Townsend, C.R., Poulin, R. & Tompkins, D.M. (2009). Has the introduction of brown trout altered disease patterns in native New Zealand fish? *Freshwater Biology*, Vol.54, 1805–1818.

Kilpatrick, A.M.; Kramer, L. D.; Jones, M. J.; Marra, P. P. & Daszak, P. (2006). West Nile virus epidemics in North America are driven by shifts in mosquito feeding behavior. *Plos Biology*, Vol.4, 606-610.

Koenig, W.D., Hochachka, W.M., Zuckerberg, B. & Dickinson, J.L. (2010) Ecological determinants of American crow mortality due to West Nile virus during its North American sweep. *Oecologia*, Vol.163, 903–909.

Kopp, K. & Jokela, J. (2007). Resistant invaders can convey benefits to native species. *Oikos*, Vol. 116, 295–301.

Kovats, R.S.; Campbell-Lendrum, D.H.; McMichael, A.J.; Woodward, A. & Cox, J.S.H. (2001). Early effects of climate change: Do they include changes in vector-borne disease? *Proceedings of the Royal Society London B*, Vol.356, 1057–1068.

Krasnov, B.R; Shenbrot, G.I.; Khokhlova, I. & Degen, A.A. (2004). Flea species richness and parameters of host body, host geography and host "milieu". *Journal of Animal Ecology*, Vol.73: 1121–1128.

Kutz, S.J.; Hoberg, E.P.; Polley, L. & Jenkins, E.J. (2005). Global warming is changing the dynamics of Arctic host-parasite systems. *Proceedings of the Royal Society London B*, Vol.272, 2571–2954.

Lee, K.A. & Klasing, K.C. (2004). A role for immunology in invasion biology. *Trends in Ecology and Evolution*, Vol.19, 523–529.

Leon-Vizcaino, L.; de Ybanez, M.R.R.; Cubero, M.J.; Ortiz, J.M.; Espinosa, J. et al. (1999). Sarcoptic mange in Spanish ibex from Spain. *Journal of Wildlife Diseases*, Vol.35, 647–659.

Li, Y.M. & Li, D.M. (1998). The dynamics of trade in live wildlife across the Guangxi border between China and Vietnam during 1993-1996 and its control strategies. *Biodiversity and Conservation*, Vol.7, 895-914.

Liang, S.Y.; Linthicum, K.J. & Gaydos, J.C. (2002). Climate change and the monitoring of vector-borne disease. *Journal of the American Medical Association*, Vol.287, 2286.

Lindenfors, P.; Nunn, C.L.; Jones, K.E.; Cunningham, A.A.; Sechrest, W. & Gittleman, J.L. (2007). Parasite species richness in carnivores: effects of host body mass, latitude, geographical range and population density. *Global Ecology and Biogeography*, Vol.1: 1–14.

Liu, J.G.; Linderman, M.; Ouyang, Z.Y.; An, L.; Yang, J. & Zhang, H.M. (2001). Ecological degradation in protected areas: The case of Wolong Nature Reserve for giant pandas. *Science*, Vol.292, 98-101.

LoGiudice, K.; Duerr, S.T.K.; Newhouse, M.J.; Schmidt, K.A.; Killilea, M.E. & Ostfeld, R.S. (2008) Impact of host community composition on Lyme disease risk. *Ecology*, Vol.89, 2841-2849.

LoGiudice, K., Ostfeld, R.S., Schmidt, K.A. & Keesing, F. (2003). The ecology of infectious disease: Effects of host diversity and community composition on Lyme disease risk. *Proceedings National Academy of Sciences USA*, Vol.100, 567–571.

Lovejoy, T.E. & Hannah, L. (2005). *Climate change and biodiversity*. Yale Univ Press, New Heaven.

Mack, R.N.; Simberloff, D.; Lonsdale, W.M.; Evans, H.; Clout, M. & Bazzaz, F.A. (2000). Biotic invasions: Causes, epidemiology, global consequences, and control. *Ecological Applications*, Vol.10, 689–710.

Marcogliese, D.J. (2001). Implications of climate change for parasitism of animals in the aquatic environment. *Canadian Journal of Zoology*, Vol.79, 1331–1352.

Marcogliese, D.J. & Cone, D.K. (1997). Food webs: A plea for parasites. *Trends in Ecology and Evolution*, Vol.12, 320–325.

Marvier, M.; Kareiva; P.; Neubert, M.G. (2004). Habitat destruction, fragmentation, and disturbance promote invasion by habitat generalists in a multispecies metapopulation. *Risk Analysis,* Vol.24, 869-878.

Materna, J.; Daniel, M. & Danielova, V. (2005). Altitudinal distribution limit of the tick *Ixodes ricinus* shifted considerably towards higher altitudes in central Europe: results of three years monitoring in the Krkonose Mts. (Czech Republic). *Central European Journal of Public Health,* Vol.13, 24–28.

McCallum, H. & Dobson, A. (1995). Detecting disease and parasite threats to endangered species and ecosystems. *Trends in Ecology and Evolution*, Vol.10, 190–194.

McIntyre, K.M., Setzkorn, C.; Baylis, M., Waret-Szkuta, A.; Caminade, C.; Morse, A.P.; Akin, S-A.; Huynen, M.; Martens, P. & Morand S. (2010). Impact of climate change on human and animal health. *Veterinary Record*: 586

MEA (2005). *Millenium Ecosystem Assessment. Ecosystems and Human Well–being: Biodiversity Synthesis*. Island Press, Washigton DC.

Meffe, G.K. (1999). Conservation medicine. *Conservation Biology*, Vol.13, 953.

Meerburg, B.G.; Singleton, G.R.; Kijlstra, A. (2009). Rodent-borne diseases and their risks for public health. *Critical Reviews in Microbiology*, Vol.35, 221-70.

Mills, J.N. (2006). Biodiversity loss and emerging infectious disease: An example from the rodent-borne hemorrhagic fevers. *Biodiversity,* Vol.7, 9-17.

Milner-Gulland, E.J. & Bennett, E.L. (2003). Wild meat: the bigger picture. *Trends in Ecology and Evolution,* Vol.18, 351-357.

Mitchell, C.E. & Power, A.G. (2003). Release of invasive plants from fungal and viral pathogens. *Nature*, Vol.421, 625–627.

Møller, A.P. & Cassey, P. (2004). On the relationship between T-cell mediated immunity in bird species and the establishment success of introduced populations. *Journal of Animal Ecology*, Vol.73, 1035–1042.

Molyneux, D.H.; Ostfeld, R.S.; Bernstein, A. & Chivian, E. (2008). Ecosystem disturbance, biodiversity loss, and human infectious disease, *In: Sustaining life : how human health depends on biodiversity*, E. Chivian & A. Bernstein (Eds), 287-323, Oxford University Press.

Morand, S. (2000). Wormy world: comparative tests of theoretical hypotheses on parasite species richness. *In Evolutionary biology of host-parasite relationships: Theory meets reality* (Poulin R, Morand S, Skorping A, eds), pp 63-79. Elsevier, Amsterdam.

Morand, S. & Harvey, P. (2000). Mammalian metabolism, longevity and parasite species richness. *Proceedings of the Royal Society London B*, Vol.267, 1999-2003.

Morand, S. & Poulin, R. (1998). Density, body mass and parasite species richness of terrestrial mammals. *Evolutionnary Ecology*, Vol.12, 717-727.

Moore, S.L. & Wilson, K. (2002). Parasites as a viability cost of sexual selection in natural populations of mammals. *Science*, Vol.297: 2015–2018.

Mouritsen, K.N. & Poulin R. (2002). Parasitism, climate oscillations and the structure of natural communities. *Oikos*, Vol.97, 462–468.

Mouritsen, K.N.; Tompkins, D.E. & Poulin, R. (2005). Climate warming may cause a parasite-induced collapse in coastal amphipod populations. *Oecologia*, Vol.146, 476–483.

Nunn, C.L.; Altizer, S.; Jones, K.E. & Sechrest, W. (2003). Comparative tests of parasites species richness in primates. *The American Naturalist*, Vol.162, 597–614.

Nunn, C.; Altizer, S.; Jones, K.E.; Sechrest, W. & Cunningham, A.A. (2005). Latitudinal gradients of parasite species richness in primates. *Diversity and Distributions*, Vol.11: 249–56.

O'Brien, S.J.; Roelke, M.E.; Marker, L.; Newman, A.; Winkler, C.A. et al. (1985). Genetic basis for species vulnerability in the cheetah. *Science*, Vol. 227: 1428–1434.

Ostfeld, R. S. & Keesing, F. (2000). Biodiversity and disease risk: The case of Lyme disease. *Conservation Biology*, 14:722-728.

Packer, C.; Holt, R.D.; Hudson, P.J.; Lafferty, K.D. & Dobson, A.P. (2003). Keeping the herds healthy and alert: implications of predator control for infectious disease. *Ecology Letters*, Vol.6, 797-802.

Patz, J.A.; Daszak, P.; Tabor, G.M.; Aguirre, A.A.; Pearl, M. et al. (2004). Unhealthy landscapes: policy recommendations on land use change and infectious disease emergence. *Environmental Health Perspectives*, Vol.112, 1092-98.

Patz, J.A.; Epstein, P.R.; Burke, T.A. & Balbus, J.M. (1996).-Global climate change and emerging infectious diseases. *Journal of the American Medical Association*, Vol.275, 217–223.

Patz, J.A.; Graczyk, T.K.; Geller, N. & Vittor, A.Y. (2000). Effects of environmental change on emerging parasitic diseases. *International Journal for Parasitology*, Vol.30, 1395–1405.

Patz, J.A.; Hulme, M.; Rosenzweig, C.; Mitchell, T.D.; Goldberg, R.A.; Githeko; A.K.; Lele, S.; McMichael, A.J. & Le Sueur, D. (2002). Climate change: Regional warming and malaria resurgence. *Nature*, Vol.420, 627–628.

Pedersen, A.B. & Fenton, A. (2006). Emphasizing the ecology in parasite community ecology. *Trends in Ecology and Evolution*, Vol.22, 133-139.

Peterson, A.T.; Sanchez-Cordero, V.; Beard, C.B. & Ramsey, J.M. (2002). Ecologic niche modeling and potential reservoirs for Chagas disease, Mexico. *Emerging Infectious Diseases*, Vol.8, 662–667.

Peterson, A.T.; Pereira, R.S. & Neves, V.F. (2004a). Using epidemiological survey data to infer geographic distributions of leishmaniasis vector species. *Revista da Sociedade Brasileira de Medicina Tropical*, Vol.37, 10–14.

Peterson, A.T., Bauer J.T. & Mills J.N. (2004b). Ecologic and geographic distribution of filovirus disease. *Emerging Infectious Diseases*, 10, 40–47.

Pounds, J.A.; Fogden, P.L.; Savage, J.M. & Gorman, G.C. (1997). Test of null models for amphibian declines on a tropical mountain. *Conservation Biology*, Vol.11, 1307–1322.

Poulin, R. (1995). Phylogeny, ecology, and the richness and parasite communities in vertebrates. Ecological Monograph, Vol. 65, 283-302.

Poulin, R. (2006). Global warming and temperature-mediated increases in cercarial emergence in trematode parasites. *Parasitology*, Vol.132, 143–151.

Poulin, R. & Morand, S. (2004). *The parasite biodiversity*. Smithsonian Institution Press, Washington.

Raymundo, L. J.; Halforda, A.R.; Maypab, A.P. & Kerr, A.M. (2009). Functionally diverse reef-fish communities ameliorate coral disease. Proc. Natl Acad. Sci. USA, Vol.106, 17067–17070.

Rayner, J.C.; Liu, W.;Peeters, M.; Sharp; P.M. & Hahn B.H. (2011). A plethora of *Plasmodium* species in wild apes: a source of human infection? *Trends in Parasitology*; Vol.27: 222-229.

Randolph, S.E. (2001). The shifting landscape of tick-borne zoonoses: Tick-borne encephalitis and *Lyme borreliosis* in Europe. *Proceedings of the Royal Society B-Biological Sciences,* Vol.356, 1045–1056.

Rogers, D.J. & Randolph, S.E. (2000). The global spread of malaria in a future, warmer world. *Science,* Vol.289, 1763–1766.

Rushton, S.P.; Lurz, P.W.W.; Gurnell, J. & Fuller, R. (2000). Modelling the spatial dynamics of parapoxvirus disease in red and grey squirrels: a possible cause of the decline in the red squirrel in the UK? *Journal of Applied Ecology,* Vol.37, 997–1012.

Saul, A. (2003). Zooprophylaxis or zoopotentiation: the outcome of introducing animals on vector transmission is highly dependent on the mosquito mortality while searching. *Malaria Journal,* Vol.2, 32-50.

Schmitz, O.J. & Nudds, T.D. (1994). Parasite-mediated comp etition in deer and moose: how strong is the effect of meningeal worm on moose? *Ecological Applications,* Vol.4, 91–103.

Scholte, E.J. & Schaffner, F. (2007). Waiting for the tiger: establishment and spread of the *Aedes albopictus* mosquitoes in Europe. In: *Emerging pests and vector-borne diseases in Europe,* W. Takken & B.G.J., Knols (Eds), 241-260, Wageningen Academic Publisher.

Shope, R. (1991). Global climate change and infectious diseases. *Environmental Health Perspectives,* Vol.96, 171–174.

Smart, S.M.; Thompson, K.; Marrs, R.H. *et al.* (2006). Biotic homogenization and changes in species diversity across human-modified ecosystems. *Proceedings of the Royal Society B-Biological Sciences,* Vol.273, 2659-2665.

Sprent, J.F.A. (1992). Parasites lost? *International Journal for Parasitology,* Vol.22, 139–151.

Stanko, M.; Miklisová, D.; Goüy de Bellocq, J. & Morand, S. (2002). Mammal density and patterns of ectoparasite species richness and abundance. *Oecologia,* Vol.131, 289-295.

Stenseth, N.C.; Mysterud, A.; Ottersen, G.; Hurrell, J.W.; Chan, K.-S. & Lima, M. (2002). Ecological effects of climate fluctuations. *Science,* Vol.297, 1292–1296.

Sutherst, R.W. (1998). Implications of global change and climate variability for vector-borne diseases: Generic approaches to impact assessments. *International Journal for Parasitology,* Vol.28, 935–945.

Sutherst, R.W. (2001). The vulnerability of animal and human health to parasites under global change. *International Journal for Parasitology,* Vol.31, 933–948.

Sutherst, R.W. (2004). Global change and human vulnerability to vector-borne diseases. *Clinical Microbiological Review,* Vol.17, 136–173.

Suzán, G.; Marcé, E.; Giermakowski, J.T., Mills, J.N.; Ceballos, G. et al. (2009). Experimental evidence for reduced rodent diversity causing increased Hantavirus prevalence. *Plos ONE,* Vol.4, e5461.

Swaddle, J.P. & Calos, S.E. (2008). Increased avian diversity is associated with lower incidence of human West Nile infection: Observation of the dilution effect. *PLoS ONE,* Vol.3, e2488.

Tersago, K.; Schreurs, A.; Linard, C.; Verhagen, R.; Dongen, S.V. & Leirs, H. (2008). Population, environmental, and community effects on local bank vole (*Myodes glareolus*) Puumala virus infection in an area with low human incidence. *Vector-Borne and Zoonotic Diseases,* Vol.8, 235–244.

Thieltges, D.W., Bordalo, M.D., Caballero-Hernandez, A., Prinz, K. & Jensen, K.T. (2008). Ambient fauna impairs parasite transmission in a marine parasite-host system. *Parasitology,* Vol.135, 1111–1116.

Thieltges, D.W.; Reise, K.; Prinz, K. & Jensen, K.T. (2009). Invaders interfere with native parasite-host interactions. *Biological Invasions,* Vol.11, 1421–1429.

Thomas, F.; Poulin, R.; de Meeüs, T.; Guégan, J.-F. & Renaud, F. (1999). Parasites and ecosystem engineering: What roles could they play? *Oikos*, Vol.84, 167–171.

Torchin, M.E.; Lafferty, K.D.; Dobson, A.P.; McKenzie, V.J. & Kuris, A.M. (2003). Introduced species and their missing parasites. *Nature*; Vol.421, 628–630.

Torres, J.; Miquel, J.; Casanova, J.-C.; Ribas, A.; Feliu, C. & Morand, S. (2006). Parasite species richness of Iberian carnivores: influences of host density and range distribution. *Biodiversity and Conservation,* Vol.15, 4619-4632

Thorne, E.T. &Williams, E.S. (1988). Disease and endangered species: the black-footed ferret as a recent example. *Conservation Biology*, Vol.2, 66–74.

Torres, J.; Miquel, J.; Casanova, J.C.; Ribas, A.; Feliu, C. & Morand, S. (2006). Parasite species richness of Iberian carnivores: influences of host density and range distribution. *Biodiversity and Conservation*, Vol.15: 4619–4632.

van Riper, C.; van Riper, S.G.; Goff, M.L. & Laird, M. (1986). The epizootiology and ecological significance of malaria in Hawaiian land birds. *Ecological Monographs*, Vol.56, 327–344.

van Riper, C.; van Riper, S.G. & Hansen, W.R. (2002). Epizootiology and effect of avian pox on Hawaiian forest birds. *The Auk*, Vol.119, 929–942.

Vitousek, P.M.; Dantonio, C.M.; Loope, L.L.; Rejmanek, M. & Westbrooks, R. (1997). Introduced species: a significant component of human-caused global change. *New Zealand Journal of Ecology*, Vol.21, 1–16.

Vittor, A.Y., Gilman, R.H., Tielsch, J., *et al.* (2006). The effect of deforestation on the human-biting rate of *Anopheles darlingi*, the primary vector of *Falciparum malaria* in the Peruvian Amazon. *American Journal of Tropical Medicine and Hygiene*, Vol.74, 3-11.

Vourc'h, G.; Marmet, J.; Chassagne, M.; Bord, S. & Chapuis, J.-L. (2007). *Borrelia burgdorferi sensu lato* in Siberian chipmunks (*Tamias sibiricus*) introduced in suburban forests in France. *Vector-Borne and Zoonotic Diseases*, Vol.7, 637-641.

Walpole, M.; Almond, R.E.; Besançon, C.; Butchart, S.H.; Campbell-Lendrum, D. et al. (2009). Ecology. Tracking progress toward the 2010 biodiversity target and beyond. *Science*, Vol.325 5947

Walther, G.-R.; Post, E.; Convey, P.; Menzel, A.; Parmesan, C. et al. (2002). Ecological responses to recent climate change. *Nature*, Vol.416, 389–395.

White, P.C.L.; Newton-Cross, G.A.; Gray, M.; Ashford, R.; White, C. & Saunders, G. (2003). Spatial interactions and habitat use of rabbits on pasture and implications for the spread of rabbit haemorrhagic disease in New South Wales. *Wildlife Research*, Vol.30, 49–58.

Wilcox, B.A. & Gubler, D.J. (2005). Disease ecology and the global emergence of zoonotic pathogens. *Environmental Health and Preventive Medicine*, Vol.10, 263–272.

Williamson, M. (1996). *Biological invasions*. Chapman and Hall, London

Wilson, M.E. (2000). Environmental change and infectious diseases. *Ecosystem and Health*, Vol.6, 8–12.

Wolfe, N. D., Daszak, P., Kilpatrick, A.M. & Burke, D.S. (2005). Bushmeat hunting, deforestation and prediction of zoonotic emergence. *Emerging Infectious Diseases*, Vol.11, 1822-1827.

14

Protected Areas: Conservation Cornerstones or Paradoxes? Insights from Human-Wildlife Conflicts in Africa and Southeastern Europe

Brandon P. Anthony and Alina Szabo
Department of Environmental Sciences & Policy, Central European University, Budapest Hungary

1. Introduction

Protected areas (PAs) are considered the cornerstone of biodiversity conservation (Chape et al., 2005), according to the International Union for Conservation of Nature (IUCN), are defined as "A clearly defined geographical space, recognised, dedicated and managed, through legal or other effective means, to achieve the long-term conservation of nature with associated ecosystem services and cultural values" (Dudley, 2008). The IUCN set the first 'target' for protection, by agreeing at the 1992 World Parks Congress in Caracas, Venezuela, that PAs should cover a minimum of 10% of each biome by 2000. There are now over 120,000 PAs worldwide, taking into account both those classified under any of the six IUCN management categories, and those not classified, including private reserves and community-conserved areas (UNEP-WCMC, 2008). The overall trend in the total surface of PAs has been steadily upward during the last few decades in all regions of the world, but has been leveling off somewhat since 2000 (Fig. 1). Recognising the importance of PAs to global efforts to halt biodiversity loss, in late 2010 at the Convention on Biological Diversity's 10th Conference of Parties in Japan, twenty targets were set for biodiversity conservation ('Aichi Targets'). These include a global increase of terrestrial and inland water PAs to 17% (from 13%) by 2020, and coastal and marine PAs from 1% to 10% in the same period (Secretariat of the Convention on Biological Diversity, 2010).

Despite this growth, however, the effectiveness of PAs in conserving habitats and species cannot simply be interpreted as the result of their number and size as it also depends on their location, structure (size, shape, connectivity) and, of equal importance, their management (Cantu-Salazar & Gaston, 2010; Rodrigues et al., 2004). Traditionally, a top-down approach was employed to PA establishment and management which excluded local participation (Ervin et al., 2010; Kiss, 1990; Stevens, 1997). As a result, people whose livelihoods chiefly involve the direct exploitation of local natural resources often come into conflict with the institutions of PAs. Frequently, communities living in and around PAs have important and longstanding relationships with these areas that embrace *inter alia* cultural identity and subsistence practices essential to sustaining livelihoods, and often contribute to maintenance of biodiversity (Anthony & Bellinger, 2007). Consequently, PAs are increasingly being recognised as 'social spaces' (Ghimire & Pimbert, 1997) and as such, cannot be decoupled from their human context in terms of management regimes (Brechin et al., 2002). Reinforcing this shift, it is now widely postulated that PAs cannot coexist in the long term with communities that are hostile to them (McNeely, 1993; Pimbert & Pretty, 1997; West & Brechin, 1991).

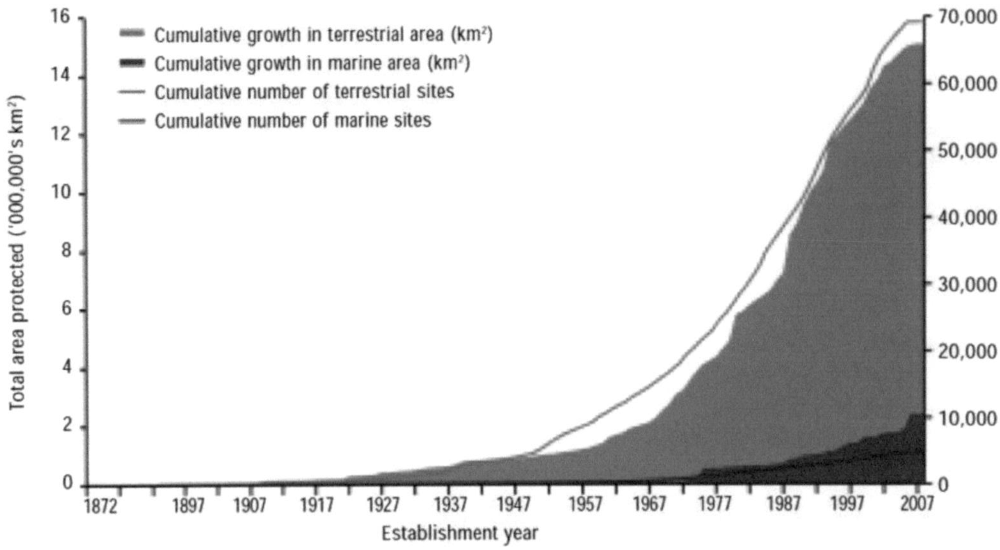

Fig. 1. Cumulative global growth in the area and number of nationally designated protected areas (1872-2007). Reproduced with permission from Coad et al. (2008)

A particular management challenge confronting PAs is human-wildlife conflict (HWC), the consequence of spatial proximity between people and their activities, and wildlife (Knight, 2000; Naughton-Treves & Treves, 2005; Treves, 2009). This poses additional challenges for biodiversity conservation as conflict tends to lead to an increase in opposition from local people and undermines political support for conservation efforts (Madden, 2004; Naughton-Treves, 1998; Newmark et al., 1994, Treves, 2009). Conflicts between humans and wildlife are the product of socio-economic and political landscapes (Graham et al., 2005) and the institutional architecture designed to manage these conflicts, and are controversial because the resources concerned have economic value and the species involved often have high profile and are legally protected (McGregor, 2005; Treves & Karanth, 2003). While humans and wildlife have a long history of co-existence, the frequency of HWC has grown in recent decades (Decker et al., 2006; Graham et al., 2005; Mishra, 1997; Wang & Macdonald 2006), mainly because of (i) extension of human presence and activities into wildlife habitat and shrinking prey populations (Linnell et al., 2001; Woodroffe, 2000; Woodroffe et al., 2005), (ii) expansion of some wildlife distributions including into the matrix surrounding PAs (Bisi & Kurki, 2005; Breitenmoser, 1998; Cozza et al., 1996; Stahl et al., 2001; Zedrosser et al., 2001), as well as (iii) a frequent inability of institutions that are meant to mediate such conflicts to respond effectively (Anthony et al., 2010; Hewitt & Messmer, 1997). Moreover, particularly in developing countries, poor and politically marginalised people frequently come into conflict with wildlife and are pitted against the state and its wildlife agencies in strongly unequal power relationships.

1.1 Definitions

To define human-wildlife conflict, Madden (2004) uses the 5th IUCN World Parks Congress recommendation which states that it 'occurs when the needs and behavior of wildlife impact negatively on the goals of humans or when the goals of humans negatively impact the needs of wildlife.' Knight (2000) offers an anthropological perspective: 'People-wildlife conflicts are relations of rivalry or antagonism between human beings and wild animals which typically arise from territorial proximity and involve reliance on the same resources or a threat to human wellbeing or safety.' There are many variations on these definitions in the literature (Conover, 2002; Graham et al., 2005; Marshall et al., 2007; Treves, 2009) but overall they can be summed up as including 'both competition and predation' (Knight, 2000) with a bidirectional character.

The conflict between people and wildlife takes many forms: attacks on people, attacks on livestock, crop-raiding, damage to forestry, competition for forage resources and for wild prey, building infestations, traffic accidents, disease, threats to other species and biodiversity, and human-induced wildlife mortality (Conover, 2002; Knight, 2000; Kruuk, 2002; Sillero-Zubiri et al., 2007; Thirgood et al, 2005). It has two dimensions, occurring *between people and wildlife* but also *between various people* with dissimilar views about wildlife (Knight, 2000; Linnell et al., 2010; Madden, 2004; Marshall et al., 2007). This latter form of conflict stems from differences in values held by those involved (Knight, 2000; Treves, 2009) as well as from distinct perceptions that people have regarding their own and others' positions (Marshall et al., 2007).

In this chapter, HWC is understood as both a clash between people and wildlife over 'space, food and life' (Treves, 2009), in agreement with the two definitions mentioned above, and a socio-political conflict. This latter dimension of conflict cannot be omitted especially taking into account that our case studies are (in the least) nationally designated PAs. As Treves (2009) emphasised, often PAs bring out 'the fundamental dilemma posed by global and national concerns for biodiversity conservation on the one hand and individual and economic motivations to safeguard human life and livelihood on the other hand.'

1.2 Factors contributing to conflict

The growing trend towards greater spatial proximity between people and wildlife and their reliance on the same resources will likely lead to an increase in the frequency and intensity of conflict. Research has shown that different areas experience dissimilar levels of conflict with some so-called conflict 'hot spots' experiencing recurring problems (Breck, 2004). In order to develop effective strategies that will allow people and wildlife to share the landscape it is necessary to understand the complexities posed by the local situation and to tease out underlying factors that lead to negative interactions, some of which we outline below.

Attacks on humans are particularly important drivers of conflict as fear of personal injury and death builds strong antagonism toward wildlife, in particular toward elephants and large carnivores (Dublin & Hoare, 2004; Löe & Röskaft, 2004; Saberwal et al., 1994). Moreover, perceptions are crucial (Naughton-Treves & Treves, 2005; Zinn et al., 2000) and any successful solution to conflict must address them directly (Madden, 2004; Treves et al., 2006). Research and practice show that perception of potential risk and of level of control over the situation are often the most important factors driving public reactions to HWC (Johansson & Karlsson, 2011; Sillero-Zubiri & Laurenson, 2001), even though perceptions are frequently incongruent with reality (Dublin & Hoare, 2004).

Furthermore, the socio-cultural context in which the conflict is embedded is significant. Such is for example the case with wolves which traditionally have elicited a strong negative reaction from the public despite the fact that they pose little danger to people (Kellert et al., 1996). Nevertheless, more recently people's attitudes toward wolves have undergone a significant transformation under the influence of changing cultural beliefs. Kellert et al. (1996) emphasised the important place of large carnivores in North American consciousness and their roles as indicators of changing attitudes toward wildlife and nature. Moreover, cultural values also inform the approach undertaken to conserving wildlife. Clark & Slocombe (2009) show how aboriginal people's respect for grizzly bears has led to the formation in southwest Yukon (Canada) of a resource management system that could impart knowledge to new strategies for managing human-bear interactions.

Another significant factor is represented by the severity of conflict between people and wildlife. This is influenced by the spatial and social distribution of damage, as well as by the ability of individuals to cope with losses inflicted by wildlife. When risk is carried at household level, material wealth is an important determinant of who is able to cope with wildlife inflicted damage (Naughton-Treves & Treves, 2005). Local people's inadequate knowledge of the ecology and behavior of wildlife (e.g. growing crops in wildlife areas) also affects the severity of conflict and, coupled with limited coping capacity, leads to increased vulnerability to wildlife damage (Naughton-Treves & Treves, 2005).

1.3 Significance

The investigation of HWC and its influence on biodiversity conservation is important for a number of reasons. First, wildlife damage represents a very tangible threat to livelihoods in terms of personal injury, crop and livestock losses, property damage, and lost opportunity costs (Choudhury, 2004; Emerton, 2001; Happold, 1995; Hill, 2004; Graham et al., 2005; Linnell et al., 2010). Second, attitudes towards PAs are often influenced by real or perceived damage caused by wildlife (Anthony, 2007; de Boer & Baquete, 1998; Els, 1995; Hill, 2004). Third, active persecution by humans following wild predator attacks on livestock has been identified as an important factor in observed carnivore declines (Hazzah et al., 2009; Mishra, 1997; Woodroffe, 2001), and may lead to increased 'edge effects' along the peripheries of PAs (Woodroffe & Ginsberg, 1998). Finally, HWC are potentially socially corrosive, creating and reflecting larger conflicts of value and class and other interests (McGregor, 2005). Especially in less developed countries and countries in transition, such conflicts have the potential to undermine both biodiversity conservation efforts and human security, and further weaken the effectiveness and legitimacy of state institutions including national parks and other PAs (Anthony et al., 2010).

These aforementioned complexities are also juxtaposed within contextual PA management regimes. As PAs establish and implement management strategies and plans which increasingly embrace socio-economic interests, they must make difficult decisions about their objectives in terms of HWC, particularly to 'identify and explicitly acknowledge the trade-offs and hard choices that are involved in advancing conservation in specific places and through specific approaches' (McShane et al., 2011). These choices have repercussions that can last for many decades and, in some cases, can affect the suite of options available for park administrations to mitigate HWC in the future. Drawing on case studies, we outline selected factors and 'choices' contributing to HWC, to what extent relevant institutions are addressing these conflicts and what the likely outcomes are for biodiversity conservation.

Finally, we show that in order to manage HWC more effectively in such contexts, five components of the conflict must be addressed: (i) baseline research, (ii) evaluation of damage, (iii) conflict management, (iv) adaptive management, and (v) identifying and acknowledging management trade-offs.

2. Methods

Our four case studies involve PAs which are currently facing HWC challenges, and with which we have personal experience. They are located in South Africa (Kruger National Park, hereafter KNP), Malawi (Vwaza Marsh Wildlife Reserve, hereafter VMWR), and Romania (Măcin Mountains National Park, hereafter MMNP; Rodna Mountains National Park, hereafter RMNP) (Fig. 2). They differ in terms of size, date of designation, ecosystems represented, and tenure (Table 1).
In order to explore the complexity of HWC issues, including the perspectives of relevant stakeholders, multi-method approaches were utilised in each of our four cases (Table 2). For more detailed descriptions of methods used, see indicated sources.

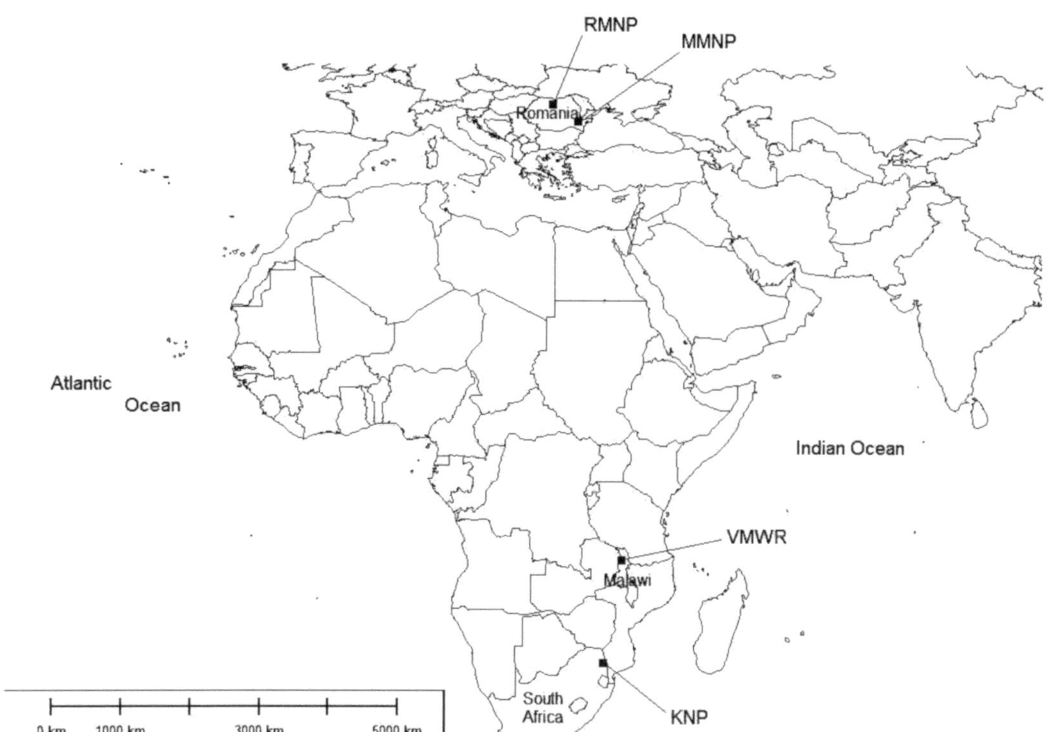

Fig. 2. Location of selected PAs in sub-Saharan Africa, and Romania in southeastern Europe

PA	Date of Designation	Area (ha)	Representative ecosystems	Tenure
KNP	1926	1,898,900	16 'ecozones' ranging from wooded hills to open plains and savannas, and riverine bushveld	state-owned, although portions subject to land claims
VMWR	1977	98,200	Mopane and miombo woodland, marshy wetlands	state-owned (100%)
MMNP	2003	11,142	Balkan-Pontic steppe and sub-Mediterranean and Balkanic forest	most (99%) state-owned; 1% municipally-owned
RMNP	1990	46,417	Eastern Carpathian Mountains including temperate forests and alpine grasslands	most (~93.4%) owned by local communities; state-owned (6.4%); private & church owned (0.17%)

Table 1. Characteristics of selected protected areas

PA	Methods	Fieldwork	Source(s)
KNP	archival analysis of policies, reports, and legislation questionnaire survey (240 households; 38 villages, C.I. = 6.28, C.L. = 95%) semi-structured interviews participant observation	Jan – Nov 2004	Anthony (2007) Anthony et al. (2010)
VMWR	archival analysis of policies, reports, and legislation village meetings (7 zones; 300+ participants) community mapping semi-structured interviews participant observation	July – Aug 2009	Anthony & Wasambo (2009)
MMNP	archival analysis of policies, reports, and legislation questionnaire survey (374 households; 14 villages + 1 town, C.I. = 5.0, C.L. = 95%) unstructured interviews	May – July 2007	Anthony & Moldovan (2008)
RMNP	archival analysis of policies, reports, and legislation semi-structured interviews participant observation	June 2010 – April 2011	Szabo, unpublished results

Table 2. Multi-method approaches used in selected case studies

3. Case studies

In this section, we provide an introduction to each of our case studies, including its history, and type and extent of human-wildlife conflict. This is followed by a description of selected factors which we believe lead to particular outcomes in terms of HWC management options for each PA.

3.1 Kruger National Park, South Africa

The KNP, established in 1926, is situated in the north-eastern region of the Republic of South Africa, and covers nearly 2 x 10^6 hectares (Mabunda et al., 2003). KNP is unequaled among South Africa's national parks system, being home to an unparalleled diversity of wildlife and maintained by one of the world's most sophisticated park management systems (Braack, 2000). Internationally, KNP functions as a major tourism destination with up to 1 million visitors annually, and serves as an important socioeconomic and ecological component of the Great Limpopo Transfrontier Park, a multi-lateral initiative involving South Africa, Mozambique, and Zimbabwe (SANParks, 2006).

Prior to 1994, as in other parts of southern Africa, the familiar approach to proclaiming PAs in South Africa was to remove (often forcefully) resident rural people and relocate them elsewhere without adequate compensation (Callimanopulos, 1984; Campbell & Shackleton, 2001; Lahiff,1997; Volkman, 1986). These and other neighbouring communities were then customarily deprived of access to PAs, any participation or input in their management, or any share of their benefits (Khan, 1994; Magome & Collinson, 1998). The result was that, despite successes gained in conserving biodiversity by producing South Africa's extensive PA network, in the process much human misery and hostile attitudes towards PAs resulted, including from communities evicted from KNP (SANParks, 2000). However, since the lifting of Apartheid and the democratic elections of 1994, the National Parks Board, whose name changed in 1997 to South African National Parks (SANP), has undergone major changes with regard to philosophy, policy and organisational structure to reflect the new economic and socio-political realities of South Africa as underpinned by the new Constitution. In addition to core objectives of conserving biodiversity and maintaining landscapes, new park management policy has moved towards integrating the interests of neighbouring communities. This includes redressing past injustices through facilitating land claims within the park.

Concomitant with these changes, KNP established its own Social Ecology Program, which facilitates seven participatory communication structures with the park's neighbouring communities, consisting of about 120 villages and private game farms with an estimated total human population of 1.5 million (SANParks, 2000). The Hlanganani Forum (representing 27 villages), in whose jurisdiction this case falls, was initiated in 1994, and meets monthly to strengthen park-neighbour relationships (Anthony, 2006). The historical background of these communities, which forms part of the former Gazankulu homeland, is characterised by a general dissatisfaction with park authorities (Els, 1994), in part due to damage to crops, livestock and property caused by wildlife, particularly lion (*Panthera leo*), hyena (*Crocuta crocuta*), elephant (*Loxodonta africana*), and buffalo (*Syncerus caffer*) which were regularly escaping from the park (Cock & Fig, 2000; Freitag-Ronaldson & Foxcroft, 2003; Tapela & Omara-Ojungu, 1999).

Until the late 1990s, KNP section rangers had authority to deal with damage-causing animals themselves, but currently their agreement with Limpopo Province stipulates that the park can only assist in controlling animals exiting the park after first obtaining permission from the provincial government *in each case*. To complicate matters, the border fence is under the responsibility of the Department of Animal Health, which is obliged to maintain it in order to prevent transmission of disease from wild animals to domestic livestock. However, despite the provincial government being the lead agency responsible for controlling animals outside the park, it currently suffers from under-funding and under-staffing, and is therefore a weak actor with little capacity to solve local problems (Anthony et al., 2010). This is exacerbated by poor communication, slow response times, weak reporting and monitoring, and vulnerability to illegal hunting practices by professional hunting outfitters tendered by the province to control animals, involving the luring of lion from the park with livestock and/or poached zebra (Anthony et al., 2010). This lion luring is a man-induced population sink, which may increase the prevalence of 'edge effects' along KNP's border (Woodroffe & Ginsberg, 1998), and subject lions to a change of prey preference if they have increased cattle-killing opportunities (van Dyk & Slotow, 2003). Tension and problems are therefore inherent in the current institutional arrangements within which HWC are dealt with. This institutional reform for controlling damage-causing animals has resulted in a piecemeal process, fraught with gaps and loopholes, which has produced increased opportunities for corruption and illegal activity. Moreover, it demonstrates the need for an improved and streamlined system of control that minimises risk and damage, an issue of acute concern for communities neighboring the park.

Community perceptions of HWC are an important aspect of KNP's interaction with its neighbors, and have great potential in shaping attitudes towards the park and its objectives. KNP is perceived by many neighbours as contributing to current injustices by harbouring dangerous animals causing extensive damage and threatening livelihoods of the very communities it seeks to empower. Based on their questionnaire, Anthony et al. (2010) reported that 12.1% of their respondents had experienced HWC between 2002 and 2004. If one considers all households within 15 km of the park border, an estimated 2,216 households had been affected. They also showed that households that had higher numbers of mammalian livestock and are closer to the park were significantly more likely to experience HWC. Moreover, those who had suffered damage were significantly less likely to believe that KNP would ever help their household economically. Negative attitudes toward KNP by community members primarily centre on HWC, including the lack of adequate maintenance of the KNP border fence, control of animals once they escape from the park and affected farmers not being financially compensated for losses, despite promises that compensation would be forthcoming. These aspects of HWC threaten, and in some cases prevent, economic diversification through sustaining or enhancing agricultural livelihoods. They also have left many community members with a sense of hopelessness.

In addition to simply influencing attitudes, due to the perceived inadequacy of control by the current institutional structure, many locals were resorting to retaliatory killing of wildlife, including the shooting of problem lions, and using snares (Anthony, 2006). In addition, carcasses have often been laced with poison to indiscriminately kill lions and hyenas, which has had lethal consequences for other species, including vultures, which are specially protected in Limpopo Province (LEMA No. 7 of 2003; Schedule 2).

3.2 Vwaza Marsh Wildlife Reserve, Malawi

Analogous to South Africa, previous conservation policies in Malawi focused solely on the conservation of wildlife resources within PAs without taking the rights and needs of the surrounding human population into account. Consequently, the primary focus of PA management was law enforcement and the relationship between PA authorities and neighboring communities was openly antagonistic (MDNPW, 2004). To change this relationship, the Department of National Parks and Wildlife embraced a collaborative management program in the early 1990s with the aim of developing a sustainable and inter-dependent relationship between PAs and the neighbouring rural populations by ensuring direct flow of benefits to these communities (MDNPW, 2004).

Vwaza Marsh Wildlife Reserve (VMWR) occupies almost 100,000 ha of diverse terrain in Northern Malawi in the Central African Plateau. The reserve comprises a region of hills and pediments in the east, and a region of wetland and alluvium in the west. It is home to a wide variety of large mammals, including buffalo, hippopotamus (*Hippopotamus amphibius*) and elephant, and an extensive range of lowland bird species. The reserve's water source is the Nyika plateau which lies to the north and east within Nyika National Park, running along the South Rukuru River. The western and part of its northern boundary coincides with the Malawi - Zambia border where wildlife can roam freely between the Reserve and the Luangwa Valley in Zambia.

Protection of the area as a group reserve began in 1941 with the proclamation of Lake Kazuni Game Reserve (Government Notice no. 166 of 1941). This included all land and water within a 5 mile distance of the centre of Lake Kazuni (Nxumayo et al., 2008). According to the Vwaza Marsh Wildlife Reserve Revised Master Plan (2004), the Vwaza Marsh was proclaimed a Wildlife Reserve in 1977 (GN 33 of 1977), which subsequently involved the eviction of most villages in the area. Bell & Mphande (1980) estimated the human population to have been approximately 2,075 immediately prior to 1977. Inhabitants were removed from the reserve through 1979 and the last group of people to leave were those of the Mowa village in the north-west of the reserve in 1984.

Despite more than a decade of active engagement between VMWR and its neighbouring communities, little is known about how those relationships have developed and what factors influence their success or failure in fostering cooperation. Previous studies have emphasised the general dissatisfaction with reserve authorities by local communities, largely subsistence farmers, in part due to village evictions and damage to crops and property caused by wildlife (Msiska 2002; Nxumayo et al. 2008; VMWR 2003). Further, VMWR Annual Reports (2003-2004 through 2008-2009) consistently highlight (i) problem animal incidents, and (ii) the ongoing challenges associated with problem animal control, alleviating damage, and adequately responding to communities' demands for compensation.

Anthony & Wasambo (2009) undertook a HWC study involving about 60 villages (organised into 7 zones) within 5 km of the reserve, in part to examine institutional roles and the effectiveness of policies and practices of VMWR and local communities in managing HWC, and offer perspectives from rural community members who live within the area. They found that of 19 identified problem taxa by local communities, elephants were perceived as the 'worst' problem animal, followed by chacma baboons (*Papio ursinus*)/monkeys (*Cercopithecus aethiops*), hyena, bushpigs (*Potamochoerus porcus*), and hippopotamus. Meeting participants were adamant in voicing their concerns over the widespread destruction of both crops and livestock in the area by problem animals (Table 3).

Zone	crop depredation	destroying food stores	chasing/killing livestock	threatening people	killing people	other
Zolokele	✓	✓		✓		
Mwazisi	✓	✓	✓	✓	✓	
Mphangala	✓	✓	✓			
Kamphenda	✓	✓	✓	✓		
Kazuni	✓	✓	✓	✓	✓	'drinking our traditional beer'
Thunduwike	✓	✓	✓	✓		'drinking our water'
Zaro	✓	✓	✓	✓		'bringing tsetse flies'

Table 3. Types of human-wildlife conflict in village zones surrounding VMWR

Participants from all 7 zones unanimously believed that incidents with problem animals had been increasing in recent years, particularly with elephants. Their reasons for saying so are based on perceived problems with the reserve border fence, increasing animal populations, insufficient buffer zone between reserve and communities, and poor control of problem animals as a consequence of low capacity within the reserve to manage the conflict.

Despite the fact that legislation and policies are in place at both the national and local reserve level concerning the control of problem animals, the implementation of these policies is weak on the ground (Anthony & Wasambo, 2009). The perceived result is that incidents with problem animals have increased in recent years, the response times by VMWR staff are variable, and the results of animal control are largely inadequate, often with animals habituating to the methods used. This has had profound consequences on local livelihoods, both directly and indirectly, including financial, social and cultural losses ranging from crop raiding and damage to water sources, to increased anxiety and loss of human life. This has led to increasingly frustrated communities who have, in some cases, over-exaggerated the extent of elephant damage (Anthony & Wasambo, 2009).

When asked about possible strategies for mitigating HWC, a wide variety of suggestions were made by participants, including adjustments in staffing, benefit sharing, increased community involvement, and improved wildlife management (Anthony & Wasambo, 2009). For our purpose here, a noteworthy suggestion was that of the role of fencing as a HWC control measure. Experience shows that where PA boundaries abut onto settlements and cultivation, especially in a densely populated country like Malawi, then HWC is inevitable. This is especially the case with species like elephants, hippopotamus, baboons and bushpigs, all of which occur in VMWR. The alignment of boundaries, as with the South Rukuru River and floodplain forming the southern boundary of Vwaza Marsh, with an inadequate buffer zone between wildlife habitat within the park, and cultivation outside, complicates the issue enormously (Hall-Martin et al., 2007). If properly designed and maintained, electric game fences have been shown to prevent unwanted movement of animals out of PAs (Hoare, 2001). Yet, communities surrounding VMWR are polarised on their use. Some would like to see the fencing extended and/or improved, whilst many others want to see the existing fence eliminated altogether, as it represents a permanent reminder of the fact that their land was illegitimately taken from them by the previous regime. This sentiment was also echoed by VMWR staff, who claim that proposals to improve the fence has always elicited mixed reactions from the community for this reason.

3.3 Măcin Mountains National Park, Romania

MMNP is a relatively small (approx. 11,000 ha) PA situated in southeast Romania. MMNP is the only PA in Europe where ecosystems typical of the Pontic-Sarmatian steppe, sub-Mediterranean and Balkan forests can be found on Hercynian mountains, the oldest in Romania and some of the earliest in Europe (Parcul National Muntii Macinului, 2006). According to Institutul de Cercetari si Amenajari Silvice (1996), steppe ecosystems once covered 16% of Romania's territory but they are now restricted to the Macin Mountains. MMNP is among the newest national parks in Romania, being designated in 2003. The park is administered by the MMNP Administration, within the National Forest Administration - Romsilva. The majority (99.6%) of MMNP is forest, along with smaller areas of pasture and arable land (Parcul National Muntii Macinului, 2006). Although comprehensive surveys regarding the biodiversity of Macin Mountains have not been undertaken, an estimated 150 vertebrate and 72 plant species classified as protected are present in the park (UNDP, 2005). Similar to our African cases, MMNP primarily treated its neighboring communities in a top-down fashion until 2006, at which time it established a Community Liaison position within its organisational structure. This person interacts with a Consultative Council, the members of which represent local municipalities.

MMNP keeps no HWC incident records, therefore no park data is available, with the exception of *ad hoc* observations provided during park staff interviews. In their MMNP-communities interaction study, Anthony & Moldovan (2008) report that 24 (6.4%) of 374 survey respondents from neighboring villages declared their household had experienced some form of damage from wildlife in the past 5 years. When the spatial reference was extended to the whole village, 18.4% declared they knew about HWC incidents in their village. Most frequently (68.1%), the identified problem was chasing or killing of livestock, followed by crop depredation (55.1%). In 13% of the cases, wildlife inflicted property damage and in four cases, human life was supposedly threatened. Wild boars (*Sus scrofa*) are thought to be the most damaging animal, followed by fox (*Vulpes vulpes*) and golden jackal (*Canis aureus*).

Given its relatively recent establishment, and the absence of radical changes in the forestry exploitation regime, local residents have not had extensive interaction with the park's administration structures and representatives. Thus, most local people tend to hold a neutral attitude towards MMNP (Anthony & Moldovan, 2008). However, those that do hold negative attitudes largely justified their opinion by stating that they believe the park plans to introduce nose-horned vipers (*Vipera ammodytes*) to the area, and this will consequently be a threat to local inhabitants. Other negative responses were derived from the perceived increase in restrictions and associated fines related to wood collection and grazing. A particularly sensitive issue is the closing of adjacent quarries due to the designation of the park, an aspect that produces mixed feelings. Some stated that quarry closure has brought about decreased air and noise pollution, while others claim it has contributed to locally high unemployment.

Reinforcing the trepidation about the park's management objectives in terms of reintroductions, although most respondents (80.2%) agreed that wild animals should be protected, a significant proportion believed that only those animals that 'do not harm people or destroy crops should be protected' (Anthony & Moldovan, 2008). Bears, wolves and, especially, snakes (vipers) were characterised as 'animals that can do harm' and, consequently, should not be protected. According to the MMNP biologist, there are no bears or wolves on the park territory, nor does the park have plans to reintroduce vipers to the area.

Due to the short history of MMNP, people have had minimal experience with the institution of the park. Thus, their attitudes are based rather on pre-existing knowledge and, therefore, on pre-constructed notions that have yet to be fully tested with prolonged interaction with the park. Therefore, MMNP must be aware of both negative and positive perceptions that its establishment and management can generate among local residents, and work to integrate the diversity of opinions, attitudes and values in order to reflect this reality. The MMNP is now at a crossroads at which its local populace can be greatly affected by future interaction with the park, including how it communicates and dispels existing rumors concerning species reintroductions.

3.4 Rodna Mountains National Park, Romania

The Rodna Mountains National Park (RMNP) is located within the Carpathian Mountain chain in northern Romania. The first attempts at protecting nature in this region occurred in 1932 when 183 ha around the Pietrosu Mare peak were declared as a PA, one of the first initiatives of this kind in Romania. This was followed by successive expansions to 2700 ha in 1971, 3300 ha in 1977, to 46,399 ha in 1990 when it was declared as a national park, and finally to its current size of 46,417.1 ha as stipulated in the park's draft management plan (Administratia Parcului National Muntii Rodnei, 2010). The park administration (APNMR) was set up only in 2004 as a structure within the National Forest Administration – RNP - Romsilva (NFA). The administration is helped by a Scientific Council which has decision-making powers and includes several experts and interacts with stakeholders through a Consultative Council which includes approximately 100 people belonging to 60 institutions. The stated purpose of the RMNP combines biodiversity conservation with encouraging and supporting local communities' traditional way of life (Administratia Parcului National Muntii Rodnei, 2010).

The importance of this PA for biodiversity conservation is demonstrated by its triple status as national park, biosphere reserve and a Natura 2000 site under both the Habitats and Birds Directives of the European Community. Forest ecosystems cover approximately 60% of the park area while alpine grasslands represent 30%. Diverse geomorphology, climate and soil types resulted in the presence of a rich biodiversity, with important habitats, and more than 2000 species each of both flora and fauna, including several endemics (Administratia Parcului National Muntii Rodnei, 2010). The internal zoning of the park currently includes the following areas (listed from highest to lowest level of protection): scientific reserves and areas of strict protection (12.2%), integral protection (43.02%), sustainable conservation (44.73%), and sustainable development (0.05%). The park's draft management plan proposes to extend the integral protection zone to 56.39% and to reduce the sustainable conservation area to 31.36% (Administratia Parcului National Muntii Rodnei, 2010). Forests and pastures included in the park are administered by several state and private structures and this fragmentation of ownership and administration poses a great challenge to the park administration and its goal of promoting integrative management in accord with biodiversity conservation.

Agriculture and forestry are the main sources of income for communities adjacent to the RMNP. Agriculture involves the cultivation of a few crops that are suitable to the harsh climate and mountainous terrain (potatoes, some corn, apples and plums) and livestock husbandry (sheep, goats, cattle, horses and pigs). Local people still practice transhumance, taking their sheep up to alpine pastures every spring and back to their villages in the autumn. In the past, mining was an important sector but this activity was decommissioned between 2004 and 2007 leading to increased unemployment in the area and greater pressure on forest resources (Administratia Parcului National Muntii Rodnei, 2010).

Previous to this research no studies were conducted on HWC and people-wildlife coexistence aspects in the area of the RMNP. Three species of large carnivores are present in the park area, namely brown bear (*Ursus arctos arctos*), gray wolf (*Canis lupus*) and Eurasian lynx (*Lynx lynx*), but data about their distribution, abundance and population dynamics are minimal, the main source of information being annual game censuses and rangers' and foresters' knowledge. No database integrating these data exists however. Moreover, habitat assessments and mapping are only partially available and more detailed assessments are planned to commence. In addition, no socio-economic investigations have been undertaken.

During this study a series of issues came to the fore. These could be grouped into conflict type (parties involved), and factors affecting large carnivores directly, for example hunting and poaching, or indirectly through habitat and prey management, and compensation for damages to livestock owners. Investigations revealed that the main direct conflict occurs between shepherds and large carnivores as a result of depredation events at sheepfolds located in alpine pastures during the summer grazing season. In this area (and elsewhere in Romania) HWC is difficult to assess and study because no records are kept of attacks and many events go unreported. No conflict monitoring system or adaptive management approaches have been implemented in the RMNP and no strategy (neither ecological nor social) for large carnivore management and for dealing with conflict has been developed so far. However, several informants stated that conflict level has decreased recently due to a decline in livestock, prey and wolf numbers and the interdiction of baiting inside the park.

The conflict between people and large carnivores is affected by several factors. Hunting plays a role despite the fact that no hunting is permitted inside the national park. However, the RMNP is surrounded by ten hunting grounds that overlap the park. Large carnivores have a protected status in Romania but hunting quotas are set yearly with the aim of bringing the populations within habitat carrying capacity or to address conflict caused by 'problem animals'. Supplementary feeding is allowed on hunting grounds and some informants stated that large carnivores are attracted outside the park where they can be hunted. Poaching of large carnivores was not perceived as a major factor although it is believed to occur. However, poaching of prey species was considered to be a more pressing issue due, at least in part, to the ease with which guns can be procured. Habitat management is also important and several respondents commented on the fact that logging operations are a source of disturbance for large carnivores. At the same time young forest regeneration areas are prime feeding grounds for bears but are also appealing to shepherds and their flocks and consequently they become sites of conflict.

The issue of compensations for damages caused to livestock by large carnivores is also an important factor influencing conflict. Although Romanian legislation specifies that compensation is to be paid in such cases, several informants stated that the procedure is complicated. Moreover, in some cases the damage cannot be assessed because livestock remains are dragged away by large carnivores. Shepherds who were interviewed declared that they generally claim compensation only when the damage is large, one stating that they make a claim only if they lose more than 10 sheep in one attack.

Regarding the socio-political dimension of HWC in the RMNP area the complexity of the issue is easily apparent. A multitude of institutions and stakeholders are involved often with unclear roles and responsibilities. Moreover, various stakeholders hold conflicting views toward conservation and resource use and often decisions are not integrated. A case in point is represented by agricultural subsidies paid to livestock owners by the National Agency for Payments and Interventions in Agriculture with a potential to encourage farmers to increase their livestock numbers, which could lead to an increased pressure on alpine pastures within the park and heightened HWC. Already several problems have occurred inside RMNP, including the illegal building of sheepfolds in the Pietrosu Mare scientific reserve where no resource use is permitted.

This situation is complicated further by the fact that the park land is owned mostly by local communities which are restricted from resource use (a significant proportion of forests and some pastures) without being compensated for this although the legislation does stipulate it. Before 1989 the area currently included in the RMNP was state owned and all resource use was carefully controlled by the communist authorities. The RMNP was established through a top-down approach without the participation of local communities which, following successive restitution laws, gained their land back but found themselves faced with restrictions imposed on natural resource use due to the park's presence. This led to a great deal of mistrust in state institutions and negative attitudes toward the park and its administration. Consequently, APNMR is faced with tremendous challenges and as yet does not dispose of adequate resources to address them (Ioja et al., 2010).

4. Discussion

PAs are considered the cornerstones of conservation. Yet, at the same time, they can also represent conservation paradoxes, particularly as sources of HWC. The very serious problem of HWC, and its mitigation, identified in our case studies demands a solution in order to improve relationships between communities and management institutions, and to arrive at better outcomes for communities and biodiversity conservation alike. Fostering communication and trust, demonstrating effort and a willingness to address the issue, and following through can lead to improved governance (Lockwood, 2010) and have a positive effect on the attitudes and actions of people in conflict with wildlife (Madden, 2004). However, with such a complex issue, employing a suite of flexible instruments and policies adapted to the local situation is required. Based on our cases, we suggest that in order to manage HWC more effectively in such contexts, five components of the conflict must be addressed: i) baseline research, ii) evaluation of damage, iii) conflict management, iv) adaptive management, and v) identifying and acknowledging management trade-offs.

4.1 Baseline research

Understanding local perceptions of affected communities, and the ecology of 'problematic' species, including their life histories and propensity for causing damage should be at the forefront of any research designed to minimise HWC. Knowledge on the spatial and temporal variation of conflicts, as well as the behaviors of involved individuals/institutions is a critical first step in planning any intervention (Treves et al., 2006). While this paucity of information might be more understandable for relatively young PAs like MMNP and RMNP, older parks such as KNP and VMWR may also suffer from a lack of baseline studies because of *inter alia* institutional reform and/or low capacity. This may also be a product of originally top-down hegemonic models of PA establishment and management where the concerns of neighboring communities were largely ignored. The trade-off here is that, while efforts to build and maintain wildlife populations were the focus, comprehensive baseline research on existing HWC (or which could potentially occur) was left wanting. Indeed, our studies were the first to systematically document HWC in all of our cases.

By conducting baseline research on what species are present, and what human activities exist (and are planned) within and adjacent to PAs, better strategies for mitigating conflict can be negotiated. Where baseline information is lacking, research capacities and efforts need to be increased, not only in culturally-sensitive social science research on and with neighboring communities (Pollard et al., 2003), but also in the areas of livestock and crop depredation (Bauer & Karl, 2001; Kolowski & Holekamp, 2006; Macandza et al., 2004), and wildlife deterrent measures (Newmark et al., 1994; Ogada et al., 2003; Sitati & Walpole, 2006).

4.2 Evaluation of damage

In addition to baseline research, systematic and effective reporting and monitoring, record keeping, and quick responses are required to ensure that the HWC is being tracked, comprehended, and adequately addressed (Treves et al., 2006). Both the design and implementation of policies formulated to manage HWC are dependent on the availability of *current, accurate,* and *long-term* information on the problem. Unfortunately, this aspect of HWC mitigation is desperately lacking in each of our cases, although there is promise that an improved system, of at least reporting, will be implemented in VMWR. In the absence of good information, the scale and nature of HWC can simply become a matter of personal opinion (e.g. elephants in VMWR, vipers in MMNP). Conflict between people and wildlife is an emotional issue and, as a result, reports and opinions can be biased, creating a false impression of the size of the problem. The systematic and objective gathering of information allows stakeholders to put the problems and threats caused by HWC into context and perspective with other problems faced by local communities. It also ensures that resources are correctly directed at solving the real issues rather than the perceived problems (Mishra, 1997), particularly where management regimes suffer from tightly constrained budgets and personnel.

In cases where record keeping is unsystematic and attending to incidents is hampered by overlapping (KNP) and/or weak institutional arrangements (VMWR), valuable data concerning the nature and extent of damage can be left wanting. Consequently, measures to minimise real or potential loss of life or livelihoods will remain unrealised and negative attitudes towards PAs from affected communities will persist (Anthony, 2007; Hazzah et al., 2009), and may eventually escalate to retaliatory killings as evidenced in KNP, and elsewhere (Mishra, 1997). Appropriate new, existing, or traditional systems and institutions need to be developed or empowered locally, and be evidence-based to ensure good management (Madden, 2004; Thirgood & Redpath, 2008). Such a system, we believe, must be mutually agreed upon and be clearly and broadly communicated to the relevant institutions, including local communities.

4.3 Conflict management

As outlined in our cases, the distributions of competencies between relevant institutions are not always aligned in structures that promote goodwill and biodiversity conservation in and around PAs. Although legislation and policies are in place to mitigate HWC in each of our four cases, their implementation on the ground is either weak or simply not carried out. We recommend that a system be created that, at least insofar as the issue of problem animals are concerned, helps establish the credibility and legitimacy of PAs. In order to achieve this objective the authority to control problem animals should be decentralised following the subsidiarity principle, in which 'the goal is to have as much local solution as possible and only so much government regulation as necessary' (Berkes, 2004). This may include local hunters to legitimately hunt valuable wildlife (KNP). Decentralisation of authority should also include allowing joint teams of qualified PA (and other institutional) staff with the authority to respond to problem animals as the need arises. These measures would go a long way in ironing out the procedural and practical difficulties now encountered in monitoring and responding to HWC incidents under the current institutional framework (KNP, VMWR, RMNP).

We believe that in situations where overlapping and/or competing institutions have a shared goal in mitigating, alleviating and eventually minimising HWC, these changes, in combination, will create a situation in which inherent institutional rivalries will be minimised due to cooperation on the ground as well as shared responsibility for oversight and low-level policy adjustments (Anthony et al., 2010). A process of social learning in which the various stakeholders understand the viewpoints of others and take some responsibility for meeting the core interests of their partners is likely to lead to greater mutual sympathy, a decrease in conflict, and more effective management in the long term.

In addition to these basic institutional arrangements, we recommend a number of other measures be taken, which are more unique to our cases. These include steps to improve relations with neighboring communities, such as maintaining and upgrading the park boundary fence (KNP), dispelling unfounded rumors about park objectives (MMNP), and following through on promises of compensation for damage(s) inflicted by wildlife (KNP, VMWR, RMNP). Although compensation schemes are generally not a good long-term solution as they may create continuing financial burdens and increase expectations (Crawshaw Jr,. 2004; Graham et al., 2005), and be counter productive to conservation by stimulating agricultural expansion (Bulte & Rondeau, 2005), the legitimacy of institutions may be enhanced where following through on long-standing promises are made. Moreover, when designed and implemented appropriately, compensation schemes can be effective (see e.g. Nyhus et al., 2003; Ogra & Badola, 2008; Schwerdtner & Gruber, 2007).

4.4 Adaptive management

Building on the data gained from comprehensive baseline research and damage evaluation, institutions responsible for mitigating HWC are in a more profitable position to adapt to changes, whether they are ecological, economic, or socio-political. Adaptive management is imperative to evaluate the effectiveness of interventions to minimise HWC (Curtin, 2002). We believe that in order to ensure performance improvement and provide a forum for timely feedback, an 'audit committee' of all relevant stakeholders where HWC challenges are faced by PAs should review cases on a regular basis and recommend changes in practice, if necessary. In the case of KNP, developing such a scheme is in the works but, to date, has not been implemented. In each of the other three cases, such a system is essential if institutions are to minimise conflicts in such dynamic ecological and socio-political landscapes, where the influences of agriculture and forestry sectors are so pronounced. However, adaptive management is superfluous, and may indeed be counter-productive, if it is not preceded by adequate evaluation of damage and conflict management.

4.5 Identifying and acknowledging management trade-offs

The goal to combine biodiversity conservation with sustainable management of natural resources and support for local communities is challenging and seemingly insurmountable (if not contradictory). This situation generates trade-offs and hard choices which are often not acknowledged and articulated and, thus, may result in unanticipated conflicts. Moreover, management decisions can affect temporal horizons in that the consequences of past management decisions can have cascading effects, leading to less-than-optimum solutions for managing current and/or future HWC. In the case of KNP, past injustices under Apartheid, and later unmet promises of compensation for wildlife damage, has led to elevated mistrust amongst local communities. Further, with the current institutional reform taking place, confusion from overlapping responsibilities coupled with a weakened response to HWC incidents is contributing to negative attitudes towards the park where, in some cases, poaching and retaliatory killing is taking place.

Similarly, in our Malawian case, past evictions from the park have created a general indignation by local communities towards the VMWR, which is limiting the options available for current park managers to mitigate HWC. Although electric fencing may be a viable option for controlling elephants in the reserve, many community members feel that such an object would represent a physical (and continual) reminder of their illegitimate removal from the area where the reserve now stands. Moreover, with the current organisational reserve structure and non-lethal techniques utilised to manage elephants, HWC are growing in the area and influencing increasingly negative attitudes towards the park and its biodiversity conservation objectives. In such situations, unless management decisions recognise these inherent trade-offs and are oriented to rectify the situation, it is inevitable that conflict will escalate beyond the capacity of the reserve to control.

Our first Romanian case is in perhaps the most enviable of all our cases. MMNP is a relatively young park, which has had limited interaction with neighboring communities. Community attitudes are mixed: some recognise and support the benefit of the park in conserving biodiversity; others feel that its restrictions on local mining operations have created heightened unemployment in an already impoverished area. Moreover, a falsely held belief of the park's intention to reintroduce dangerous vipers to the area is prevalent amongst many community members. How explicitly, and when, the MMNP communicates its position on these issues to local communities is of vital importance as it will affect outcomes and can sway the attitudes of local communities towards the park and its conservation objectives. Fortunately, unlike our African cases, it does not also have to address past injustices in doing so.

Finally, concerning RMNP, the proposal to include most alpine pastures in the integral protection area of the park has the potential to encourage ecotourism development in the area, drive local livestock owners to reduce their stock and could result in a considerable reduction in HWC inside the park but, at the same time, it could deprive local people and communities of significant agricultural subsidies and generate a series of law suits from disenfranchised land owners. On the other hand, if the park does not follow up on its intention of extending the integral protection area it could be faced with increased pressure on grasslands and probably increased levels of HWC as a result of farmers being stimulated by subsidies to increase their livestock numbers. These choices are not simple as, in addition to biodiversity outcomes, they also involve issues of power and justice with respect to the distribution of their consequences (e.g. developing ecotourism could promote a local elite to the detriment of poorer shepherds).

5. Conclusion

Although protected areas are hailed as the primary mechanism of conserving the planet's biodiversity, they can also be sources of conflicts with communities living in and around these areas, particularly in terms of contributing to human-wildlife conflict. Incidents of HWC that are not adequately resolved assure the maintenance of tense relationships between PAs and communities, which has undesirable social consequences and poses risks for PAs and biodiversity conservation in the longer-term. Thus, developing adequate responses to HWC should be a high priority for PA authorities and other governmental bodies.

Our cases indicate that addressing baseline research, evaluation of damage, conflict management, and adaptive management are tantamount if conserving biodiversity is to persist where HWC exists. Moreover, recognising and articulating inherent management trade-offs amongst diverse actors are requisite if HWC is to be fully understood, and mitigated. Our case studies individually differ in terms of their attention to these components (Table 4), which is likely a reflection of principles of 'scale, context, pluralism and complexity' (McShane et al., 2011). Yet, there are opportunities for cross-learning here, as relatively young and inexperienced PAs (e.g. MMNP) can avoid making decisions that have had negative (and unexpected) consequences for PAs elsewhere.

Our chapter re-emphasises that HWC are complex, dynamic, and driven not only by ecological factors, but by economic as well as socio-political forces. Embedded within this framework is the need for conservation agencies to encourage the wise and sustainable use of natural resources, which in some cases, are becoming increasingly threatened. The question remains as to whether strategies developed by PAs (and others) to effectively integrate these identified components will gain normative weight so that local institutions will be able to meet both their biodiversity conservation and socio-economic objectives. We have shed some light on these complexities and it is hoped that this will contribute to a more stable and sustainable future for both PAs and their neighbours, where HWC continues to be a challenge. In a world in which biodiversity is under increasing pressure from human encroachment, and in which people's rights to justice and secure livelihoods must be respected, investments in addressing HWC more holistically should be treated as mandatory.

	KNP	VMWR	MMNP	RMNP
Baseline Research	minimal; expertise available	none	none	minimal
Evaluation of Damage	better in past; poor now	poor, efforts in place to improve	*ad hoc* (poor)	*ad hoc* (poor)
Conflict Management	legislation & policies exist, but implementation lacking; compensation a concern	legislation & policies exist, but weak implementation; compensation a concern	legislation & policies exist, but uncertain realization	legislation & policies exist, but weak realization; compensation a concern
Adaptive Management	potential exists	lacking capacity	potential exists	potential exists
Trade-offs	PAST Apartheid > mistrust + land claims CURRENT institutional reform > confusion; negative attitudes; poaching and retaliatory killings (direct + indirect); possible edge effects	PAST evictions > mistrust; ambivalence towards fencing CURRENT poor animal control > negative attitudes	CURRENT a) mining > mixed attitudes b) viper reintroduction (false beliefs) > trepidation & negative attitudes	PAST land nationalization; strict resource control > forests and game thrived CURRENT land ownership change; negative attitudes; & fragmented management

Table 4. Summary of HWC components across selected case studies

6. References

Administratia Parcului National Muntii Rodnei (2010). *Planul de management al Parcului National Muntii Rodnei* (The Management Plan of Rodna Mountains National Park), Rodna, Romania

Anthony, B.P. (2006). *A view from the other side of the fence: Tsonga communities and the Kruger National Park, South Africa*, Ph.D. thesis, Department of Environmental Sciences and Policy, Central European University, Budapest, Hungary

Anthony, B. (2007). The dual nature of parks: attitudes of neighbouring communities towards Kruger National Park, South Africa. *Environmental Conservation*, Vol.34, No.3, (September 2007), pp. 236-245, ISSN 0376-8929

Anthony, B.P. & Bellinger, E. G. (2007). Importance value of landscapes, flora and fauna to Tsonga communities in the rural areas of Limpopo province, South Africa. *South African Journal of Science*, Vol.103, No.3/4, (March / April 2007), pp. 148-154, ISSN 0038-2353

Anthony, B.P. & Moldovan, D. (2008). Poised for engagement: local communities and the Măcin Mountains National Park, Romania. *The International Journal of Biodiversity Science & Management* Vol.4, No.4, (December 2008), pp. 230-241, ISSN 1745-1590

Anthony, B.P.; Scott, P. & Antypas, A. (2010). Sitting on the fence? Policies and practices in managing human-wildlife conflict in Limpopo Province, South Africa. *Conservation & Society*, Vol.8, No.3, (July-September 2010), pp. 225-240, ISSN 0972-4923

Anthony, B.P. & Wasambo, J. (2009). *Human-Wildlife Conflict Study Report: Vwaza Marsh Wildlife Reserve, Malawi*, Report prepared on behalf of Malawi Department of National Parks and Wildlife, Central European University, Budapest, Hungary

Bauer, H. & Karl. S. (2001). Assessment of people-predator conflict through thematic PRA in the surroundings of Waza National Park, Cameroon. *PLA Notes*, Vol.41, (June 2001), pp. 9-13, ISSN 1357 938X

Bell, R.H.V. & Mphande, J.N.B. (1980). Vwaza Marsh Game Reserve: A Report on a Survey in Aug - Sept 1979, and Recommendation for Management and Development, Wildlife Research Unit, Department for National Parks and Wildlife, Kasungu, Malawi

Berkes, F. (2004). Rethinking community-based conservation. *Conservation Biology*, Vol.18, No.3, (June 2004), pp. 621-630, ISSN 0888-8892

Bisi, J. & Kurki, S. (2005). *Wolf discourse in Finland: provincial and national expectations and objectives for the management of the Finnish wolf population [in Finnish]*. Julkaisuja 3. Helsingin Yliopisto Maaseudun tutkimus- ja koulutuskeskus.

Braack, L.E.O. (2000). *Kruger National Park* (2nd edition) New Holland Publishers, ISBN 1859744710, London, UK

Brechin, S.R.; Wilshusen, P.R.; Fortwangler, C.L. & West, P.C. (2002). Beyond the square wheel: toward a more comprehensive understanding of biodiversity conservation as social and political processes. *Society & Natural Resources*, Vol.15, No.1, pp. 41-64, ISSN 0894-1920

Breck, SW. (2004). Minimizing Carnivore-Livestock Conflict: The Importance and Process of Research in the Search for Coexistence, In: *People and predators: from conflict to coexistence*, Fascione, N., Delach, A. & Smith, M.E. (Eds.), pp. 13-27, Island Press, ISBN 9781559630832, Washington DC, USA

Breitenmoser, U. (1998). Large predators in the Alps: the fall and rise of man's competitors. *Biological Conservation*, Vol.83, No.3, (March 1998), pp. 279-289, ISSN 0006-3207

Bulte, E.H. & Rondeau, D. (2005). Why compensating wildlife damages may be bad for conservation. *Journal of Wildlife Management* Vol.69, No.1, pp. 14-19, ISSN 0022-541X

Callimanopulos, D. (1984). Relocating blacks in South Africa, In: *Cultural Survival Quarterly*, Vol.8, No.1, 03.3.2011, Available from: http://www.culturalsurvival.org/publications/cultural-survival-quarterly/south-africa/relocating-blacks-south-africa

Campbell, B. & Shackleton, S. (2001). The organizational structures for community-based natural resource management in Southern Africa, In *African Studies Quarterly*, Vol.5, No.3, 12.4.2010, Available from: http://www.africa.ufl.edu/asq/v5/v5i3a6.htm

Cantu-Salazar, L. & Gaston, K.J. (2010). Very large protected areas and their contribution to terrestrial biological conservation. *Bioscience*, Vol.60, No.10, pp. 808-818, ISSN 0006-3568

Chape, S.; Harrison, J.; Spalding, M. & Lysenko, I. (2005). Measuring the extent and effectiveness of protected areas as an indicator for meeting global biodiversity targets. *Philosophical Transactions of the Royal Society B: Biological Sciences*, Vol.360, No.1454, (February 2005), pp. 443-455, ISSN 1471-2970

Choudhury, A. (2004). Human-elephant conflicts in northeast India. *Human Dimensions of Wildlife*, Vol.9, No.4, pp. 261-270, ISSN 1087-1209

Clark, D.A. & Slocombe, D.S. (2009). Respect for grizzly bear: an Aboriginal approach for co-existence and resilience. *Ecology and Society*, Vol.14, No.1, (June 2009), art. 42, ISSN 1708-3087

Coad, L.; Burgess, N.; Fish, L.; Ravillious, C.; Corrigan, C.; Pavese, H.; Granziera, A. & Besançon, C. (2008). Progress towards the Convention on Biological Diversity terrestrial 2010 and marine 2012 targets for protected area coverage. *Parks*, Vol.17, No.2, (December 2008), pp. 35-42, ISSN 0960-233X

Cock, J. & Fig, D. (2000). From colonial to community based conservation: environmental justice and the national parks in South Africa. *Society in Transition*, Vol.31, No.1, pp. 22–35, ISSN 1028-9852

Conover, M. (2002). *Resolving human-wildlife conflicts: the science of wildlife damage management*, Lewis Publishers CRC Press Company, ISBN 9781566705387, Boca Raton, Florida, USA

Cozza, K.; Fico, R.; Battistini, M.L. & Rogers, E. (1996). The damage-conservation interface illustrated by predation on domestic livestock in central Italy. *Biological Conservation*, Vol.78, No.3, (December 1996), pp. 329-336, ISSN 0006-3207

Crawshaw, P.G. Jr. (2004). Depredation of domestic animals by large cats in Brazil. *Human Dimensions of Wildlife*, Vol.9, No.4, pp. 329-330, ISSN 1087-1209

Curtin, C.G. (2002). Integration of science and community-based conservation in the Mexico-U.S. borderland. *Conservation Biology*, Vol.16, No.4, (August 2002), pp. 880-886, ISSN 0888-8892

de Boer, W.F. & Baquete, D.S. (1998). Natural resource use, crop damage and attitudes of rural people in the vicinity of the Maputo Elephant Reserve, Mozambique. *Environmental Conservation*, Vol.25, No.3, (September 1998), pp.208-218, ISSN 0376-8929

Decker, D.J.; Jacobson, C.A. & Brown, T.L. (2006). Situation-Specific "Impact Dependency" as a Determinant of Management Acceptability: Insights from Wolf and Grizzly Bear

Management in Alaska. *Wildlife Society Bulletin*, Vol.34, No.2, pp. 426-432, ISSN 1938-5463

Dublin, H.T. & Hoare, R.E. (2004). Searching for solutions: the evolution of an integrated approach to understanding and mitigating human-elephant conflict in Africa. *Human Dimensions of Wildlife*, Vol.9, No.4, pp. 271-278, ISSN 1087-1209

Dudley, N. (Ed.) (2008). *Guidelines for Applying Protected Area Management Categories*, IUCN, ISBN 978-2-8317-1086-0, Gland, Switzerland

Els, H. (1994). Werknermers van die Nasionale Krugerwildtuin se waarde-oordele oor aspekte van die natuur en natuurbewaring [Value judgements of personnel of the Kruger National Park on aspects of nature and nature conservation]. *South African Journal of Ethnology*, Vol.17, No.4, pp. 109-120, ISSN 0379-8860

Els, H. (1995). Conflicts between man, agriculture, and wildlife, In: *Integrating people and wildlife for a sustainable future*, Bissonette, J.A. & Krausman, P.R. (Eds.), pp. 355-359, The Wildlife Society, ISBN 0933564120, Bethesda, Maryland, USA.

Emerton, L. (2001). The nature of benefits and the benefits of nature: why wildlife conservation has not economically benefitted communities in Africa, In: *African wildlife & livelihoods: the promise & performance of community conservation*, Hulme, D. & Murphree, M. (Eds.), pp. 208-226, David Philip Publishers, ISBN 0325070598, Cape Town, South Africa

Ervin, J.; Sekhran, N.; Dinu, A.; Gidda, S.; Vergeichik M, & Mee, J. (2010). *Protected areas for the 21st century: Lessons from UNDP/GEF's Portfolio*, United Nations Development Programme and Convention on Biological Diversity, ISBN 92-9225-274-7, New York and Montreal

Freitag-Ronaldson, S. & Foxcroft, L.C. (2003). Anthropogenic influences at the ecosystem level, In: *The Kruger Experience: Ecology and Management of Savanna Heterogeneity*, du Toit, J.T., Rogers, K.H. & Biggs, H.C. (Eds.), pp. 391–421, Island Press, ISBN 1559639822, Washington DC, USA

Ghimire, K.B. & Pimbert, M.P. (1997). Social change and conservation: an overview of issues and concepts, In: *Social change and conservation: environmental politics and impacts of national parks and protected areas*, Ghimire, K.B. & Pimbert, M.P. (Eds.), pp. 1-45, Earthscan Publications Ltd., ISBN 1853834106, London, UK

Graham, K.; Beckerman, A.P. & Thirgood, S. (2005). Human–predator–prey conflicts: ecological correlates, prey losses and patterns of management. *Biological Conservation*, Vol.122, No.2, (March 2005), pp. 159-171, ISSN 0006-3207

Hall-Martin, A.; Nzima, H.E.; Myburgh, W. & van Riet Jr., W. (2007). *Establishment and Development of Malawi-Zambia Transfrontier Conservation Areas: Nyika Transfrontier Conservation Area: Joint Management Plan*. June 2007, Peace Parks Foundation.

Happold, D.C.D. (1995). The interaction between humans and mammals in Africa in relation to conservation: a review. *Biodiversity and Conservation*, Vol.4, No.4, (June 1995), pp. 395-414, ISSN 0960-3115

Hazzah, L.,; Borgerhoff-Mulder, M., & Frank, L. (2009). Lions and Warriors: Social factors underlying declining African lion populations and the effect of incentive-based management in Kenya. *Biological Conservation*, Vol.142, No.11, (November 2009), pp. 2428-2437, ISSN 0006-3207

Hewitt, D.G. & Messmer, T.A. (1997). Responsiveness of agencies and organizations to wildlife damage: policy process implications. *Wildlife Society Bulletin*, Vol.25, No.2, pp. 418-423, ISSN 1938-5463

Hill, C.M. (2004). Farmers' perspectives of conflict at the wildlife-agriculture boundary: some lessons learned from African subsistence farmers. *Human Dimensions of Wildlife*, Vol.9, No.4, pp. 279-286, ISSN 1087-1209

Hoare, R.E. (2001). *A decision support system for managing human–elephant conflict situations in Africa*. IUCN/SSC African Elephant Specialist Group, Nairobi, Kenya

Institutul de Cercetari si Amenajari Silvice (1996). *National Strategy and Action Plan for Biodiversity Conservation and Sustainable Use of its Components in Romania*. Institutul de Cercetari si Amenajari Silvice [Forest Research and Management Institute], Bucureşti, Romania

Ioja, C.I.; Patroescu, M. ; Rozylowicz, L. ; Popescu, V.D. ; Verghelet, M. ; Zotta, M.I. & Felciuc, M. (2010). The efficacy of Romania's protected areas network in conserving biodiversity. *Biological Conservation*, Vol.143, No.11, (November 2010), pp. 2468-2476, ISSN 0006-3207

Johansson, M. & Karlsson, J. (2011). Subjective experience of fear and cognitive interpretation of large carnivores. *Human Dimensions of Wildlife*, Vol.16, No.1, pp. 15-29, ISSN 1087-1209

Kellert, S.R.; Black, M.; Rush, C.R. & Bath, A.J. (1996). Human culture and large carnivore conservation in North America. *Conservation Biology*, Vol.10, No.4, (August 1996), pp. 977-990, ISSN 0888-8892

Khan, F. (1994). Rewriting South Africa conservation history - the role of the Native Farmers Association. *Journal of Southern African Studies*, Vol.20, No.4, pp. 499–516, ISSN 0305-7070

Kiss, A. (Ed.) (1990). Living with wildlife: wildlife resource management with local participation in Africa. World Bank Technical Paper 130. World Bank, Washington, DC, USA

Knight, J. (2000). Introduction, In: *Natural enemies: people-wildlife conflicts in anthropological perspective*, Knight, J. (Ed.), pp. 1-35, Routledge, ISBN 9780415224413, New York, USA

Kolowski, J.M. & Holekamp, K.E. (2006). Spatial, temporal, and physical characteristics of livestock depredations by large carnivores along a Kenyan reserve border. *Biological Conservation*, Vol.128, No.4, (April 2006), pp. 529-541, ISSN 0006-3207

Kruuk, H. (2002). *Hunter and hunted: relationships between carnivores and people*. Cambridge University Press, ISBN 9780521891097, New York, USA

Lahiff, E. (1997). Land, water and local governance in South Africa: a case study of the Mutale River Valley. Rural Resources Rural Livelihoods Working Paper Series, Paper No. 7, IDPM, Manchester, UK

Linnell, J.D.C.; Swenson, J.E. & Andersen, R. (2001). Predators and people: conservation of large carnivores is possible at high human densities if management policy is favourable. *Animal Conservation*, Vol.4, No.4, (November 2001), pp. 345-349, ISSN 1367-9430

Linnell, J.D.C.; Rondeau, D.; Reed, D.H.; Williams, R.; Altwegg, R.; Raxworthy, C.J.; Austin, J.D.; Hanley, N.; Fritz, H.; Evans, D.M.; Gordon, I.J.; Reyers, B.; Redpath, S. & Pettorelli, N. (2010). Confronting the costs and conflicts associated with biodiversity. *Animal Conservation*, Vol.13, No.5, (October 2010), pp. 429-431, ISSN 1367-9430

Lockwood, M. (2010). Good Governance for Terrestrial Protected Areas: A Framework, Principles and Performance Outcomes. *Journal of Environmental Management,* Vol.91, No. 3, (January-February 2010), pp. 754–766, ISSN 0301-4797

Löe, J. & Röskaft, E. (2004). Large carnivores and human safety: a review. *Ambio*, Vol.33, No.6, (August 2004), pp. 283-288, ISSN 0044-7447

Mabunda, D.; Pienaar, D.J. & Verhoef, J. (2003). The Kruger National Park: a century of management and research, In: *The Kruger Experience: Ecology and Management of Savanna Heterogeneity,* du Toit, J.T., Rogers, K.H. & Biggs, H.C. (Eds.), pp. 3–21, Island Press, ISBN 1559639822, Washington DC, USA

Macandza, V.A.; Owen-Smith, N. & Cross, P.C. (2004). Forage selection by African buffalo in the late dry season in two landscapes. *South African Journal of Wildlife Research,* Vol.34, No.2, pp. 113-121, ISSN 0379-4369

Madden, F. (2004). Creating Coexistence between Humans and Wildlife: Global Perspectives on Local Efforts to Address Human–Wildlife Conflict. *Human Dimensions of Wildlife,* Vol.9, No.4, pp. 247–257, ISSN 1087-1209

Magome, D.T. & Collinson, R.F.H. (1998). *From Protest to Pride: a Case Study of Pilanesburg National Park, South Africa.* World Bank/WBI's CBNRM Initiative, 06.6.2010, Available from: http://srdis.ciesin.org/cases/south_africa-003.html

Malawi Department of National Parks & Wildlife [MDNPW] (2004). Nyika National Park Master Plan, 2004. Final draft. January 2004.

Marshall, K.; White, R. & Fischer, A. (2007). Conflicts between humans over wildlife management: on the diversity of stakeholder attitudes and implications for conflict management. *Biodiversity and Conservation*, Vol.16, No.11, (October 2007), pp. 3129-3146, ISSN 0960-3115

McGregor, J. (2005). Crocodile crimes: people versus wildlife and the politics of postcolonial conservation on Lake Kariba, Zimbabwe. *Geoforum*, Vol.36, No.3, (May 2005), pp. 353-369, ISSN 0016-7185

McNeely, J. (Ed) (1993). *Parks for life: Report of the IVth World Congress on National Parks and Protected Areas,* IUCN, ISBN 2831701627 Gland, Switzerland

McShane, T.O.; Hirsch, P.D.; Trung, T.C.; Songorwa, A.N.; Kinzig, A.; Monteferri, B.; Mutekanga, D.; Thang, H.V.; Dammert, J.L.; Pulgar-Vidal, M.; Welch-Devine, M.; Brosius, J.P.; Coppolillo, P. & O'Connor, S. (2011). Hard choices: Making trade-offs between biodiversity conservation and human well-being. *Biological Conservation,* Vol.144, No.3, (March 2011), pp. 966-972, ISSN 0006-3207

Mishra, C. (1997). Livestock depredation by large carnivores in the Indian trans-Himalaya: conflict perceptions and conservation prospects. *Environmental Conservation*, Vol.24, No.4, (December 1997), pp. 338-343, ISSN 0376-8929

Msiska, H.G. (2002). *Guidelines on community involvement in problem animal control with vermin animals (draft)*. February 2002. Vwaza Marsh Wildlife Reserve and Department of National Parks and Wildlife

Naughton-Treves, L. (1998). Predicting patterns of crop damage by wildlife around Kibale National Park, Uganda. *Conservation Biology*, Vol.12, No.1, (February 1998), pp. 156-168, ISSN 0888-8892

Naughton-Treves, L., & Treves, A. (2005). Socio-ecological factors shaping local support for wildlife: crop-raiding by elephants and other wildlife in Africa, In: *People and wildlife: conflict or co-existence?* Woodroffe, R., Thirgood, S. & Rabinowitz, A. (Eds.), pp. 252-277, Cambridge University Press, ISBN 9780521532037, Cambridge, UK

Newmark, W.D.; Manyanza, D.N.; Gamassa, D.G.M. & Sariko, H.I. (1994). The conflict between wildlife and local people living adjacent to protected areas in Tanzania: human density as a predictor. *Conservation Biology*, Vol.8, No.1, (March 1994), pp. 249-255, ISSN 0888-8892

Nxumayo, G.Z.; Msiska, H.G. & Kataya, A. (2008). *Reflection on the relationship between Department of National Parks & Wildlife (DNPW) and bordering communities: a case study in the border zone areas of Nyika National Park and Vwaza Marsh Wildlife Reserve in the Northern Division*. October 2008. Department of National Parks and Wildlife, and Nyika-Vwaza Association

Nyhus, P.; Fischer, H.; Madden, F. & Osofsky, S. (2003). Taking the bite out of wildlife damage. *Conservation in Practice*, Vol.4, No.2, (April 2003), pp. 37-40, ISSN 1936-2145

Ogada, M.O.; Woodroffe, R.; Oguge, N.O. & Frank, L.G. (2003). Limiting depredation by African carnivores: The role of livestock husbandry. *Conservation Biology*, Vol.17, No.6, (December 2003), pp. 1521-1530, ISSN 0888-8892

Ogra, M. & Badola, R. (2008). Compensating human-wildlife conflict in protected area communities: Ground-level perspectives from Uttarakhand, India. *Human Ecology*, Vol.36, No.5, (October 2008), pp. 717-729, ISSN 0300-7839

Parcul National Muntii Macinului (2006). *Raport general de activitate pentru anul 2006 [General Activity Report for 2006]. Internal report*. Available from Parcul National Muntii Macinului [Măcin Mountains National Park], Tulcea, Romania

Pimbert, M.P. & Pretty, J.N. (1997). Parks, people and professionals: putting 'participation' into protected area management, In: *Social change and conservation: environmental politics and impacts of national parks and protected areas*, Ghimire, K. & Pimbert, M., (Eds.), pp. 297-330, Earthscan Publications Ltd., ISBN 1853834106, London, UK

Pollard, S.; Shackleton, C. & Carruthers, J. (2003). Beyond the fence: People and the lowveld landscape, In: *The Kruger experience: Ecology and management of savanna heterogeneity*, du Toit, J.T., Rogers, K.H. & Biggs, H.C. (Eds.), pp. 422-446, Island Press, ISBN 1559639822, Washington DC, USA

Rodrigues, A.S.L.; Andelman, S.J.; Bakarr, M.I.; Boitani, L.; Brooks, T.M.; Cowling, R.M.; Fishpool, L.D.C.; da Fonseca, G.A.B.; Gaston, K.J.; Hoffmann, M.; Long, J.S.; Marquet, P.A.; Pilgrim, J.D.; Pressey, R.L.; Schipper, J.; Sechrest, W.; Stuart, S.N.; Underhill, L.G.; Waller, R.W.; Watts, M.E.J. & Yan, X. (2004). Effectiveness of the global protected area network in representing species diversity. *Nature*, Vol.428, No.6983, (April 2004), pp. 640-643, ISSN 0028-0836

Saberwal, V.K.; Gibbs, J.P.; Chellam, R. & Johnsingh, A.J.T. (1994). Lion-human conflict in the Gir forest, India. *Conservation Biology*, Vol.8, No.2, (June 1994), pp. 501-507, ISSN 0888-8892

SANParks (2000). *Visions of Change: Social Ecology and South African National Parks.* Development Communications Co. in association with South African National Parks, Johannesburg, South Africa.

SANParks (2006). Kruger National Park Management Plan (Version 1, 31 October 2006), 10.04.2011, Available from: http://www.sanparks.org/conservation/park_man/kruger.pdf

Schwerdtner, K. & Gruber, B. (2007). A conceptual framework for damage compensation schemes. *Biological Conservation* Vol.134, No.3, (January 2007), pp. 354-360, ISSN 0006-3207

Secretariat of the Convention on Biological Diversity (2010). Press Release: A new era of living in harmony with Nature is born at the Nagoya Biodiversity Summit. Nagoya, Japan 29 October 2010, 03.11.2010, Available from: www.cbd.int/doc/press/2010/pr-2010-10-29-cop-10-en.pdf

Sillero-Zubiri, C. & Laurenson, M.K. (2001). Interactions between carnivores and local communities: conflict or co-existence?, In: *Carnivore Conservation*, Gittleman, J.L., Funk, S.M., Macdonald, D.W., & Wayne, R.K. (Eds.), pp.282-313, Cambridge University Press, ISBN 0 521 66232 X, Cambridge, UK

Sillero-Zubiri, C.; Sukumar, R. & Treves, A. (2007). Living with wildlife: the roots of conflict and the solutions, In: *Key topics in conservation biology*, Macdonald, D.W., & Service, K. (Eds.), pp. 266–272, Oxford University Press, ISBN 978-1-4051-2249-8, Oxford, UK

Sitati, N.W. & Walpole, M.J. (2006). Assessing farm-based measures for mitigating human-elephant conflict in Transmara District, Kenya. *Oryx*, Vol.40, No.3, (July 2006), pp. 279-286, ISSN 0030-6053

Stahl, P.; Vandel, J.M.; Herrenschmidt, V. & Migot, P. (2001). The effect of removing lynx in reducing attacks on sheep in the French Jura Mountains. *Biological Conservation*, Vol.101, No.1, (September 2001), pp. 15-22, ISSN 0006-3207

Stevens, S. (1997). Introduction, In: *Conservation through cultural survival: indigenous peoples and protected areas,* Stevens, S. (Ed.), pp. 1-8, Island Press, ISBN 1559634499, Washington, D.C., USA.

Tapela, B.N. & Omara-Ojungu, P.H. (1999). Towards bridging the gap between wildlife conservation and rural development in post-apartheid South Africa: the case of the Makuleke community and the Kruger National Park. *South African Geographical Journal*, Vol.81, No.3, pp. 148–155, ISSN 0373-6245

Thirgood, S., & Redpath, S. (2008). Hen harriers and red grouse: science, politics and human-wildlife conflict. *Journal of Applied Ecology*, Vol.45, No.5, (October 2008), pp. 1550-1554, ISSN 1365-2664

Thirgood, S.; Woodroffe, R. & Rabinowitz, A. (2005). The impact of human-wildlife conflict on human lives and livelihoods, In: *People and wildlife: conflict or co-existence?*

Woodroffe, R., Thirgood, S. & Rabinowitz, A. (Eds.), pp. 13-26, Cambridge University Press, ISBN 9780521532037, Cambridge, UK

Treves, A. (2009). The human dimensions of conflicts with wildlife around protected areas, In: *Wildlife and Society: The Science of Human Dimensions*, Manfredo, M.J., Vaske, J.J., Brown, P.J., Decker, D.J., & Duke, E.A. (Eds.), pp. 214-228, Island Press, ISBN 9781597264082, Washington, DC., USA

Treves, A. & Karanth, K.U. (2003). Human-carnivore conflict and perspectives on carnivore management worldwide. *Conservation Biology*, Vol.17, No.6, (December 2003), pp. 1491-1499, ISSN 0888-8892

Treves, A.; Wallace, R.B.; Naughton-Treves, L. & Morales, A. (2006). Co-managing human-wildlife conflicts: a review. *Human Dimensions of Wildlife*, Vol.11, No.6, pp. 383-396, ISSN 1087-1209

UNDP (2005). *Romania: Strengthening Romania's Protected Area System by Demonstrating Best Practices for Management of Small Protected Areas in Măcin Mountains National Park.* United Nations Development Programme (UNDP) Project No. 00047111

UNEP-WCMC (2008). *State of the world's protected areas: an annual review of global conservation progress*, UNEP-WCMC, Cambridge, UK

van Dyk, G. & Slotow, R. (2003). The effects of fences and lions on the ecology of African wild dogs reintroduced to Pilanesberg National Park, South Africa. *African Zoology*, Vol.38, pp. 79-94, ISSN 1562-7020

Vwaza Marsh Wildlife Reserve (VMWR) (2003). *Vwaza Marsh Wildlife Reserve Master Plan.*

Volkman, T.A. (1986). The hunter-gatherer myth in Southern Africa, In: *Cultural Survival Quarterly*, Vol.10, No.2, 22.03.2011, Available from: http://www.culturalsurvival.org/ourpublications/csq/article/the-hunter-gatherer-myth-southern-africa

Wang, S.W. & Macdonald, D. W. (2006). Livestock predation by carnivores in Jigme Singye Wangchuck National Park, Bhutan. *Biological Conservation*, Vol.129, No.4, (May 2006), pp. 558-565, ISSN 0006-3207

West, P.C. & Brechin, S.R. (Eds). (1991). *Resident peoples and national parks: social dilemmas and strategies in international conservation*, University of Arizona Press, ISBN 0816511284, Tucson, USA

Woodroffe, R. (2000). Predators and people: using human densities to interpret declines of large carnivores. *Animal Conservation*, Vol.3, No.1, (February 2000), pp. 65-173, ISSN 1367-9430

Woodroffe, R. (2001). Strategies for carnivore conservation: lessons from contemporary extinctions, In: *Carnivore conservation.*, Gittleman, J.L., Funk, S.M., Macdonald, D.W., & Wayne, R.K. (Eds.), pp. 61-92, Cambridge University Press, ISBN 0 521 66232 X, Cambridge, UK

Woodroffe, R. & Ginsberg, J.R. (1998). Edge Effects and the Extinction of Populations Inside Protected Areas. *Science*, Vol.280, No.5372, (June 1998), pp. 2126-2128, ISSN 0036-8075

Woodroffe, R.; Thirgood, S. & Rabinowitz, A. (Eds). (2005). *People and wildlife, conflict or co-existence?* Conservation Biology, No. 9., Cambridge University Press, ISBN 0521825059, Cambridge, UK.

Zedrosser, A.; Dahle, B.; Swenson, J.E. & Gertsl, N. (2001). Status and management of the brown bear in Europe. *Ursus*, Vol.12, pp. 9-20, ISSN 1537-6176

Zinn, H.C.; Manfredo, M.J. & Vaske, J.J. (2000). Social psychological bases for stakeholder acceptance capacity. *Human Dimensions of Wildlife*, Vol.5, No.3, pp. 20-33, ISSN 1087-1209

15

Human Wildlife Conflicts in Southern Africa: Riding the Whirl Wind in Mozambique and in Zimbabwe

Sébastien Le Bel[1], Amon Murwira[2], Billy Mukamuri[2], René Czudek[3], Russell Taylor[4] and Mike La Grange[5]
[1]*Cirad (Centre de coopération internationale en recherche agronomique pour le développement)*
[2]*University of Zimbabwe*
[3]*FAO Subregional Office for Southern Africa*
[4]*WWF-Nambia*
[5]*AWMC (African Wildlife Management and Conservation)*
[1]*France*
[2,3,5]*Zimbabwe*
[4]*Namibia*

1. Introduction

Human Wildlife Conflict (HWC) exists when the needs and behaviour of wildlife impact negatively on the goals of human beings (Cumming et al., 2007). It tends to manifest itself in scenarios where human strategies affect free movement of wild animals and vice versa. Thus, HWC can be considered inevitable in all communities where human and wildlife co-exist and share the same habitat (Fig. 1).

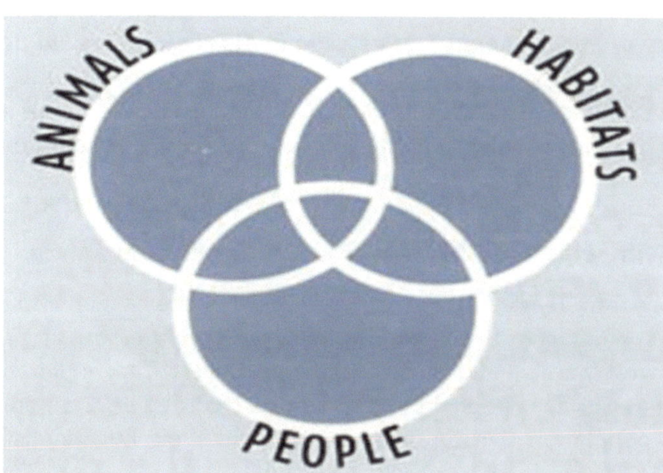

Fig. 1. People and Wildlife co-existence: a difficult equation

HWC has an international magnitude even in urban areas (Breitenmoser et al., 2005; McGinnis, 2008) and a significant social impact which depends on the capacity of a community to support a certain level of conflicts (Woodroffe et al., 2005). In Africa, HWC is not restricted to a particular geographical region or climatic condition but is common in all areas where wildlife and human populations coexist and have limited resources (Distefano, 2004). The February 2010 meeting of the Southern African Development Community (SADC) Technical Committee on Wildlife pronounced that HWC was one of the main problems for Africa's rural populations in terms of personal security and economic loss, and the situation is getting worse (LeBel et al., 2010). The population of the African continent, which has the world's largest reserves of wildlife, is expected to double from 0.8 billion to 1.8 billion people in the next 40 years (ILRI, 2009). Africans will not only be packed more tightly into cities, they and their crops will also increasingly impinge upon territory populated by wildlife. The tragedy of HWC is in its dynamic; it can only be stopped by suppressing one of the two antagonists. All too often, affected communities and support agencies are engaged in a headlong rush with little visibility of what should be done and how.

The objective of this chapter is not to give a recipe of devices to solve all HWCs (e.g. problems of crop raiding elephants) or to give a roll map to NGO's in an attempt to reconcile hungry communities and free ranging mega-herbivores. As no blueprint or panacea exists, our philosophy is to explore options which will help rural communities to improve their capacity to live with problem animals. The principles developed though this chapter aim to increase human tolerance of wildlife species and to decrease negative interactions with them.

To achieve this, we will be referring to recent works conducted in Mozambique and in Zimbabwe, both countries who decided with the assistance of FAO (Food and Agriculture Organization of the United Nations) and AFD (Agence Française de Développement) to develop a national strategy to manage HWC.

The first section will point out key principles of HWC based on the example of Mozambique and specifically the case of Limpopo National Park. A focus on communities living in areas adjacent to national parks in Zimbabwe will help explain the depth of the HWC problem. A new approach, based on disease monitoring, is to invite communities to develop their own prevention strategies with the support of a HWC toolkit developed by BIO-HUB[1] and FAO. The damages and problems associated with HWC have been high in areas adjacent to major National Parks. The improvement of alert systems allowing a quick and efficient control of problem animals will conclude this chapter.

2. Co-existing with wildlife in Mozambique, a national problem

Mozambique is a country located on the southeast coast of Africa with a total surface area of 801,590 km^2. It is bound by Swaziland to the south, South Africa to the southwest, Zimbabwe to the west, Zambia and Malawi to the northwest, Tanzania to the north and the Indian Ocean to the east. The national population was recorded at 22 million in 2009; 80% of the Mozambican population live in rural areas where their livelihoods rely heavily on

[1] a consortium of CIRAD, IGF, IUCN-ROSA & WWF-SARPO

subsistence agriculture or fishing and on the use of natural resources; 45% of them (i.e. around 8 million people) have no access to water supply equipment and have to collect river water to meet their daily needs. In addition, about two hundred thousand people live in close contact with wildlife, either in the buffer zones or even within the protected areas. This explains the increasing incidence of HWC whose impact is exacerbated by the overall poverty prevailing in one of 10 poorest developing countries in the World.

2.1 Human-wildlife conflict facts: National trends since 2006

2.1.1 National context

HWC is not a new issue in Mozambique, but a dramatic increase in HWC cases has been noted in recent years with 265 people killed between July 2006 to September 2008 mostly by crocodile (79%), 1,116 ha destroyed in 2008 mainly by elephant (86%) and hundreds of problem animals killed each year (Ministério da Agricultura, 2009).

In a response to the social impact of HWC, a national strategy to reduce HWC was developed and recently approved by the Government of Mozambique with the support of FAO. The Ministry of Agriculture (MINAG) is responsible for wildlife management outside protected areas and, since HWCs are more prone to succeed there, is in charge of the overall HWC management. More precisely HWC is managed by the Wildlife Department of the Directorate of Lands and Forests (DNTF) which has specifically appointed officers to manage this problem.

2.1.2 Materials and methods

DNTF started to collect HWC incident information from 1997. Between 1997 and 2003, a small data set of 34 records reported a few HWC cases in 9 of the 10 Mozambican provinces. During this period (1997-2003) 240 persons and 304 wild animals were killed; predominantly crocodile related attacks in Tete and Sofala provinces and man-eating lion attacks in Cabo Delgado Province (Anderson & Pariela, 2005). The lack of data prevents any statistical analysis. Another set of 66 records was collected between 2004 and 2006. The absence of information on the timing (month), the location of the event (district) and the species involved in HWC make any analysis futile.

The available database is a set of HWC monthly reports compiled by DNTF from 2006 to 2010; data was only partially recorded during 2006 and 3 months of 2008 are missing (October, November and December).

The following analyses are based on 1,141 monthly incidents recorded. The XLSTAT software package was used to analyse the data. Data are presented as mean ± SE. In order to test for independence between percentages of measured variables, the Pearson's Chi-square test was used. The null-hypothesis between classes was rejected at $P<0.05$ and results with $P<0.001$ were considered highly significant. With the assumption that HWC events are strongly related to the presence of specific wild species in certain area at a certain time, a Multiple Correspondence Analysis (MCA) was conducted to study the association between spatial and temporal factors.

2.1.3 National trends

2.1.3.1 HWC everywhere but targeted sites

The most affected provinces in Mozambique are Cabo Delgado and Gaza which experienced 18% of HWC incidents recorded between 2006 and 2010 (201 and 205 cases) followed by

Manica, Maputo, Tete and Zambezia where 10% of incidents occurred during the same period (Fig. 2).

Fig. 2. HWC frequency per Province between 2006 and 2010

2.1.3.2 An increasing problem all year long

HWC appears to be an escalating problem, with reported cases increasing by a magnitude of 4.3 between 2006 and 2010 (Fig. 3). The frequency of HWC per month varies between 6 to 10% (Fig. 4) and occurs all year long with a similar incidence during the dry (51%) and the rainy seasons (49%). The risky months appear to be September and July (10% of incidents reported) followed by March, April, June and October (8% of HWC cases).

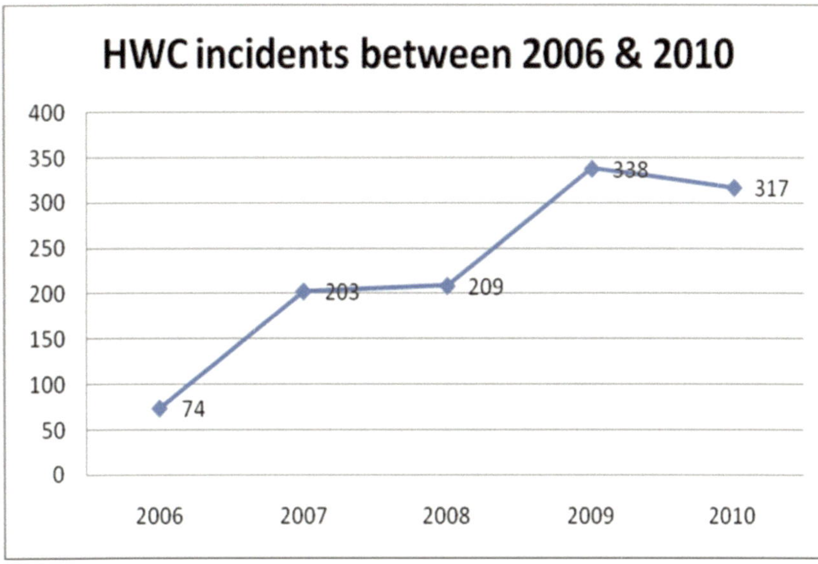

Fig. 3. Increase of HWCs between 2006 and 2010 in Mozambique

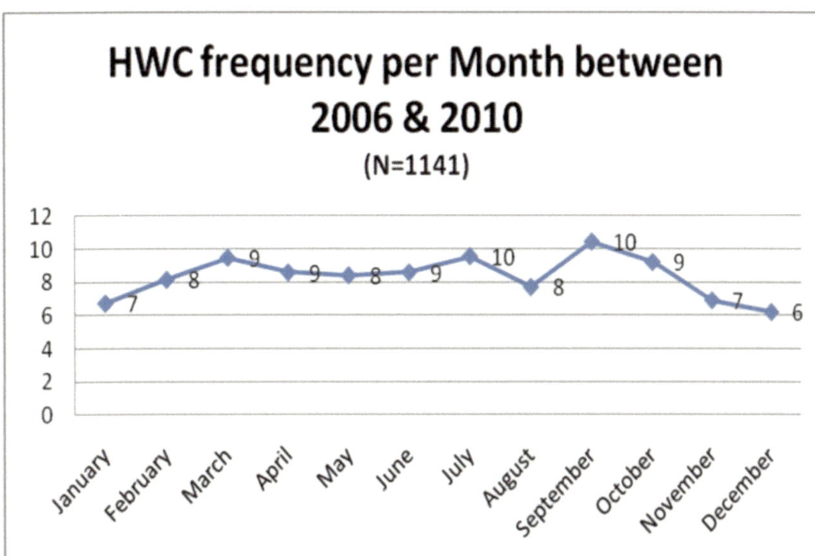

Fig. 4. Monthly frequency of HWC in Mozambique

2.1.3.3 Mainly with elephant and crocodile

Among the thirteen species involved in HWC between 2006 and 2010, four species predominated: elephant (39% of the incidents) and crocodile (29%) showed significantly more cases of attacks than hippopotamus and lion with respectively 16% and 10% of HWC records. The remaining species were occasionally involved in HWC with a few cases reported per year (Fig. 5).

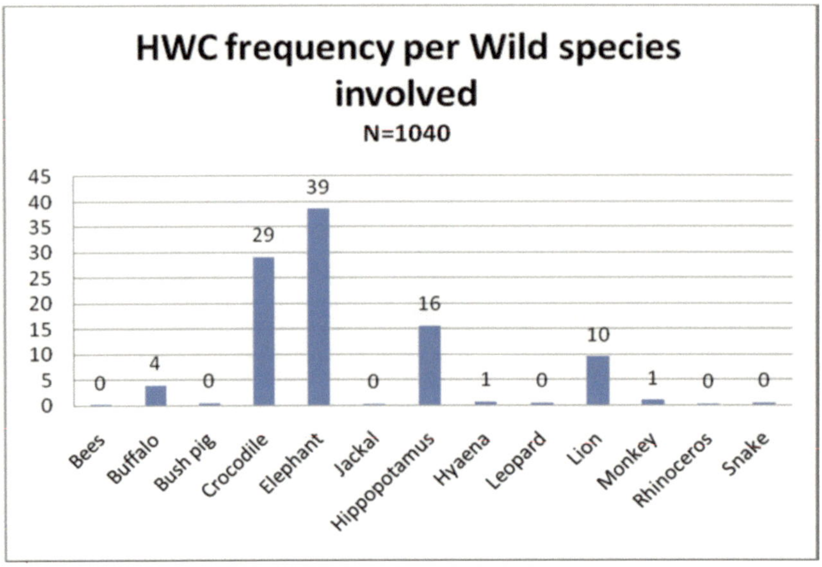

Fig. 5. Problem animals in Mozambique

A focus on the most affected six provinces reveals that elephants are the main problem animal with 41% of HWC recorded mainly in Cabo Delgado Province with 72% of HWC (Table 1). Species involvement is more balance in the other provinces except for Tete where crocodiles were the predominant species (54%). Lion seem to be only a problem in Maputo Province with 24% of HWC. Observed differences are highly significant (Pearson Chi-Square=171, df=15, p=0.001, n=755).

Wildlife involved in HWC		Province						
		Cabo Delgado	Gaza	Manica	Maputo	Tete	Zambezia	Total
Crocodile	Count (%)	20 (12)	49 (28)	37 (35)	25 (25)	44 (39)	51 (54)	226 (30)
Elephant	Count (%)	121 (72)	69 (40)	39 (37)	27 (27)	39 (35)	14 (15)	309 (41)
Hippopo-tamus	Count (%)	3 (2)	45 (26)	12 (11)	26 (26)	24 (21)	23 (25)	133 (18)
Lion	Count (%)	24 (14)	11 (6)	17 (16)	24 (24)	5 (5)	6 (6)	87 (12)
Count		168	174	105	102	112	94	755

Table 1. HWC frequencies per wildlife species and province between 2006 and 2010

2.1.3.4 A significant impact

During this period 281 reports were generated for a total of 431 people being killed by a wild animal. This represented 25% of HWC incidents (281/1141). The number of people being injured during the same period was less, with 106 reports (9% of HWC incidents) for a total of 169 injured. When grouping people killed and injured as human casualties, the crocodile results in the greatest number of incidents with 61% of them (216), followed by the elephant with 21% (73), hippopotamus with 7% (25) and lion with 5% of the people being killed or injured (17). Observed differences are highly significant (Pearson Chi-square= 349, df=13, p=0.001, n=1146).

Between 2006 and 2010, 97 attacks against various domestic animals were recorded: cattle (44%), goat (35%), sheep, donkey, dog and chicken. A total of 807 individual animals were preyed upon, mainly by lion (58%), crocodile (33%) and hyena (4%). Observed differences are highly significant (Pearson Ch-square=433, df=13, p=0.001, n=1146).
Almost 25 km² of crop were destroyed between 2006 and 2010 with an average of 11 ± 2 ha per monthly incident reported (n=238). A total of 60 ha were destroyed in 2006, 962 ha in 2008 followed by 620 ha in 2010 (Fig. 6).

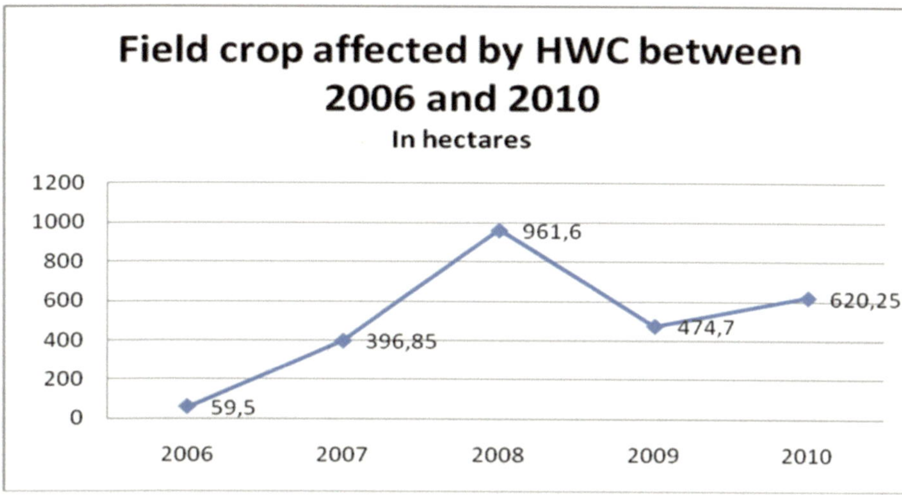

Fig. 6. Crops raids impact from 2006 to 2010

Four hundred and thirty five (n=435) cases of field crop destruction were recorded but for 44% of them the levels of damage were not estimated. In two thirds of the cases (65%) elephant was the species involved, followed by hippopotamus (26%) and buffalo (3%). Differences observed are highly significant (Pearson Chi-square=532, df=13, p=0,001, n=1146).
During the study period, a few observations (n=29) indicate that houses were targeted (42%), granaries (19%), canoes (10%) and so-called properties (10%). In most cases (68%) elephants were responsible for the damages, followed by hippopotamus (13%) and crocodile (10%). Differences observed are highly significant (Pearson Chi-square=35, df=13, p=0,001, n=1146).

2.1.3.5 Killing problem animals as the main mitigation strategy

Monitoring (38%) and killing (36%) were the two most popular methods employed to deal with HWC (Table 2). Killing problem animals occurred mainly in Maputo (55%), Zambezia (42%) and Cabo Delgado (39%) provinces. This usually relates to a single animal (75%, n=365) and these were listed as: elephant (38%, 137), hippopotamus (24%, 86), crocodile (23%, 82), lion (7%, 26) and buffalo (7%, 24). Between 2006 and 2010, 574 problem animals were destroyed, mostly elephants (33%), crocodiles (29%), hippopotamus (20%) and lions (7%). Note that the "No reaction" strategy significantly occured in 18% of cases and that the "Awareness" is a marginal strategy only implemented in Cabo Delgado (1,2%). Observed differences are highly significant (Pearson Chi-Square=65, df=25, p=0.001, n=755).

Mitigation measures		Province						
		Cabo Delgado	Gaza	Manica	Maputo	Tete	Zambezia	Total
Awareness	Count (%)	2 (1)	0	0	0	0	0	2 (<1)
Chasing	Count (%)	4 (2)	10 (6)	4 (4)	4 (4)	4 (4)	6 (6)	32 (4)
Shooting	Count (%)	12 (7)	5 (3)	0	0	6 (5)	1 (1)	24 (3)
Killing	Count (%)	66 (40)	50 (29)	25 (24)	56 (55)	39 (35)	39 (42)	275 (36)
Monitoring	Count (%)	54 (32)	74 (43)	59 (56)	24 (24)	46 (41)	32 (34)	289 (38)
No reaction	Count (%)	30 (18)	35 (20)	17 (16)	18 (18)	17 (15)	16 (17)	133 (18)
Count		168	174	105	102	112	94	755

Table 2. Frequencies of mitigation measures implemented per Province between 2006 and 2010

2.1.3.6 HWC typology

With the 5 main species involved in HWC and the 6 most affected provinces, a Multiple Correspondence Analysis (MCA) was conducted to study the association between spatial and temporal factors. The first four factors are predominating represent 72% of the variance. The major axis F1 explaining 49% of variation is characterised by elephant being a major problem animal and Cabo Delgado in one direction, versus Zambezia Province and with crocodile and hippopotamus as the main dangerous species. The second axis F2 explaining 10% of the variation is characterized on one direction by Tete with elephant and crocodile, versus Maputo and lion (Table 3).

	F1	F2
Cabo Delgado	20,463	1,673
Gaza	-4,745	2,404
Manica	0,454	-3,062
Maputo	-3,260	-18,955
Tete	-3,762	12,497
Zambezia	**-12,778**	**4,207**
DryS	**5,017**	**6,066**
RainyS	**-5,017**	**-6,066**
Crocodile	**-14,965**	**6,132**
Elephant	**20,419**	**8,367**
Hippopotamus	**-11,991**	**-2,281**
Lion	**4,327**	**-18,957**

Table 3. Test values (Variables) The values displayed in bold are significant at the level alpha=0.05

The explanatory weight of the Variable Season is limited. Four types can be identified (Fig. 7.). The first two are very distinct:

- The Cabo Delgado type with elephant conflict and a slight influence of the dry season.
- The Maputo type with lion conflicts.

The last two are more complex, with more than one species involved:

- Zambezi type with crocodile and hippopotamus conflict.
- Tete type with crocodile and elephant conflict.

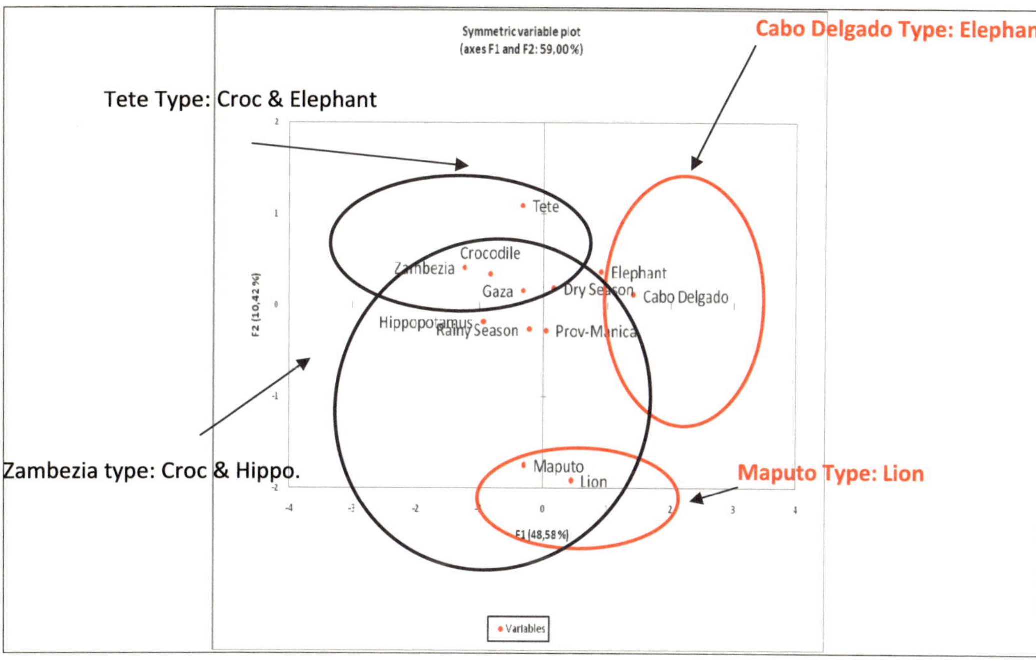

Fig. 7. Factorial plan for HWC data

2.2 From fact to perception: The case of Limpopo National Park

2.2.1 A new born park: The Limpopo National Park

The Coutada 16 (a hunting block) was proclaimed as the Limpopo National Park (LNP) in November 2001. The status of NP protects the land and prohibits human activities including prospecting and mining (Magane et al., 2003). On 9th December 2002, the presidents of South Africa, Mozambique and Zimbabwe signed the agreement on the establishment of the Great Limpopo Transfrontier Park (GLTP), and two days later the first 45 km of fencing between South Africa and Mozambique was removed. LNP is in the west of Gaza Province, adjacent with the international border with South Africa and to the south of the international border with Zimbabwe. Covering an area of 11,233 km^2, LNP is part of the GLTP which includes the Kruger NP and Gonarezhou NP (Fig. 8).

The climate in LNP is subtropical with hot, wet summers and mild, dry winters (360 mm in the north and 500 mm in the south). Rain occurs from September to April. The hydrology is characterized by 3 rivers which dictate the distribution of wildlife: the Limpopo and Elefantes rivers predominate, and to a lesser degree the Singwedzi which dries up in winter (Magane et al., 2003).

Fig. 8. Limpopo National Park limits

Mozambique's wildlife population was all but destroyed during the 14 years of war preceding 1990, and the country's parks have been battling to restore their game reserves ever since. About four thousand wild animals (3,885) were translocated from Kruger National Park to LNP between 2001 and 2006 (Whyte & Swanepol, 2006). Planes game included buffalo, hartebeest, impala, roan, waterbuck, wildebeest and zebra; mega-herbivores included elephant (111 in 2001 and 2003), giraffe (61) and white rhino (12). From an area survey undertaken in 2006 covering most of the Shingwedzi River basin (3,000 km^2), 630 elephant in sixty groups of 10.5 animals (mean size) were observed to be widely dispersed. This number greatly exceeded the number of elephants translocated to PNL suggesting that a natural re-colonization process had begun. All key problem animals exist in the area including the habitual crop raiders (elephant, buffalo, hippopotamus), livestock predators (lion, leopard, hyena, jackal) and man eaters (lion and crocodile). Densities are low but have been and will continue to be fueled by migration from Kruger National Park. No practical tool is in place or has been tested to monitor this population dynamic (density and movement).

The predominant tribe is the Shangaan. Approximately 4,350 inhabitants are living along the Shingwedzi River with about 5,200 head of cattle. A further 20,000 people live along the

Limpopo and Elefantes rivers within the Buffer Zone of the Park. These people are distributed across 44 villages which consist of a total of 5,530 households. The support zone roughly extends on a 260 km straight line from Pafuri up to the junction with the Elefantes river and 65 km from the junction up to Chibotane; this gives an average distance of 7.4 km between each village. The population is expected to increase at an average rate of 2.6% per annum (Magane et al., 2003). Zoning is a management tool that delineates the Park into areas where developments of different type and scale are allowed, and where the levels of resource use and conservation inputs vary.

Communities are concerned with the Buffer Zone (BZ) of the park. The BZ should support both existing traditional subsistence livelihoods including crop and livestock agriculture, and sustainable economic development including opportunities for community-based tourism. The BZ extends westwards from the Limpopo River for a distance of 8 - 10 km and includes settlements and agricultural lands (crop and stock). Combining the land north of the confluence of the Limpopo and the Elefantes rivers and that between Massingir and Mabelane, this represents about 3,707 km^2 (33% of the LNP). The main activity is rain fed agriculture (maize, pumpkin and beans) complemented by raising livestock (Woodburne et al., 2002). Excepting the alluvial river banks, soils are poor and low unpredictable rains limit crop production. Trees are utilized for firewood, construction and production of charcoal and bushmeat is taken as an important source of protein. Cash income from employment is a rare occurrence (Magane et al., 2003). The dependence of the people on the natural resource base is increasing the potential conflicts between the people and the wild animals (Woodburne et al., 2002).

2.2.2 Materials and methods

A four day HWC training course was conducted in the south of LNP at a training centre in Massingir from the 5th - 8th of April 2011 then at the same period in the central of LNP at Mapai from the 11th - 14th of April. A total of 48 game rangers attended, 9 of them were dedicated HWC workers and 3 extension officers. The first day of the training was focused on gaining the game rangers perception of the current HWC situation. Their perceptions were extracted from a series of 13 questions using simple cards where they were asked to write their answers using one card per idea produced. The return given was good with a total of 1,116 cards produced (2 to 3 responses per question), giving an idea of the level of participation during the course. In parallel, the analyses of 384 incidents recorded by LNP between 2007 and 2010 allowed us to compare the perceptions of the game rangers with facts. The XLSTAT software package was used to analyze the data. Data are presented as mean ± SE.

2.2.3 Main findings

2.2.3.1 Key species involved in HWC

The 384 HWC incidents between 2007 and 2010 reached a peak in 2008 with 144 cases (mean=96 per year). From HWC reports, the elephant was the main species involved in HWC with 83% of incidents. Game rangers also identified the elephant as the main problematic species but also mentioned the lion, buffalo, hippopotamus, crocodile and baboon as problem animals (Fig. 9).

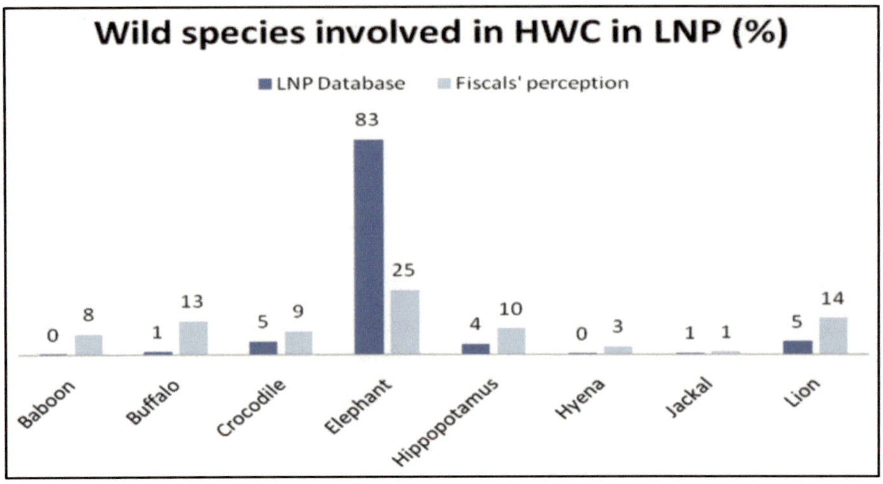

Fig. 9. Problem animals of LNP[2]

2.2.3.2 Risky months

When questioned "what are the risky months?" most of parks' staff mentioned January and March with 19% of HWC incidents, followed by February and June with 13% and 14% of incidents. January to May, and also August are perceived as periods of elevated risk with a high prevalence of human-elephant conflict. June and July are dominated by lion conflicts. During the dry season (September to November), crocodiles are the main species involved and in December human-hippo conflicts predominate. When compared with the perception of the game rangers (fiscals), one can see that they are overestimating the occurrence of HWC during the first quarter and under-estimated during the second quarter of the year (Fig. 10).

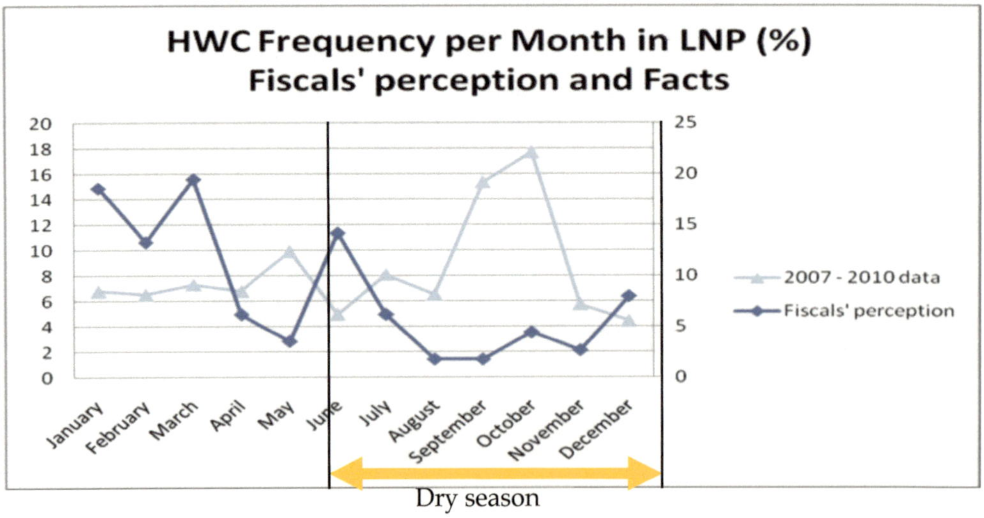

Fig. 10. HWC has a strong seasonal pattern with 2/3 of the incidents happening during the dry season (n=384)

[2] "Fiscal" is the Portuguese name given to game scouts in Mozambique

2.2.3.3 Period of the day at most risk

Night is perceived by the game rangers to be the most risky period of the day with a high frequency (67%) of encountering dangerous animals. Human activity in the mornings and afternoons are also perceived as risky periods but with a smaller percentage of HWC (14%). During the night the most dangerous animal is the elephant, during day time it is the crocodile.

2.2.3.4 Localization of the hot spots of HWC in the Limpopo NP

Using pins on a map the game rangers were asked to localize places mostly affected by HWC. Both groups from the North and South training courses indicated the triangle between the Limpopo and Elefantes Rivers (1), groups of villages close to the two pickets of Mapai and Pafuri (2) and villages along the Shingwedzi River (Fig. 11). The analysis of historical HWC incidents confirms that the southern part of the park with the District of Massingir is the most affected by HWC with ¾ of incidents recorded between 2007 and 2010 (Fig. 12). More precisely 55% of all HWC of LNP between 2007 and 2010 (212 incidents) were concentrated in 10 villages of this district.

Fig. 11. South course HWC spots localization

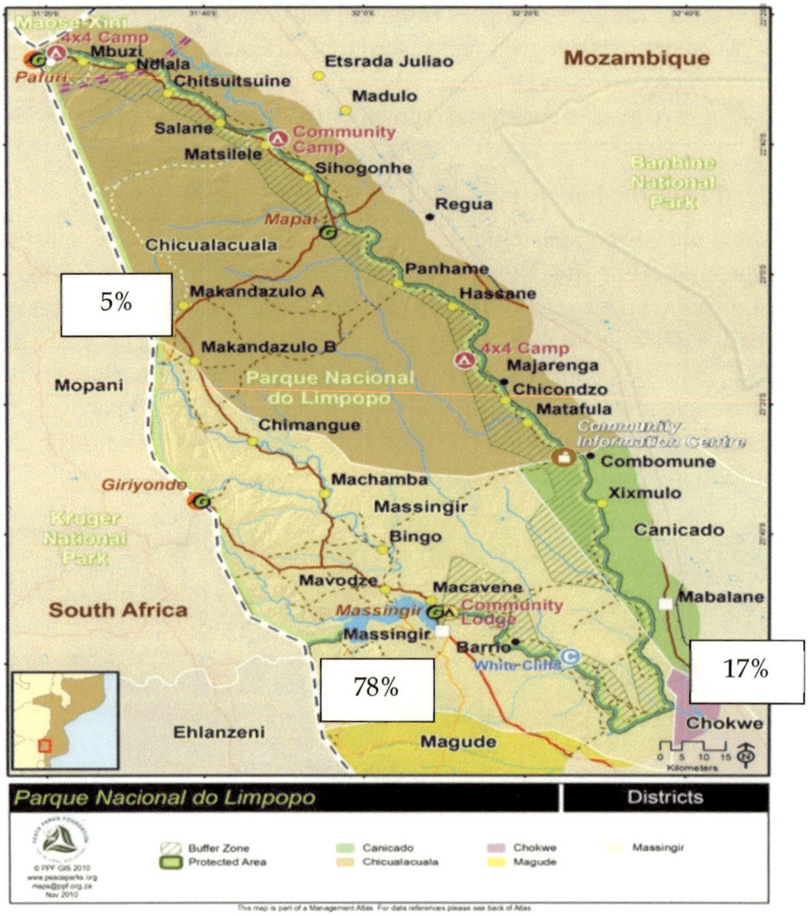

Fig. 12. Spatial records of HWC in LNP between 2007 and 2010

2.2.3.5 Situations in which HWC occurs and categories of people in danger

When asked "In which circumstances does HWC appear?" game rangers mentioned farming (31%) and river based activities (27%). Forests were mostly associated with the "lion risk" (71%) and walking on a path had equal risk of encountering an elephant or a snake. Farming activity increased the risk of human-elephant conflicts (68%) and the river represented a risky place for crocodile attacks (82%) or problems with hippopotamus (14%). The home is not necessary a safe place as it appears to be a site of conflict with snakes (36%), lion (24%) and hyena (16%).

According to game rangers, the categories of people more likely to be involved in HWC are men (40%) followed by women (37%) and children (24%). Children are most likely to be attacked by a crocodile (41%), followed by a lion or an elephant (18%). Men have conflicts predominantly with elephant (54%) or lion (30%) and women with crocodile (56%) or elephant (32%). These results, based on the perception of game rangers, are in line with human activities which could increase wildlife encounters. Nevertheless, information kept in the HWC data could not confirm this statement.

2.2.3.6 Main impacts of HWC

The impact of HWC is perceived by game rangers mainly as crops destroyed (49%) followed by livestock being eaten (27%) and people being killed or injured (24%). Crops are mainly destroyed by elephant (57%) and hippopotamus (14%) and livestock are attacked by lion (65%), crocodile (22%) and hyena (14%). People are mostly frightened and attacked by crocodile in the river (42%), and lion (27%) and elephant (15%) in the field or in the forest.

The analysis of LNP data base revealed that human casualties are low at 3% (12/384) and occurred mainly in Massingir (83%). Crocodile were the biggest problem (67% of human casualties), followed by hippopotamus at 17% and then buffalo and elephant (both 8% of human casualties). Most of the incidents recorded were related to crop destruction (82%, 316/384), mostly maize (48% pure maize, 39% maize and vegetable, maize, vegetable and fruit trees 12%), with an average of 75 ± 6 crop raids reported per year of which 93% were by elephant. During the same period, 723 fields were reported to be destroyed with an average of 2.3 ± 0.3 fields per incident (n=384) but in 94% of the HWC no evaluation of destruction was conducted. The few assessments done (n=18) gave an average of 0.11 ± 0.03 ha of crops destroyed.

Livestock predation represented a small volume of HWC (8%, 32/384) with the same proportion of cattle (44%) or goats (38%) preyed on in small numbers (5 ± 1, n=32). Lion were involved in cattle (79%) and donkey (100%) predation, and crocodile targeted sheep (100%) or goats (55%). The latter was also attacked by jackal (18%) and baboon (9%). In the last 4 years only one case of equipment destruction (irrigation pipe) by an elephant was recorded.

2.2.3.7 Preventive measures taken by the communities

Making noise and having ones property or field protected by a fence was perceived as the best way to prevent HWC (24%). Other strategies mentioned were making fire (20%), making noise (17%) or fencing without making noise (12%). The interesting point is the diversity of solutions reported by the game rangers, indicating (i) the importance of HWC and (ii) the motivation of local farmers to cope with such a problem (Fig. 13). The lack of detail information in the HWC database prevented any comparison.

For the elephant the main mitigation measures were to make fire (32%), to beat drums (30%) and combine fencing with making noise (19%). Other techniques were used at lower frequencies. In order to minimize lion predation, farmers were firstly herding (40%) and also making fire (20%), preventing the access to the krall (corral) with strips (20%) or stopping the lion with snares (20%). Fencing (33%), making noise (33%) or combining both (17%) with in some cases utilization of a flashlight (17%) were the main techniques to repel buffalo. For the hippopotamus, farmers were mainly making fire (36%) or combining fencing with beating drums (36%), while the use of scarecrows was also reported. Fencing the water point and making noise (40%) were the most successful deterrents for the crocodile, or simply to fence them out (20%). The use of chili as a deterrent was also reported (20%). Baboon were involved in crop raiding and in small livestock predation, methods reported by the game rangers varied from herding (38%), using plastic strips to prevent access to crops (33%) and also using dogs to alert (13%).

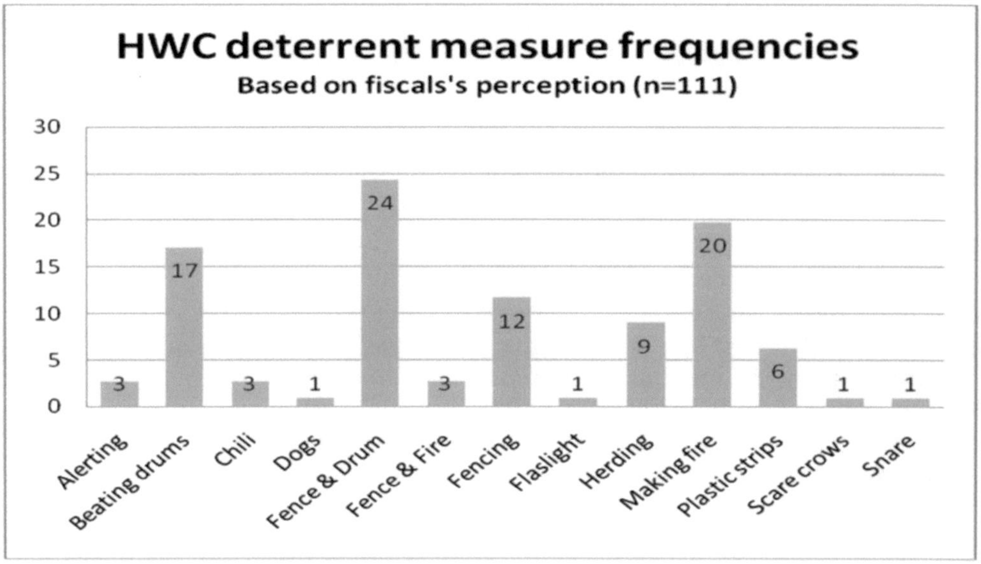

Fig. 13. Communities' mitigation measures

2.2.3.8 Types of reaction taken by the fiscals

The main game rangers intervention was to chase the animals involved in HWC (64%) followed by killing them (19%). Little was done to monitor or evaluate the situation (9%) or provide awareness to farmers (8%). Killing is used as a reprisal to all species (57%) but especially to crocodile (14%), buffalo (14%), elephant (7%) and lion (7%). This attitude was partly confirmed by the HWC data base with 53% of game rangers reporting chasing problem animals and 36% evaluating the situation. Killing seems to be rarer than was perceived with an average of one animal killed per year (1.3 ± 0.2, 1 to 4, n=21) between 2007 and 2010, mainly elephant (50%), crocodile (25%) and lion (25%).

3. Understanding Human Wildlife Conflict (HWC) in areas adjacent to national parks and conservancies in Mbire, Chiredzi and Hwange districts, Zimbabwe

3.1 Background

Zimbabwe is endowed with a stock of natural resources that could support the development of a diversified economy, thus enabling the country to address social development issues. To reach its objectives of food security and sustainable environment preservation, the country faces important challenges, the most important being the existence of acute conflicts between rural communities and wildlife. Recently conflicts between human and wildlife became one of the biggest obstacles for community-based natural resource management in Zimbabwe, this situation been exacerbated by the 1999 Land Reform which resulted in Africans settling on former white owned commercial farms, as well as game safari land and sections of National Parks.

When examining causes, several interlinked factors could be identified. The post colonialism period was marked by profound changes in landscape arrangement. Firstly restructuring of the agricultural sector and general human population growth had resulted in human encroachment on wildlife habitat and natural movement corridors.

Changes in individual land use strategies involving the switch from large fields to scattered cultivated lands likely contributed to increased conflicts in rural lands adjacent to Protected Areas. Meanwhile, because of the high rate of unemployment and increasing poverty, the area of subsidence farming of cultivated lands expanded. Growing poverty has led to the over-exploitation of natural resources and the increase of illegal activities including poaching. Inevitably an augmentation in conflicts between poor communities living side-by-side with Protected Areas and wildlife has been experienced. In addition, the situation is exacerbated by insufficient revenue from wildlife to communities resulting in their decreased tolerance level towards wildlife. The direct costs to local communities encompass threat to human life and economic losses with a decrease in agriculture performances. HWC are also extremely costly in term of wildlife conservation. Revenge killing of problem animals leads to the death of non-targeted animals. The use of snares, traps, poisoned water and poisoned carcasses may affect the entire biodiversity chain. Conflicts can represent a real threat to endangered and protected species. Local revenue generated through hunting tourism can also be reduced when there is excessive removal of trophy animals under problem animal control (PAC) activities. Ultimately, people tend to develop a negative attitude towards wildlife management and conservation initiatives proposed by the government or conservation authorities. This in turn can lead to non-cooperation of local communities and increased instances of poaching and other illegal activities.

The main objectives of this baseline survey were to: determine the drivers of HWC, characterize the spatial and temporal dynamics of HWC, and explore the relationships between the frequency of HWC and environmental factors at community level.

3.2 Materials and methods

3.2.1 Study sites

The study was conducted in three sites located in Hwange, Mbire and Chiredzi districts of Zimbabwe (Fig. 14). Each study site was adjacent to conservation areas. In Hwange district, the study site comprised wards 11, 14, 15, 16 and 17. In Mbire district, the study was conducted in wards 1, 2, 3, 9 and 12, while in Chiredzi district, the study site comprised wards 1, 22, 24, 26 and 32.

3.2.2 Sampling and data collection

To determine the levels of human-wildlife conflict in each study site, we administered questionnaires to randomly selected households to solicit information on the type and frequency of occurrence of HWCs. In each study site, households were selected following several steps. Firstly, we digitized and mapped all homesteads in the selected wards based on 2010 SPOT satellite image in Google Earth (http/google.earth.com). The satellite image had a spatial resolution of 2.5 m which made it possible to detect homesteads. Secondly, in Chiredzi district, we randomly selected homesteads within each sampled ward from a population of homesteads mapped using the method described above. In Hwange and Mbire districts, we generated and overlaid at least four transects in a Geographic Information System (GIS). Transects originated at conservation boundaries and passed through human settlements (homesteads) and were oriented in such a way that some homesteads were closer to the boundaries of protected areas (mostly National Parks) while others were farther away. Along each transect, homesteads were selected at random. This sampling strategy was employed to gather data for testing whether the frequency of occurrence of HWC is related to distance from conservation areas.

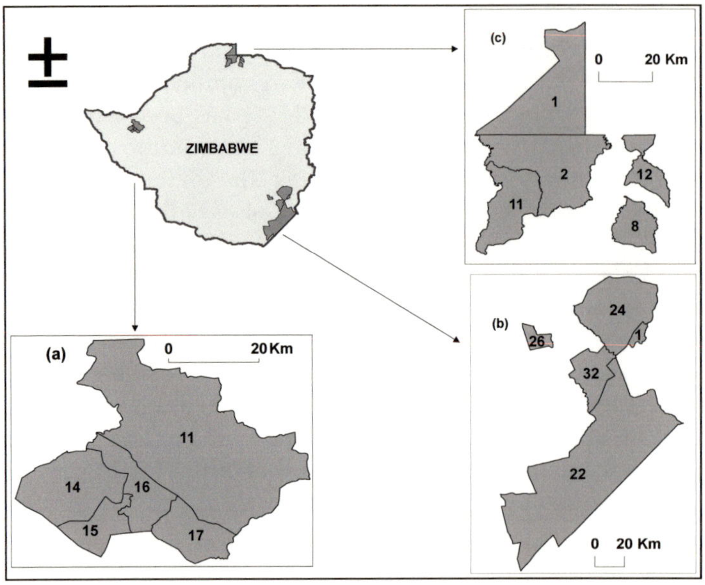

Fig. 14. Location of (a) Hwange, (b) Chiredzi and (c) Mbire study sites in Zimbabwe

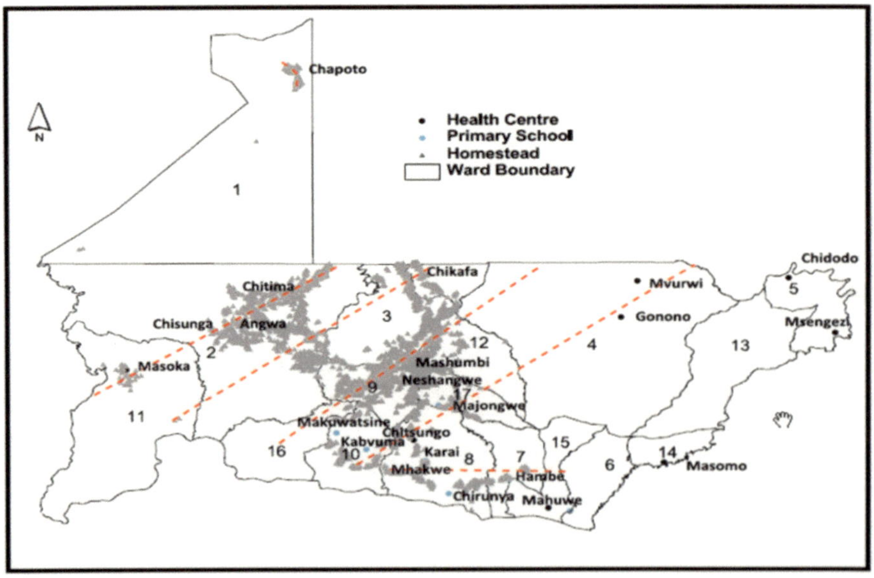

Fig. 15. The distribution of sampled homesteads along transects in Mbire district, Zimbabwe

The coordinates of the selected homesteads were stored in a Global Positioning System (GPS) unit, which was used to locate the homesteads in the field. Data on the location of sampled homesteads was used to aid analysis of the spatial patterns of HWC in each study site. Figure 15 shows the distribution of the homesteads in selected wards along transects in Mbire districts. At each of the selected homesteads, the most senior household member belonging to that homestead was interviewed.

3.2.3 Data management and analysis

Data collected during the questionnaire surveys were captured and managed in a database that we specifically designed for analyzing HWC in Microsoft Access (MS Access). The main goal for designing and developing a database for HWC was to enable an understanding of the changing nature of HWC in space and time; in order to achieve this the database was equipped with a simple data entry interface which enables capture of data on HWC in the future.

To facilitate statistical analysis of the HWC across the three study sites, questionnaire data were also captured in a statistical package (Statistical Package for the Social Sciences, SPSS). SPSS was also used to explore and understand the relationships between the spatial patterns of HWC, their seasonal dynamics and several household characteristics that include sources of livelihoods, time of settlement, ethnicity and perception of HWC, as well as ecosystem services in the study sites. The main objectives of these analyses were to determine and explain the geographic distribution of HWC in relation to key household characteristics such as time of settlement and location of settlement, and to ascertain whether the level ecosystem services in a site was a key driver of the spatial variations in the perceptions and occurrence of HWC. To map the geographic distribution of HWC and isolate hotspots of conflict, we first mapped and plotted as point data all households who either reported or did not report that they had experienced any HWC such as crop raids by wildlife. We then plotted the intensity of HWC by fitting a kernel function on point data indicating the locations of households that reported conflict in a GIS. This analysis was only performed for study sites in Hwange district because the data was suitable for this type of spatial point pattern analysis. The plotted function enabled us to visually explore patterns in HWC.

For Mbire and Chiredzi districts we used logistic regression to relate the occurrence/non-occurrence of HWC with environmental factors such as the distance from the conservation area boundary and vegetation cover. Vegetation cover was estimated remotely from Landsat Thematic Mapper (TM) satellite image of May 2010 using the normalized difference vegetation index (NDVI). The purpose of this analysis was to test whether the observed spatial pattern in HWC was significantly related to environmental factors. $P < 0.05$ was used as the critical level of significance. Where logistic regression functions relating environmental factors to HWC were significant, the functions were in turn used to map the spatial distribution of the probability of occurrence of HWC in each of the three study sites in a GIS. All statistical analyses were performed using SPLUS version 8 (Insightful Inc.) while Arc View GIS and ILWIS GIS were used for spatial analysis and mapping.

3.3 Results

3.3.1 Spatial aspects of HWC in the study sites

Figure 16 shows that in Hwange and Chiredzi districts, a significantly higher proportion of households reported that they experienced an HWC incident but in Mbire, the difference between households that reported HWC and those who did not is small.

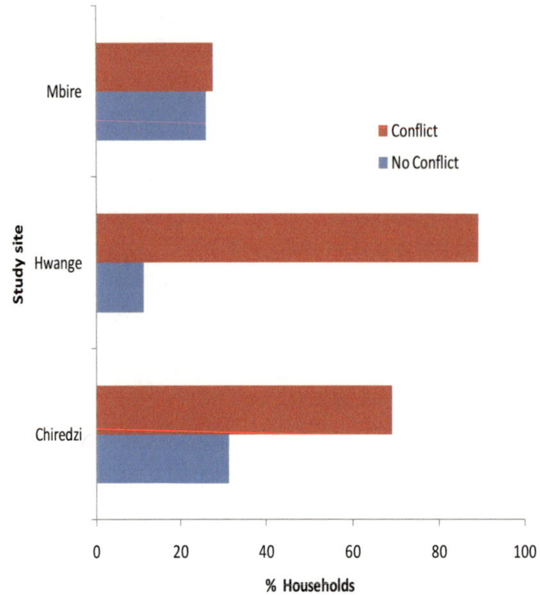

Fig. 16. Occurrence of HWC in the study sites (n=613)

Figures 17 and 18 show the spatial distribution HWC in Hwange and Mbire study sites, respectively. In Hwange, HWC were mainly concentrated in wards 15, 16, and 17 which are adjacent to Hwange National Park. In Chiredzi, HWC were also concentrated inside and closer to Gonarezhou National Park. In Mbire, the situation is different. There are no observable clusters of HWC.

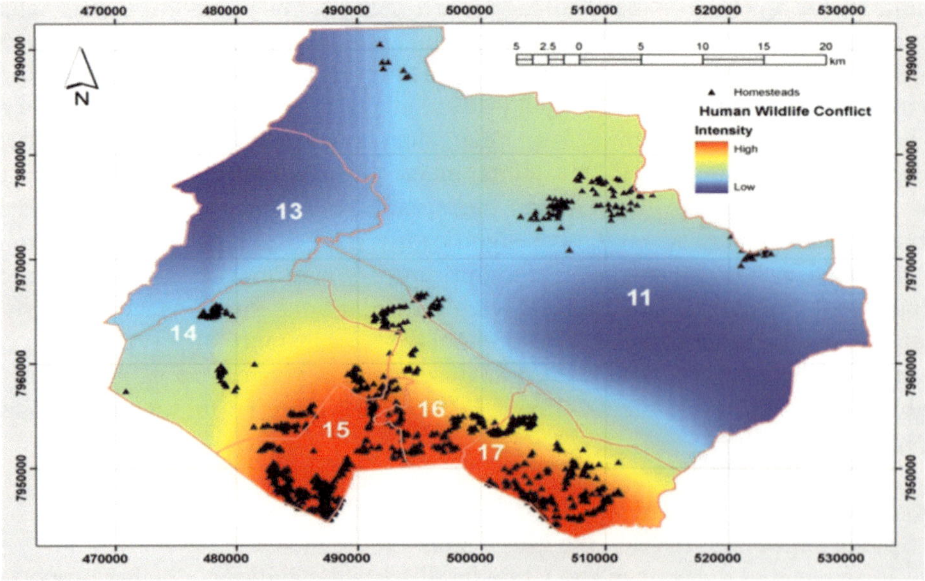

Fig. 17. The distribution of the intensity of HWC in the study wards in the Hwange study site based on fitting a kernel function on the occurrences of HWC

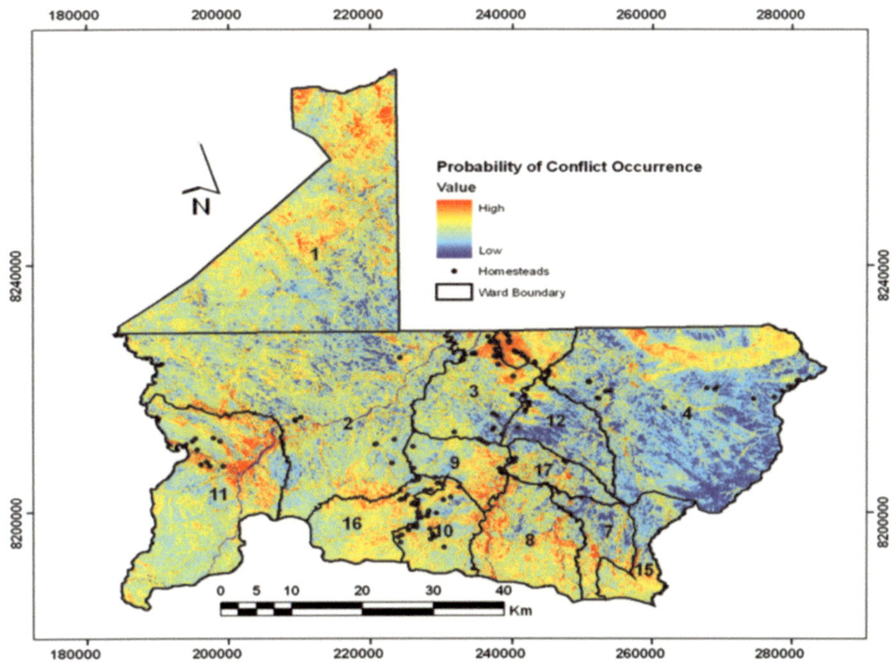

Fig. 18. The distribution of the intensity of HWC in the study wards in the Mbire study site

3.3.2 Biophysical characteristics in the HWC sites neighbourhood

The probability of occurrence of HWC in the study sites was significantly related to vegetation density (Fig. 19). This indicates that areas that have high vegetation cover are

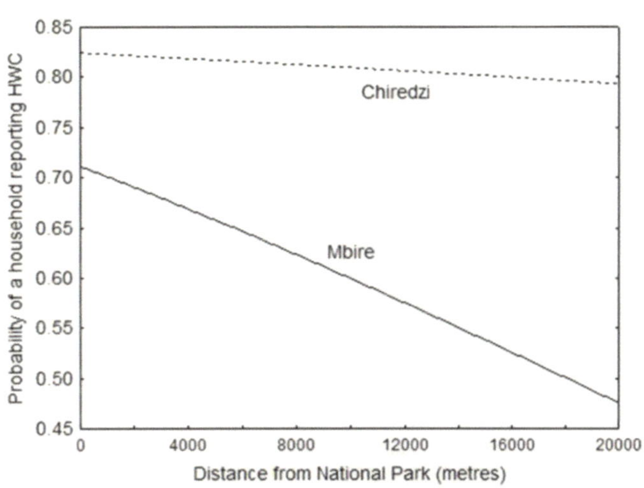

Fig. 19. The probability of HWC in Mbire and Chiredzi correlated to distance from conservation areas

associated with high incidents of HWC. Figure 20 illustrates that there is a significant negative relationship between increasing distance from the National Parks and the probability of occurrence of HWC, specifically for Mbire and Chiredzi study sites. For the Hwange study site, there is also a negative relationship between the probability of HWC occurrence and distance from Hwange National Park but this holds only when the vegetation density is high (Fig. 21). This suggests that areas that are closer to Hwange National Park but have low vegetation density are less likely to experience elevated levels of HWC than those which are closer and more densely vegetated.

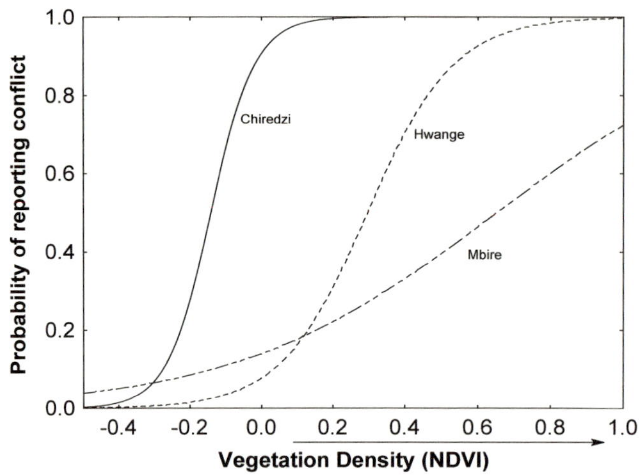

Fig. 20. The probability of HWC in Hwange, Mbire and Chiredzi in the study sites as a significant ($P < 0.05$) function of NDVI (vegetation density)

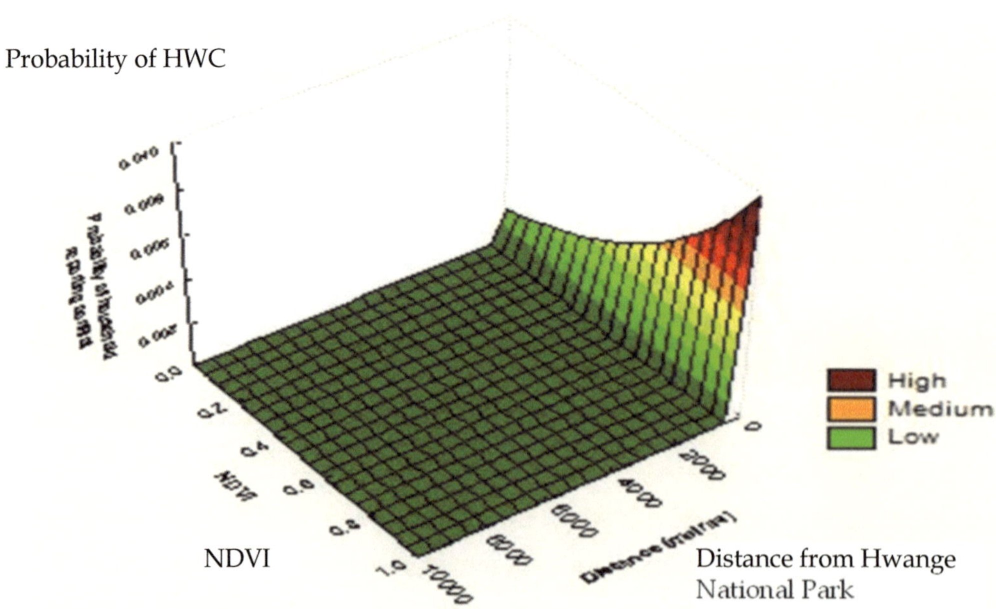

Fig. 21. The probability of HWC in Hwange as a function of the interaction of distance away from Hwange National Park boundary and vegetation density (NDVI)

Moreover, results from specific sites where HWC incidents take place reveal that most HWC incidents occur in the fields (85%, n=387). Observed differences are highly significant (Pearson Chi-Square=49, df=8, p=0.001, n=387) with 93% of HWC occuring in the field in Chiredzi, 85% in Hwange and 78% in Mbire.

3.3.3 Temporal aspects of HWC in the study sites

3.3.3.1 Diurnal dynamics of HWC

The survey investigated temporal aspects of HWC in all three study sites these included seasonal as well as diurnal distribution of raids or attacks. In all the three study sites, most of the attacks or raids occurred during the night with 86% of HWC. Chiredzi district had the highest number nocturnal attacks (94%), followed by Hwange (84%) and Mbire (82%). Mbire accounted for the highest number of attacks happening during the afternoons (12%) followed by Hwange (8%) and Chiredzi (2%). Observed differences are significant (Pearson Chi-Square=10, df=4, p=0.034, n=410).

3.3.3.2 Seasonal dynamics of HWC

The survey investigated respondents' experiences regarding the seasonal movements of wild animals into human settlements and arable fields. Results show that in all the three study sites most HWC occurred in summer with 89% of incidents observed. Differences between the three districts are highly significant (Pearson Chi-Square=23, df=4, p<.000, n=426) with Hwange having the greatest number of HWC during the summer (94%) and Mbire a significant number of raids during winter (17%).

3.3.3.3 Movements of wild animals into human settlements and arable lands in the winter season

Elephants are the main problem animals, involved in half of HWC incidents (predominating in summer vs. 27% in winter) with a high prevalence in Chiredzi (78% of incidents reported). Hyenas are perceived as the second most problematic animal by the farmers, involved in 17% of HWC in summer and 28% in winter. Baboons and wild pigs are mentioned as a problem in summer (7% of HWC), lions and buffalos are more active in winter with respectively 17% and 11% of HWC incidents reported by the informants. Differences between the three districts are highly significant (Pearson Chi-Square=111, df=18, p<.000, n=358) and can be summarized as followed in descending order:

- Mbire District:
    - In winter: Elephant (24%) > buffalo (18%) > hyena (18%) > baboon (16%) > lion (15%).
    - In summer: Elephant (47%) > wild pig (17%) > baboon (12%) > buffalo (8%) > hippopotamus (7%).
- Chiredzi District:
    - In winter: Elephant (34%) > lion (29%) > hyena (22%).
    - In summer: Elephant (78%).
- Hwange:
    - In winter: Hyena (51%) > elephant (26%) > lion (12%).
    - In summer: Elephant (50%) > hyena (32%) > baboon (8%)

3.3.3.4 Socio-economic characteristics and HWC including cropping and grazing practices

We also assessed the probability of reporting HWC as a function of the time a homestead was established. It is clear that recent settlers are more likely to report incidents of HWC than early settlers. This finding implies that recent settlers are allocated land closer to park boundaries and therefore buffer the older settlements from HWC. Logistic regression confirmed that the negative relationship between the probability of a household (respondent) reporting conflict and the time the household's homestead was established was significant ($p<0.05$). A further investigation on HWC reporting patterns by households shows that the level of education has an influence on HWC (Fig. 22). Respondents with only primary education constituted the highest number of those who reported HWC.

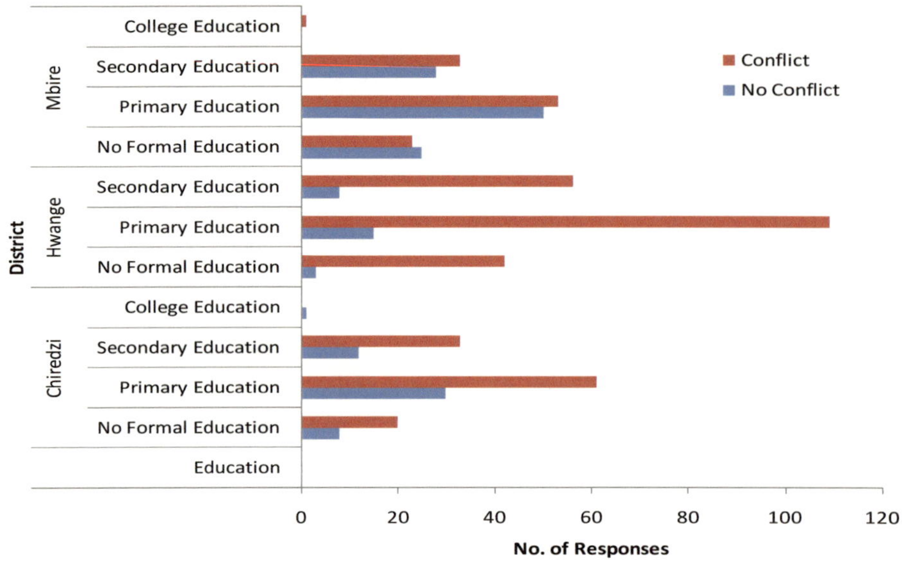

Fig. 22. Number of households (respondents) who reported the occurrence and non-occurrence of HWC in the three different study sites (n=613)

3.3.4 Consequences and responses to HWC by households

3.3.4.1 Effect of wild animals on crops, animals and persons

The major impact of HWC is crop destruction (77% of HWC reported by informants) followed by animal predation (17%) and human casualties (6%). The differences observed between the three districts are highly significant (Pearson Chi-Square=78, df=4, $p<.000$, n=407) with Chiredzi mostly affected by crop raiding (94%), Hwange having the highest percentage of livestock attacked (25%) and Mbire having the greatest number of persons being attacked by wildlife (20%).

The most vulnerable crops to raiding by wild animals include maize (43%), cotton (35%) and sorghum (16%) but significant variations exist in terms of crops raided by wild animals across the study sites (Pearson Chi-Square=232, df=18, $p<.000$, n=537). Maize which is a staple food in all three study sites is the most raided crop, especially in Hwange (56%) and in Chiredzi (50%). Second on the list of crops mostly raided by wild animals is sorghum, which is a traditional staple crop mainly in Hwange (28%). Results also show that cotton raids are largely specific to Mbire (67%) and Chiredzi (31%), and less common in Hwange (2%). In Mbire, cotton started being introduced after the eradication of tsetse fly in the mid 1980s and is fast expanding. With the support of local and multinational companies, such as Cottco, and Cargill and Olam, cotton has also been introduced in Chiredzi by newly resettled farmers.

3.3.4.2 Responses to wild animal attacks/raids by local people

Human wildlife conflict calls for action at different levels. In all the three study sites households report wild animal attacks to traditional leaders (33%), Problem Animal Control (PAC) unit (16%), or a councillor (13%) but one third of informants are not making reports. However the reporting system even in the same site differs (Pearson Chi-Square=54, df=14, p<.000, n=443). In Mbire, households are mainly reporting to traditional leaders (41%) or to the Problem Animal Control unit (24%). In Chiredzi complainants report either to traditional leaders (36%) or chose not to report (39%). In Hwange, most respondents (40%) indicated that they do not report problem animals to anyone.

In addition to reporting problem animals, affected households also undertake activities to lessen the impact of wild animal attacks. Most households tried to chase or scare the animal away (70%) and a significant proportion did not react (18%). Mitigation measures differ significantly between the three sites (Pearson Chi-Square=69, df=6, p<.000, n=492). Apart from trying to chase the animals away, in Mbire, a certain proportion of households call National Park's personnel (16%) or even to kill problem animals (7%) while in the two other districts a significant proportion (20% in Hwange and 30% in Chiredzi) are doing nothing.

The study investigated satisfaction levels of respondents in relation to responses by responsible authorities to problem animals across the three study sites. All three districts reflected negativity with the least satisfied respondents being in Hwange (97%), Chiredzi (92%) and Mbire (82%). The interesting finding here is that respondents living close to a National Park are the least satisfied, followed by those living close to a local authority controlled Safari Hunting areas. In terms of this survey, though not very statistically significant, respondents living close to a privately controlled Conservancy are more satisfied than their counterparts.

3.3.5 Assessment of compliance with regulations and perceptions on natural resources

3.3.5.1 Community resource assessment and HWC

Results show that several factors contribute to the HWC in the study sites. Most respondents in the study sites attribute the occurrence of HWC to the decline of the natural resource base. Climate change and human activities such as the opening of new fields and homesteads are also viewed as important drivers of the decline in the status of the natural resource base. Observed differences between the 3 sites are highly significant (Pearson Chi-Square=37, df=4, p<.000, n=544).

3.3.5.2 Compliance to wildlife regulations

Rules are in place to regulate the human benefits derived from the environment. However, the benefits can only accrue if people know and comply with the rules set. Compliance depends on whether the affected individuals perceive such rules as beneficial. Two third of households felt that when rules are observed, the community is likely to benefit; they are

70% in Hwange compared with 50% and 67% in Mbire and Chiredzi, respectively. Observed differences are highly significant (Pearson Chi-Square=19, df=2, $p<.000$, n=519).

3.3.5.3 Perceptions on wildlife

Overall, most respondents perceive wildlife as a threat to both people and domestic animals but perceptions differ as to whether disease transmission, destruction of crops and human and livestock predation are the key threats within and across study sites. Chiredzi has the highest percentage of respondents who perceive wildlife to be a threat to both humans and livestock (94%), followed by Mbire (87%) and Hwange (85%). Transmission of diseases is perceived to be a major threat by 73%, 20% and 7% of the respondents in Mbire, Chiredzi and Hwange study sites, respectively. Destruction of crops by wildlife is also a threat in all study sites identified by 37%, 34% and 28% of the respondents in Chiredzi, Mbire and Hwange study sites, respectively. Similarly, predation on humans and domestic animals is also considered as a threat in all three study sites by 35% of respondents in Chiredzi, 33% in Hwange and 32% in Mbire.

4. HWC, living with uncertainty in Southern Africa

4.1 An on-going problem with various expressions

Human population growth, agricultural expansion into wildlife areas, and habitat loss contribute to increasing HWC. In Mozambique, the prevalence of HWC has been multiplied by 4.3 between 2006 and 2010. In Zimbabwe, between 2002 and 2006, more than five thousand cases of Human-Elephant Conflicts (HEC) were recorded of which around three thousand cases were attended to, resulting in 774 elephants being killed during subsequent problem-animal-control operations (Campfire Programme, 2007). The survey conducted in 2010 in the three districts echoes to this finding with a significantly higher proportion of respondents reporting the occurrence of HWC.

Nevertheless, observations suggest that records of HWC are largely underestimating the occurrence of the problem. The lack of information is linked either to the fact that, like in Hwange District in Zimbabwe, a significant proportion of respondents (40%) feel helpless to report problem animals to anyone, or from the absence of a good recording system. As an example in Mozambique, DNTF recorded 200 HWC between 2006 and 2010 in Gaza Province while the staff of the Limpopo National Park in the same province reported the doubling of incidents for a shorter period (2007 - 2010).

The study of the temporal pattern acknowledged the reality that HWC is occurring all year long even if in some places significant differences are perceived or recorded by Wildlife Authorities. It is difficult to identify specific seasons or months more at risk as it depends on social and ecological factors (see above) but it is evident that most attacks or raids occurred during the night with the exception of crocodile accidents which are linked to specific sites (rivers) and water based activities. According to Samu (2010) in Hwange and Mbire districts, night attacks occurred while people are busy guarding their crops; Cumming et al. (1982) linked gender to human attacks by wild animals and observed that most people who are attacked and killed by wild animals are males mainly because they herd cattle, forage for bush meat and walk alone during the night. A study by Kock (2003) corroborates the above by mentioning that activities around the settlements such as the collection of wild fruits, fishing and poaching increased the inhabitants risks of wildlife attacks.

Cumming et al. (2007) pointed out that human deaths and injuries are less common than crop damages. The low occurrence of human casualties observed in Zimbabwe, or in the buffer zone of the Limpopo National Park with 3% of HWC, must not hide the fact that it can be much more significant at local (Mbire District) or country level (Mozambique) with, for the latter, up to 25% of HWC reporting human casualties mainly by crocodile (61%).

This impact is of serious political concern (Lamarque, 2010); however, HWC expression is more likely to include crop or livestock losses, asset destruction and the less mentioned spread and dissemination of wildlife borne pathogens (WWF, 2005). In the two countries, crop destruction is a greater issue than livestock destruction, although the latter can be a specific problem (e.g. hyenas) in certain areas (Hwange buffer zone). Elephant appear to be the main species (80% of cases) involved in crop raiding with destruction in some areas reported to be more than half the anticipated yield. In addition to this direct cost of HEC, indirect impacts include restrictions on movement of people, access to key resources, such as water, firewood and thatching grass as well as the transaction costs, of guarding crops and property against wildlife degradation, resulting in negative attitudes towards wildlife and increases in unsustainable and unregulated hunting (WWF-SARPO, 2005). A number of studies on crop and livestock predation reported have a negative impact on the local economy (Katerere, 1997; Barnes, 1998; Murphree, 2005; FAO, 2009). This last point is reflected by informants both in Mozambique and in Zimbabwe but is not supported by a good evaluation system; most of them, like the case of LNP 94% of crop destruction, have not been assessed.

4.2 Fuelled by the increase of human and wildlife population

4.2.1 From crops raiding elephants...

With elephant populations in southern Africa increasing at 5% per annum (Blanc et al., 2005; Cumming & Jones, 2005) together with the expansion of human settlement into wildlife areas, local communities living in marginal land adjacent to protected areas are faced with increasing occurrences of HEC (Nelson et al. 2003). The aetiology of HEC differs depending on the area but is linked by space requirements for human development which clashes with the ecological needs of elephants. In that sense, it is predictable that human growth (Fig. 25) will progressively increase the number and intensity of HEC in Africa as about 80% of elephants range lies outside protected areas. The potential range expansion in southern Africa may be able to accommodate up to about 75,000 elephants (Fig. 26), the equivalent of 15% of the regional population of elephants (500,000 elephants) in the next 12 to 15 years (Cumming & Jones, 2005).

4.2.2 To crocodile crimes

With a third of HWC recorded mainly in the Zambezia and Tete provinces, crocodiles are the second main problem animal in Mozambique responsible for two thirds of human casualties. The main reason is the lack of natural food due to over-fishing, and risky behaviour (Anderson & Pariela, 2005). In addition, conservationists are implementing the successful re-introduction of crocodile in big head waters such as Lake Kariba in Zimbabwe (McGregor, 2005). Considering the nature of the accident, the numbers of casualties are likely to be under-reported (Anderson & Pariela, 2005).

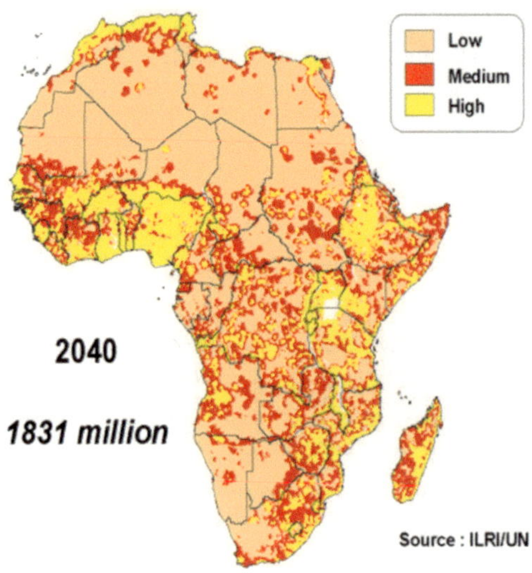

Fig. 23. Human growth in Africa

Fig. 24. Potential expansion range for elephant populations in Southern Africa

4.3 Spatial distribution and HWC dynamics

According to Steinfeld et al. (2006) competition for space and available resources between people and wildlife is the major driver of HWC. It is also generally assumed that HWC used to be low when resources (land, food and water) were in abundance for both people and wildlife. The competition reduced the 'once' peaceful co-existence of people and wildlife. In the particular case of these districts in Zimbabwe, the spatial distribution of the reported human wildlife conflict incidents show evidence of clustering especially in areas closer to conservation areas. We therefore conclude that human wildlife conflict incidents are not randomly distributed but form clusters (hotspots) in relation to distance from conservation areas, as well as, vegetation density. In Mozambique, the roots of HWC have been analysed by the National Strategy for HWC mitigation (FAO, 2009). The lack of land use plan which leads to anarchic human settlements in wildlife areas and to the destruction of the natural habitat is a major explanation for HWC between people and elephant (Araman, 2009). As mentioned above, the poor access to water supplies is another common cause of HWC, notably with crocodile (Ministério da Agricultura 2009).

This finding echoes events within the buffer zone of the Limpopo National Park where 10 villages out of 44 are experiencing most of the conflicts with wildlife. It also opens another dimension when wildlife conservation policies are extending land targeted for wildlife into grazing and arable land which is highly valued by pastoralists and agriculturalists (Kock, 2003). With the Coutada 16 proclaimed as the Limpopo National Park in November 2001, all the ingredients for the emergence of a typical HWC hotspot were combined: (i) an important human population (25,000) living along the Elephant and Limpopo rivers; (ii) a scattered, but increasing wildlife resource (Blanc et al., 2007), due to the translocation of animals and the lowering of fences between Kruger and Limpopo National Parks; and (iii) a water surface system dictating the distribution of wildlife during the dry season (Magane et al., 2003). With elephant occurring at low densities, Limpopo National Park is already facing crop damage (BRL, 2006) with an average of 32 incidents per month. It is not only a finding from the analysis of the HWC data base which indentified Gaza and Cabo Delgado provinces as elephant conflicts hotspots, but the perception of LNP game scouts who scored elephant as their main problem animals. The problem is getting worse if we consider that a systematic survey conducted in 2006 (Whyte, 2006) gives an estimation of 0.06 elephant per km^2 which is less than the density observed in the two neighbouring parks (1 elephant per km^2 in Gonarezhou; 0.63 elephant per km^2 in Kruger) (Blanc et al. 2007). With the free movement from Kruger and natural population growth which may mean a population doubling in less than 15 years, there may soon be up to 2,000 elephants living in the Buffer Zone of LNP, undermining the development of LNP and compromising the support to the Great Limpopo Transfrontier Conservation Area.

Competition for water resources especially during the dry season is a classic source of HWC (Butler, 2000) when permanent settlements are depriving wildlife access to water. Murphree (2005) and Steinfied et. al. (2000) cite conflict over access to water bodies as a major source of HWC, especially with crocodiles attacking both humans and livestock (Anderson & Pariela, 2005; McGregor, 2005).

The reporting patterns of respondents in Zimbabwe shows that recent settlers (1980s-2000s) have more problems with wildlife attacks compared with their earlier settlers. Coupled with evidence that attacks were occurring closer to conservation areas and that these attacks were mainly on crop fields, we can conclude: that (1) crop based land use practices of recent settlers attract wildlife attack and (2) that these settlers are located in densely vegetated areas close to the wildlife frontier, which mainly explains the high incidents of human wildlife conflict involving recent settlers. Human migration is a major driver of HWC (McGregor, 2004). Harsh living conditions characterised by droughts, floods, civil unrest and sometimes wars disrupt the livelihoods of rural people and force them to move to wildlife areas such as National Parks, where resources like fuelwood tend to be more readily available (Hulme & Murphre, 2001). It is important to note that in many parts of Africa, protracted civil wars and civil unrest force people to seek shelter in Protected Areas where they exert pressure on the natural resources and compete with wildlife resulting in increased cases of HWC (Mizutani et al., 2005; Steinfeld et al., 2006). As discussed above, immigrants are less likely to protect wildlife which they view as a competitor.

4.4 Competing attitudes and perceptions

The majority of the respondents perceive wildlife as a threat and, in addition, natural resources as dwindling. However, most respondents also agree that if rules are observed, wildlife will be beneficial to communities. Thus, we conclude that there is room for sustainable management of natural resources including wildlife among the communities that live close to protected areas, but this still has a long way to go. In the meantime, HWC results in negative impacts on human social, economic or cultural life, on the conservation of wildlife populations, and on the environment, and play a major role in the perception of local communities neighbouring protected areas and ultimately, in the success of conservation strategies. Murombedzi (1999) reported that people living with wildlife and who encounter problems on a daily basis often display negative attitudes towards wildlife. They often perceive wildlife as 'good', only for their meat and feel insecure in terms of both quality of life and food insecurity and this sense of insecurity may exacerbate HWC (Murombedzi, 1999). The situation is exacerbated by insufficient revenue from wildlife to the community resulting in decreased tolerance levels from communities towards wildlife and the public or private owners in charge of their management (Mombeshora & LeBel 2009; Mupamhadzi et al., 2009). This attitude is highlighting a new dimension, the existence of a Social Carrying Capacity (SCC) which measures the sensitivity of local human populations to the presence of wildlife (McGinnis, 2008). The SCC names the Wildlife Acceptance Capacity (WAC: Decker & Purdy, 1988) as the population size of a species that the human community is willing to tolerate; it is a very subjective measure which needs to be taken into-account in HWC mitigation strategies (Woodroffe et al., 2005).

The negative attitudes of humans towards wildlife impact (Struhsaker, 1997) can become engrained in the minds of many people (Kock, 2003). For example, in Mbire district of Zimbabwe it has been reported that baboons and monkeys 'steal' food from people's houses, raid crops, 'destroy' granaries, kill small livestock like chickens and yet, both do not have any perceived value to locals (Samu, 2010). In other parts of the country, baboons and monkeys draw 'sympathy' from locals who associate them with autochthonous spirits which are believed to bring rainfall. Ethnicity plays a big part in shaping local attitudes towards wildlife. In Zimbabwe, wildlife areas are characterised by ethnic mosaics resulting from settlement patterns that were created by the colonial regime between 1950 and the late 1970s. For example, in the South East Lowveld (SEL) of Zimbabwe, people of different ethnic origins were forcibly moved from their areas and settled in these areas following creation of white owned large scale commercial farms (Mombeshora & LeBel, 2009). These ethnic groups pursue diverse livelihoods. For example, Shangaan people living in the SEL predominantly rear livestock while the Karanga and Ndebele ethnic groups prefer crop production. Most Shangaan people do not like the idea of living with wildlife because the latter attack their livestock, compete with livestock for forage and sometimes spread diseases to livestock. On the other hand, the Karanga and Ndebele peoples dislike wildlife because some wild animals raid crop fields. Similarly, in Mbire district the Karanga people who recently moved into the area grow cereals and cotton, and the Doma people, who are indigenous to the area and are traditional hunter-gatherers, only value wildlife as meat (Dzingirai, 1999). Linked to the above, landowners and users and even wildlife managers still sometimes deliberately kill wildlife they consider a threat with the view of reducing the problem or exterminating the problem animals species from their neighbourhood (Kock, 2003).

Some societies also advocate for the total eradication of lions which are viewed as useless and largely responsible for killing people and their livestock. Such views have contributed to the setting up of Problem Animal Control (PAC) programmes in CAMPFIRE (Communal Areas Management Programme for Indigenous Resources) areas of Zimbabwe. PAC entails killing or capturing of problem animals by officials from National Parks. However, people living in Campfire areas often complain about the way PACs are handled. Officials are often blamed for responding too late, especially during the off-hunting season or when the problem animal does not have a high trophy value. The manner in which PAC and Campfire dividends are handled by officials helps in creating negative attitudes towards wildlife by local people. An ongoing study being implemented by the Centre for Applied Social Sciences (CASS - UZ, Zimbabwe) has documented cases where locals in the South East Lowveld (SEL) have negative attitudes towards Campfire and wildlife in general. Local communities indicated that they have not received dividends from Campfire proceeds over the last ten years. Local authorities, especially rural district councils (RDCs) have been blamed for withholding proceeds from Campfire. This is understandable because of the hardships RDCs have been facing during the inflationary environment from 2004 to 2009.

4.5 Mitigation measures

Human-wildlife conflict is a complex problem, requiring a combination of approaches to manage the conflict (Fig. 25), including wildlife barriers, protecting property, traditional methods and removal of the specific problem animals (Nelson et al., 2003; WWF-SARPO, 2005; LaGrange, 2006; Parker et al., 2007; FAO, 2009). For any human-wildlife conflict management strategy to succeed, it must be sustainable and is therefore ideally administered by the local community itself (WWF-SARPO, 2005).

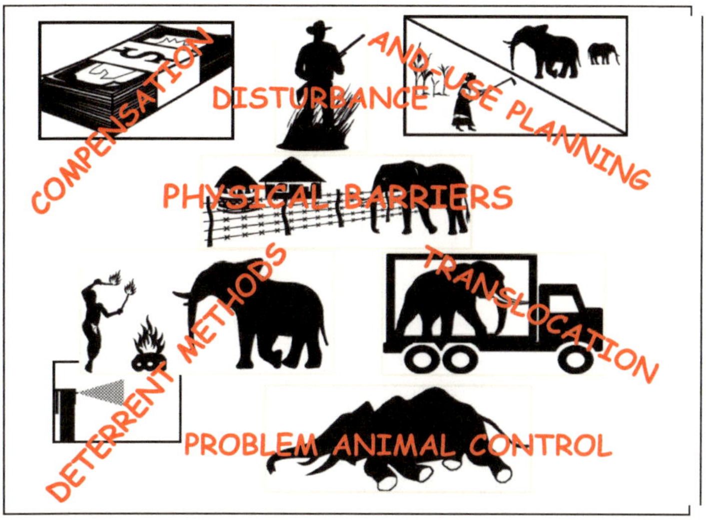

Fig. 25. Mitigation tools for HWC

Farmers' groups have been trained to use non-lethal methods, separated into three separate categories (Osborn & Parker, 2002; WWF-SARPO, 2005), including vigilance methods aimed at alerting farmers to the presence of approaching wildlife, and passive methods aimed at impeding the passage of potential crop-raiding animals using simple physical barriers and deterrents. The third method, an active one, is implemented to scare-off intruders using various forms of disturbance measures such as fires, noisemakers and chemical deterrents.

Most respondents of the Zimbabwean study were not satisfied with the response of local authorities to incidents of human wildlife conflict. In Mozambique, while farmers are deploying a range of solutions, game scouts are monitoring and killing rather than trying to prevent. In this regard, measures of PAC may need to be improved as they are presently perceived as unsatisfactory or weak.

4.6 Ways forwards: A new PAC approach

A number of initiatives aimed at reducing HWC and its related negative perceptions by humans towards wildlife have been proposed by governments and wildlife authorities and conservation groups (Katerere, 2005). The CAMPFIRE programme launched in Zimbabwe was considered as one of the key initiatives adopted to deal with HWC (Mukamuri et al., 2009). But the failure by Campfire to reduce poaching has led decision makers to suggest construction of buffer zones or erecting fences (Samu, 2010), ensuring, like the south part of the Limpopo National Park, that wildlife areas will not extend into agricultural land and vice versa (Magane et al., 2003). Another suggestion involves settling people away from wildlife areas such as National Parks and other protected areas (Brockington, 2002). However, a study by Brockington & Igoe (2006) documented failures of such resettlement policies as many villagers would return to graze their livestock as well as to have continual access to the wildlife meat. This finding illustrates the fact that no quick-fix or one-time solution to the problem of HWC exists (Duffy, 2000) and it requires a multipronged approach, a sustained effort and commitment from local residents and officials. In other words, the art of living with this problem is by increasing the human tolerance of wildlife and to decrease the negative interactions with it.

4.6.1 Learning from disease control

Based on strategies developed in response to the threat of emerging zoonotic diseases (Fromenty, 2011), we suggest a coping strategy developed by implementing an improvement of HWC data management as this is the backbone of any attempts to control and prevent HWC incidents.

This PACL approach will be based on a system combining Prevention measures, early Alert mechanisms, application of Control solutions with the objective of Learning from past experiences (Fig. 26). The early detection of HWC generates information, which alerts decision makers timeously, therefore allowing a quick and adequate reaction to control dangerous animals. By providing a complete set of explanatory variables, the flow of information will feed the HWC database. The analysis of it will help to improve the understanding of HWC incidents and improve their mitigation by targeted prevention campaigns. This last point is crucial if we are aiming to reduce the global prevalence of HWC. Previous studies have demonstrated that an adequate prevention campaign, enhancement of early warning and guarding efforts on previously raided farms can reduce incidents of crop raiding by 90% (Sitatai et al., 2005). The monitoring of this cycle will aid and improve the learning from past experience.

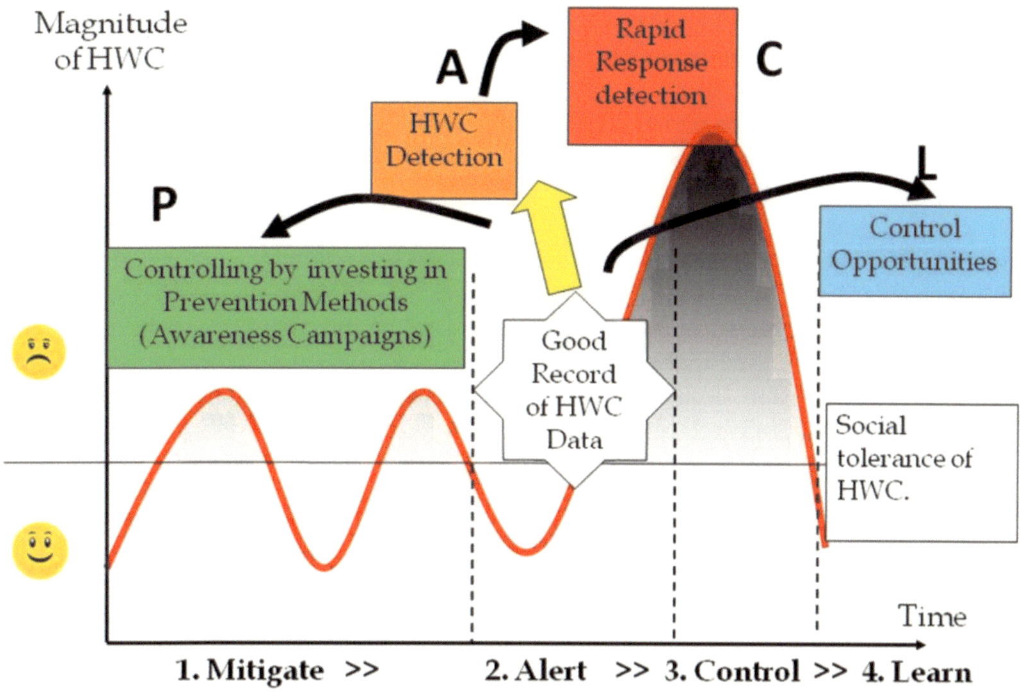

Fig. 26. A new PACL approach

4.6.2 Alerting and controlling in time

Monitoring of HWC and conflict mitigation measures has been weak to date, with unclear policies and lack of appropriate responses, this being largely the responsibility of wildlife authorities. Two initiatives are currently developed to address this gap. The first one is a user-friendly educational, management tool, named Management Oriented Monitoring System (MOMS) which was successfully developed in Namibia and introduced in Zimbabwe to monitor illegal offtake in the South East Lowveld of Zimbabwe. This tool, utilized for HWC monitoring in Namibia, allows immediately visualization of the consequences of certain management decisions which makes them also an excellent tool for conflict avoidance or mitigation involving stakeholders and decision makers (Stuart-Hill et al., 2005).

The second one is adding value on the latter in facilitating a quick flow of information, so reducing delays of reaction. Providing an immediate, direct, reliable, simple to use and relatively cheap method of data flow, it will inform local decision makers when urgent reactions have to be taken and increment local and national HWC data base. The proposed solution based on the following assumptions is using FrontlineSMS www.frontlinesms.com, free open source software used by a variety of organizations to distribute and collect information via text messages (SMS).

- Game rangers generate the information based on direct assessment of any HWC incident and utilize various means to communicate including mobile phone.

- The mobile phone is the nearest thing to a computer that anyone has. SMS, which is cheaper than traditional voice calls, is the dominant form of communication.
- From all district offices, game rangers have the ability to access a GSM network and to send and receive SMS.
- The local level of managing the information is either the head quarters of a National Park or local government focal point. The final destination of all HWC information is the National Data Base.

The case of Quirimbas National Park in Mozambique will help to explain how this approach can be implemented (Fig. 27). Data management needs to combine the management of data from Protected Areas by MITUR (Ministry of Tourism) and from non Protected Areas by MINAC (Ministry of Agriculture). The complex flow of information will be eased with FrontlineSMS and the automatic creation of feedback messages to game rangers, and forward messages alerting decision makers, while raw information will feed Park data bases at provincial level by SMS and a national data base by E-mail.

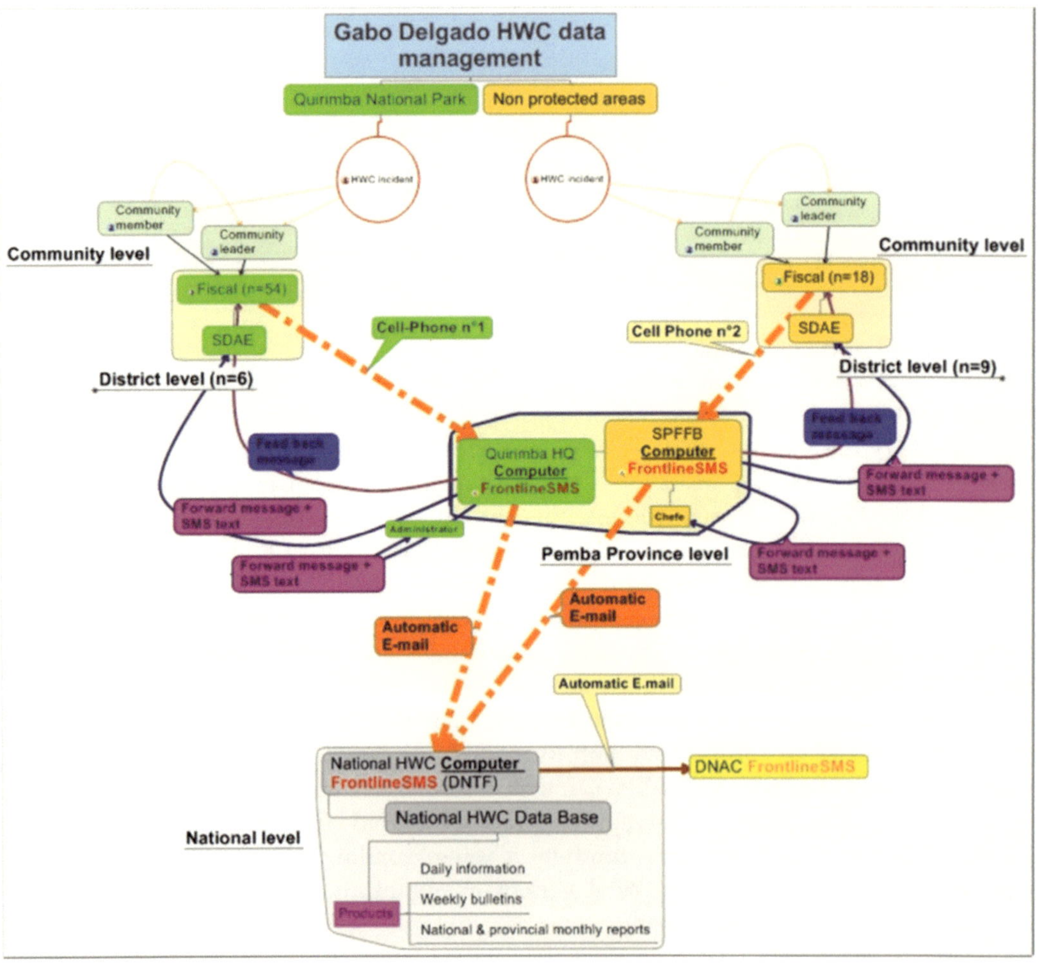

Fig. 27. Theoretical HWC data management in Cabo Delgado Province

4.6.3 Investing on prevention measures

Current methods of managing this conflict have been mostly external, requiring outside intervention, and centralized, requiring authority from a district or central authority. They have been both expensive (use of ammunition, guns, vehicles, etc.) and ineffective (e.g. using firearms to scare-off wildlife). For any HWC management strategy to succeed, it must be sustainable and should therefore be administered by the local community. Mediation and conflict resolution need to take into account the lack of connection between the official framing of conflicts and local responses, efficacy of traditional mechanisms of conflict resolution, competing and contradictory policies, local institutional complexity and its significance to conservation debates and interventions. A number of conservation initiatives aimed at reducing HWC and its related negative perceptions by humans towards wildlife have been proposed by governments, individuals and wildlife authorities have been put into practice with success and failures.

Fig. 28. The PAC options show the different solutions offered from Awareness to Lethal measures

To address some gaps in implementing mitigation measures, BIO-HUB and FAO has recently developed a HWC toolkit. This practical tool, developed for the reinforcement of local capacities, acknowledges that human-wildlife conflict is multifaceted and that some of the mitigation practices proposed to date are ineffective on their own over time. Therefore the toolkit presents tools and practices that can have great success when used in combination (Fig. 28). It is designed not only to help protect people, their livestock and their crops from wild animals, but also to safeguard wild animals from people. The HWC toolkit

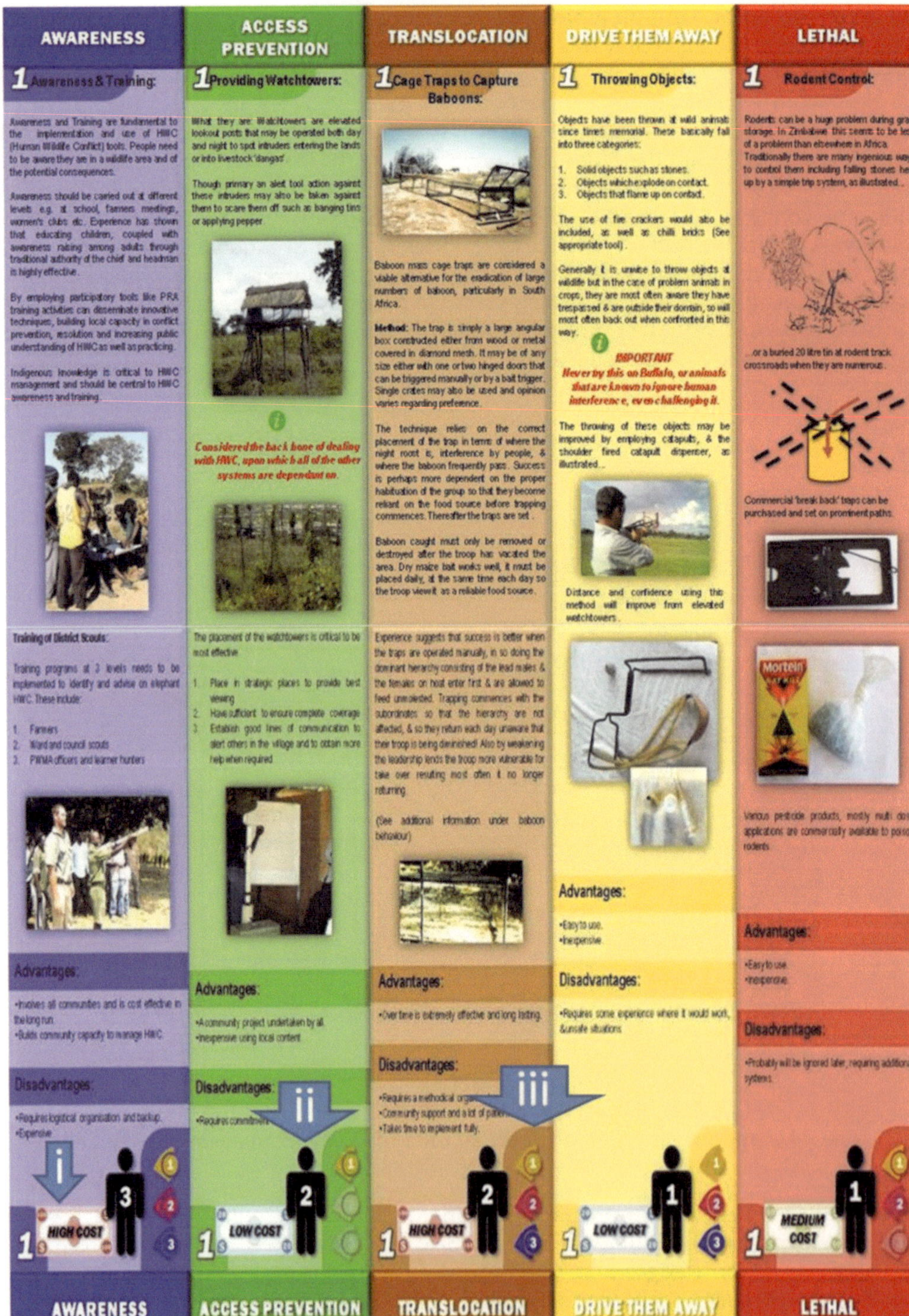

Fig. 29. First set of Tools. Information given at the bottom of each tool indicates (i) cost, (ii) manpower requested or (iii) level of implementation (1=village, 2=district, 3=national).

is designed for use by extensionists working with communities. In addition to presenting tools and practices that can have great success when used in combination, it guides users through HWC situations from incidence to resolution. By intervening in the conflict, a range of solutions become possible using simple to High-Tech devices. The objective is to show that simple and low cost solutions can be implemented at grass root levels before opting for lethal options (LeBel et al., 2010).

Each HWC toolkit is divided in 8 different sections: a Manual indicates how to use the toolkit. The Policy section summarizes what measures are allowed according to the national wildlife legislation which will vary between countries. The Reporting section teaches the recording of HWC incidents. The Environment booklet looks at the environmental situation: People, Village, Livestock, River/Waterhole and Crops/Fields, as the main areas of conflict. In each situation, the toolkit highlights factors of risks linked to human activities. The Wildlife booklet examines sixteen animals (predators, big and small herbivorous, birds, etc) considered predominant in HWC. Understanding the behaviour of each animal helps to pin-point potential conflicts and gauge requirements for mitigation measures. A combination of tools is put forward for mitigation in 5 colour-coded categories (awareness, access prevention, translocation, driving animals away and lethal control). The Tools section is the largest with more than 90 solutions proposed to reduce and control HWC incidents. Clear explanation is given on how to implement the control options in the five colour-coded categories (Fig. 29).

5. Conclusion

Even with low game density, HWC has a significant social and economic impact on communities living at the edge of Protected Areas and it is inevitable that HWC on the one side will increase with the free movement and growing populations of wildlife species in Southern Africa while on the other, the increasing demand for land for agriculture. This is even most evident in respect to elephant populations that have the ability to double in numbers within 15 years. If this situation is not addressed, so that humans and wildlife coexist, wildlife as a natural resource will be condemned to survive only in fenced protected areas.

In an attempt to improve the situation, we suggest an approach based on three pillars:

1. Appreciating the wildlife-domestic interface demands the knowledge of how wildlife move and utilize their home range in comparison to the use of this land for agriculture and wildlife perception by communities living there. Records of HWC activity documented using simple technology such as SMS will help to monitor this interface while determining the dynamics and magnitude of the problem.
2. Minimizing the negative impact of wildlife has to be based on the understanding of animal behaviour. The dissemination and utilisation of the HWC toolkit in its different versions will help communities to be more sympathetic and adopt less riskless attitude when confronting dangerous animals. Also it enables target intervention of specific problem animals through the concept of '*memory fences*' to facilitate wild animals respecting human activities and settlement (e.g. *discipline* of crop raiding elephant employing chilli pepper).
3. Improving community tolerance toward wildlife must start by capacitating the local communities to protect themselves. Building on existing *traditional approaches* through the HWC toolkit will promote a range of applicable solutions adaptable to their situation. Promoting and improving wildlife based-revenue ventures are considered essential to provide for long term mitigation and human-wildlife coexistence.

6. References

Ministério da Agricultura (2009). *Informe Sobre a implentaçao da Estratégia de Gestao de Conflito Homem-Fauna Bravia, de Janeiro a Dezembro de 2009.*

Anderson, J. and Pariela, F. (2005). *Strategies to mitigate human-wildlife conflicts in Mozambique. Wildlife Management Working Paper n°8,* FAO, Rome.

Araman, A. (2009). Human-Elephant Conflict in Northern Mozambique. *Human Wildlife Conflict Training of Trainers Workshop,* BIO-HUB, Harare: 14.

Barnes, J.I.T. (1998). *Wildlife conservation and utilizations. Complements to Southern Africa development.* DEA Research Discussion Paper 27. Windhoek, Namibia: 18.

Blanc, J.J., Barnes, R. F. W., Craig, G. C., Douglas-Hamilton, I., Dublin, H. T., Hart, J. A. and Thouless, C. R. (2005). *Changes in elephant numbers in major savanna populations in eastern and southern Africa.* Pachyderm, Vol. 38, 19-28.

Blanc, J.J., Barnes, R., Craig, G.C., Dublin, H.T., Thouless, C.R., Douglas-Hamilton, I. and Hart, J.A. (2007). *African Elephant Status Report 2007: An Update from the African Elephant Database.* Occasional Paper Series of the IUCN Species Survival Commission N°33. IUCN/SSC. African Ele. Spe. Group. IUCN Gland, Switzerland, 276 pages. : 276.

Breitenmoser, U., Angst, C., Landry, J.M., Breitenmoser-Würsten, C., Linnell, J.D.C. and Weber, J.M. (2005). *Non-lethal techniques for reducing depredation. People and Wildlife. Conflict or Coexistence?* S. T. A. R. Rosie Woodroffe. Cambridge, Cambridge University Press. 9: 49-71.

BRL ingénerie (2006). *Development of the Limpopo National Park and its Support Zone: Feasibility study Draft Report.* DBSA, AFD, Ministry of Tourism of Mozambique, Nepad Project Proposal and Feasibility Fund: 291.

Brockington, D. (2002). *Fortress conservation .The preservation of the Mkomazi game reserve Tanzania.* Oxford London.

Campfire Programme (2007). *PAC statistics 2002-2006.* Harare, Zimbabwe, Campfire Association.

Cumming, D. and Jones, B. (2005). *Elephants in Southern Africa: Management issues and options.* WWF-SARPO Occasional Paper.

Decker, D. J. and Purdy, K. G. (1988). *Toward a concept of wildlife acceptance capacity in widllife management.* Wildlife Society Bulletin 16(1): 53-57.

Distefano, E. (2004). *Human-Wildlife Conflict worldwide: collection of case studies, analysis of management strategies and good practices.* FAO: 30.

Duffy, R. (2000). *Killing for conservation: Wildlife policy in Zimbabwe.* Oxford, James Currey.

Dzingirai, V. (1999). *Human migration and natural resources management in communal lands: the case of Binga in Zimbabwe:* CASS. Harare, University of Zimbabwe. DPhil.

Food and Agriculture Organization (FAO) (2009). *Human-wildlife conflict in Africa: causes, consequences and management strategies.* Rome, FAO.

Fromenty, P. (2011). *From forecasting to control of zoonotic diseases - Linking animal and human systems.* OIE Global Conference on Wildlife Animal Heallth and Biodiversity - Preparing the Future. . Paris (France). Page 22.

Hulme, D. and Murphre, M. Eds. (2001). *African wildlife and livelihoods .The promise and performance of community conservation.* Oxford. James Currey.

International Livestock Research Institute (ILRI) (2009). *Climate livestock and poverty - challenges at the interface.* C. R. 2008-9. Nairobi, Kenya, International Livestock research Institute.

Katerere, J. (1997). *Parks without peace.* International Parks without peace conference on trans-boundary protected areas as a vehicle for international co-operation, Somerset West, South Africa.

Kock , R. A. (2003). *What is the famous wildlife disease interface ? A review of current knowledge for the African continent.* South and East African experts panel on designing successful conservation and development interventions at the wildlife /livestock interface and humans health (AHEAD) forum IUCN 5TH world parks conference, Durban, South Africa.

LaGrange, M. (2006). *The Capture, Care and Management of Wildlife.* Pretoria, South Africa, Van Schaik.

Lamarque, F. (2010). *Memorandum of the consultancy mission carried out in Maputo, Mozambique by Francois Lamarque, Program Officer - IGF Foundation.* IGF/FAO: 7.

LeBel, S., Mapuvire, G. and Czudek, R. (2010). *Human-Wildlife Conflict Toolkit: comprehensive solutions for farmers and communities.* Unasylva 236(61): 12-13.

Magane, S., Soto, B., Munthali, S., Schneider, M., Vicente, G., van Wyk, A., Nhalidede, A., Rode, P., Grossman, D., Holden, P., Kleibl, T. and Maluleke, L. (2003). *Limpopo National Park: Management & Development Plan.* F. Edition, Ministerio Do Turismo, Direcçao Nacional de Areas de Conservaçao: 96.

McGinnis (2008). Background on Urban Deer Management. Grand Haven, Urban Deer Management Plan: 5.

McGregor, J. (2005). *Crocodile crimes: people versus wildlife and the politics of postcolonial conservation on Lake Kariba, Zimbabwe.* Geoforum 36: 353-369.

Mizutani, F., Muthiani, E., Kristjanson, P. and Recke, H. (2005). *Impact and value of wildlife in pastoral livestock production in Kenya .Possibilities for health ecosystem conservation and development for the poor.* South and East experts panel on designing successful conservation and development interventions in the wildlife/livestock interface and human health(AHEAD) forum IUCN 5TH world parks conference, Durban, South Africa.

Mombeshora, S. and Le Bel, S. (2009). "Parks-people conflicts: the case of Gonarezhou National Park and the Chitsa community in south-east Zimbabwe." Biodivers Conserv 18: 2601-2623.

Mukamuri, B. B., Manjengwa, J. M. and Anstey, S. (2009). *Beyond proprietorship :Murphree laws on community based natural resources management in Southern Africa.* Harare, Weaver press.

Mupamhadzi, I. T., Matema, C. and Muvengwi, J. (2009). *Observations and Informal Findings on Wild Meat Trade: Socio-economic research with local communities in and around the Save Valley Conservancy, Zimbabwe.* Harare, PARSEL Project: 32.

Murombedzi, J. (1999). *Capacity Building: The Theoretical Relevance and Practical Importance to the Development of Common Property Institutions for Management of Natural Resources: A Southern African Perspective.* A paper presented on a workshop on Proceedings of Regional Natural Resource Management, Johannesburg, South Africa.

Murphree, M. J. (2005). *Counting sheep. The comparative advantages of wildlife and livestock, a community perspective.* South and East African experts panel on designing e successful conservation and development interventions at the wildlife /livestock interface and human health (AHEAD) forum. 5th world parks conference, Durban South Africa, IUCN.

Nelson, A., Bidwell, P. and Sillero-Zubiri, C. (2003). *A review of human-elephant conflict management strategies.* P. W. Initiative. Oxford University, Wildlife Conservation Research Unit: 25.

Osborn, F. V. and Parker, G. E. (2002). *Living with elephants II: A manual for implementing an integrated program to reduce crop loss to elephants and to improve livelihood security of small-scale farmers.* Mid Zambezi Elephant Project.

Parker, G. E., Osborn, F. V., Hoare, R. E., and Niskanen, L. S. (2007). *Human Elephant Conflict Mitigation: A Training Course for Community-Based Approaches in Africa. Trainer's Manual.* Livingstone Zambia & Nairobi Kenya, Elephant Pepper Development Trust & IUCN/AfESG.

Samu, N. (2010). *Securing farming systems and livelihoods in communal lands adjacent protected areas through human wildlife conflict management.* CASS. Harare, University of Zimbabwe. Masters Degree thesis.

Sitatai, N. W., Walpole, M. J. and Leader-Williams, N. (2005). *Factors affecting susceptibility of farms to crop raiding by African elephants: using a predictive model to mitigate conflict.* Journal of Applied Ecology 42: 1175-1182.

Steinfeld, H., Gerber, P., Wassenaar T. D. and Castelet, V. (2006). *Livestock long shadow.* Environmental issues and options, FAO.

Struhsaker, T. T. (1997). *Ecology of an African rainforest: Logging in Kibale and the Conflict between Conservation and Exploitation.* Univ. Press of Florida, Gainesville.

Stuart-Hill, G., Diggle, R. , Munali, B., Tagg, J. and Ward, D. (2005). *The Event Book System: a community-based natural resource monitoring system from Namibia.* Biodiversity and Conservation 14: 2611-2631.

Whyte, I. J. and W. Swanepol (2006). *An aerial census of the Shingwedzi Basin area of the Parque Nacional do Limpopo (Limpopo National Park) in Mozambique.* Scientific Report 02/2006, South African National Park.

Woodburne, S., Prangley, A. and Mabuza, L.M. (2002). *Socio-economic, Demographic, land-use and Attitudinal Survey of the communities residing in the Singuedzi River Bassin, Limpopo National Park, Gaza Province, Mozambique.* Peace Park Foundation, Ministry of Tourism: 201.

Woodroffe, R., S. Thirgood and Rabinowitz, A. (2005). *The impact of human-wildlife conflict on natural systems. People and Wildlife. Conflict or Coexistence?* S. T. A. R. Rosie Woodroffe. Cambridge, Cambridge University Press. 9: 1-12.

World Wide Fund for Nature Southern Africa Regional Programme Office (WWF-SARPO) (2005). *Human wildlife conflict manual.* Harare, Zimbabwe, WWF-SARPO.

16

The Cultural Weight of Nature: The Intra and Inter-Institutional Conflicts About Biodiversity and Ethnicity in Chile and Mexico

E. Silva Rivera[1], B. Contreras-Ruiz Esparza[1] and E. Parraguez-Vergara[2]
[1]*Centro de Investigaciones Tropicales*
University of Veracruz
[2]*Pontificia Universidad Católica de Chile*
Santiago
[1]*Mexico*
[2]*Chile*

1. Introduction

The rapid chemical and mechanical technification of production systems has greatly contributed to the current social, environmental and economic crises. Especially over the past decade, it has become even more evident that these transformations are putting at serious risk the integrity of the planet's social and ecological systems (MEA, 2005). At the beginning of the new millennium, the need to reflect about better ways to regulate and organize the current management practices, as well as analyzing how their effects can be buffered, has come consistently to the fore (Chapin et al 2006; Orstrom, 2007).

The negative effects of such technification are due greatly to the fact that there is not a specific set of laws and regulations that considers the planet's Socio-Ecological Systems (SSE) as a whole. In addition, technology has been applied within a competition based, market economy framework. Since the seventies it has become increasingly evident that this model favors unequal power relations resulting in a few individuals maintaining control over a few, jeopardizing the wellbeing of other social groups. This is the case of native people in Latin America that for diverse historic and cultural reasons have maintained their social, economic and political organization practices. Connected with this scenario, the market economy model has not been able to prevent the implications of overexploiting natural resources. The result has been irreversible biological and cultural diversity losses, as well as the dramatic reduction of survival possibilities for the current and future human populations (MEA, 2005).

In Latin America, this problem is acute. One of the central reasons is rooted in the history of Iberic colonization of American territories. During this process, the Spanish imposed their own land management strategies, as well as their ways to regulate the society over American societies and ecosystems that were radically different from European ones. After the new nations became independent, a large part of the new governments maintained segregation policies implemented during the colony. Only until 1957, native people were recognized internationally as citizens with the same fundamental rights as other human

beings, when the 107 Tribal People's Convention was signed. This was superseded in 1989 by the 169 Tribal People's Convention that says among other things that:
"Considering that the developments which have taken place in international law since 1957, as well as developments in the situation of indigenous and tribal peoples in all regions of the world, have made it appropriate to adopt new international standards on the subject with a view to removing the assimilationist orientation of the earlier standards, and
Recognising the aspirations of these peoples to exercise control over their own institutions, ways of life and economic development and to maintain and develop their identities, languages and religions, within the framework of the States in which they live, and
Noting that in many parts of the world these peoples are unable to enjoy their fundamental human rights to the same degree as the rest of the population of the States within which they live, and that their laws, values, customs and perspectives have often been eroded…"
The perception has taken a long time to permeate through Mexican and Chilean states, which despite signing the above agreements, still uphold practices that marginalize, weaken and fragment those cultures entailing valuable ancestral socio-ecological knowledge. Within the global change framework, and with the possibility of a collapse for many socio-ecological systems, it has become an overriding task to develop a solid, diagnostic method that will allow analyzing diversity processes and multiplicity of potential solutions about social and biophysical aspects. These methods are to tackle the large scale changes approaching (Orstrom, 2007). This chapter analyses two case studies in Mapuche (Chile) and Totonac (Mexico) communities. Both were Spanish colonies, however, their socio-ecological systems developed contrasting systems of governance. Along their history of interaction with the new government systems, we identify some of the socio-ecological factors that allowed them to resist the challenges that a radically different culture, and political system imposed.

2. Parallel historical processes

Traditional land management practices are the result of accumulated knowledge and experiences over hundreds of years, including data on useful species, their habitat, different types of soil, phenology, and the interactions amid different plants, and between them and animals. This knowledge allowed many cultures to develop highly complex agricultural systems (i.e. Gadgil and Berkes, 1991; Toledo et al., 2003). It has been suggested that studying these practices can be very useful in the design of the present management practices (Gadgil et al, 1993, Gómez-Pompa and Kaus, 1999).
Hundreds of species were managed for various purposes; for food, as fuel, for medicines, for construction work and artcrafts, among others. Agricultural cycles were coupled to the seasons, and marked by festivities and rituals that also maintained the sense of community and belonging, that fostered an interdependent connection with the ecosystems. In this context, human populations saw themselves as another element that interacted and contributed to vital cycles. Aside from having to endure the imposition of a management system completely different from theirs, clearly inadequate to the native system dynamics, a vast number of native people lost the right to use the land (Bengoa, 2003).
Currently, native people in Latin America share a common history of colonization and the consequent acculturation. Being such a large and diverse region, the ability of ethnic groups and *campesino* communities (rural farmers) to adapt and respond, has been determined by

the biophysical context in which political and social events have occurred. Despite having been marginalized and exploited, many resisted change and managed to conserve their culture and traditions at least partially. This trust, along with collective decision making and social networks of reciprocity, frequently based on family links, has been defined as 'social capital' (Bourdieu, 1985; Coleman, 1990). Putnam (1993, 1996) established some elements in social organization practices, for example networks of trust which facilitates action and cooperation among members from the same community. The emergence of this concept contributed to a more detailed understanding of power relationships among people living in vulnerable conditions, like *campesinos* or ethnic groups, and government and private sectors that possess greater economic and political power over the land and natural resources. We argue here that this has been mainly due to the existence of social organization practices based on trust.

'Social capital' also helped to define those elements that contribute to maintain cohesion, or those that cause fractures within the same community members. This discussion is central for understanding social and political conflicts in Latin America, as ethnic groups and *campesinos* that resisted change brought by colonists, have been historically marginalized, and have had restricted access to land and natural resources *de facto*. The concept of social capital (Bebbington, 2005) is also useful as it raises issues that are traditionally invisible to the dominant economic theories and policies.

'Social capital' is articulated in the cultural domain, where community members learn at an early age that there are expectations about their behavior as individuals that have a role in a community. The contributions that they make in the different social and productive areas are also determined by this sense of collectivity. The global economy has a strong influence on the ways in which social capital is strengthened or weakened, accordingly. Therefore, interactions among members of a society, for instance rural communities and government, become complex, due to different perceptions on what is important in the negotiation codes, language and cultural barriers.

Beside 'social capital', there are other key factors that help to maintain the integral wellbeing of rural communities, such as productivity, predictability, and ecosystem health. The fact that an ethnic group has its basic food, shelter, housing and health needs covered means that they are in a more advantageous position to negotiate land or economic and political power. Productivity in a system relates to carrying capacity, in this case, it would refer to the number of people that can be maintained per unit area. Another important element is predictability, which is very important for communities with an agricultural vocation, since it allows to plan when to plant and when to harvest, or, when to establish hunting and gathering seasons. Another determinant factor is extreme weather. It is not the same; to be able to produce food all year long than to only have the summer for covering this basic need. When an ecosystem keeps healthy, it has the capacity to be resilient. The endurance of this ability is vital, because a highly fragmented ecosystem that has been repeatedly disturbed will hardly absorb new disturbances.

With the approach of Socio-Ecological Systems, we make a comparative study of two case studies, one based in Chile, where 17 Mapuche communities live, in the margins of two natural protected areas. The other case study is based in Mexico, where 5 Totonac communities live inside a polygonal Archaeological Monuments Zone, located in lowlands mainly, with tropical rain forest vegetation. This site was declared World Heritage site by

UNESCO (ICOMOS, 14 December, 1992). Both case studies are considered touristic attractions for their scenic beauty. The relationship between the state and ethnic groups and the capacity to counteract change are compared (see Table 1 for important characteristics for both SSE).

	Totonac	Mapuche
Ecosystems where culture was developed	Weather. Warm and humid. Summer and Fall rains. Strong northern winds in the Winter. Temperature. 18.5° - 27.5. °C, Both limits tend to become more extreme (6° and 35°C). Tropical forest in the lowlands, and in the mountains, subtropical mountain forests. Topography. Broad valleys, with streams and mountainous chains. Agriculture viability. Fertile soils. Easy access to several ecological floors (e.g. the same farmer could maintain crops in lowlands and highlands). Products' availability all year long.	Weather. Temperate and humid. Rains all year long. January and February are the driest months. 1 m snowfalls. Temperature. 0º - 20º C Highest superior peak (30º C). Coniferous mountain forest, with wild pasture and lava covered soils. Topography: More than 3000 m volcanoes with valleys, lakes and rivers. During the rainy season, around 50% floods. Agriculture viability. Due to the steep terrain, only those areas closer to the valley are suitable for cultivation. Crops are cultivated in Spring and Fall only.
Main products	Subsistence: maize, beans, squash Economic: orange, banana	Subsistence: pinion (*Pinus cembroides*) Economic: *mosqueta*, berries, *murta, digueñes* (wild mushrooms).
Species uses	More than 300 edible species (endemic and introduced)	More than 100 edible species (endemic and introduced)
Land tenure	Precolonia: comunal Republic: private property and *ejidos*. Current situation: When the Archaeological site was decreed, land was expropriated and people were relocated in congregations.	Private. Chilean State: Reserves and Villarica National Park. Current situation: Communities recognised by the indigenous law. (private territories).

	Totonac	Mapuche
Population statistics and land use	Land use Forest cover ~10% Wild pastures and agricultural areas 75% Population density Precolonial[1] 56 hab/km^2 Current 160 hab/km^2 Totonacs represent 24% of municipal population (INEGI 2005).	Land use 40% forest cover (30% forest plantations). Agricultural area 30% Prairies and bushes 22.4% Bare soils or permanent snows 10.5%. Population density Precolonial[2] 1 hab/km2 Current: 6.3 hab/km^2 Mapuche represent 50% of Curarrehue population (2002). Population with moderately poor level 43,1%, Population living in extreme poverty 22,3%.
Ecosystem management	Integrated management system. Landscape units were managed in ~40 yrs cycles (Family vegetable gardens with consumption animals, maize, managed wild pasture (wood), regenerating acahual, incipient forest, mature forest). Vanilla and orange plantations, and wild pastures were added later on. The rotation systems implied that lands were communal but harvested products were private.	Economic agricultural system in forest clearings or riverbanks, with abundant gathering products (pinion, berries, fungus). Subsistence cattle ranching. Added to this are small scale agriculture and fruit trees. *Chilihueques* (llamas) were substituted by cows, goats and pork.
Impacts in the ecosystem impacts	High degree of deforestation mainly due to cattle ranching and soil pollution by PEMEX. 3,220 oil wells and many more in prospect. Chemical pollution.	In lands occupied by Mapuche and non Mapuche population, the main problem is deforestation. Soil erosion and nutrient lixiviation are caused by rain and floods.

	Totonac	Mapuche
Important events in the State conformation	Independence 1810 Revolution 1910 Neoliberalism 1982 N° 169 Indigenous and Tribal People's Convention year 1989. Zapatista Rebellion 1994 (Revindication of indigenous peoples rights) Indigenous law	Independence 18 September 1810. Nuevo Imperial Agreement 1989 (basis for the creation of the indigenous law). Indigenous Law 1993. N° 169 Indigenous and Tribal People's Convention, year 2008.
Important regional events	Three declarations that recognize the importance of Totonac culture: • UNESCO Declaration, 1992. Human Heritage and Monuments Zone. • Presidential declaration. 2000. Federal Law for Archaeological, Artistic and Historical Monuments. • UNESCO Declaration. 2010. Flyers ritual. Intangible Human Heritage. When the ZMAT obtained the decree, territories were expropriated, many people were relocated and the generated tension remains until today.	No events recognize the importance of the Mapuche culture. 1912 Villarrica National Reserve is created. 1940 Villarrica National Park is created.
Totonac social organizations	• CAPIT • Popular and *Campesino* Organizations Central (Cocyp) • Flyers Organizations • Juan Simbrón Cooperative (for pepper and maize leaves' commerce) • 3 sewers' and embroiderers' organizations • Artcraft sellers	Indigenous communities with juridical parsonality (mechanism to relate with the State according to the Indigenous Law). Ponuwemanke Association (10 communities to coopt the tourism bidding at the Villarica National Park). Artcraft workshops for men and women.

	Totonac	Mapuche
Institutions	A large number of social scientists from different Mexican universities apply research in the region. • Instituto Veracruzano de Cultura. • Intercultural University of the State of Veracruz • National Institute of Anthropology and History • Takilhsukut Thematic Park • Center for Indigenous Arts	Very few research studies, only in focal points, such as the Catholic University of Chile through the Villarica campus.

Table 1. Main characteristics and events in the Totonac and Mapuche SES

3. The Chilean and Mexican state

The following section presents an overview of public policies in Chile and Mexico, as in general terms both governance forms have much in common in how they relate with the different ethnic groups and their traditions, habits and values. Public policies have not been able to incorporate native people's cosmovision within the plans, norms, programs, and development projects. Neither the State has developed a coordinated and planned strategy over time. It seems that a central aim is to integrate native people to the State, without recognizing their identity. An example is those issues associated with development and quality of life of the Mapuche. On the one hand is the State offer from a Western perspective, and on the other are cultural identity demands (Arce, 2006). The result is a cultural clash, where in the end the power of decisions and the government's economic system impose.

On December 1994, in Chiapas, Mexico, the Zapatista uprising brought to the public eye fundamental social and political issues that affect American Indians. Themes such as autonomy, marginalization, discrimination and constant abuses from the landlords and the state itself were brought to the surface. Until now, these demands have not been solved. The Mexican government opted for a low intensity war (Proceso, 2004). On the other hand, due to national and international public demands, and funded by development organizations, the Mexican government has dedicated to promote clientelist projects. Government projects in the discourse aim to improve living conditions for indigenous communities. However, in practice the great majority have essentially failed to hamper or even reduce poverty.

In general, decision making and action implementation from public institutions is extremely complex. The first obstacle is function duplicity, not only among institutions but also within them. One example is conventional tourism initiatives against rural tourism. Secretaries or ministries frequently develop local programs that clash with other institutions connected with indigenous groups that have similar interests. These examples evidence the sectorial character and functions' duplicity by which public institutions work. Institutional interventions cover sectorial aims, without coordinating with the rest of public institutions.

At the local level, invested resources end up unstated, and with few positive outcomes. Since results lack follow up and assessment, they are recurrently replaced by new development initiatives. At the same time, local development strategies homogenize broad regions, blocking each region's specific potentials. Political and administrative decisions respond to the centralized form in which the state functions. Undoubtedly this represents an obstacle in communication channels with local actors; hence policies are designed in a top-down manner: there are no evaluation or feed-back mechanisms. The latter causes more problems than those that were supposed to be solved originally.

In Mexico, indigenous groups have organized and acquired legal and organizational knowledge to defend their rights. They have had support from human rights groups, from the organized civil society, and from broad *mestizo* population sectors. However, with regards to access to natural resources, there have been diverse conflict points. The basic interest of ethnic groups is to conserve natural resources and use them in the same way they have done for centuries. This management form is not considered viable for the market economy, neither for development plans. Therefore, for indigenous people and their descendants, defending the right to the land has had high social costs, causing jailing and even murder of their members.

This lack of power for decision making is another factor that accentuates vulnerability of indigenous people (Westra, 2008). The meagre representation lies in the fact that the elaboration and development of public instruments linked with indigenous people is reduced to public consultations that only include the public institutions component. Surprisingly, the right to participate in decision making is not considered in Chile's indigenous law. There is a National Commission for Indigenous Development (CONADI) and it is through its Advisor Council how the interests of indigenous people are veiled. In Mexico, the issue of representation is very similar. The directives of the National Commission for Indigenous people are defined by the president, and are not of direct ethnic origin.

4. The Mapuche

The Mapuche live in mountainous ecosystems, in valleys and near the sea. Due to the extreme seasonal conditions with harsh winters and floods, they did not develop a complex agrarian system, they just used the riverbanks and forest clearings to plant potatoes and '*pallar*', a native potatoe which required little soil preparation. The *piñon* (a pinenut, *Araucaria araucana*) was, and still is the most important food product, that used to be stored during the winter, and would be complemented with berries, mushrooms, game and fish. *Chilihueques* (llamas) were another food source for self consumption. Because they were not connected with an established agrarian system, they had mobility, migrating seasonally, a characteristic of hunters and gatherers. Historic records describe them as strong, healthy and bellicose, which reveals sufficient food availability conditions. The "bellicose" adjective derives from the defense they made of their territories against the Spanish for more than three centuries (Bengoa, 2000).

The Mapuche occupied the whole Araucania region until the "Pacification" process that ended in 1883; when they lost the war against the colonizers and were forced to live in reservations. The criteria for the establishment of these reservations were defined by the

Chilean state. Many Mapuche ranaway from the central valley to the mountains that were not populated due to the extreme weather. Some asked for citizenship permits. This is how our study site, Curarrehue, was born. Their traditional organization consisted of a *lof*; and the hierarchy was a political chief (*lonko*), another figure in charge of health and religion (*machi*, mainly) and a chief is case of war (*toqui*). In 1993, with the promulgation of the Indigenous Law, it was established that community representatives would lose their historic ways of organization and replace them with the Western hierarchies: president, treasurer and secretary. The law determined that the new hierarchy should renew every two years. This shift has caused division among communities, stopping them from maintaining the juridical personality in the National Comission for Indigenous Development. Hence, they struggle to access public development plans promoted by the State. Since the return to democracy in Chile, in 1990, the Indigenous Law and the CONADI, have set the regulatory basis for various initiatives with the objective to restitute lands and to launch local development plans. Understandably, this meant the introduction of new productive activities, different from the practices traditionally performed by the Mapuche.

Sixty percent of the political-administrative area of Curarrehue belongs to protected areas. This is the case of the Villarrica National Reserve and Park, and this explains why development projects are tourism oriented. Productive activities are subsistence agriculture, subsistence cattle ranching, and forestry. However, as the native forests in Araucania, have been reduced, and there are restrictions in the present normative, the latter activity has lost importance. In the *comuna* there are not many differences among the Mapuche and the *mestizo* population, although Mapuche have to live in the reservations. Nevertheless, some even have moved to cities like Curarrehue, Temuco (regional capital), Santiago (the capital), and the frontier area with Argentina, in the search for work opportunities. Soil erosion and biodiversity loss, in addition to roads in bad condition are synergistic factors that make poverty more acute during winter, increasing financial dependency from the State. Both Mapuche and *campesino* population, classified with a high poverty indexes work their fields in the traditional way, while those considered to be moderately poor, have introduced new productive systems, and have accessed the municipality's equipment for which they have to pay. Production remnants are sold, and this activity is complemented with informal jobs that are also badly paid. Another sector of Mapuche and non Mapuche have jobs as technicians and other informal jobs in Pucón, Villarrica or Temuco. In the field they do minimum work, or are hired by other members of the community.

The Mapuche belonging to the lower economical tier receive State benefits, through a series of social assistance programmes. The ones considered moderately poor also have access to State social development programs, but additionally, they have implemented activities connected to tourism, artcraft production, honey, preserves and other products. To obtain these benefits, they have constituted different social organization forms. Through social organization, seven communities have gotten access to natural resources from Villarrica National Reserve, such as *piñon* (pinenut) recollection and *veranadas* (sheep ranching). However, in 2007 a conflict arose due to the exploitation of 12,8 has of Oregon pine, in the same reserve. The government proposed 60% of the benefits for local communities, and 40% for the State, but local communities were asking for a 70%. Finally the discussion scaled to a regional level and a political decision was made. At the beginning, communities felt marginalized, but when they started receiving benefits, they conceived further basic demands, however most of them have not been possible to fulfill given the present protected area regulations. Another conflict aspect is the management of the Villarrica National Park. The state called the park for licitation with touristic purposes. For this reason, ten communities came together as a legal community organization and now they are in direct competition with the private enterprise for the park bidding.

5. The Totonac

The Totonac lived in several towns in the mountains and the lowlands in northern Veracruz, they shared language, cosmovision and culture. The Lords (*señores*) ruled in each town and collected tributes in form of work and spices. When the Spanish arrived, the region was dominated by the Aztecs that collected taxes. The Totonacs allied with the Spanish against the Aztecs; hence the Totonac did not have the 'defeated' status that other native groups had. This alliance was broken afterwards, and a new tax and control system was established by the *criollos* (Spanish born in America). Nevertheless, many Totonac families remained in power and established new alliances allowing traditional productive systems and their government forms to endure without great changes. Many new crops were added to the system. Until today, citric and banana plantations have an important economic value in the region. During this time, community leaders were restricted to religious and political posts, but these positions served as an ethnic and political resistance strategy. Religion became syncretic, as in many other areas of Mexico. Catholic festivities were adjusted to the planting and harvesting soil rituals.

However, during the XIX century, these posts were disbanded and reduced to the sacramental spheres (Masferrer, 2005). Also a series of rebellions related with landlord abuses and religious prohibitions unleashed. Between 1830 and 1910, vanilla commerce hit the highest peak, exporting to Europe, which meant economic bonanza, and the configuration of new power groups, both Totonac and *mestizo,* that came to dominate and transform social, economic and political life in the region (Kouri, 2000). The latter had deep implications for the Totonac, because the vanilla boom meant a reorganization of the old land tenure models. Vanilla growers associated with the wealthier Totonac landlords and were able to fragment and militarize the communal Totonac lands (called *'condueñazgos'*); in this way they took possession of the vanilla plantations (Kouri, 2000; Masferrer, 2005). The loss of land and the alliance of some Totonac with the *mestizos* produced the social fabric rupture that had resisted for so many years (Kouri, 2000). After the Mexican Revolution, *ejidos* and communal lands were legally established. Nevertheless, in the sixties, a great deal of the vanilla plantations, still privately owned, were sold to cattle ranchers, that represented the new economic alternative.

For centuries, community representatives were elected under the uses and customs system, where the community representative was assigned by common agreement. However, this election has been lately politicized, and currently, several candidates affiliated to political parties are proposed and voted. This has contributed to fracture the social fabric even more, as many do not recognize them as their representatives and therefore do not respond to their calls. Changes in land tenure had deep implications in the ecosystem. Traditionally, crop lands where open by slash-burn, used for 3 years and then left to recover to forest in a 40-50 years rotational cycle. But currently, as paddocks are left out of the landscape rotation units, large areas of land are permanently open. With land use intensification, and plot reduction, *campesinos* were forced to apply a non-rotational system. As a result only some tropical forest fragments remain. Aside from being strongly degraded, these fragments are surrounded by paddocks and occasionally, maize fields.

Over the last decades, Totonac have been receptors of development programs that provide limited resources, which coincided with a boom in oil exploitation in the region since the seventies. Currently, thousands of new oil exploration platforms are being opened, and each one requires at least a hectare for their exploitation. Proprietors are receiving significant payments for allowing the oil company to use the land, which is cleared and polluted. Situations like these bring new instability in the area: land prices rise, and food and basic goods also become more expensive due to the arrival of many workers from the oil companies, that live there on a temporary basis, but cause an intensive social, economic and environmental impact as long as the oil wells are producing.

Despite this problematic, in El Tajín, the majority of the land was privately owned until 2001, when it was expropriated by the government after the Archaeological Monument Zone was decreed. People were relocated in congregations with 300 m^2 plots in average. The Totonac lost once more their lands, a situation that currently generates greater conflicts within local people and the National Institute of Anthropology (Instituto Nacional de Antropología, INAH, in Spanish), the institution responsible of keeping and maintaining those sites of archaeological importance in the country. This is another example of intra-institutional conflicts, as another function of INAH, is to steward the living cultural heritage. However, on the positive side, efforts are taking place to implement a management plan, lead by INAH and CITRO, that looks for ways to better communicate with local people's needs by incorporating different disciplinary perspectives.

With the lack of land, the majority of Totonac people and their descendants have looked for other subsistence forms, as paid workers for other landowners, as construction workers, bus and taxi drivers, shop and oil company employees, or as art craft sellers in the archaeological site. This last occupation has allowed an income increase for many families living inside the archaeological site polygon. However, it has broadened a gap within the agro-ecological system that in the past used to be their subsistence means and deeply rooted in their cultural identity. On the other hand, the opening of the archaeological site to the general public, aside from becoming an important income generating activity, has revived interest for the Totonac culture, bringing international recognition (Table 1). It was mentioned that since the 1994 Zapatista uprising took place, indigenous cultures were projected in the global scenario. This revival has had a direct local impact, motivating private and governmental initiatives for the establishment of programs focused on promoting the knowledge preservation and, in this case, Totonac values and traditions. Currently, young Totonacs are at a crossroad, between the two worlds, that of their ancestors, and the present communications era. In front of them is the challenge to preserve the cultural and natural legacy inherited from their parents.

Despite the multiple historic and political vicissitudes, basic organization forms have remained, and the ethnic-political chiefs have been able to maintain very discreet but effective mechanisms to consolidate their leaderships. In nearby *ejidos,* agricultural practices subsist, and family vegetable gardens in the studied communities are still rather diverse, despite their small size (Contreras and Silva, 2010). The socio-ecological system is once more in a re-structuring stage, but this time, the internal social structure and the ecosystem have been greatly fragmented. Hence, there is a need to find the appropriate means to establish connectivity within the elements in the system.

6. The past and present of ethnic groups in the context of environmental governance

The historic process that has taken the Mapuche and the Totonac to the current situation, has been broadly similar in terms of the great events that influenced them and modified their socio-ecological systems. They both received the impact of the Spanish colonization, however, in different historical moments. In Totonac territories, there were Spanish settlements since 1530, but it was until the end of 1800 that the Mapuche lands were conquered by the Spanish. Although Mexico's and Chile's independence took place in the same year, 1810, the way in which each of these groups has responded, first to colonization and later to the imposition of a development model centered in private property, competitiveness and efficiency, has not been the same (see Table 1).

In the following section, we will compare some aspects of the biophysical environment in which both ecosystems evolved. The Totonac, geographically located in the neotropical belt, lived in a lower montane forest ecosystem. The weather is warm and humid, favoring the presence of fertile soils and water. The Totonac had easy access to several ecological floors that allowed them to grow crops both in lowlands and highlands. In this way a complex and very productive agricultural subsistence system emerged. Whereas the Mapuche, living also in a mountainous ecosystem, at higher altitudes, face extreme weather conditions and a temperate climate, with a very steep terrain, floods and heavy snowfalls which reduce the area suitable for cultivation. Therefore, they developed an economic system based on a variety of seasonal activities, fruit and wild plants recollection, fishing, hunting and nomadic cattle ranching.

The Mapuche society had reached a high degree of complexity in its political and social organization. They had access to abundant resources and a natural and biological ordinance that was regulated to the inside of large extended families, through a specialized power system, that regulated conflicts for the establishment of alliances, and consisted of an advisor and main judge called *ulmén* (Bengoa, 2000).

Development policies are based mainly on land tenure, however, the traditional property concept used to be different. Groups maintained hunting and collecting territories (Mapuche), or communal lands for agricultural management (Totonac) over which they implemented their hunting, collecting and cultivating rights. This type of tenancy allowed flexibility and adaptation capacity, as groups could freely move within their territories. The Mapuche would do it seasonally, according to the fructification months, and animal migration. Agricultural activity for Totonacs allowed the rotation of landscape units within a 40-50 years cycle, to let the land rest and obtain resources from vegetation at different regeneration stages. In literature, the importance of establishing management agreements has been constantly debated (Agrawal, 2002). The efficiency of these agreements is essential because groups can evaluate how actors comply with the agreed conventions, and the kind of response given by each resource to this type of management. In practice, this calibration between management rules and the ecosystem was facilitated by mobile management systems where actors moved through the different biogeographical regions, and under a temporary scale that allowed them to observe the consequences of their actions.

Today, as they are being restricted to a fixed territory, it is more difficult to relate the causes of the management forms with the consequences, since they are using fast changing new technologies and share space with other management forms (for example, forestry or cattle ranching). Another aspect is the delayed effect of agrochemicals in the impoverishment of the soil, as they break the structure and the biotic community. The first harvests are rather abundant, but afterwards, inputs need to be increased to maintain sufficient production levels. However, there is a conscience based on the experience of many years of using agrochemicals, that chemicals 'kill the soil', and that many species that would flourish spontaneously in the maize fields, are now disappearing (Silva-Rivera, 2004).

Ecosystem productivity and type of weather are other important aspects to consider when designing management strategies. Despite the current systems are much more connected within them - as many foods come from other regions-, the system resilience to economic or energetic crises, depends greatly from the regional food sustainability. The amount of food that can be produced is limited by the carrying capacity of the system. El Tajin region, is more densely populated than Villarrica (160hab/km^2 versus 6 hab/km^2), has a higher carrying capacity, producing hundreds of kilograms of maize, beans and squash per hectare, as well as tons of orange or bananas, generating food and securing income each year. The Mapuche population density, despite being very low, has increased 6 times with regards to density in the pre-colonial era. In the case of Totonacs, although having a higher population density, there has been only a 2.7 increase, keeping the system closer to the carrying capacity limits. Even though the Mapuche today have vegetable gardens and poultry, a lot of their collecting and production practices are still done traditionally. In addition, although the collection of berries and *mosqueta* roses generate income for mapuche families, as they are not being processed, they imply lower sales prices. This fact, added to the exponential increase in population density, has meant that the carrying capacity of the territory has been taken close to its limits. This is how the high poverty index is explained, among other motives. The latter makes them more vulnerable and dependent of governmental assistance hence less able to negotiate than the Totonacs.

Despite the majority of people living in El Tajin polygon lost their land, they are surrounded by productive ejidos, which leaves them with the ability to reach a great diversity of food products at very accessible prices that they occasionally can have through barter or gifts. In both productive systems, new species have been introduced. The utilization of endemic and native species is associated with beliefs, practices and traditional cultural knowledge on the environment that surrounds them. However, the use of introduced species shows the transculturation process that both groups have gone through since the colonial times. Species assimilation has meant new innovations, knowledge and commercialization processes. In the Totonac case, the assimilation of more than 41 new species has taken 500 years, and began with the insertion of orange and banana Spanish plantations. During this time, species were tested and incorporated into the productive systems. This assimilation reveals a general knowledge on how to domesticate and adapt plants to the system, a characteristic that distinguishes tropical areas, recognized as many species' domestication and diversification centers.

In the Mapuche case, climate is much more extreme and the state has been the main mediator of change through recent local development projects. The latter has caused strongly adapted and resistant species and varieties to be left aside. The extreme weather is an important factor that makes them dependent on state subsidies and benefits, limiting their possibilities to maintain more resilient and autonomous livelihoods.

Dependence on subsidies is also shared with the Totonacs, as they require government support for the installment of basic services like water, electric energy, sewage and streets. The State has not created the regulatory policies for more efficient and democratic governance; as a consequence, the dialogue with local actors becomes more difficult. This model should be based upon a communication system that flows in both directions, however, it is reduced to environmental assessments and literature reviews. For instance, despite the extensively documented evidences pointing at how inefficient it is to maintain bovine cattle in the tropics, every year, governments support and encourage such initiatives. The simple ratification of agreements and international treaties is insufficient if they do not go along internal norms that generate communication mechanisms within these groups and the State.

These regulation mechanisms should also consider other actors in the society; such would be the case of private businesses. On the other hand, is the process of competition for natural and financial resources between actors. This is what happens with the Mapuches, and the various associated projects with ethnotourism, or special interests tourism. The inexistence of regulatory mechanisms not only generates unequal competence among actors, it also increases dependency of Mapuches from the State financial resources.

The process turns even more complex when it comes to a resource that is a nation's patrimony, as in the case of the Villarrica National Park and Reserve. Currently, land overexploitation by these groups means new erosion processes, biodiversity losses, etc. In consequence they have no choice but to look for new strategies, increasing pressure over vital local resources. Biodiversity valuation mechanisms imposed by the State only consider economic factors such as profitability or income. However, ethnic groups' cosmovision values them in ways that go beyond the Western utilitarian ones, recognizing human populations as another component in the larger ecosystem. As long as alternative visions on the relationship of humans and nature are not acknowledged, the State clientelist mechanisms will continue to be perpetuated and local development projects will remain alien to the environmental conditions and cultural patterns of a given territory. The consequences add on to the multiple causes that will continue to put in risk the integrity of social and ecological systems.

7. References

Arce, M. (2006). La dimensión del conflicto mapuche. Políticas y estrategias (102): 110-119.

Agrawal, A. (2002). Common Resources and Institutional Sustainability. In: Drama of the Commons, ed. Committee on the Human Dimensions of Global Change, National Research Council. National Academies Press, Washington, DC. P:41-86. URL: http://www.nap.edu/books/0309082501/html/.

Bebbington, A. (2005). Estrategias de vida y estrategias de intervención: el capital social y los programas de superación de la pobreza. In: Arriagada Irma, ed. Aprender de la experiencia. El capital social en la superación de la pobreza. CEPAL, Santiago. pp:19-57.

Bengoa, J. (2000). Historia del pueblo mapuche: Siglo XIX-XX. 6ª. Edición. Biblioteca del Bicentenario, Santiago. 423 p.

Bengoa, José. (2003). Historia de los antiguos mapuches del sur: desde antes de la llegada de los españoles hasta las paces de Quilín; siglos XVI y XVII. Santiago: Catalonia. 524 p.

Bourdieu, P. (1985). The Forms of Capital. In: Richardson, ed. Handbook of Theory and Research for the Sociology of Education. Grenwood, New York. P:241-258.

Chapin, S., A. Lovecraft, E. Zavaleta, J. Nelson, M. Robards, G. Kofinas, S. Trainor, G. Peterson, H. Huntington and R. Naylor. (2006). Policy Strategies to Address Sustainability of Alaskan Boreal Forests in Response to a Directionally Changing Climate. Proceedings of the National Academy of Sciences 103(45):16637–16643.

Coleman, J. (1988). Social capital in the creation of human capital. American Journal of Sociology : 95-120.

Coleman, (1990). Foundation of Social Theory. The Belknap Press of Harvard University Press.

CONADI. Comisión Nacional para el Desarrollo Indígena. Retrieved from: http://www.conadi.gob.cl/. Last accessed: 14 July, 2011.

Contreras, B. y E. Silva-Rivera. (2010). The diversity, economic and nutritional value of Totonac homegarden produce. Conference Proceedings of The 51th Annual Meeting of the Society for Economic Botany. Xalapa, México. p 45

Gadgil, M. and Berkes, F. (1991). Traditional Resource Management Systems. Resource Management and Optimization 18 (3-4): 127-141.

Gadgil, M., Berkes, F. and Folke, C. (1993). Indigenous knowledge for biodiversity conservation. Ambio, XXII (2-3), 151-156.

Gómez-Pompa, A. and A. Kaus. (1999). From prehispanic to future conservation alternatives: lessons from Mexico. Proceedings of the National Academy of Sciences 96: 5982-5986.

ICOMOS. Comité Nacional Mexicano del Consejo Internacional de Monumentos y Sitios. Retrieved from: http://www.icomos.org.mx/cultural.php. Last accessed: 14 July, 2011.

Kouri, E. (2000). La vainilla de Papantla: Agricultura, comercio y sociedad rural en el siglo XIX. Signos Históricos II 3: 105 -130.

MEA. (2005). Millennium Ecosystems Assessment. Ecosystems and Human Well-being. Synthesis Resport. Washington, D.C.

Masferrer Kan, E. (2005). Los dueños del tiempo. Los tutunakú de la Sierra Norte de Puebla. Fundación Juan Rulfo, México.D.F. 319 p.

Orstrom, E. (2007). Sustainable Social-Ecological Systems: An Impossibility? Annual Meeting of the American Association for the Advancement of Science, "Science and Technology for Sustainable Well-being," 15-19 February, 2007. San Francisco, CA. Retrieved from http://ssrn.com/abstract=997834. Last accessed: 15 May, 2011.

Proceso (2004). Los muertos no se olvidan (Editorial). *Proceso*, Año 28, Edición Especial 13, 6-10

Putnam, R. D. (1993). The prosperous community: social capital and public life' in the *American Prospect*, 4:13

Putnam, R. D. (1996). The Strange Disappearance of Civic America, in the *American* Prospect, 7: 24

Silva-Rivera, E. (2004). Campesino producer organisations and sustainability in Mexican coffee production. PhD Dissertation. University of East Anglia, Norwich, UK. 403 p.

Toledo, V., B. Ortiz-Espejel, L. Cortés, P. Moguel, and M. Ordoñez. (2003). The multiple use of tropical forests by indigenous peoples in Mexico: a case of adaptive management. Conservation Ecology 7(3): 9.

Tribal People's Convention. C169 Indigenous and Tribal Peoples Convention, 1989. Retrieved from: http://www.ilo.org/ilolex/cgi-lex/convde.pl?C169. Last accessed: 14 July, 2011.

Westra, L. (2008). Environmental justice & the rights of indigenous peoples. International & domestic legal perspective. Earthscan, London. 315 p.

17

Biodiversity and the Human Factor – The Need to Overcome Humankind's Addiction to Growth

Philip Lawn
Flinders University
Australia

1. Introduction

It is often believed that biodiversity loss is principally due to poor environmental management or the failure to preserve critical ecosystems. Although these factors are critical, the greatest threat to biodiversity conservation is unquestionably humankind's addiction to growth. Economies are subsystems of the larger ecosphere and dissipative structures in the sense that they must continuously digest low-entropy resources and excrete high-entropy wastes to maintain their physical order (Georgescu-Roegen, 1971; Perrings, 1986; O'Connor, 1991; Daly, 1996; Common & Stagl, 2005). Because of thermodynamic limits to both materials recycling and the technical efficiency of production (Ayres & Miller, 1980), the continued growth of economic systems requires an ever-rising rate of resource throughput that must eventually exceed the ecosphere's regenerative and waste assimilative capacities. Whether we like it or not, all attempts to continuously grow our economic systems must inevitably deplete the natural capital that supports them as well as the critical ecosystems that contain much of the planet's biodiversity (Jansson et al., 1994; Lawn, 2000, 2007; Victor, 2008).

The increase in the rate of throughput, as the global economy expands, and its degenerative impact on natural capital and biodiversity can be represented by way of a comparison between the Earth's biocapacity and humankind's ecological footprint (see Figure 1). The global ecological footprint constitutes the area of land *required* to service humankind's production and consumption desires, whereas the Earth's biocapacity constitutes the area of land that is *available* to service them (Wackernagel et al., 1999). Between 1965 and 2005, humankind's ecological footprint continuously rose. Worse still, humankind's ecological footprint began to exceed the Earth's biocapacity in the mid-1980s (Global Footprint Network, 2008). That is, humankind's demands on the biosphere eventually surpassed the Earth's capacity to support them on a long-term sustainable basis. In effect, since the mid-1980s, humankind's ability to meet its production and consumption desires has only been possible by running down stocks of natural capital and, in the process, eroding the Earth's biodiversity.[1]

---

[1] Natural capital can still be depleted, even when humankind's ecological footprint is less than the Earth's biocapacity, if natural capital is exploited imprudently.

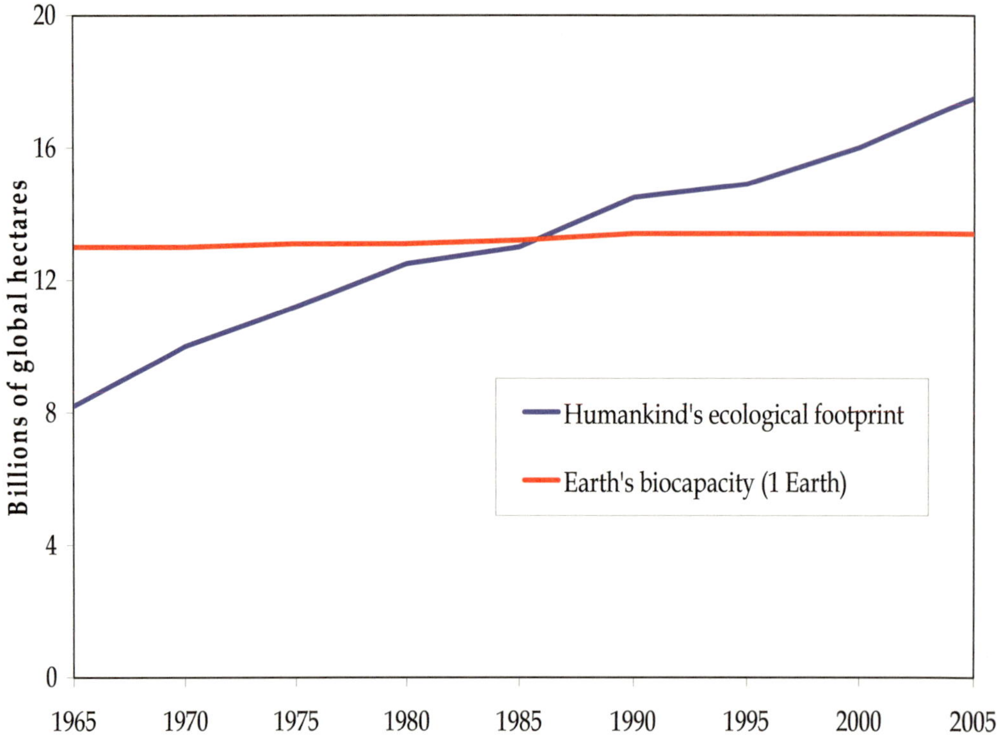

Fig. 1. Humankind's ecological footprint versus the Earth's biocapacity - 1965-2005 (Source: Global Footprint Network, 2008)

The important connection between growth of the global economy and biodiversity loss is reflected by the fact that the rise and trend change in humankind's ecological footprint between 1965 and 2005 corresponds almost exactly with the rise in real Gross World Product over the same period. Indeed, the more temperate rate of increase in the ecological footprint in the early-1980s and early-1990s coincided with global output recessions when Gross World Product also increased at more modest rates.

Clearly, in spite of efficiency gains, conserving biodiversity will require all nations to eventually make the transition to a steady-state economy - that is, a physically non-growing economy maintained by an ecologically sustainable rate of resource throughput (Daly, 1991). Whilst, for many impoverished countries, some further growth is both possible and desirable, for wealthy nations, the need to make the transition to a steady-state economy is required immediately and likely to involve the physical shrinking of their economies (Latouche, 2007; Martinez-Alier, 2009; Kerschner 2010; Martinez-Alier et al. 2010). There is little doubt that the need to make such a transition will encounter considerable political and institutional barriers. However, this only highlights the urgency of the problem at hand and the extent to which humankind has failed miserably to deal with the problems of excessive growth.

Putting aside political realities for a moment, there are four critical questions that need to be answered in relation to macroeconomic policy and its connection with biodiversity:

- To what extent does contemporary macroeconomic policy contribute to the growth of economic systems and the subsequent loss of biodiversity?
- Can macroeconomic policy be implemented to ensure the scale of economic systems remain within the limits imposed by the regenerative and waste assimilative capacities of the ecosphere and the need to preserve critical ecosystems?
- If the answer to the above question is no, what basic policy instrument needs to be implemented to achieve the condition of ecological sustainability which is necessary to preserve the planet's biodiversity?
- Where does this leave the role of macroeconomic policy?

2. Contemporary macroeconomic policy and its relationship with growth

Before answering the first question, let me say a few things about contemporary macroeconomic policy.

Macroeconomic policy exists at two levels. The first level involves 'internal' fiscal and monetary policies, while the second level involves the 'external' policy area of international trade. Fiscal policy primarily relates to the expenditure decisions of central governments as well as the impact of taxation on private-sector spending. Monetary policy, which involves manipulatory intervention in money and bond markets, is decreasingly being conducted in terms of money-supply targets. It is now almost exclusively conducted by way of interest rate adjustments to either facilitate or dampen private-sector spending. In most countries, monetary policy is carried out by central banks acting independently of the central government. Central banks usually perform their monetary policy role with the aim of achieving a desired inflationary target band inscribed in the central bank's charter (Dornbusch & Fischer, 1990).

Whilst international trade policy can involve the central-government impositions of tariffs, import quotas, and foreign exchange regulations, contemporary international trade policy is increasingly characterised by minimal government intervention in global markets and the relatively fluid movement of international financial capital across international borders. As a consequence, international trade outcomes are determined primarily by private-sector agents engaged in international transactions. This is not to say that central governments do not implement policies to influence international trade outcomes. Central-government policy is often aimed at increasing the international competitiveness of domestic industries in order to boost net exports. Unfortunately, policies designed to increase competitiveness are not always desirable and invariably reflect the detrimental impact that globalisation forces are having on domestic policy (Daly, 1996). I shall return to this issue later in the chapter.

Whether it is internal fiscal and monetary policies or the external policy area of international trade, macroeconomic policy is essentially directed towards achieving the maximum 'inflation acceptable' growth rate of real GDP (Gross Domestic Product). By 'inflation acceptable' I mean a growth rate which does not lead to price-inflation exceeding the upper end of a desired target band (usually 3% per annum). Although central governments prefer a low unemployment rate, it is because of the desire to avoid an excessive rate of price-inflation that full employment ceased to be a major objective of macroeconomic policy in most countries during the 1970s. Thus, by stopping short of encouraging a rate of growth that leads to excessive inflation, most governments are content to accept a higher unemployment rate than what would generally be regarded as full employment.[2]

A minority of economists are critical of this 'fight inflation first' approach to macroeconomic policy insofar as it is results in a 'sacrificial' pool of unemployed labour (Forstater, 2000; Mitchell & Watts, 2002). Advocates of the mainstream position respond by asserting that macroeconomic policy aimed at achieving full employment leads to an unacceptably high inflation rate that, in turn, leads to macroeconomic instability and prolonged periods of even higher rates of unemployment. Hence they argue that the current approach is the lesser of two evils. Critics of the mainstream position disagree and believe that a non-inflationary full employment outcome can be achieved through the implementation of an appropriately designed and government-financed 'employer of last resort' programme (Wray, 1998; Mitchell & Watts, 2002; Mitchell & Muysken, 2008).[3]

You may well ask what the employment implications of macroeconomic policy have got to do with biodiversity conservation. I believe a great deal since, in the case of industrialised nations at least, economies organised to operate sustainably will need to be much smaller in scale than they are at present and this could lead to potential difficulties if associated with very high rates of unemployment. Should society regard the ensuing unemployment rate too repugnant to accept, it could inhibit the transition to ecological sustainability and thus undermine conservation efforts.

Irrespective of the economic debate surrounding employment and inflationary issues, the general flavour of all mainstream macroeconomic policy is essentially the same - it is unashamedly pro-growth. Given the link between growth and natural capital depletion, there is little doubt that contemporary macroeconomic policy contributes significantly to the growing pressure being exerted on the ecosphere and the subsequent loss of biodiversity. To make matters worse, the lack of efficient 'cost internalisation' policies - which is partly due to the globalisation forces arising from the international trade aspect of contemporary macroeconomic policy - is further increasing environmental pressure by allowing damaging activities to proceed without the perpetrators incurring a financial penalty commensurate with the spillover costs they impose on the rest of society.

I might also point out that ecological economists believe there is an 'economic' limit to growth that is likely to be arrived at prior to reaching any ecological limits. The reasoning for this is that the costs of growth - in particular, the costs associated with the depletion of natural capital and any accompanying loss of biodiversity - typically rise at an increasing rate (principle of increasing marginal costs). Conversely, the benefits of growth typically increase at a declining rate (principle of diminishing marginal benefits). Thus, well before the physical scale of the economy becomes ecologically unsustainable, a point is reached where the additional costs of growth exceed the additional benefits. As a consequence, growth beyond this point reduces a nation's economic welfare.

Studies using a Genuine Progress Indicator (GPI), which is a recently established indicator of economic welfare, appear to support this contention (Redefining Progress, 1995; Lawn & Clarke, 2006). The studies reveal that the economic welfare enjoyed by wealthy nations has

---

[2] A full-employment rate is not 0%. Because there are always people moving in and out of jobs, there is always some 'frictional' unemployment. Full employment is usually regarded as being somewhere near a 2% unemployment rate.

[3] A good example of an 'employer of last resort' programme is the Job Guarantee described in detail in Mitchell & Muysken (2008).

either plateaued or been in decline for some 30-40 years despite the almost continuous rise in GDP (Daly & Cobb, 1989; Max-Neef, 1995; Jackson & Stymne, 1996) (see Figure 2).

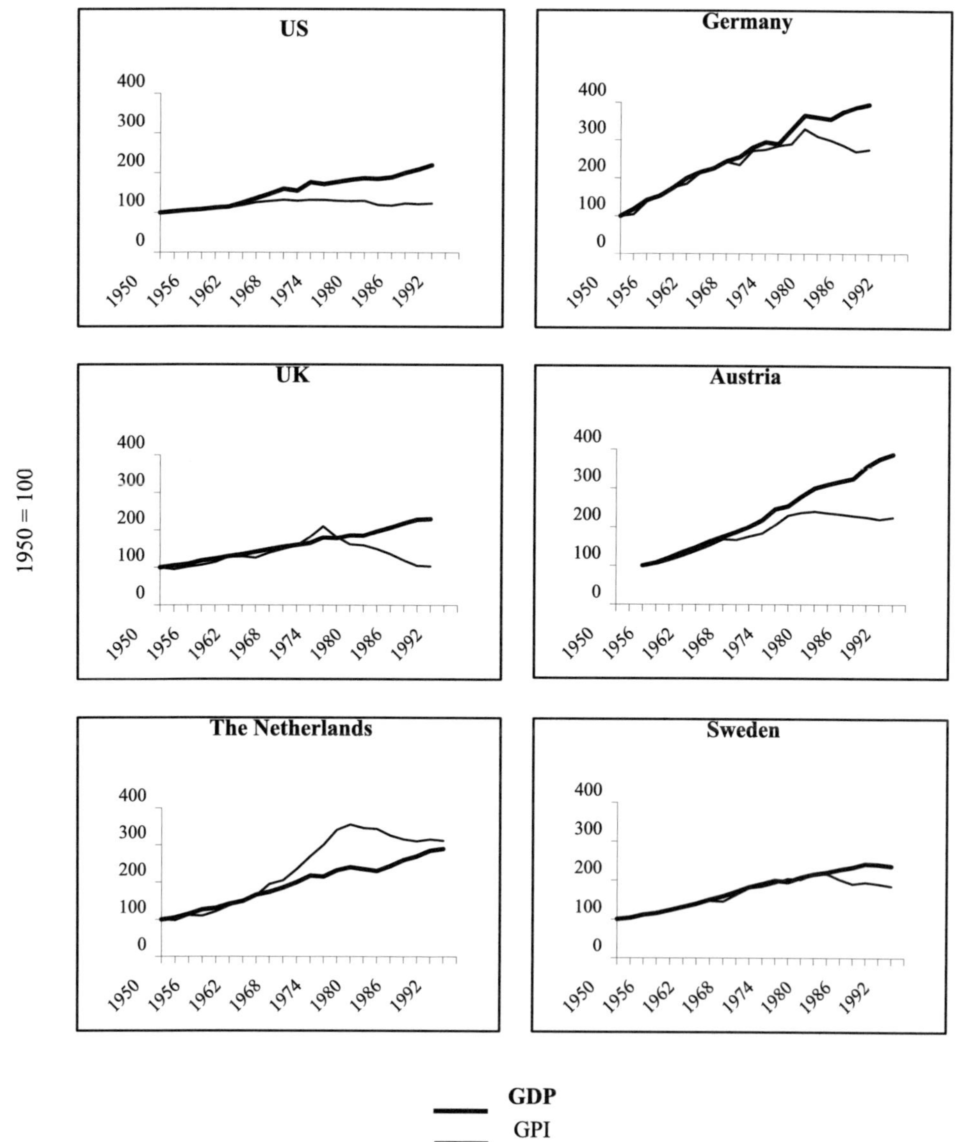

Fig. 2. A comparison of the per capita GDP and per capita GPI of six industrialised nations (Source: Jackson & Stymne, 1996)

Although the timing of the peak of the GPI varies from country to country, what doesn't change, but is not evident in Figure 2, is that the per capita GPI almost always starts declining when the per capita GDP of a nation reaches somewhere around US$20,000. When this trend was first recognised in the mid-1990s, it led Manfred Max-Neef (1995) to put forward a *threshold hypothesis* regarding the growth of a nation's economy and the economic welfare it generated. Max-Neef argued that a per capita GDP of approximately US$20,000 constituted a threshold income level at which point continued growth of a nation's economy would reduce the economic welfare enjoyed by the average citizen.

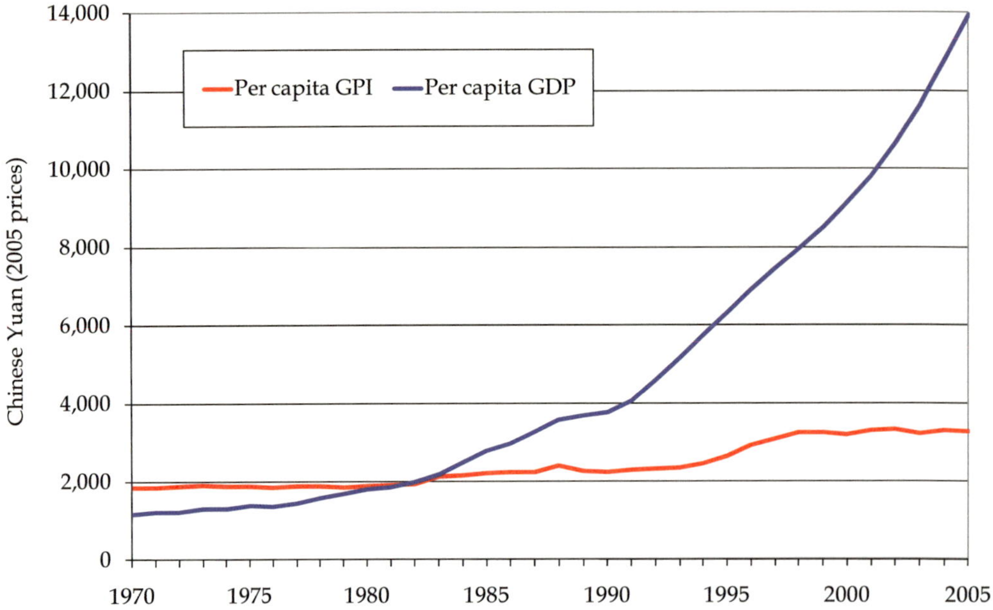

Fig. 3. Per capita GPI versus per capita GDP - China, 1970-2005 (Source: Wen et al., 2008).

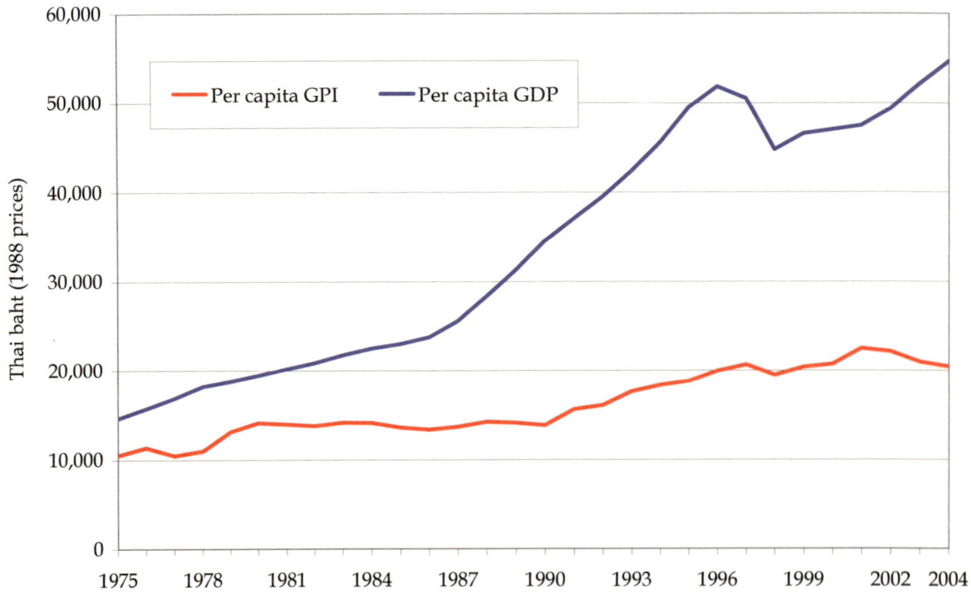

Fig. 4. Per capita GPI versus per capita GDP - Thailand, 1975-2004 (Source: Clarke & Shaw, 2008).

As disconcerting as the threshold hypothesis seemed for the rich countries at or beyond the threshold income level, it appeared to offer great comfort to the world's poorer nations. Since all poor countries had a per capita GDP well below US$20,000, the threshold hypothesis suggested that the growth of their per capita GDP would increase the economic welfare of its citizens for some considerable time. Distressingly, a recent GPI study of seven countries in the Asia-Pacific region - four of which are relatively poor nations - casts doubt over this prognostication. Consider Figures 3 and 4, which show that the per capita GPIs of China and Thailand have begun to fall despite the per capita GDP levels of both countries being well short of the US$20,000 envisaged by the proponents of the threshold hypothesis. Indeed, China's per capita GPI peaked when its per capita GDP was around US$5,000. For Thailand, its per capita GPI peaked at a per capita GDP of around US$7,500.

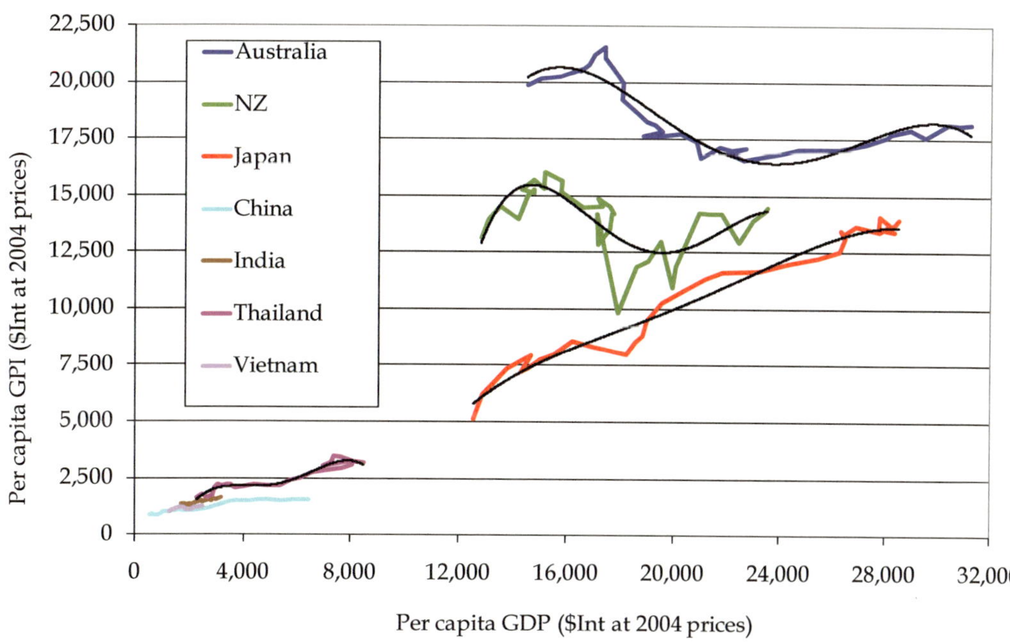

Fig. 5. Per capita GPI versus per capita GDP of seven Asia-Pacific countries (Source: Lawn and Clarke, 2008)

Now consider Figure 5, where the annual per capita GPI values of the seven countries included in the Asia-Pacific study have been plotted against their corresponding per capita GDP values. The figure reveals that the three wealthy countries of the region - Australia, New Zealand, and Japan - along with China and Thailand, have all reached a threshold level of per capita GDP. Although India and Vietnam, with a per capita GDP of around US$3,000 and US$2,500 respectively, have yet to reach a threshold point (i.e., their per capita GPI is still rising), both countries are experiencing substantial increases in environmental and social costs (Lawn, 2008; Nguyet Hong et al., 2008). It is not unreasonable to assume that both countries are likely to cross a threshold point at a lower per capita GDP than Thailand and China. From the results of this GPI study at least, there is little doubt that the later a nation experiences an initial expansion phase of its economy, the lower is its per capita GDP when its per capita GPI begins to decline. This is no better exemplified than by the tunnelling of each country's per capita GPI-GDP curve below that of their growth predecessor.

Lawn and Clarke (2008) believe that the phenomenon revealed in Figure 5 can be mostly explained by: (a) low consumption levels in poor nations relative to their domestic production – a consequence of their large net exports of goods and services; (b) the migration of manufacturing operations to poor countries where wages are low and environmental regulations are weak; and (c) growth late-comers having to contend with GDP expansion in a world full of human beings and human-made capital, yet one with much less natural capital and many fewer ecosystems. Whereas (a) reduces the marginal benefits enjoyed by poor countries and (c) increases the marginal cost of an increment of GDP growth, (b) results in poor nations having to bear a disproportionately large share of the world's social and environmental costs.

It is because of the above factors that Lawn and Clarke (2008) have proposed a new *contracting threshold hypothesis*. The hypothesis is essentially this: as the economies of the world collectively expand in a globalised economic environment, there is a contraction over time in the threshold level of per capita GDP. As such, growth late-comers (poor nations) face the prospect of never attaining the level of economic welfare enjoyed by the early growth-movers (rich nations).

Despite this new hypothesis, Lawn and Clarke still believe it is possible for poor nations in the Asia-Pacific region and the world generally to experience higher levels of economic welfare. However, they argue that progress will only occur if an extension can be made to the threshold point at which the per capita GPI of poor countries begins to decline. This, according to Lawn and Clarke, will necessitate dramatic policy changes on the part of poor nations. Just as importantly, it will require rich nations to cease growing their economies (i.e., make the transition to a steady-state economy) in order to provide the 'ecological space' that poor nations need to enjoy a phase of welfare-increasing growth.

All in all, apart from having to make the transition to a steady-state economy to achieve ecological sustainability, it is becoming increasing apparent that wealthy nations must stabilise their economies at a much smaller physical scale in order to maximise the economic welfare enjoyed by their citizens. Once this scale has been attained – referred to by ecological economists as the optimal macroeconomic scale – the emphasis of economic activities should shift from quantitative expansion (growth) to qualitative improvement (development), where the latter would involve improvements in the stock of physical goods, a more equitable distribution of income and wealth, minimisation of the rate of resource throughput via increases in the rate of recycling and the technical efficiency of production, and reorganisation of the production process to increase job satisfaction and reduce the cost of unemployment, crime, and family breakdown. All such advances would enable economic welfare to increase without the need for further growth. What's more, they would allow businesses to increase profits without the need for continuous businesses expansion, thus allaying any concerns that a steady-state economy is incompatible with capitalism (Lawn, 2011a).

3. Achieving ecological sustainability and biodiversity conservation

If contemporary macroeconomic policy can be largely blamed for the growth that threatens biodiversity conservation, it ought to follow that growth can be slowed or halted via a radical alteration of the fiscal, monetary, and international trade policy stances of central governments. Although international trading arrangements warrant a serious overhaul, a radical redirection of fiscal and monetary policy stances is not the solution. Certainly, the growth process could be reversed by implementing macroeconomic policy in such a way as to totally destabilise the economy. But sabotaging a growth economy is hardly the appropriate means of dealing with the dilemma that humankind confronts.

Why, then, is macroeconomic policy unable to halt the biodiversity-eroding growth of economic systems? It has already been argued that the scale of economic systems must remain within the limits imposed by the need to preserve critical ecosystems and the regenerative and waste assimilative capacities of the ecosphere. These limits are determined by biophysical criteria, yet macroeconomic policy is designed to meet economic criteria and thus has no capacity to achieve biophysically-based targets. Macroeconomic policy cannot, therefore, directly solve the biodiversity erosion crisis.

Mainstream economists will be the first to admit that macroeconomic policy cannot satisfy biophysical criteria. However, they will usually respond by arguing that there is no need to align macroeconomic policy with ecological constraints because an alignment of this nature is automatically achieved at the microeconomic level. According to mainstream economists, if there is any possibility that the growth of the economy may lead to an excessive rate of resource throughput, it will be adequately registered in individual resource markets and will induce appropriate remedial action. That is, if we value ecological sustainability and the biodiversity conservation that comes with it, the increased scarcity of natural resources will cast a shadow in the form of higher natural resource prices. This will induce greater resource use efficiency and lessen our demands on the stock of natural capital per unit of economic activity. In doing so, the rise in natural resource prices accompanying any increase in natural resource scarcity will automatically ensure that the necessary stock of natural capital is maintained, including the critical ecosystems containing a great deal of the planet's biodiversity. Hence, regardless of the macroeconomic policy being undertaken, it will only end up being expansionary if the increase in efficiency induced by rising natural resource prices permits the production of a larger quantity of goods and services from what the market will always ensure is a sustainable rate of resource throughput.

It is at this point that mainstream economists come unstuck. The belief that increased resource scarcity will immediately result in a rise in all resource prices is erroneous (Hall & Hall, 1984; Reynolds, 1999; Lawn, 2007, 2010). Resource prices are very good at reflecting the *relative* scarcity of resources (i.e., how scarce a particular resource is relative to a substitute resource). But resource prices are woefully inadequate at reflecting the *absolute* scarcity of each resource type (i.e., how much remains of each particular resource) and are even worse at reflecting the services provided by nature's waste sinks and life-supporting ecosystems. This is because ecological sustainability is based on the need to meet ecological criteria and markets are only capable of satisfying economic criteria (i.e., the efficiency criteria). What's more, even if resource prices do eventually rise and induce greater efficiency, there is nothing inherent in markets to prevent the percentage increase in efficiency from being overwhelmed by the percentage increase in economic output - a phenomenon known more widely as the 'Jevons' Paradox' (Jevons, 1865; Blake, 2005). Hence, there is nothing to prevent the rate of throughput from rising and therefore nothing to prevent the stock of natural capital from subsequently declining. Finally, market decisions are made by currently existing people who have a natural tendency to discount the future ramifications of their present actions. This leaves future generations - the people who will suffer most in a resource-poor world - unable to partake in the current resource bidding process. Thus, market decisions are always biased against future generations.

Put simply, markets have no ears, noses, and eyes to sense an ecologically sustainable rate of resource throughput just as they cannot sense an equitable distribution of income and wealth. As such, no economic forces exist to ensure economic systems operate sustainably.

There are, of course, many actions that need to be taken to ensure economic activity is ecologically sustainable and the need for sound environmental management, ecosystem preservation, and the preclusion of human-beings from the direct use of a relevant portion of the total ecosphere have already been mentioned as important requirements. As I have argued, all will prove futile unless the rate of resource throughput is kept within the limits imposed by the ecosphere's regenerative and waste assimilative capacities. Since markets and economic instruments cannot satisfy ecological criteria, the policy instrument required to resolve the resource-limiting aspect of sustainability dilemma must exist in the form of society-imposed restrictions on the annual rate of resource throughput.

A policy instrument of this nature would be more complex than first imagined. To begin with, it would require a unique resource extraction regime for each resource type. Furthermore, the extraction regime for each resource would need to differ from one geographical location to another (e.g., tree species 'X' could thrive in both dry and moist environments but regenerate more slowly in the former). As for many non-renewable resources, restrictions would need to be determined by the rate at which renewable resource substitutes can be cultivated to keep intact a combined stock of resource-providing natural capital. In other instances, limits on the incoming resource flow would be dictated by a paucity of environmental sink capacity.

Although the imposition of throughput constraints involves the establishment of a policy instrument external to the market domain, there is no reason why it cannot be combined with the policy instrument of relative prices to maximise efficiency. This can be achieved by introducing a system of tradeable resource use permits, or what is often referred to as a 'cap-auction-trade' system. There is insufficient space to outline the full details of such a system except to say that the cap on the number of permits sets a throughput constraint on economic activity (see Lawn, 2007, Chapters 11 & 13). This ensures the economy at all times operates sustainably. Meanwhile, the initial auctioning and subsequent trading of the permits by a government authority ensures that permit prices, by reflecting ecological limits, serve as an absolute scarcity tax. The tax-like feature of the permit prices enables a central government to achieve two things. Firstly, it provides a central government with an opportunity to reduce the marginal tax rate on low incomes, which not only shields low-income citizens from the impact of higher resource and energy prices, but also narrows the income gap between rich and poor. Secondly, it encourages the efficient allocation of the capped (sustainable) resource flow under existing technology while also facilitating the development of new resource-saving technology.

I should also point out that a cap-auction-trade system can also be extended to various forms of pollution to ensure certain types of waste do not exceed the ecosphere's capacity to absorb them. Generally speaking, the system need not be applied to all forms of pollution because, in keeping with the first law of thermodynamics (the law of conservation of matter and energy), limits imposed on the input of low-entropy resources automatically imposes limits on the output of high-entropy wastes. However, a cap-auction-trade system is ideally suited to certain types of waste that are more difficult to regulate via caps on resource inputs. A good example of this are greenhouse gas emissions, where not only can a cap prevent emissions from exceeding dangerous levels – unlike a carbon tax – the price paid for emissions permits can encourage the development and uptake of pollution-reducing technology. Permit prices can also facilitate the gradual shift from non-renewable resources to renewable resources.

4. The role of macroeconomic policy in a throughput-constrained economy

One of the major ramifications of imposing throughput constraints on the economy is its impact on expansionary macroeconomic policy. Let's assume that the central government is of the view that the prevailing unemployment rate is too high. Believing that an economic stimulus would not lead to an excessive rate of price-inflation, it implements an expansionary monetary policy. Such a policy would immediately bring about a rise in the demand for low-entropy resources. However, with a cap-and-trade system in place, there would be no corresponding increase in the low-entropy resources supplied in resource markets. The resultant excess demand for low-entropy resources would almost certainly drive up the price of resource permits and raise the cost of resource use in production. This, in turn, would increase the general price level of all goods and services and therefore reduce real income. To cut a long story short, it is possible, as a consequence of implementing an expansionary monetary policy in a throughput-constrained economy, for real GDP to decline and for unemployment to rise - the opposite outcome to the one intended (see Heyes, 2000; Lawn, 2007, Chapters 13 & 14).

The altered dynamics surrounding the eventual fall-out of macroeconomic policies implemented in a throughput-constrained economy is not something to be taken lightly. I alluded earlier to the idea that bad macroeconomic policy can undermine conservation efforts if society rejects the ensuing deterioration of economic outcomes and reverts to the growth objective. It is therefore critical that we gain a greater understanding of the impact of macroeconomic policy in a throughput-constrained economy. To date, very few researchers have undertaken work in this area and this includes ecological economists who have long stressed the need for nations to make the transition to a steady-state economy.

From the small amount of work conducted on this issue, there are three things worthy of note. Firstly, except for impoverished nations that require a dose of efficient and equitably-distributed growth, fiscal and monetary policy should not be expansionary. Nor should it necessarily be contractionary since any economic contraction required to align the rate of resource throughput with the ecosphere's regenerative and waste assimilative capacities is best left to the introduction and gradual tightening of a cap-auction-trade system. In addition, healthy government expenditure levels are necessary to maintain public goods and critical infrastructure as well as to keep unemployment low during downturns of the business cycle, all of which are necessary to soften the social impact of transitioning to a steady-state economy.

Of course, if there is a gradual tightening of a cap-auction-trade system and a consequent reduction over time in the rate of resource throughput, there is no need to be concerned that the maintenance of strong expenditure levels would promote the undesirable growth of the economy. Indeed, all that healthy government expenditure levels would effectively do is alter the macro-allocation of resources towards the public sector (i.e., increase the public sector's share of real GDP relative to the private sector). Again, since public sector output tends to involve the provision of public goods and the redistribution of services towards the poor, this maybe necessary to ensure that the transition to a steady-state economy is socially acceptable. Overall, the importance of maintaining government expenditure indicates how potentially disastrous it could be to use contractionary macroeconomic policy as means of discouraging growth – even more so given that it does not guarantee ecological sustainability and biodiversity conservation.

Secondly, as a fiscal instrument, taxes are best used to alter the nature of consumer behaviour and economic activities rather than as a means of controlling the aggregate level of private-sector spending.[4] There are two reasons for this. In the first instance, the level of aggregate private-sector spending is best modified at the margin by the price-influencing impact of a cap-auction-trade system (i.e., as it alters the purchasing power of household disposable income). In the second instance, the tax burden should be redirected towards such 'bads' as resource depletion and pollution and away from such 'goods' as income and profit (a policy commonly referred to as *ecological tax reform*). Whilst the former discourages environmentally destructive activities, the latter rewards and thus facilitates increases in efficiency and value-adding in production (Gale et al., 1995; Daly, 1996; O'Riordan, 1997; Lawn, 2007). Incidentally, the imposition of a tax on depletion and pollution would be achieved to a large extent by the cap-auction-trade system given that resource permit prices are, as already mentioned, the equivalent of an absolute scarcity tax.

Thirdly, international trade policy should be focused on restricting the international mobility of financial capital. The massive increase over the past thirty-five years in the fluidity of financial capital movements has transformed the global economy from a federation of independent, trade-linked, national economies into one large economy devoid of national economic boundaries where production-location decisions are often determined by the desire of transnational corporations to avoid high wages and stringent environmental and workplace regulations. This globalisation phenomenon has emerged because the increase in financial capital mobility has shifted the principle governing international trade from 'comparative advantage' (which is based on the relative cost of production) to 'absolute advantage' (which is based on the absolute cost of production) (Daly and Cobb, 1989; Ekins et al., 1994; Daly, 1996; Lawn, 2007). In a world where comparative advantage is the dominant principle, lower absolute costs arising from lower standards confer little if any trade advantage because competing nations can continue to enjoy a relative cost advantage in the production and trading of most goods even if they suffer an absolute cost disadvantage in all goods. But in a world dominated by the principle of absolute advantage, lower absolute costs confer a significant trade advantage because the ability of corporations to take full advantage of a low-cost location is no longer impeded by a lack of international capital mobility.

Apart from the detrimental impact of global capital gravitating towards low-wage and environmentally damaging production locations, globalisation is increasing the pressure on governments in wealthy nations to lower environmental and workplace standards to prevent an exodus of capital and an associated loss of industries. It is for this reason that many ecological economists believe that globalisation forces are perhaps the most important to deal with since they severely undermine domestic attempts to introduce policies, such as cap-auction-trade systems, to achieve ecological sustainability, biodiversity conservation,

[4] This does not mean that taxes do not and should not be used to reduce private-sector spending power. Central-government taxation, by reducing private-sector spending power, plays an important role in nullifying most of the inflationary effect of a central government's own spending. Taxes should not, however, be used to control private-sector spending power at the margin in order to achieve an acceptable rate of price-inflation. I believe this is best left to variations in central-government spending via an 'employer of last resort' programme (see Lawn, 2011b).

and greater resource use efficiency. Ecological economists are therefore calling for urgent institutional reform at the international level - in particular, reform of the Bretton Woods institutions, such as the World Bank, International Monetary Fund, and World Trade Organisation (Daly, 2007).

5. Conclusion

Biodiversity conservation is threatened by many things but none more so than humankind's addiction to growth. Since contemporary macroeconomic policy is geared to maximise the growth rate of real GDP, it contributes significantly to the growing pressure being exerted on the ecosphere and the accelerating loss of biodiversity. However, turning macroeconomic policy on its head is not a viable solution to the dilemma we confront. This is because ecological sustainability, which is necessary to achieve biodiversity conservation, is a biophysical problem and macroeconomic policy involves the use of economic instruments which are designed to deal with economic problems. Macroeconomic policy settings cannot, therefore, be fine-tuned to directly solve the biodiversity erosion crisis.

To achieve ecological sustainability, it is necessary to impose resource throughput restrictions to ensure the rate of throughput does not exceed the ecosphere's regenerative and waste assimilative capacities. Given, also, the important role that efficiency plays in ensuring economic systems operate effectively within sustainable limits, cap-auction-trade systems should be widely introduced. However, cap-auction-trade systems will have major implications for macroeconomy policy setting. It is therefore important that more is known about the likely impacts of various macroeconomic policy stances in a throughput-constrained economy. Despite the lack of research in this area, it is safe to conclude that macroeconomic policy setting should cease to be expansionary, except in impoverished nations, where, for a limited period of time, a dose of efficient and equitable GDP growth would be clearly beneficial. Instead, macroeconomic policy should be directed towards: (a) maintaining public goods and critical infrastructure - necessary to ensure an appropriate macro-allocation of the sustainable incoming resource flow; (b) engaging in counter-cyclical increases in government expenditure to keep unemployment rates low; (c) ecological tax reform - necessary to discourage environmentally destructive activities and reward efforts to increase efficiency and value-adding in production; and (d) restoring comparative advantage as the principle governing international trade - necessary to promote a better international allocation of resources and to put a halt to standards-lowering competition at the international level.

Instituting policies of this nature would almost certainly reduce the real GDP of many wealthy countries, but it is our best chance of achieving sustainability and biodiversity conservation. Of course, as we have seen, and as I have been at pains to argue, there is no reason why appropriately designed and instituted policies that reduce the physical scale of wealthy economies should lead to a decline in human well-being. Indeed, if ecological sustainability is achieved in the manner outlined in this chapter, I believe that economic welfare, as reflected by the GPI, would rise. But if we sabotage the growth economy through the blunt use of contractionary macroeconomic policy, the GPI is almost certainly likely to decline along with real GDP. And that, I believe, would put at risk any genuine attempt to conserve the world's biodiversity.

6. References

Ayres, R. & Miller, S. (1980). The role of technological change. *Journal of Environmental Economics and Management*, Vol.7, pp. 353-371.

Blake, A. (2005). Jevons' Paradox. *Ecological Economics*, Vol.54, pp. 9-21.

Clarke, M. & Shaw, J. (2008). Genuine Progress in Thailand: A Systems-Analysis Approach, In: *Genuine Progress in the Asia-Pacific: Studies Using the Genuine Progress Indicator*, P. Lawn & M. Clarke (Eds.), pp. 260-298, Edward Elgar, Cheltenham, UK.

Common, M. & Stagl, S. (2005). *Ecological Economics: An Introduction*, Cambridge University Press, Cambridge, UK.

Daly, H. (1991). *Steady-State Economics: Second Edition with New Essays*, Island Press, Washington DC.

Daly, H. (1996). *Beyond Growth: The Economics of Sustainable Development*, Beacon Press, Boston.

Daly, H. (2007). *Ecological Economics and Sustainable Development: Selected Essays of Herman Daly*, Edward-Elgar, Cheltenham, UK.

Daly, H. & Cobb, J. (1989). *For the Common Good: Redirecting the Economy Toward Community, the Environment, and a Sustainable Future*, Beacon Press, Boston.

Dornbush, R. & Fischer, S. (1990). *Macroeconomics*, Fifth Edition, McGraw-Hill, New York.

Ekins, P.; Folke, C. & Costanza, R. (1994). Trade, environment, and development: the issues in perspective. *Ecological Economics*, Vol.9, pp. 1-12.

Forstater, M. (2000). Full Employment and Economic Flexibility, In: *The Path to Full Employment*, W. Mitchell & E. Carlson (Eds.), pp. 49-88, University of NSW Press, Sydney.

Gale, R.; Barg, S. & Gillies, A. (Eds.) (1995). *Green Budget Reform*, Earthscan, London.

Georgescu-Roegen, N. (1971). *The Entropy Law and the Economic Process*, Harvard University Press, Cambridge, MA.

Global Footprint Network (2008). *The Ecological Footprint Atlas 2008*, Global Footprint Network, Oakland, CA.

Hall, D. & Hall, J. (1984). Concepts and measures of natural resource scarcity with a summary of recent trends. *Journal of Environmental Economics and Management*, Vol.11, pp. 363-379.

Heyes A. (2000). A proposal for the greening of the textbook macro: IS-LM-EE. *Ecological Economics*, Vol.32, pp. 1-7.

Jackson, T. & S. Stymne (1996). *Sustainable Economic Welfare in Sweden: A Pilot Index 1950-1992*, Stockholm Environment Institute, Stockholm.

Jansson, A.; Hammer, M.; Folke, C. & Costanza, R. (Eds.) (1994). *Investing in Natural Capital: The Ecological Economics Approach to Sustainability*, Island Press, Washington DC.

Jevons, S. (1865). *The Coal Question*, Macmillan & Co., London.

Kerschner, C. (2010). Economic de-growth vs. steady-state economy. *Journal of Cleaner Production*, Vol.18, pp. 544-551.

Latouche, S. (2007). Sustainable consumption in a de-growth perspective, In: *Sustainable Consumption, Ecology, and Fair Trade*, E. Zaccai (Ed.), pp. 178-185, Routledge: London.

Lawn, P. (2000). *Toward Sustainable Development: An Ecological Economic Approach*, CRC Press, Boca Raton.

Lawn, P. (2007). *Frontier Issues in Ecological Economics*, Edward Elgar, Cheltenham, UK.

Lawn, P. (2008). Genuine Progress in India: Some Further Growth Needed in the Immediate Future but Population Stabilisation Needed Immediately, In: *Genuine Progress in the Asia-Pacific: Studies Using the Genuine Progress Indicator*, P. Lawn & M. Clarke (Eds.), pp. 191-227, Edward Elgar, Cheltenham, UK.

Lawn, P. (2010). On the Ehrlich-Simon bet: Both were unskilled and Simon was lucky. *Ecological Economics*, Vol.69, pp. 2045-2046.

Lawn, P. (2011a). Is steady-state capitalism viable? – A review of the issues and an answer in the affirmative, In: *Ecological Economic Reviews*, R. Costanza & I. Kubiszewski (Eds.), pp. 1-25, New York Academy of Sciences Press, New York.

Lawn, P. (2011b). Wake up economists! – Currency-issuing central governments have no budget constraint. Munich Personal RePEc Archive paper No. 28224, http://mpra.ub.uni-muenchen.de/28224

Lawn, P. & Clarke, M. (2006). *Measuring Genuine Progress: An Application of the Genuine Progress Indicator*, Nova Science Publishers, New York.

Lawn, P. & Clarke, M. (Eds.) (2008). *Sustainable Welfare in the Asia-Pacific*, Edward Elgar, Cheltenham, UK.

Martinez-Alier, J. (2009). Socially sustainable economic de-growth. *Development and Change*, Vol.40, pp. 1099-1119.

Martinez-Alier, J.; Pascual, U.; Vivien, F.-D. & Zaccai, E. (2010). Sustainable de-growth: mapping the context, criticisms, and future prospects of an emergent paradigm. *Ecological Economics*, Vol.69, pp. 1741-1747.

Max-Neef, M. (1995). Economic growth and the quality of life. *Ecological Economics*, Vol.15, pp. 115-118.

Mitchell, W. & Watts, M. (2002). Restoring full employment: the Job Guarantee, In: *The Urgency of Full Employment*, E. Carlson & W. Mitchell (Eds), pp. 95-114, University of New South Wales Press, Sydney.

Mitchell, W. & Muysken, J. (2008). *Full Employment Abandoned: Shifting Sands and Policy Failures*, Edward-Elgar, Cheltenham, UK.

Nguyet Hong, V.-X.; Clarke, M. & Lawn, P. (2008). Genuine Progress in Vietnam: The Impact of the Doi Moi Reforms, In: *Genuine Progress in the Asia-Pacific: Studies Using the Genuine Progress Indicator*, P. Lawn & M. Clarke (Eds.), pp. 299-330, Edward Elgar, Cheltenham, UK.

O'Connor, M. (1991). Entropy, structure, and organisational change. *Ecological Economics*, Vol.3, pp. 95-122.

O'Riordan, T. (Ed.) (1997). *Ecotaxation*, Earthscan, London.

Perrings, C. (1986). Conservation of mass and instability in a dynamic economy-environment system. *Journal of Environmental Economics and Management*, Vol.13, pp. 199-211.

Redefining Progress (1995). *The Genuine Progress Indicator: Summary of Data and Methodology*. Redefining Progress, San Francisco, CA.

Reynolds, D. (1999). The mineral economy: how prices and costs can falsely signal decreasing scarcity. *Ecological Economics*, Vol.31, pp. 155-166.

Victor, P. (2008). *Managing Without Growth*, Edward-Elgar, Cheltenham, UK.

Wen, Z.; Yang, Y. & Lawn, P. (2008). From GDP to GPI: quantifying thirty-five years of development in China, In: *Genuine Progress in the Asia-Pacific: Studies Using the*

Genuine Progress Indicator, P. Lawn and M. Clarke (Eds.), pp. 228-259, Edward Elgar, Cheltenham, UK.

Wackernagel, M.; Onisto, L.; Bello, P.; Callejas Linares, A.; López Falfán, S.; Méndez García, J.; Suárez Guerrero, A. I. & Suárez Guerrero, M. G. (1999), National natural capital accounting with the ecological footprint concept. *Ecological Economics*, 29 (3), pp. 375-390.

Wray, R. (1998). *Understanding Modern Money: The Key to Full Employment and Price Stability*, Edward Elgar, Cheltenham, UK.

18

Mobilizing Community Capitals to Support Biodiversity

Cornelia Butler Flora
Iowa State University
USA

1. Introduction

The mobility of living species is increasing dramatically. When these species reproduce without control, they are viewed as invasive, crowding out native species and reducing biodiversity. While humans are often instrumental in knowingly or unknowingly introducing these invaders, increasing temperatures and extreme weather events provide environments that enhance the growth of these pests. Climate change creates more extreme weather events, and rising temperatures and increased rainfall provide favorable conditions for pests to thrive while endemic plants and animals are threatened. Human commerce and policy is accompanied by the introduction of new pests, both consciously (as the introduction of kudzu in the South of the United States by the Soil Conservation Service to restore cover to land deprived of nutrients by over-production of cotton, or rabbits in Australia introduced by "gentlemen farmers" to promote recreational hunting) or inadvertently, as pests were part of shipments of (as with Russian thistle, also known as tumble weed, in the U.S. introduced to South Dakota in the US by Russian migrants as a contaminant in flax seed) or pests on packing pallets or in shipping containers.

Biodiversity contributes to ecosystem health, which is a part of the triple bottom line that also includes economic security and social inclusion (Flora, 2001, Blewett, 2008). When there is social inclusion and economic security, it is easier to mobilize collective action to promote biodiversity. By investing in all seven of the community capitals (natural, cultural, human, social, political, financial and built), local communities can collaborate with scientists to enhance all three aspects of the triple bottom line. The community capitals framework has been successfully mobilizing in maintaining and increasing endemic biodiversity (Flora, 2001, Cepeda, et al. 2008) and ecosystem health (Flora 2004a; Gutierrez-Montez, 2005, Flora and Delaney, forthcoming). Capitals such as human, social and built (technological) are particularly important in adaptation to climate change, according to Gardener, et al. 2009).

Multiple capitals have been used by development practitioners to foster a holistic approach (Bebbington, 1999; DIFD 1999, 2000, 2001; Carney, 1998; Pretty, 2000, 1998). Building on their experience and our practice and scholarship in the U.S. and Latin America, we have found that consideration of these seven capitals is critical in making sure that programs are both sustainable and effective. (Figure 1)In our analysis, we separate individual facts from social facts Durkheim (1902, 2001). While individuals may have access and control of each of the seven community capitals, the community's stocks and flows of these capitals these are more than the sum of individual attitudes and possessions. Thus, you explain how individuals behave based on their attitudes and characteristics, and you explain how communities behave by looking at their structures and their collective histories.

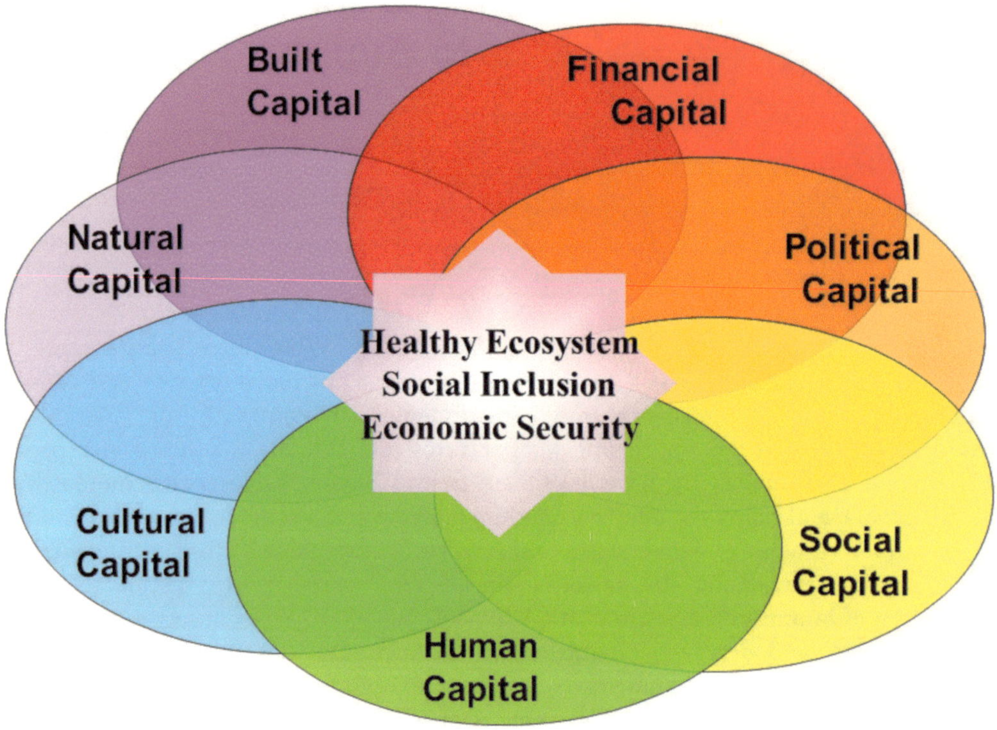

Fig. 1.

2. Community capitals

No matter how poor or how rich, all communities have resources that can be depleted, saved for future use, or invested to create new resources. Those resources, when invested to create new resources over a long time, are referred to as "capital". Stocks and flows of the seven capitals are ends in themselves and means to achieve desired future conditions. A dynamic balance among the capitals and investments in them support sustainable strategies to address the emergent threats of invasive species and decline of biodiversity in a global economy and a changing climate that lead to ecological regime shifts (Lin, 2011; Flora and Flora, 2008).

2.1 Natural capital

The natural assets of a location, including weather, geographic isolation, natural resources, amenities, and natural beauty, make up natural capital. Water, soil and air - their quality and quantity - are a major building block of natural capital, as is biodiversity (Costanza, et al. 1997; Altieri, M.A. 1999). Biodiversity may increase with invasive species, but invasive species may damage endemic biodiversity, which then impacts the landscape, water quality and even water quantity. Urban and rural forests in the U.S. are impacted by several recently introduced invasive species, including the Asian Longhorned Beetle (ALB). The eradication program for ALB has greatly impacted the local areas where this beetle has been found because of the removal of thousands of trees, which has cost millions of dollars. The United States has implemented stricter trade regulations to prevent further introductions. If the established populations of ALB are not eradicated, the beetle could threaten the maple sugar industry, fall-foliage tourism, natural ecosystems, recreational areas, and many beloved backyard and street trees (Sawyer, et al. 2010). And as the climate warms, maple trees are more susceptible to ALB.

The dying out of native tree species in the U.S. where the emerald ash beetle is decimating ash forests, contribute to global warming. Rural development activities influence natural capital, often negatively, decreasing long term development. Because different groups in our obesogenic society (Parkin, 2010) have different access to and understanding of natural capital, they have important firsthand knowledge of changes that are occurring on the landscape. For example, new immigrants are more likely to fish and gather wild foods than are native born populations. Linking to their daily experiences of linking to the natural world can help determine subtle changes in the flora and fauna of a place where they hunt, fish and gather.

A community-based approach to natural capital develops healthy ecosystems with multiple community benefits, where human communities act in concert with natural systems, rather than simply to dominate these systems for short term gain (Ostrum, 1990; Ostrum et al. 1994). As climate changes, conflicts over natural capital increase (Sondorp and Patel, 2003; Barnett and Adger, 2007). Utilizing climate science and local knowledge allow adaptations which include increasing biodiversity (Ensor and Berger, 2009).

2.2 Cultural capital

Cultural capital reflects the way people "know the world" and how to act within it as well as their traditions and language. It includes *cosmovisión* (spirituality and how the different parts are connected), ways of knowing, food and language, ways of being, and definition of what can be changed (Flora and Flora, 2008). Very often local people will have multiple terms for stages of plant and forest growth which reflect an intimate understanding and close observation of these growth stages and enable these people to give an early warning that a situation is abnormal and an exotic pest may be present. That ability to see differences is very important in identifying the possible presence of a new pest or condition (Flora, 2008).

It is often easy to discount local voices in discussions of biodiversity. Hegemony privileges the cultural capital of dominant groups (Bourdieu 1986, Flora, *et al.*, 2004). An important part of cultural capital involves the collective belief that despite changing climate conditions and seemingly unconquerable invasive species, things can change. Generally, that comes from working across groups to increase endemic biodiversity through controlling invasive species. The increase temperatures affect plant species and invasive exotic species move farther North, including the multiflora rose in the river in Chico, California. The Friends of Bidwell Park in Chico, California are a group that cares about a specific place and the biodiversity it supports. They periodically gather together wearing closed-toe shoes and long pants to remove the multiflora rose. They are convinced that by working together they can at least contain this spreading pest that threatens their endemic biodiverse park. They build on cultural differences, are patient with those who originally were enchanted by the "pretty flower" and show reverence for endemic biodiversity around Chico.

2.3 Human capital

Human capital has long been identified as critical for positive social change by scholars (Becker, 1964; Schultz, 1961, 1963, 1964). However, human capital champions often focus on formal education. To preserve and enhance biodiversity, the concept must be broadened to include the skills and abilities of people to develop and enhance their resources, and to access outside resources and bodies of knowledge in order to increase their understanding, identify promising practices, and to access data to enhance community capitals. Human capital also includes health and leadership.

Age and gender impact the skills and abilities available to a community seeking to enhance biodiversity. Boys and girls are sometime taught about quite different aspects of their environment. In the highland of Peru, women know about where to take livestock to graze the best mixture of forbs and grasses and which potatoes taste the best and cook with less fuel. Men know the amount of manure to put into the soil and which soils are best for each type of potato (Flora and Kroma, 1998).

Endemic biodiversity enhances human capital by providing a varied landscape and potential plants and animals for human curing and consumption, cultural capital by maintaining native species of sacred uses, social capital by providing pleasant places to gather together, and political capital to justify maintaining habitat.

Scientists often think of themselves as the major creators of high level human capital, transferring knowledge from scientists to beneficiaries. However, addressing human capital to enhance natural capital, especially biodiversity, requires more nuance than the expert transferring knowledge that has been generated from the scientists or their faculties. Partnerships aimed at enhancing natural capital and biodiversity, such as the Raccoon River Watershed Association, use capital framework to identify the motivations and abilities of each individual to improve natural capital, increase the skills and health of individuals to act to enhance natural capital, and recombine the skills and motivation of the community to a more sustainable collective future. Partnering with local communities to enhance biodiversity requires transferring knowledge about how to combat invasive species and enhance endemic based on what is already in place: local knowledge its complementarity with scientific knowledge in order to promote biodiversity (Gasteyer and Flora, 2000).

2.4 Social capital

Social capital is increasingly recognized as critical for ecosystem health (Triglia, 2001; C. Flora, 1995, 1998a, 2000; J. Flora, 1998). It reflects the connections among people and organizations or the social glue to make things, positive or negative, happen (Coleman, 1988; Portes and Sessenbrenner, 1993). It includes mutual trust, reciprocity, groups, collective identity, sense of a shared future, and working together (Putnam, 1998, 1995; 1993a, b).

Bonding social capital refers to those close ties that build community cohesion. Bridging social capital involves loose ties that bridge between organizations and communities (Narayan, 1999; Daasgupta and Serageldin (2000). A specific configuration of social capital - entrepreneurial social capita1 (ESI) is related to community economic development (Flora and Flora, 1989; 1993; Flora, et al., 1997). ESI includes inclusive internal and external networks, local mobilization of resources, and willingness to consider alternative ways of reaching goals. Entrepreneurial social capital can be invested in enhancing ecosystem health.

An example of use of social capital to fight an invasive plant is the group, Fighting Phragmites, in Oakland County in central Michigan. Phragmites has been present in North America for over 3,000 years; however, over the past century, it has come to dominate many mid-Atlantic marshes. Scientists attribute the plant's rapid expansion to an aggressive, competitively superior genetic phragmites strain introduced from Eurasia in the 19th century via dry ballast from ships. They destroy native species in shallow lakes and marshes. Michigan State University Extension used bridging and bonding social capital to successfully combat the invasive. The Michigan Conservations Stewards brought together scientific knowledge from Michigan State University, local ingenuity of volunteer engineers employed by the automobile industry and volunteer monitors and workers to harvest and pull out the roots of these plants in local marshes.

It is critical that local people who most feel the pressure of the invasive species provides a portion of the knowledge as well as the work to address the issues of invasive species and biodiversity. When outsiders provide resources to a community without the community determining how those resources can contribute to sustainable development, any traditional dependency on political parties and politicians, based on personal relations of one or two people, may be simply transferred to the scientists involved. The group already working conservation stewardship demonstrated bonding social capital through local action for the common good. Through bridging social capital, help build flexible and porous boundaries that increased the human capital to address the issue. Social capital can be a key vehicle of cutting transaction costs in linking local and scientific knowledge to control and eradicate invasive species.

In this case and others of citizen participation to decrease the presence of invasive species and maintain endemic biodiversity, the long term institutional presence in the community through Extension served as links to other institutional actors can be key to project success. Indicators of a balanced increase in bridging and bonding social capital include a shared vision (which takes time and trust to develop), building first on internal resources (which means the community has together determined existing assets that can be turned into capitals for participatory rural development, looking for alternative ways to respond to constant changes (rather than the one solution of a certain investment to solve everything), the loss of the victim mentality (feeling overwhelmed by globalism or climate change), and loss of a cargo cult mentality (where the community waits for an outside investment to rescue them) (Flora et al. 1999; Flora, et al. 1996).

2.5 Political capital

Political capital is the ability of a group to mobilize their norms and values to influence standards, regulations and enforcement of those regulations that determine the distribution of resources and the ways they are used (Flora and Flora, 2008). When a community has high political capital, its people have the collective ability to find their own voice and to engage in actions that contribute to the well being of their community. In Fighting Phragmites, the norms and values surrounding invasive species not only mobilized human and social capital, but got the local county and the state Department of Natural Resources to invest in the effort.

Political capital has been commonly operationalized as the power to disrupt or stop something from happening, or to get specific goods from a central authority (Aiger, *et al.* 2001). By increasing bridging and bonding social capital to all segments of the community, the community's power of negotiation, particularly in identifying allies that share their vision for a sustainable future with increased biosecurity, grows.

Indicators of increased political capital that can enhance endemic biodiversity include organized groups working together for biodiversity and a healthy ecosystem, local people knowing and feeling comfortable around powerful people, including scientists and government functionaries, and local concerns becoming part of the agenda in the regulation and distribution of resources related to increasing ecosystem health. Focusing only on the technical or mechanistic means to maintain biodiversity serves to mask or obscure the fundamentally political nature of remediation and adaptation processes. Engaging a broad range community residents in the discussion of policy issues around biodiversity and a healthy ecosystem builds political capital.

2.6 Financial capital

Financial resources available to invest in community capacity building, to underwrite invasive pest management and business development, to support civic and social entrepreneurship, and to accumulate wealth for future enhancement of the other capitals make up financial capital. Financial capital is often privileged as an indicator of progress and dominates evaluation and monitoring when looking at changes in natural capital. There is a tendency to monetize other capitals, rather than considering the reverse: money gained from natural resource destruction results in the decline of natural and often human capital as well. In an era when profitability for private actors is viewed as the most important measure of success, biodiversity should be creatively linked to financial capital to help legitimize the activities undertaken by local people as scouts for invasive species and as actors to control them.

Bridging social capital can enhance financial capital (Grannovetter, 1973; 1985). Appropriate investment in financial capital can create an appropriately diverse and healthy economy and to increase biodiversity if that social capital includes biodiversity considerations. Otherwise, bridging social capital can foster more rapid decline in biodiversity as water quality declines, as is happening in areas in the states of Pennsylvania and New York where hydraulic fracturing (fracking) is used to mine natural gas. When there are public investments in increasing biodiversity, poverty can be reduced, business efficiency can be enhanced, economic diversity can increase, and the assets of those who live in the community can grow. A number of projects of the USDA Forest Service in New Mexico were aimed at maintaining forest health and biodiversity through the harvesting of small timbers by local people and the creation of small saw mills to process that wood.

2.7 Built capital

Physical and technical infrastructures make up built capital. It includes such diverse human-made objects and systems as sewers, water systems, electronic communication, soccer fields and processing plants. For control of invasive species, it can be as simple as hand tools such as picks and shovels and wheel barrows to physically remove the plants. Or it can include chemical substances that can make pests sterile

Built capital can enhance or decrease biodiversity. Indeed landscaping for new subdivisions is often a source of major invasive plants, such as privet, used decoratively for hedges and which can readily invest a hardwood forest. An example of such potential for both enhancement and degradation is a rural water system, which can run sewage into the stream the community depends on for drinking water, or provide for appropriate and cost-effective sewage treatment (Flora, 2004).

Built capital can enhances other community capitals when it serves multiple users (human capital), it can be locally maintained and improved (human and financial capital), it links local people together equitably (bonding social and cultural capital), and it links local people, institutions and businesses to the outside (bridging social and financial capital). And appropriate investments in infrastructure can discourage the growth of invasive species, while other build capital decreases it.

2.8 Spiraling down

Loss of one capital can lead to disinvestment in other capitals (Waquant, 1997). Decline in natural capital, which is often triggered by invasive species, has been linked to decreased financial capital (lowered productivity and income), decreased human capital (out-migration and illness), decreased cultural capital (loss of bio-diversity and village rituals), and decreased social capital (as there is increased inequality). Responses to invasive species often are based on technological fixes (a form of built capital) in hopes of at least stemming the downward spiral. By addressing only the immediate threat of invasive species, sustainable actions to increase the community's ability to respond to future ecosystems changes will not occur.

2.9 Spiraling up

Built capital is not the first investment necessary to reverse the downward spiral of biodiversity decline (Cepeda, 2008; Gutierrez, 2005). Investment in both bonding and bridging social capital, including accepting the time it takes to build trust and reciprocity, is often a key entry point. Maintaining biodiversity and combating invasive species is a way to build trust and reciprocity through giving all members of the community to do what they say they will do and to give local people a way to reciprocate in a way that maintains their dignity and meaning.

3. Conclusions

The increasing vulnerability of local landscapes to invasive species requires on-going mobilization of all community capitals in order to have the agility of prompt and effective responses. Bridging social capital among communities can serve as an early alert that can mobilize local residents and scientists to work together to experiment with and implement appropriate remedial action. But when all the capitals are invested in, including cultural capital so that local populations understand that they do not have to accept crop and forest destruction as inevitable, scientist-community partnerships can increase biosecurity for a healthy ecosystem and a vital economy with a high level of social well-being (Flora, 2003).

4. References

Aigner, S.M., C.B. Flora and J.M. Hernandez. Fall, 2001. "The Premise and Promise of Citizenship and Civil Society for Renewing Democracies and Empowering Sustainable Communities." Sociological Inquiry. 71 (4): 493-507. http://www.ag.iastate.edu/centers/rdev/pubs/flora/empower.htm

Altieri MA. 1999. "The ecological role of biodiversity in agroecosystems." Agriculture, Ecosystems and Environment 74: 19–31.

Barnett, J. and W. N. Adger. 2007. Climate change, human security, and violent conflict. Political Geography. 26:639-655.

Bebbington, A. 1999. "Capitals and Capabilities: A Framework for Analyzing Peasant Viability, Rural Livelihoods and Poverty." World Development. 27: 2021-2044.

Becker, Gary. 1964. Human Capital: A Theoretical and Empirical Analysis with Special Reference to Education. New York: Columbia University Press.

Blewitt, J. (2008). Understanding Sustainable Development. London: Earthscan

Carney. D. (ed.) 1998. Sustainable Rural Livelihoods: What Contribution Can We Make? London: Department for International Development.

Cepeda Gómez, C. I. Gutiérrez-Montes, A. Imbach, F. Alpizar y N. Windevoxhel. 2008. Tiburón ballena y bienestar comunitario en Holbox, Quintana Roo, México. Recursos Naturales y Ambiente 55:109-117

Coleman, James C. 1988. "Social capital in the creation of human capital." American Journal of Sociology 94 (Supplement S95-S120): 95-119.

Costanza, R., R. d'Arge, R. de Groot, S. Farber, M. Grasso, B. Hanson, K. Limburg, S. Naeem, R.V. O'Neil, J. Parvelo, R.G. Raskin, P. Sutton, and M. van den Belt. 1997. "The Value of the World's Ecosystem Services and Natural Capital." Nature 387:253-260.

Dasgupta, P. and I. Serageldin, eds. 2000. Social Capital: A Multifaceted Perspective. Washington, D.C.: World Bank.

DFID. 1999, 2000, 2001. Sustainable Livelihoods Guidance Sheets. Available from: http://www.livelihoods.org/info/info_guidancesheets.html

Durkheim, E. 1902, 1947. The Division of Labor in Society. Glencoe, IL: Free Press.

Ensor, Jonathan and Rachel Berger. 2009. Understanding Climate Change: Lessons from Community-Based Approaches. The Schumacher Centre for Technology and Development, Bourton on Dunsmore, Rugby, Warwickshire, CV23 9QZ, UK:. Practical Action Publishing

Flora, C.B. 2001. "The Creation of Social Incentives for the Conservation of Biodiversity." Pp. 135-144 in Integrated Conservation and Development in Tropical America: Experiences and Lessons in Linking Communities, Projects, and Policies. R.E. Rhoades and J. Stallings (eds.) Athens, GA.: SANREM and CARE.

Flora, C.B. 2004a. "Community Dynamics and Social Capital." Pp. 93-107 in Agroecosystems Analysis. D. Rickert and C. Francis (eds.) Madison, Wisconsin: American Society of Agronomy, Inc., Crop Science Society of America, Inc., Soil Science Society of America, Inc.

Flora, C.B. 2003. "Comprehensive Community Initiatives: Evaluating Multi-Level Collaboration for Systems Change." Rural Development News. 26:3: 1-4.

Flora, C.B. 2001. "Shifting Agroecosystems and Communities." Pp. 5-14 in Interactions Between Agroecosystems and Rural Communities. C. Flora (ed.) Boca Raton: CRC Press.

Flora, C.B. 2008. "Social capital and community problem solving: Combining local and scientific knowledge to fight invasive species." Learning Communities: International Journal of Learning in Social Contexts *(Australia), &* Kritis: Journal of Interdisciplinary Development Studies *(Indonesia). BIOSECURITY BILINGUAL MONOGRAPH,* pp. 41-54.

Flora, C.B. 2004b. "Social Aspects of Small Water Systems." Journal of Contemporary Water Research and Education 128: 6-12.

Flora, C. B. 1995. "Social Capital and Sustainability: Agriculture and Communities in the Great Plains and the Corn Belt." Research in Rural Sociology and Development 6: 227-246.

Flora, C.B. 1998a. "Sustainable Production and Consumption Patterns: Community Capitals." Pp. 115-122 in The Brundtland Commission's Report - Ten Years. G. B. Softing, G. Benneh, K. Hindar, L. Walloe, and A. Wijkman (eds.) Oslo: Scandinavian University Press.

Flora, C.B. 2000. "Sustainability in agriculture and rural communities." Pp.191- 207 in Nachhaltigkeit in de Landwirschaft: Landwirtschaft im Spannungsfeld zwischen Ökologie, Ökonomie und Sozialwissenschaften. (Sustainability in Agriculture: Tensions between Ecology, Economics and Social Sciences). M. Härdtlein, M. Kaltschmitt, I. Lewandowski, and H.N. Wurl (eds.) Berlin, Germany: Erich Schmidt Verland.

Flora, C.B. and J. Flora. 1989. "Characteristics of Entrepreneurial Communities in a Time of Crisis." Rural Development News. 12 (2): 1-4.

Flora, C.B. and J. Flora. 1996. El Desarrollo Comunitario Basado en los Bienes Comunitarios: Como Mobilizar a una Comunidad Entera. Quito: SANREM CRSP Ecuador Documentos de Trabajo.

Flora, C.B. and J. L. Flora. 1993. "Entrepreneurial Social Infrastructure: A Necessary Ingredient." The Annals of the Academy of Social and Political Sciences 529:48-58.

Flora, C.B. and J. L. Flora, 2003. "Social Capital." Pp. 214-227 in Challenges for Rural America in the Twenty-First Century. D. L. Brown and L. E. Swanson (eds.) University Park, Pennsylvania: The Pennsylvania State University Press.

Flora, C.F. and J.L. Flora. 2008. Rural Communities: Legacy and Change. 3rd edition. Boulder, CO: Westview Press.

Flora, C.B., J. Flora, K. Wade. 1996. "Measuring Success and Empowerment." Pp. 57-74 in Community Visioning Programs: Practices and Principles. N. Waltzer (ed.) Westport, CT: Greenwood Press.

Flora, C.B. and M. Kroma. 1998. "Gender and Sustainability in Agriculture and Rural Development." Pp. 103-116 in Sustainability in Agricultural and Rural Development. G. E. D'Souza and T. G. Gebermedhin (eds.) Brookfield, VT: Ashgate Publishing Company.

Flora, C.B., M. Kinsley, V. Luther, M. Wall, S. Odell, S. Ratner, and J. Topolsky. 1999. Measuring Community Success and Sustainability. (RRD 180). Ames, IA: North Central Regional Center for Rural Development

Flora, J. 1998. "Social Capital and Communities of Place," Rural Sociology 63: 481-506.

Gardner J, Dowd A-M, Mason C, Ashworth P. 2009. A Framework for Stakeholder Engagement on Climate Adaptation. Commonwealth Scientific and Industrial Research Organisation Climate Adaptation Flagship Working Paper no. 3. CSIRO.

Gasteyer, S. and C.B. Flora. 2000. "Measuring PPM with Tennis Shoes: Science and Locally Meaningful Indicators of Environmental Quality." Society and Natural Resources. 13: 589-597.

Granovetter, M. S. 1973. "The strength of weak ties." American Journal of Sociology 78: 1360-1380.

Granovetter, M. S. 1985. "Economic Action, Social Structure and Embeddedness." American Journal of Sociology 91:481-510.

Gutierrez-Montes, I.A. 2005. Healthy Communities Equals Healthy Ecosystems? Evolution (and Breakdown) of a Participatory Ecological Research Project Towards a Community Natural Resource Management Process, San Miguel Chimalapa (Mexico). Ames, IA: Iowa State University PhD dissertation.

Landolt, P. 2001. "Salvadoran Economic Transnationalism: Embedded Strategies for Household Maintenance, Immigrant Incorporation, and Entrepreneurial Expansion," Global Networks 1: 217-241.

Lin, B. B. 2011. "Resilience in Agriculture through Crop Diversification: Adaptive Management for Environment Change." BioScience 61:183-193.

Lorenz, E. 1999. "Trust, Contract and Economic Cooperation," Cambridge Journal of Economics 23: 301-15.
Mollering, G. 2002. "Perceived Trustworthiness and Inter-firm Governance: Empirical Evidence from the UK Printing Industry," Cambridge Journal of Economics 26: 139-160.
Nayaran, D. 1999. Bonds and Bridges: Social Capital and Poverty. Washington, DC: World Bank Poverty Group.
Ostrum, E. 1990. Governing the Commons: The Evolution of Institutions for Collective Action. New York: Cambridge University Press.
Ostrum, E., R. Gardner, and J. Taylor. 1994. Rules, Games and Common-Pool Resources. Ann Arbor, MI: University of Michigan Press.
Parkin, S. *The Positive Deviant: Sustainability Leadership in a Perverse World.* London, UK: Earthscan, 2010.
Portes, A. and J. Sensenbrenner. 1993. "Embeddedness and Immigration: Notes on the Social Determinants of Economic Action." American Journal of Sociology 98:1320-1350.
Pretty, J. N. 2000. "Conditions for Successful Implementation of Sustainable Agriculture". Pp. 323-343 in Nachhaltigkeit in de Landwirschaft: Landwirtschaft im Spannungsfeld zwischen Ökologie, Ökonomie und Sozialwissenschaften. M. Härdtlein, M. Kaltschmitt, I. Lewandowski, and H.N. Wurl (eds.) Berlin, Germany: Erich Schmidt Verland.
Pretty, J. N. 1998. The Living Land: Agriculture, Food and Community Regeneration in Rural Europe. London: Earthscan Publications Ltd.
Putnam, R. D. 1995. "Bowling Alone: America's Decline in Social Capital." Journal of Democracy. 6: 65-78.
Putnam, R. D. 1993a. Making Democracy Work: Civic Traditions in Modern Italy. Princeton, NJ: Princeton University Press.
Putnam, R. D. 1993b. "The Prosperous Community: Social Capital and Public Life." The American Prospect 13:35-42.
Sawyer, A. J.; Panagakos, W. S.; Horner, A. E.; Freeman, K. J. 2010. "Asian longhorned beetle, over the river and through the woods: habitat-dependent population spread." In: McManus, K. A; Gottschalk, K. W., eds. Proceedings. 21st U.S. Department of Agriculture interagency research forum on invasive species 2010; 2010 January 12-15; Annapolis, MD. Gen. Tech. Rep. NRS-P-75. Newtown Square, PA: U.S. Department of Agriculture, Forest Service, Northern Research Station: 52-54.
Schultz, T. 1963. The Economic Value of Education. New York: Columbia University Press.
Schultz, T. 1961. "Investment in Human Capital," American Economic Review 51: 1-17.
Schultz, T. 1964. Transforming Traditional Agriculture. New Haven: Yale University Press.
Siles, M., S. D. Hanson, and L. J. Robison. 1994. "Socio-Economics and the Probability of loan Approval," Review of Agricultural Economics 16: 363-372.
Sondorp, E. and P. Patel. 2003. "Climate Change, Conflict and Health." Transactions of the Royal Society of Tropical Medicine and Hygiene. 97:139-140.
Talmud, I., and G. S. Mesch. 1997. "Market Embeddedness and Corporate Instability: The Ecology of Inter-Industrial Networks," Social Science Research 26: 419-441.
Triglia, C. 2001. "Social Capital and Local Development," European Journal of Social Theory 4: 427-442.
Wilson, P. N. 2000. "Social Capital, Trust, and the Agribusiness of Economics," Journal of Agricultural and Resource Economics 25: 1-13.

Part 3

Interactions Between Living Organisms and Abiotic Factors

19

Biodiversity on Stone Artifacts

Oana Adriana Cuzman[1], Piero Tiano[1],
Stefano Ventura[2] and Piero Frediani[1]
[1]ICVBC-CNR – Istituto per la Conservazione e la Valorizzazione dei Beni Culturali
[2]ISE-CNR – Istituto per lo Studio degli Ecosistemi
Italy

1. Introduction

The biological colonization of stone artifacts exposed under outdoor conditions (e.g. artistic fountains, statues in parks, monuments in urban areas, archeological sites, etc.) is always influenced by the microclimate and the high biodiversity of the opportunistic airflora/airfauna. There is a strength relation between biological development and different factors such as the type of rocks, meteorological phenomena and urban or rural environment. Many organisms, such as phototrophs, may only use the stone as a support, but they can favour more complex colonization, with possible biodeterioration effects. In fact, these organisms are able to growth using the stone mineral components or the superficial deposits of it, exerting damages of stone material by their metabolic activity. Organisms causing biological decay effects are called biodeteriogens. Knowledge of the biodiversity, ecological and physiological aspects of the biological colonization on monumental stones is essential to maintain and preserve the stone cultural heritage for further generations.

2. Bioreceptivity of stone material

Natural and artificial stones have been used from ancient times as building and art materials. Most of the stone artifacts are located outdoor and consequently are frequently subjected to epilithic, chasmolithic and endolithic biological colonization. These biological micro-ecosystems can have different colours and aspects such as pustules, crusts, patinas, biofilms or carpets. Microorganisms may colonize a stone substratum in a variety of ways, especially if they are motile, whereas not motile microorganisms (many cyanobacteria, diatoms) and propagules (spores, seeds, etc.) are presumably deposited on a substratum by gravity from air or water flows. The development of stone biocenosis depends on the combination of environmental location and climatic conditions, in addition to the chemical-physical and petrographic features of stones. A synergistic biodeteriorative action on stone substratum can be started by their concomitant growth of phototrophic and heterotrophic populations during the ecological succession. The whole properties that contribute to biological (flora and/or fauna) colonization has been defined as "bioreceptivity" by Guillitte in 1995, who further defined different types, such as primary, secondary and tertiary bioreceptivity.

2.1 Characteristics of stone substrata

Once extracted from the quarry, the stone material is subjected to irreversible meteoric alteration induced by the synergic action of external and intrinsic factors, with different deteriorating levels, and therefore, with different compositional and geometrical changes. The natural or artificial stones are characterized by their petrographic structure, texture, colour, mineralogical and chemical composition. Among these, the surface roughness, porosity, hygroscopicity, chemical composition, as well as the state of conservation of the stone, are the most important features, which favour the biocolonization and that could, in various ways, lead to deterioration of artworks (Caneva & Ceschin, 2008). Sedimentary (limestones, sandstones, travertine, volcanic tuffs), metamorphic (marble), igneous (basalt) and plutonic rocks (granite, diorite) have been commonly used both as construction materials and artistic compositions. Even artificial stones such as ceramic (bricks, tiles, terracotta), plasters, mortars, stuccos and frescoes, have been used. These have different characteristics and in general a larger porosity with respect to natural stones (Pinna & Salvadori, 2008).

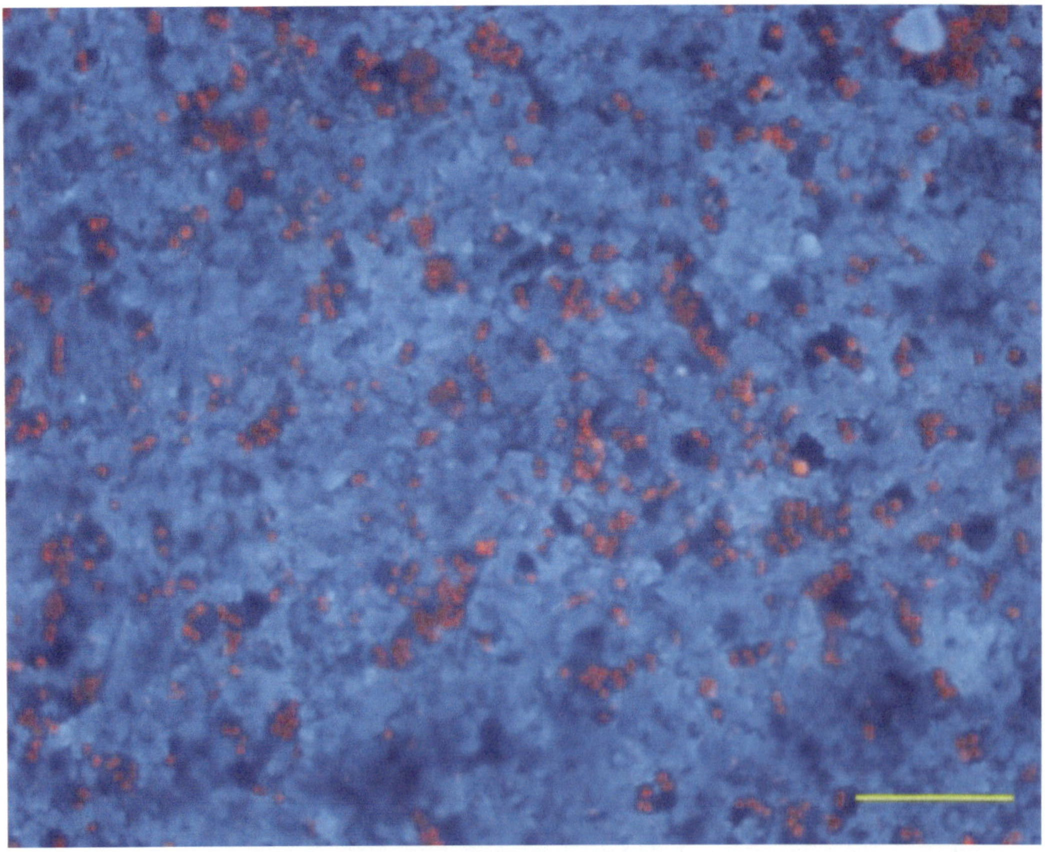

Fig. 1. Phototrophic presence within the micro-irregularities of the surface of a Carrara marble slab under UV light. Scale bar 100 μm

Porosity represents the "empty spaces" in a total volume of a stone material. It affects the permeability and molecular diffusion phenomena. The permeability phenomenon needs usually big pores, while the molecular diffusion can occur even through small interstitial spaces (Amoroso, 2002). Only parts of the void fraction are communicating among them and with the exterior constituting the "open and partial-open pores", while the non-communicating parts are the "closed pores". The distribution, shape and pores size, influence the water absorption and the alterations connected with it (freeze-thaw, salt crystallisation and hydration, biological colonization). Phototrophic micro-organisms, such as coccoid cyanobacteria and algae, start to inhabit in niches formed by its surface porosity, in the primary colonization phase of a stone (Fig. 1). Also, the spores and seeds from air may deposit, by chance, in the pores or crevices and start to develop when microenvironmental conditions are favourable. Some lithotypes are more easily colonized than others, such as highly porous materials (tuffs, limestones, sandstones).

Fig. 2. Different phototrophic colonization in accordance with moisture availability in a stone microhabitat (wall from historical center of Gavorrano, Italy)

Beside the external factors involved in a biological dwelling and growth, the stone intrinsic properties are very important. These are related to the porosity and water transport, retention and diffusion. Therefore, *hygroscopicity* is a significant characteristic to be considered. It is related to the porous structure and capillary pressures of the water present in the stone. The water, in vapour or liquid phase, may follow different penetration mechanisms, by imbibition or condensation processes. The water vapours are strictly correlated with the surrounding environment tending to reach an equilibrium i.e. higher is the relative humidity, higher is the moisture absorption. The liquid water is absorbed by capillarity and moves by diffusion and/or osmosis inside the stone. The water availability is essential for the life development, but it is crucial how long the moisture is retained on/into the stone. Therefore, another parameter to be considered is the *water evaporation process* – a low evaporation usually leads to a higher biological colonization. In Fig. 2 can be seen how different stones show different colonization levels due to their different properties, even if they are located in the same micro-environment. A long lasting living microbial establishment can occur in fine-grained stones that hold moisture for long time, while coarse-grained stones, with high water permeability, can favour only temporary biological colonisations (Warscheid & Braams, 2000; Tomaselli et al., 2002; Prieto & Silva, 2005; Miller et al., 2009).

Surface roughness represents the complex vertical deviations in time of the topography of a stone surface. In fact, this is changing due to the climatic action, soiling processes, eutrophication and biological succession. Many authors (Tiano et al., 1995; Tomaselli et al., 2000; Miller et al., 2009) confirm that higher microbial colonisation occurs when the roughness increases, because this enhances the total surface area, diminishing the shear forces and increasing water absorption.

Chemical components such as macro and microelements, can influence the types of biological growth allowing the availability of inorganic micronutrients. For example, Miller et al. (2006) observed that *Gloeocapsa alpina* and *Sticochoccus bacillaris* were able to grow on carbonate substrates, but showed very limited growth on silicate substrate. Also, many lichen species are growing on a specific type of substratum, such as on siliceous (*Acarospora fuscata, Caloplaca flavovirescens, Lecanora rupicola*) or calcareous stone (*Acrocordia conoidea, Lecanora spadicea, Protoblastenia incrustans*). When the stone surface is eutrophicated, and therefore the chemical composition of the stone become less important, other microorganisms can develop, such as microfungi, heterotrophic bacteria and nitrophilus lichens (*Xanthoria calcicola, X. parietina, Diplocacia canescens*) (Nimis et al., 1992).

2.2 Environment, stone material and biodiversity

Each stone material is part of an ecosystem, being interconnected with surrounding abiotic and biotic factors. At a global level, an ecosystem is part of a biome, and the latter one is part of the ecosphere - the planetary system consisting of atmosphere, geosphere, hydrosphere and biosphere, which are continuously exchanging matter and energy in order to reach a dynamic equilibrium. The climate is a term used to describe the above mentioned interlinked systems powered by the sun, in time and space scale. Macro-climate includes a large area, such as a region or a country, while micro-climate includes only a small area (still part of the macro-climate) such as a single building or a statue (Fig. 3). The artwork ecosystem is a complex self-regulating system, composed by the biotope (stone material), the biocenosis (organisms) and the surrounding environment (micro-climate). The climate is the principal factor controlling the distribution and dynamics of ecosystems (Levéquê, 2003). Pollution is a consequence of industrialization and urbanization of the society and is the main cause of climate changes. This has an impact not only on human health and natural environment, but also on the built environment where, chemical, physical and biological weathering are influenced by climatic factors. The effect of climatic and weather changes on all living organisms (bioclimate) is evidenced especially by loss of biodiversity and increasing of uniformity, both on macro- and micro-ecosystem scale (Caneva, 2010). Studies on biodeterioration of monuments in relation to climatic changes (Caneva et al., 1995; Ariño et al., 2010) stressed the floristic diversity reduction and disappearance of the most sensitive species, such as many lichens. In the same

time, other species among algae, cyanobacteria, fungi and lichens developed resistance to air pollutants, becoming invasive species. It has been reported that introduction of building materials, not common in a certain geographical area, increase biodiversity by colonization of specific microorganisms (Ariño & Saiz-Jimenez, 1996).
Within the general stone cultural heritage sector, it can be considered several types of environments (Table 1), which play a considerable role in the development of specific and common biological agents (Albertano et al., 2008).

Environment	Characteristics	Biological colonizers
TERRESTRIAL ENVIRONMENTS		
A. Enclosed environments:	*reduced exchange with the outdoor environment*	
A1. museums	inside microclimate conditions (usually controlled and monitored) are strongly influenced by location and geography of the site, characteristics of the building and museum management	rarely heterotrophs (bacteria and fungi)
A2. churches and crypts	a certain stability of the internal microclimate conditions, not much influenced by the daily and seasonal variations, but influenced by their function (number of visitors, heating), structural and architectural characteristics	actinomycetes, cyanobacteria, microalgae, fungi
A3. tombs, catacombs and hypogean environments	located below the ground level, with a high and constant relative humidity level (>70%) and low and relative constant temperature (between 10-18°C); the stability of microclimate conditions is related with the size and depth of hypogean space, characteristics of the terrain above, the degree of pollution in the atmosphere, and absence of visitors	subaerial microflora – archaeobacteria, bacteria, actinomycetes, cyanobacteria, microalgae, fungi, phototrophic biofilms, mosses, insects, bats
B. *Outdoor environments:*	*different characteristics according to the climate context, geographical and topographical location*	
B1. urban environments	high population density, with human features expressed by urbanization processes, with a microclimate becoming warmer and drier, with increasing atmospheric pollution	black crust (bacteria, cyanobacteria, algae, fungi) and vascular plants, less lichens and mosses, insects and avifauna

Environment	Characteristics	Biological colonizers
B2. parks and rural environments, archaeological remains	microclimate conditions induced by forest and vegetation cover, with lower temperatures, lower chemical pollution, higher relative humidity, higher nutrients availability and biological contamination, with respect to urban environment	abundant and varied biological colonization forms, bacteria, cyanobacteria algae, fungi, lichens, mosses, vascular plants micro and macrofauna
B3. coastal environments (sandy coastline and maritime rocks)	influenced by the action of the sea, with high concentrations of marine salts both in the air and on the substrate, according to distance from the sea, with an urban or rural microclimate characteristics	halotolerant and halophytic species, algae, cyanobacteria, fungi, lichens, mosses, plants
B4. fountains and nymphea	usually located outdoor, in parks or urban environments, with a constantly or sporadically water supply which can wet totally, partly or not the stone surfaces, often with incrustations and eutrophication areas	aquatic communities, poikilohydric microorganisms, subaerial microflora, lichens, mosses, vascular plants, fishes
C. *Semienclosed environments:*	*partially circumscribed by natural materials (walls, rocks, soil) or by a roof cover*	
C1. loggia and porticoes	microenvironmental conditions can be uniform or with strong fluctuations, depending on the architectural structure, distance from the ground, height from the pavement, inclination, extensiveness of the roof cover	photoautotrophic microorganisms (cyanobacteria, algae), heterotrophic microflora, lichens, bryophytes, weeds
C2. rupestrian environments and caves	located in a natural environment, dug into the rock, generally in close contact with the soil, the microclimate conditions vary greatly in relation to the distance from outside, with similarities with hypogean environments	sciaphilic cyanobacteria, bacteria, microfungi, algae, insects
C3. sheltered archaeological sites	microclimate conditions vary according to size and type of the shelter, the moisture level usually is lower with respect of non-sheltered sites but it could generate a green-house effect	phototrophs, lichens, mosses, vascular plants, micro and macrofauna
MARINE AND FRESHWATER ENVIRONMENTS	microclimate conditions afforded by the specific aquatic environment	biofouling organisms

Environment	Characteristics	Biological colonizers
EDAPHIC ENVIRONMENTS	microclimate conditions afforded by the specific edaphic environment (e.g. clay seems to create an anoxic environment, sandy soils create a low water content environment, while permafrost a low temperature one)	microflora - bacteria (heterotrophic and chemoautotrophic), yeasts, actinomycetes, fungi, cyanobacteria

Table 1. Types of environments and their impact on biocolonization of art and building stone materials

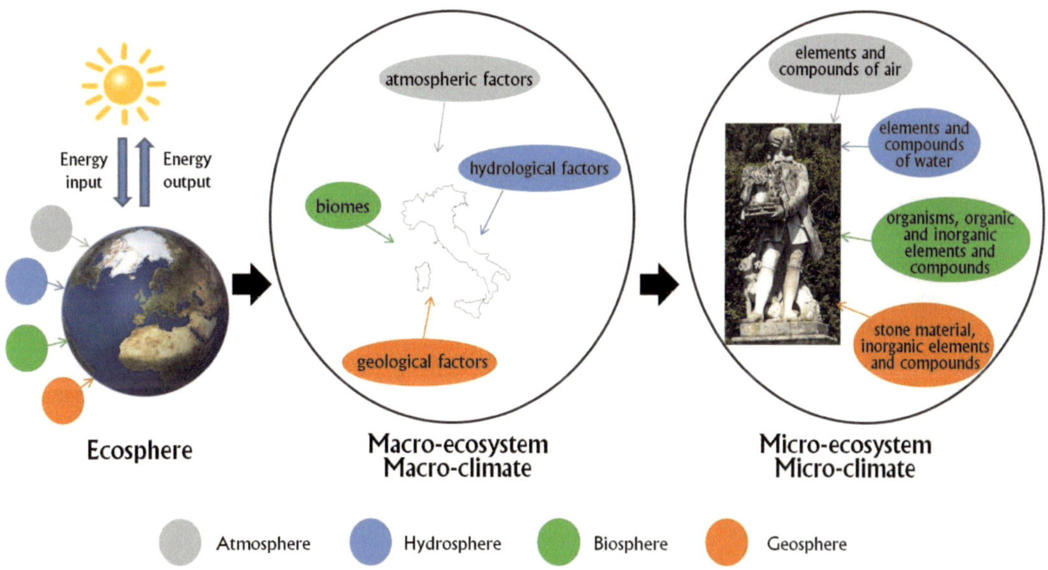

Fig. 3. Ecological hierarchy of an artwork ecosystem as part of the ecosphere

2.3 Mechanisms of colonization

The colonisation mechanism starts when one organism finds the optimal ecological conditions for dwelling, growth and multiplication. In fact, microorganisms may settle on the stone material during its entire life cycle, before and after its processing, from the quarry to the finished work of art (Garcia-Vallès et al., 2000; Cámara et al., 2011). This initial contact may result in a transitory chemical attraction that is difficult to characterize. Cell control attachment and release, are depending on the suitability of the microhabitat.

The photosynthetic microorganisms develop easily on the stone surfaces and, once established, allow the growth of more complex microbial consortia formed by heterotrophic microroganisms which can exercise stronger deteriorating activity (Tiano, 1993; Tomaselli et al., 2000; Crispim et al., 2003; Peraza Zurita et al., 2005). These phototrophic microrganisms are building up a so-called *biofilm*, enriched with organic and inorganic compounds.

As regard the stone colonization by lichens, their role of primary colonizers seems doubtful, especially for the crust like endolithic forms. According to Savoye & Lallemant, 1980, lichens start to develop on the substrate after it was partially enriched with nutrients from air and previous colonizers - bacteria (Lisci et al., 2003). The prime colonizers are in fact the single components (algae and microfungi) and if they are compatible can develop in a new symbiotic organism, the lichen (Joneson & Lutzoni, 2009).

At more advanced stage of stone colonization, the bryophyte communities and vascular plants can develop (Warscheid & Braams 2000). An important role for their growing is the presence of a protosoil (Gómez-Alarcón et al., 1995a), which is favoured by previously colonizations (cell debris) and by airborne particles retained on biofilms or accumulated in cracks or holes.

Biofilms occur on all solid surfaces in habitats where exists a constant availability of moisture. This complex biocenosis consist principally of water (70 to 95% of the fresh weight), extracellular polymeric substances (Flemming, 1993) and microorganisms such as phototrophs (algae, cyanobacteria, diatoms) and heterotrophs (bacteria, fungi, protozoa, nematods), which are embedded in this hydrated matrix. The proportion of each group is varying seasonally and it is influenced by different habitats (Underwood, 1984; Anderson, 1995; Roeselers et al., 2007). Furtermore, the biofilm contains cells debris, airborne particles and spores together with inorganic material adsorbed from the substratum (Warscheid, 2008). This complex structure also contains biopolymers such as exopolysaccharides (EPS) with adhesive properties, which are very relevant in the early development stage of a biofilm, since they facilitate the attachment of cells to the substrate (Decho, 2000, Barranguet et al., 2005).

The organization of a biofilm on a solid surface consists of three main levels: molecular fouling, micro-settlement and a macro-settlement. It has been noted that the formation of an organic molecular layer should be realized before the attachment of the microorganisms on a solid substrate. After that a reversible stage of adhesion of the primary colonisers starts, followed by their fixing and propagation, with an auto-organised three-dimensional structure. The adhesion of secondary colonizers (filamentous cyanobacteria, fungi) with further colonization of invertebrates, led to a mature biofilm with an expressed specific micro-ecosystem structure. This is a homeostatic phase of the biofilm, with a continuous growth and detachment of small parts. The mature biofilm has a complex genetic heterogeneity which confers stability and resistance to the biocenosis (Nikolaev & Plakunov, 2007). Biofilm evolution is also dependent by stochastic and mechanical processes, deterministic phenomena and temporal changes (Wimpenny et al., 2000). The formation of a visible biofilm on a stone surface in natural environment is a quite longer process (even months), and its proliferation is strongly dependent on the environmental conditions (especially light, temperature and water content) and on the concentration of the microorganisms that can adhere on it.

The observed patina composition on a marble specimen with a confocal microscope (CSLM) contains are the green algae, cyanobacteria and diatoms. These phototrophic pioneers are observed, as first stone colonizers in a young biofilm just after 6 days of incubation in contaminated water (Fig. 4) and it is possible to see that cyanobacteria prefer to colonize deeper irregularities of the stone substrate. The presence of the EPS and heterotrophs can be observed just after 22 days of immersion of the marble specimen in the contaminated water, under indoor laboratory conditions (Fig. 5).

3. Biological heterogeneity on stone artifacts

Biological development on stone monuments is strongly influenced by the whole complex interaction between biotic and abiotic factors. All together contribute to different biological developments. In fact, when it occurs, may give rise to a great variety of colonization patterns following dynamic phenomena and/or ecological succession.

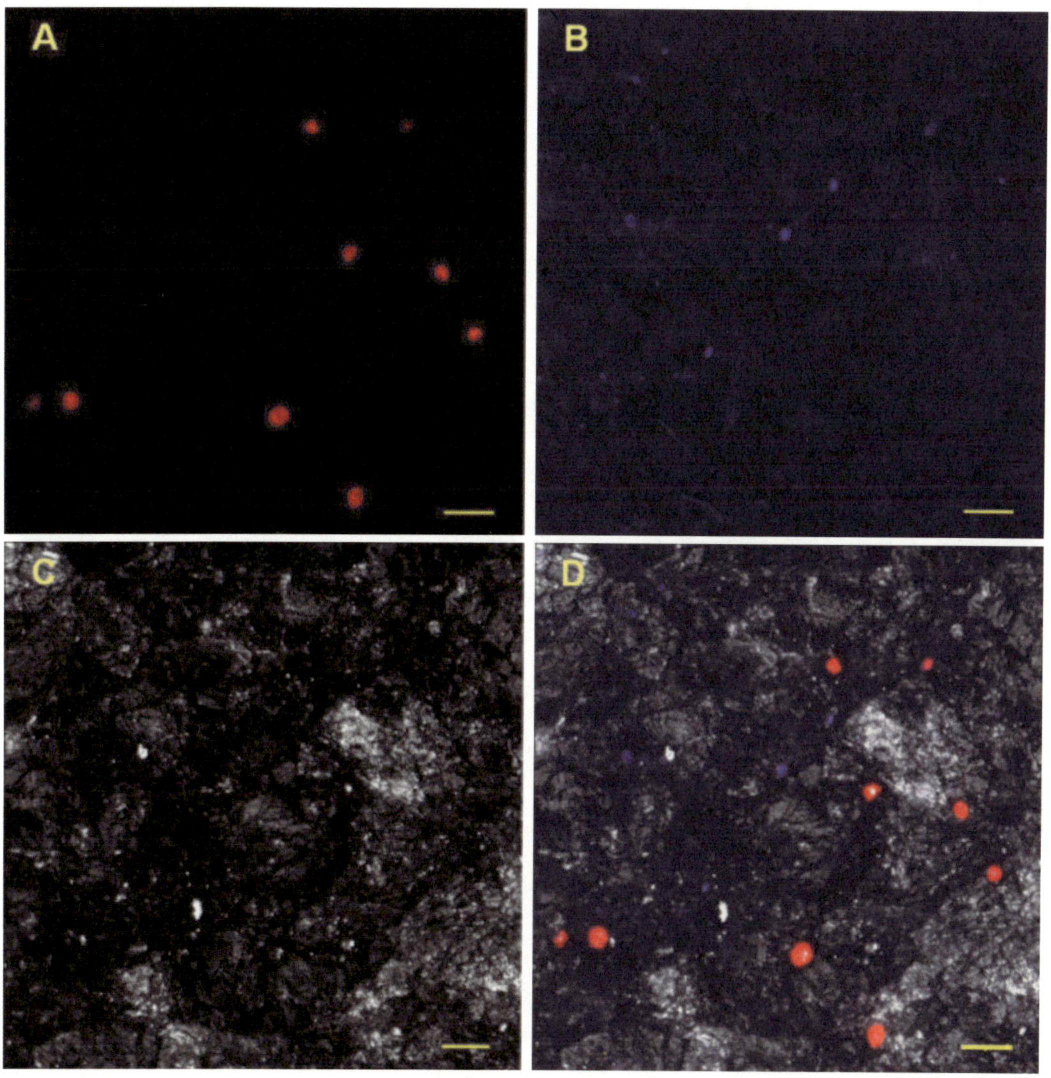

Fig. 4. Maximum intensity projection images of 6 days old biofilm developed on the marble sample examined by CLSM. Images are showing (A) the autofluorescence of individual algae in the red channel, (B) unidentified microorganisms in the blue channel, (C) reflection of the substratum in the blue channel, and (D) the resulting overlay of the four channel (in the green channel was not recorded any signal). Scale bar 20 µm. Colors are false colors

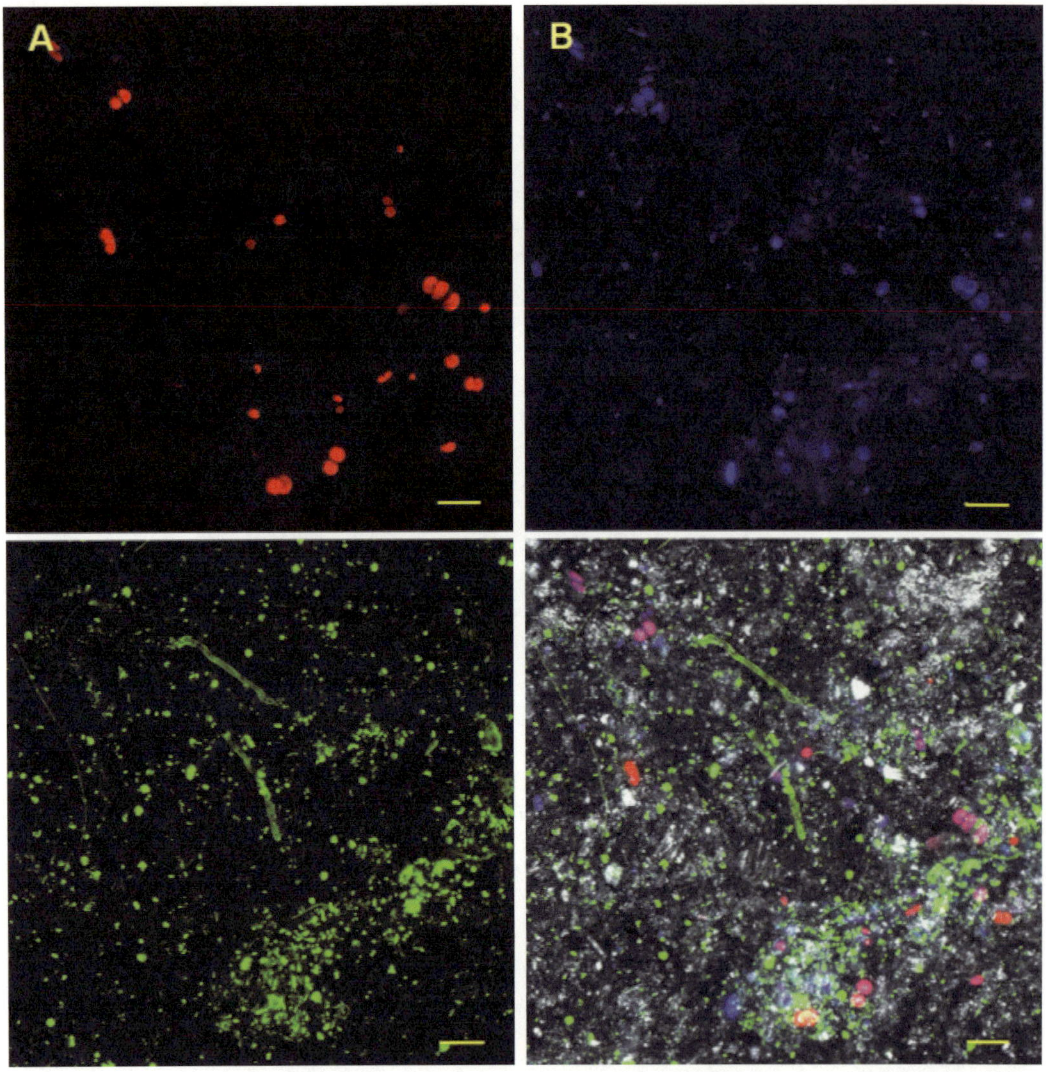

Fig. 5. A 22 days old biofilm developed on a marble sample. The single channel signals are showing (A) algae and diatoms in division - autofluorescence in the red channel, (B) algae and cyanobacteria - autofluorescence in the green channel, (C) EPS and fungal hyphae stained with concanavaline A cojugated with Alexa Fluor 488 - blue channel, and (D) the resulting overlaying of all four channels (including the channel used for capturing the substratum signal, in reflectance mode). Scale bar 20 μm

The biological growth can present different colours, morphologies and agents. The bio-patina or patina is the most general term for defining a epilithic microbiota spread on a stone surface. This, different from the "artistic patina", alters the aesthetical, physical and chemical aspects (Krumbein, 2003) of the substrate. When this patina exerts a protective role for the stone substrata, it can be called bioderma. It must be also considered that in the initial phase some biological agents can develop inside the stone with a concealed pattern. The type of organisms can be subdivided in chasmoendoliths (colonize cracks, fissures or pores, being in contact with the stone surface), cryptoendoliths (colonize pores inside rocks and/or develop in strata inside parallel to the stone surface) and euendoliths (actively dissolve and penetrate the stone surface, forming cavities with different morphologies). The bio-patina can support the development of other organisms such as protozoa, nematodes, insects and arthropods.

According to different typologies found in the literature, biological colonization can be divided in: **A.** *micro-colonial epilithic growths*, **B.** *epilithic formations with patina aspect* and **C.** *chasmo-endolithic developments*; each of these categories can be mono- or multispecies associations (Figs. 6 and 7). The epilithic organisms may be superficially attached to substratum, through synthesized exopolysaccharides (EPS), as the phototrophs, while fungi, lichens, mosses or plants, are owning specific structure (hyphae, rhizines or rhizoids, roots) that allow them to mechanically fix into the stone porosity. Mature biofilms or crusts, demonstrate great resistance to environmental stresses and conservation interventions.

Biological colonisations on monumental stone surface essentially assume the following forms:

A. Epilithic growth with micro-colonial aspect

A1. *Pustules* (Fig. 6a,b)

A pioneer stage of colonization, with hemispheric morphology, with a verrucose and gelatinous aspect, due to the EPS secretion, being formed by one or more phototrophic biotypes, with a compact pseudo-parenchimatic structure (Sánchez Castillo & Bolívar, 1997).

Organisms: green algae (*Palmella miniata*) and cyanobacteria (*Pleurocapsa minor, Chamaesiphon polonicus*) (Sánchez Castillo & Bolívar, 1997), diatoms (*Navicula, Achnanthes*).

Ecology: in wet or very humid areas, in the monumental fountains and nymphea, in hypogean environments.

A2. *Microbial concealed growth*

This microbial contamination (up to 5 cm deep), gives an eroded and/or pulverized aspect to the stone surface (Warscheid & Braams, 2000). Microbiological development difficult to observe were found on mural paintings. Some pigments can change colour due to their oxidation, reduction or metal ions transfer after chemical processes induced by bacteria (Petushkova & Lyalikova, 1986; Nugari et al., 2008).

Organisms: chemiotrophic bacteria (*Thiobacillus, Nitrobacter*) (Caneva et al., 1991; Warscheid & Braams, 2000), heterotrophic bacteria (*Bacillus, Arthrobacter*) (Petushkova & Lyalikova, 1986; Nugari et al., 2008).

Ecology: on coarse grained porous stones, sandstones, man-made stones and frescoes, in urban environments, churches and crypts, tombs and catacombs.

A3. *Irregular stains*

The stains are related to the pigments belonging to the stone colonizers: *black stains* (melanin, melanoidins, products of chlorophyll degradation) (Fig. 6f); *green and greenish*

Fig. 6. Various typologies of biological colonization found on monumental stones: (a) green algae pustules and (d) microbial mat on artistic fountain from Alhambra complex, Granada (Spain); (b) diatoms brown pustules and (c) phototrophic biofilm developed on Tacca's fountains from Florence (Italy); (e) irregular whitish stains produced by actinomycetes in an Etruscan tomb from Chiusi (Italy); (f) microcolonial fungal growth on marble stone, from Trespiano monumental cemetery, Florence (Italy)

stains (photosynthetic pigments from algae and cyanobacteria) (Fig. 6a); *yellow-orange-brownish stains* (carotens and carotenoids and degradation products of chlorophyll such as phycobiliproteins) (Fig. 6b); *bright orange, pink and red stains* deriving from pigments of chemoorganotrophic (halophilic) bacteria and degradation products of cyanobacteria and algae with iron enrichment (Warscheid & Braams, 2000). They may have a powdered, mucous or compact aspect, according to environmental conditions and constituting organisms.

Organisms: black fungi (*Ulocladium, Alternaria, Scolecobasidium, Phytomyces*); yeasts (*Rhodotorula minuta*); heterotrophic pigmented bacteria (*Micrococcus roseus, Flavobacterium*) (Tiano & Tomaselli, 1989), actinomycetes (*Streptomyces, Nocardia*) (Fig. 6e); chemoorganotrophic bacteria (*Bacillus, Arthrobacter, Rhodococcus, Brevibacterium*) (Warscheid & Braams, 2000; Pinna & Salvadori, 2008);

Ecology: in tombs, catacombs and hypogean environment, in artistic fountains and nymphea

B. Epilithic formations with patina aspect

B1. *Films*

Films are formed by being thin unistratified sheets fairly homogeneous, generally a further development of the pustules, without an increase in species complexity or represented biotypes. It can become a more complex community throughout its development under favourable environmental conditions (Sánchez Castillo & Bolívar, 1997). They are located superficial or along natural crackers and fissures, mostly unilamellar thin structure of mono- or multispecies (Warscheid & Braams, 2000).

Organisms: bacteria, phototrophic microflora (*Chamaesiphon, Pleurocapsa, Pleurastrum, Chlorogea*), microfungi.

Ecology: in wet or very humid areas, in the monumental fountains and nymphea.

B2. *Biofilms, subaerial biofilms*

Biofilms are aggregates of microorganisms embedded in a mucilaginous organic matrix (EPS) found in aqueous environments, usually intended as a spread microbial growth at the water-substratum interface, such as the basins and artistic fountains pools or the architectural elements under a continuous water flow (Cuzman et al., 2010). According to the quantity, turbulence and water spraying the stone surface can partially or totally in direct contact with the air as well. The *subaerial biofilms* develop on surfaces, at the atmosphere-rock interface, but in very damp environmental conditions. They are reported in hypogean monuments, catacombs (Gorbushina, 2007), on the walls or statues with north orientation and surrounded by vegetation, or areas with water percolations. Subaerial biofilms if colonize not only the irregularities of the stone surface but actively penetrate deeper into the substratum, are called *endolithic subaerial biofilms* (see C3).

These formations are characterized by patchy growth, usually in presence of high humidity conditions, being more complex than a film or pustules grouping, but thinner and less complex than a microbial mat. They may have a various coloured patinas aspect, staining the stone surface with organic pigments (e.g. chlorophylls, carotenoids, melanins) (Warscheid & Braams, 2000). The colour is depending on the type of biocenosis, its development stage and of the growth phase of the prevailing species. Individuals within these communities avoid sexual reproduction, but cooperate extensively with one another especially to avoid loss of energy and nutrients (Gorbushina, 2007). They can survive as

vegetative cells or tissue-like structure for long periods in case of stressing conditions (Gorbushina & Broughton, 2009), being very resistant to control treatments.
Organisms: mixed association of different biological groups, such as coccoid cyanobacteria (*Gloeocapsa, Chlorococcum, Myxosarcyna*), filamentous cyanobacteria (*Phormidium, Leptolyngbya*), coccoid algae (*Chlorella, Muriella terestris*), filamentous algae (*Trentepohlia, Mougeotia*), heterotrophic bacteria (*Rubrobacter radiotolerans*), and fungi (*Acremonium, Fusarium, Penicillium, Aspergillus*).
Ecology: in wet, very humid, with water condensation or percolation areas, in the monumental fountains and nymphea, in tombs, catacombs and hypogean environment, crypts, marine and fresh water environments

B3. *Microbial mats*

Phototrophic formations with a considerable thickness, characterized by prevailing filamentous species, developed at both water or atmosphere–stone interface. They may present filamentous fungi or lichens as well (Sánchez Castillo & Bolívar, 1997). These aggregations are a type of stratified and complex biofilm, laminated or forming stromatolites, being bound to the substrate and held together by slimy EPS. They are found especially in extreme environments (Riding, 2000). Usually the uppermost layers are dominated by aerobic phototrophs, while the lowest layers are dominated by anaerobic bacteria. The microbial mat may be *mucous* (made of microorganisms with sheath) or *fibrous* (made of microorganisms without sheets) (Sánchez Castillo & Bolívar, 1997).
Organisms: mucous mats (Fig. 6d): *Phormidium, Diadesmis, Apatococcus;* fibrous mats: *Cladophora, Gongrosira, Pleurastrum, Melosira* (Fig. 7e).
Ecology: in areas with long dry-wet alternation cycles, in the monumental fountains and nymphea, in the coastal environments.

B4. *Crusts*

The formation of biogenic crust is due to a chemical-physical process (calcite precipitation) mixed with the biological growth present on stone surface, or can be constituted by the spreading of the organisms that form crust-like structures, such as the crustose lichens and some type of mosses.
Crusts can be found in the uppermost layers of the stone, up to 1 cm depth, being composed by a complex and stable microflora (Warscheid & Braams, 2000). This mineralization process can occur in biofilms, subaerial biofilms or mats. It offers to subaerial biofilms protection from the environment and allows the accumulation and agglutination of airborne particles and deposition of minerals (Gorbushina & Broughton, 2009). The encrusted biofilm is able to penetrate inside the rock by hyphal growth and by biocorrosive activity (Warscheid & Braams, 2000). In monumental fountains the mineralization formations (calcareous deposits) may present a flat or nodular aspect (Bolívar & Sánchez Castillo, 1997; Sánchez Castillo & Bolívar, 1997).
Lichenic crust is closely adhering to the substrate through the hyphae of inferior cortex that are penetrating to varying degrees into the substratum. The lichenic crust can have different aspects, according to morphological variation of visible part of the lichens: leprous, continuous, areolate, verrucose, peltate, or placodiomorphic. (Piervittori et al., 2008).
Mosses crusts are not so often found on monumental stone surfaces. It is well known that many mosses are involved with the formation of calcareous deposits, being usually

associated with other microorganisms, such as cyanobacteria, green microalgae and diatoms. Crusts forming mosses play an important role in artistic fountains degradations, and can led to incrustation several decimetres thick, completely obliterating the legibility of the work, as was observed on Fountain of the Dragons in Villa d'Este in Tivoli (Ricci, 2008).
Organisms: encrusted biofilms: many cyanobacteria and algae (*Schizothrix, Symploca, Chlorosarcinopsis* minor, *Apatococcus, Scenedesmus, Trentepohlia; Scytonema julianum, Geitleria calcarea, Loriella osteophila, Herpizonema pulverulentum*) (Ortega-Calvo et al., 1995; Sánchez Castillo & Bolívar, 1997); chemoorganotrophic fungi (*Exophiala, Penicillium, Cladosporium, Aspergillus, Phoma, Ulocladium)* (Warscheid & Braams, 2000); crust forming lichens (*Caloplaca, Dirinia massiliensis, Lecanora, Pertusaria*); crust forming mosses (*Eucladium verticillatum, Barbula tophacea, Cratoneuron, Philonotis*).
Ecology: may occur in areas with dry-wet alternation cycles which favour the mineralization process, but with enough humidity, in monumental fountains and nymphea, in parks, rural and urban environments, in coastal environments and in hypogeal monuments.

B5. *Black crusts (Fig. 7a,c)*

The black sulphated crust, due to the interaction between calcareous substrate and the polluted atmosphere, differs from the biogenic one as being usually formed in areas sheltered from rainfall. The biological black crusts originate in a humid environment and in stone areas with running water (Gómez-Alarcón et al., 1995b; Lewin, 2006). Many heterotrophic and phototrophic organisms can adhere on the gypsum-rich black crusts. These biological agents are able to metabolize the organic (phenanthrene) or inorganic compounds (sulphur) found on it (Ortega-Calvo et al., 1995; Saiz-Jimenez, 1997). A current study reports in unpolluted environments, the presence of thin black crusts with biogenic origin, composed by a massive presence of filamentous cyanobacteria (Gaylarde et al., 2007).
Organisms: in non-biogenic black crust: bacteria (*Bacillus licheniformis, B. subtilis, B. brevis, Corynebacterium glutamicum, Actinomyces, Flavobacterium breve, Pseudomonas stutzeri, Nocardia*) (Turtura et al., 2000), fungi (*Papulaspora-like, Engyodontium album, Aureobasidium pullulans, Cladosporium sphaerospermum*) (Saiz-Jimenez, 1997; Frank-Kamenetskaya et al., 2009), and cyanobacteria (*Gloeothece, Chlorosarcinopsis*) (Ortega-Calvo et al., 1994); in biogenic black-crust: cyanobacteria (*Gloeocapsa, Nostoc flagelliforme, Chloroglea microcystoides, Scytomena, Oscillatoria*), black lichens (*Verrucaria*), and black fungi (*Ulocladium, Phoma, Alternaria*), meristematic fungi (*Aureobasidium, Sarcinomyces petricola*).
Ecology: in drier and occasionally wetted habitats and exposed to sunlight, in urban and rural environments.

B6. *Whitish efflorescence (Fig. 6e)*

Irregular heterotrophic colonial growth with puffy (due to white fungal mycelia) or powdering aspect (due to the bacterial sorted colonies). The actinomycetes development give rise to extensive, thin and pulverulent whitish patinas or to thicker and more localized forms which closely resemble saline efflorescences (Nugari et al., 2008).
Organisms: fungi (*Fusarium solani*) (Bastian & Alabouvette, 2009), actinomycetes (*Streptomyces, Micromonospora, Nocardia*) and bacteria (*Pseudomonas fluorescens*) (Nugari et al., 2008; Nugari et al., 2009).
Ecology: in habitats with organic matter, in shadow and humid areas, on wall paintings in tombs, catacombs, hypogean and rupestrian environment.

Fig. 7. Different patterns of biological growths found on monumental stones: (a) homogenous black crust with a detail (b) of interganular development of threadlike microcolonies containing cyanobacteria and black fungi, as revealed by microscopic insight, Carrara Marble Museum, Italy; (c) heterogenous lichenic black crust with a wasp nest on a limestone tomb, Trespiano monumental cemetery; (d) bryophite extended carpets, Fiesole archeological site, Italy; (e) fibrous microbial mat with a filamentous diatom (*Melosira* sp.) as a dominant species, Lindaraja Fountain from Alhambra complex, Granada (Spain); (f) micropits containing the apotheci of an endolithic lichen, Trespiano monumental cemetery

B7. *Carpet and cushion forms (Fig. 7d)*

These formations are composed by bryophyte associations, and rarely by foliose lichens developing in carpets-like form. The *carpets* are extended on wide stone surfaces, with individuals growing parallel to one another, either with an uniaxial or a branched growth habit. The *cushions* can be found in sheltered areas of a monument, where soil particles easily accumulate, with individuals placed radiate and having lateral ramifications with a hemispheric structure (Ricci, 2008). These biological formations can favour the seeds germination of small size vascular plants.
Organisms: saxicolous mosses (*Tortula muralis, Eucladium verticillatum*), foliose lichens (*Parmelia, Xanthoria*).
Ecology: in damp habitats, in archaeological sites, parks and rural environments.

C. Chasmo-endolithic growth

C1. *Micropits*

Small cavities created by active substrate dissolution. The euendolithic cyanobacteria may have different boring patterns, according to the various species (Caneva et al., 1991; Hoppert et al., 2004). The endolithic lichens on carbonate matrix can colonize to a depth of few millimetres and in exceptional cases deeper than one centimeter (Piervittori et al., 2008). The biopitting phenomena can be easily confused with abiotic alterations, such as soiling, due to the fact that appears as a simply variation of the stone's natural colour, often green or black (Pinna & Salvadori, 2008).
Organisms: blue-green algae (*Hyella balani, Hyella caespitose, Hormatonema paulocellulare, Kyrtuthrix dalmatica*), endolithic fungi (*Lichenothelia*), endolithic lichens (Fig. 7f) (*Lecidea, Caloplaca*) (Caneva et al., 1991).
Ecology: in humid and dim habitats, sometimes on surfaces sporadically wet and exposed to direct sunlight, in urban, rural and coastal environments.

C2. *Microcolonial structure*

The microcolonial development can have three main patterns: (a) inter-crystalline growth (Fig. 7b), (b) biopitting growth (see *C1*) and (c) growth in already formed cracks and fissures (Urzì et al., 2000). The biologic agents forming this kind of development are usually cyanobacteria and/or black meristematic fungi. It seems that only the fungal filaments closest to the rock surface are pigmented, whereas filaments grown deeper in the stone are colourless (Hoppert et al., 2004).
Organisms: lichen-forming fungi (*Verrucariales, Lichenothelia*), black yeast and meristematic fungi (*Sarcynomyces,Exophiala, Aureobasidium, Capnobotryella, Phaeococcus, Trimmatostroma*); cyanobacteria (*Gloeocapsa*).
Ecology: usually on surfaces exposed to direct sunlight, but with enough humidity when phototrophs are present, in urban, rural and coastal environments.

C3. *Endolithic subaerial biofilms*

In the long-established subaerial biofilms, the biofilm communities are not only on the stone surface but they have crept into deeper layers of the substrate and offered themselves an endolithical niche in which the environmental extremes are better buffered. It has a complex spatial pattern of mineral grains, pores and fissures (Gorbushina & Broughton, 2009). Sometimes this biocenosis has a sandwich aspect, with compact and distinct green layer beneath the stone surface (Gorbushina, 2007; Saiz-Jimenez et al., 1990).

Organisms: cyanobacteria (*Chroococcidiopsis, Synechococcus*) (Saiz-Jimenez et al., 1990), meristematic fungi, endolithic lichens.

Ecology: in humid and/or sporadically dry areas, in urban, rural and coastal environments.

4. Consequences of biological presence on artistic stone material

Beside the weathering agents, a synergistic biodeteriorative effect on stone surfaces can be started by the concomitant growth of phototrophic and heterotrophic populations. The biodamages produced on stone substrata are especially related from one side to the metabolic activity of living organisms and from the other side to aesthetical changes of the stone surface. The biological agents dwelling on stone monuments are involved directly or indirectly in the weathering of stones and constituent minerals (Warscheid & Braams, 2000), and can induce various alterations forms, as can be seen in Table 2.

Cause	Alteration type	Potential biological development
mechanical stresses such as swelling-shrinking and physical penetration inside the rock	decohesion fissures fractures	biofilms and subaerial biofilms, lichenic crusts, physical penetration of fungal hyphae or specialized organs such as rhizines (lichens), rhizoids (mosses) or roots (vascular plants)
reaction with atmospheric pollutants, correlated with condensation phenomena	peeling swelling pitting	black crusts
solubilization due to metabolic processes, acid attacks	disintegration pulverization flakes	microbial concealed growth (bacteria), lichens
deposits of different kinds of materials, crystallization of minerals	encrustation, concretions, film peeling	biofilms, subaerial biofilms, microbial mats, lichenic and mosses crusts
active penetration by chemical dissolution	pitting	endolithic cyanobacteria and lichens
presence of various natural pigments and degradation products	chromatic alteration spotting	phototrophic microorganisms black fungi chemoorganotrophic bacteria

Table 2. Different weathering processes associated with biological colonization and related alterations on artistic stone material

The main types of biodeterioration processes, often intercorrelated, can be classified as (Amoroso, 1995; Tiano, 1998):

- *physical-mechanical*: abrasion, detachments, disaggregation, fissures and crevices formation due to organisms growth or movement.
- *chemical*: chemical changing of the stone substrata due to interaction with metabolic products, solubilisation, biopitting, and formation of new-reaction products.
- *aesthetical*: coloured patches, patinas, crusts.

Generally speaking, besides the discoloration and staining processes produced by biogenic pigments, the microflora leads to the change of materials characteristics with regard to their mechanical properties, superficial absorbency/hydrophobicity, stability, density, diffusivity and thermal-hydric behavior (Warscheid, 1996, 2008). For example, the biological formations with high water content (films, biofilms, subaerial biofilms, mats) can induce combined alteration processes, due to the fluid retention, to the repeated wetting and drying cycles with the subsequent expansions and contractions of the biocenosis, to the aggressive action of both metabolic products and to atmospheric pollutants entrapped in the EPS, by increasing chemical corrosion process. Therefore, the stony structure can present pitting, ion transfer, leaching processes and dwindling (Tiano, 1998).

A particular aspect of biodeterioration of stone monuments is linked to endolithic organisms colonizing the interior of rocks (Salvadori, 2000). The presence of endolithic formations (crusts, micropits, endolithic subaerial biofilms, plant root systems), which can penetrate up to depths of several millimetres, with the diffusion of their excreted products into the intergranular matrix, enhance the weathering reactions and decrease mechanical properties of the stone material.

The fauna has also an important role in stone biodeterioration. Microfauna, such as various arthropods which are feeding fungi and lichens, contributes in their diffusion with the transport of propagules and spores. Macrofauna, especially the birds, have a decay action due both their movement and organic material accumulation, which favors the growth of heterotrophic organisms.

5. Acknowledgements

The authors are grateful for funding to ICVBC-CNR, through the TeCon@BC Project (Regione Toscana – Italy, POR FESR 2007-2013, cod 57476) and to Mariona Hernández-Mariné (Facultat de Farmàcia, Universitat de Barcelona, Spain) and Mónica Roldán (Servei de Microscòpia, Universitat Autònoma de Barcelona, Spain), for their kindly assistance with the CLSM analysis.

6. References

Albertano, P.; Altieri, A.; Caneva, G.; Ceschin, S.; Maggi, O.; Nugari, M.P.; Pasquariello, G.; Persiani, A.M.; Piervittori, R.; Pietrini, M.; Pinna, D.; Ricci, S.; Roccardi, A.; Urzì, C.; Tomaselli, M.L. & Valenti, P. (2008). Problems of biodeterioration in relation to particular types of environments, In: *Plant biology for cultural heritage*, The Getty Conservation Institute, pp. 171-218, ISBN 978-0-89236-939-3, Los Angeles, USA

Amoroso, G.G. (1995). Incrostazioni, patine e alterazioni delle superficie lapidee, In: *Il restauro della pietra nell'architettura monumentale*, Dario Flaccovio Editore, pp. 111-144, ISBN 88-7758-244-8, Italy

Amoroso, G.G. (2002). *Trattato di scienza della conservazione dei monumenti*, Alinea Editrice, ISBN 88-8125-410-7, Firenze, Italy

Anderson, M.J. (1995). Variations in biofilm colonizing artificial surfaces: seasonal effects and effects of grazers. *Journal of Marine Biology Association of United Kingdom*, Vol.75, No.3, (August 1995, online May 2009), pp. 705-714, ISSN 0025-3154

Ariño, X. & Saiz-Jimenez, C. (1996). Biological diversity and cultural heritage, *Aerobiologia*, Vol.12, No.4 (December 1996), pp. 279-282, ISSN 0393-5965 (print) 1573-3025 (online)

Ariño, X.; Llop, E.; Gómez-Bolea, A. & Saiz-Jimenez, C. (2010). Effects of climatic change on microorganisms colonizing cultural heritage stone materials, In: *Climate change and cultural heritage*, Edipuglia, pp. 193-198, ISBN 978-88-7228-601-2, Bari, Italy

Barranguet, C.; Veuger, B.; van Beusekom, S.A.M.; Marvan, P.; Sinke, J.J. & Admiraal, W. (2005). Divergent composition of algal-bacterial biofilms developing under various external factors. *European Journal of Phycology*, Vol.40, No.1, pp.1-8, ISSN 1469-4433 (electronic) 0967-0262 (paper)

Bastian, F. & Alabouvette, C. (2009). Lights and shadows on the conservation of a rock art cave: the case of Lascaux cave. *International Journal of Speleology*, Vol.38, No.1, (January 2009), pp. 55-60, ISSN 0392-6672

Bolívar, F.C. & Sánchez Castillo, P.M. (1997). Biomineralization processes in the fountains of Alhambra, Granada, Spain. *International Biodeterioration & Biodegradation*, Vol.40, No.2-4, pp. 205-215, ISSN: 0964-8305

Cámara, B.; De los Rios, A.; Urizal, M.; Álvarez de Buergo, M.; Varas, M.J.; Fort, R. & Ascaso, C. (2011). Characterizing the microbial colonization of a Dolostone quarry : implications for stone biodeterioration and response to biocide treatments. *Microbial Ecology*, DOI 10.1007/s00248-011-9815-x, ISSN: 0095-3628 (print) 1432-184X (online)

Caneva, G.; Nugari, M.P. & Salvadori, O. (1991). Biodeterioration of inorganic materials, In: *Biology in the conservation of works of art*, ICCROM, pp. 87-112, Sintesi Grafica, ISBN 92-9077-101-1, Rome, Italy

Caneva, G.; Gori, E. & Montefinale, T. (1995). Biodeterioration of monuments in relation to climatic changes in Rome between 19-20th centuries. *The Science of the Total Environment*, Vol.167, No.1-3 (May, 1995), pp. 205-214, ISSN 0048-9697

Caneva, G. & Ceschin, S. (2008) Ecology of biodeterioration, In: *Plant biology for cultural heritage*, The Getty Conservation Institute, pp. 35-58, ISBN 978-0-89236-939-3, Los Angeles, USA

Caneva, G. (2010). Cultural landscapes and climate change, In: *Climate change and cultural heritage*, Edipuglia, pp. 181-192, ISBN 978-88-7228-601-2, Bari, Italy

Crispim, C.A.; Gaylarde, P.M. & Gaylarde, C.C. (2003). Algal and cyanobacterial biofilms on calcareous historic buildings. *Current Microbiology*, Vol.46, No.2 (February 2003), pp. 79-82, ISSN 0343-8651 (print) 1432-0991 (online)

Cuzman, O.A.; Ventura, S.; Sili, C.; Mascalchi, C.; Turchetti, T.; D'Acqui, L. & Tiano, P. (2010). Biodiversity of phototrophic biofilms dwelling on monumental fountains. *Microbial Ecology*, Vol. 60, No.1, (July 2010), pp. 81-95, ISSN 0095-3628 (print) 1432-184X (online)

Decho, A.W. (2000). Microbial biofilms in intertidal systems: an overview, *Continental Shelf Research*, Vol.20, No.10-11, (July 2000), pp. 1257-1273, ISSN 0278-4343

Flemming, H.C. (1993). Biofilms and environmental protection. *Water Science Technology*, Vol.27, No.7-8, pp. 1-10, ISSN 0273-1223

Frank-Kamenetskaya, O.V.; Vlasov, D.Y; Zelenskaya, M.S.; Knauf, I.V. & Timasheva, M.A. (2009). Decaying of the marble and limestone monuments in the urban environment. Case of studies from Saint Petersburg, Russia. *Studia UBB, Geologia*, Vol. 54, No.2, pp. 17-22, ISSN 1221-0803 (print) 1937-8602 (online)

Gaylarde, C.C.; Ortega-Morales, B.O. & Bartolo-Pérez, P. (2007). Biogenic black crusts on buildings in unpolluted environments. *Current Microbiology*, Vol. 54, No.2, (February 2007), pp. 162-166, ISSN 0343-8651 (print) 1432-0991 (online)

Garcia-Vallès, M.; Urzì, C.; De Leo, F.; Salamone, P. & Vendrell-Saz, M. (2000). Biological weathering and mineral deposits of the Belevi marble quarry (Ephesus, Turkey). *International Biodeterioration & Biodegradation*, Vol.46, No.3, (October 2000), pp. 221-227, ISSN 0964-8305

Gómez-Alarcón, G.; Muñoz, M.; Ariño, X. & Ortega-Calvo, J.J. (1995a). Microbial communities in weathered sand stones: the case of Carrascosa del Campo church, Spain. *The Science of the Total Environment*, Vol.167, No.1-3, (May 1995), pp. 249-254, ISSN 0048-9697

Gómez-Alarcón, G.; Cilleros, B.; Flores, M. & Lorenzo, J. (1995b). Microbial communities and alteration processes in monuments at Alcala de Henares, Spain. *The Science of the Total Environment*, Vol.167, No.1-3, (May 1995), pp. 231-239, ISSN 0048-9697

Gorbushina, A.A. (2007). Life on the rocks. *Environmental Microbiology*, Vol.9, No.7, (July 2007), pp. 1613-1631, ISSN 1462-2912 (print), 1462-2920 (online)

Gorbushina, A.A. & Broughton, W.J. (2009). Microbiology of the atmosphere-rock interface: how biological interactions and physical stresses modulate a sophisticated microbial ecosystem. *Annual Review of Microbiology*, Vol.63, ISSN 0066-4227

Guillitte, O. (1995). Bioreceptivity : a new concept for building ecology studies. *The Science of the Total Environment*, Vol.167, No.1-3 (May 1995), pp. 215-220, ISSN 0048-9697

Hoppert, M.; Flies, C.; Pohl, W.; Günzl, B. & Schneider, J. (2004). Colonization strategies of lithobiontic microorganisms on carbonate rocks. *Environmental Geology*, Vol.46, No.3-4, (August 2004), ISSN 0943-0105 (print), 1432-0495 (online)

Joneson, S. & Lutzoni, F. (2009). Compatibility and thigmotropism in the lichen symbiosis: a reappraisal. *Symbiosis*, Vol. 47, No.2, pp. 109-115, ISSN 0334-5114

Krumbein, W.E. (2003). Patina and cultural heritage - a geomicrobiologist's perspective. *Proceedings of the 5th EC Conference Cultural Heritage Research: A Pan-European Challenge*, pp. 39-47, ISBN 92-894-4412-6, Krakow, Poland, May 14-15, 2003

Levéquê, C. (2003). *Ecology – from ecosystem to biosphere*, Science Publishers, Inc., pp.85-200, ISBN 1-57808-294-3, Enfield, NH, USA

Lewin, R. A. (2006). Black algae. *Journal of Applied Phycology*, Vol.18, No.6, (December 2006), pp. 699-702, ISSN 0921-8971 (print), 1573-5176 (online)

Lisci, M.; Monte, M. & Pacini, E. (2003). Lichens and higher plants on stone: a review. *International Biodeterioration & Biodegradation*, Vol.51, No.1, (January 2003), pp. 1-17, ISSN 0964-8305

Miller, A.; Dionisio, A. & Macedo, M.F. (2006) Primary bioreceptivity : a comparative study of different portuguese lithotypes. *International Biodeterioration & Biodegradation*, Vol.57, No.2, (March 2006), pp. 136-142, ISSN 0964-8305

Miller, A.Z.; Dionisio, A.; Laiz, L.; Macedo, M.F. & Saiz-Jimenez, C. (2009). The influence of inherent properties of building limestones on their bioreceptivity to phototrophic microorganisms. *Annals of Microbiology*, Vol.59, No.4, pp. 705-713, ISSN 1590-4261

Nikolaev, Y.A. & Plakunov V.K. (2007). Biofilm–"City of microbes" or an analogue of multicellular organisms?. *Microbiology*, Vol.76, No.2, pp. 125-138, ISSN 0026-2617

Nimis, P.L.; Pinna, D. & Salvadori, O. (1992). *Licheni e conservazione dei monumenti*, Editrice Club Bologna, Bologna, Italy

Nugari, M.P.; Pinna, D. & Salvadori, O. (2008) Artificial stone. In: *Plant biology for cultural heritage*, The Getty Conservation Institute, pp. 144-149, ISBN 978-0-89236-939-3, Los Angeles, USA

Nugari, M.P.; Pietrini, A.M.; Caneva, G.; Imperi, F. & Visca, P. (2009). Biodeterioration of mural paintings in a rocky habitat: The Crypt of the Original Sin (Matera, Italy). *International Biodeterioration & Biodegradation*, Vol.63, No.6, (September 2009), pp. 705-711, ISSN 0964-8305

Ortega-Calvo, J.J.; Ariño, X, Stal, L.J. & Saiz Jimenez, C. (1994). Cyanobacterial sulfate accumulation from black crust on a historic building, *Geomicrobiological Journal*, Vol. 12, No. 1, pp. 15-22, ISSN 0149-0451 (print) 1521-0529 (online)

Ortega-Calvo, J.J.; Ariño, X, Hernandez-Marine, M. & Saiz Jimenez, C. (1995). Factors affecting the weathering and colonization of monuments by phototrophic microorganisms. *The Science of Total Environment*, Vol.167, No.1-3 (May 1995), pp. 329-341, ISSN 0048-9697

Peraza Zurita, Y.; Cultrone, G.; Sánchez Castillo, P.M.; Sebastián, E. & Bolívar, F.G. (2005). Microalgae associated with deteriorated stonework of the Fountain of Bibatauín in Granada, Spain. *International Biodeterioration & Biodegradation*, Vol.55, No.1, (January 2005), pp. 55-61, ISSN 0964-8305

Petushkova, J.P. & Lyalikova, N. (1986). Microbiological degradation of lead-containing pigments in mural paintings. *Studies in Conservation*, Vol. 31, No.2, (May 1986), pp. 65-69, ISSN 0039-3630

Piervittori, R.; Nimis, P. & Tretiach, M. (2008) Lichens. In: *Plant biology for cultural heritage*, The Getty Conservation Institute, pp. 77-81, ISBN 978-0-89236-939-3, Los Angeles, USA

Pinna, D. & Salvadori, O. (2008). Stone and related materials, In: *Plant biology for cultural heritage*, The Getty Conservation Institute, pp. 128-149, ISBN 978-0-89236-939-3, Los Angeles, USA

Prieto, B. & Silva, B. (2005). Estimation of the potential bioreceptivity of granitic rocks from their intrinsic properties. *International Biodeterioration & Biodegradation*, Vol.56, No.4, (Dicember 2005), pp. 206-215, ISSN 0964-8305

Ricci, S. (2008). Bryophytes. In: *Plant biology for cultural heritage*, The Getty Conservation Institute, pp. 81-87, ISBN 978-0-89236-939-3, Los Angeles, USA

Riding, R. (2000). Microbial carbonates: the geological record of calcified bacterial-algal mats and biofilms. *Sedimentology*, Vol.47, No. Suppl.1, (February 2000), pp. 179-214, ISSN 1365-3091

Roeselers, G.; van Loosdrecht, M.C.M. & Muyzer, G. (2007). Heterotrophic pioneers facilitate phototrophic biofilm development. *Microbial Ecology*, Vol.54, No.4, (November 2008), pp. 578-585, ISSN 0095-3628 (print) 1432-184X (online)

Saiz-Jimenez, C.; Garcia-Rowe, J.; Garcia del Cura, M.A.; Ortega-Calvo, J.J.; Roekens, E. & Van Grieken, R. (1990). Endolithic cyanobacteria in Maastricht limestone. *The Science of the Total Environment*, Vol.94, No.3, (May, 1990), pp. 209-220, ISSN 0048-9697

Saiz-Jimenez, C. (1997). Biodeterioration *vs* biodegradation: the role of microorganisms in the removal of pollutants deposited on historic buildings. *International Biodeterioration & Biodegradation*, Vol.40, No.2-4, pp. 225-232, ISSN 0964-8305

Salvadori, O. (2000). Characteristisation of endolithic communities of stone monuments and natural outcrops, In: *Of microbes and art – The role of microbial communities in the degradation and protection of cultural heritage*, pp. 89-101, Cifferi O., Tiano. P, Mastromei G., ISBN 0-306-46377-6, Kluwer Academic/Plenum Publishers, New York, Boston, Dordrecht, London, Moscow

Sánchez Castillo, P.M. & Bolívar, F.C. (1997). Caracterización de comunidades algales epilíticas en fuentes monumentales y su aplicación a la diagnosis del biodeterioro. *Limnetica*, Vol.13, No.1, pp. 31-46, ISSN 0213-8409 (print), 1989-1806 (online)

Savoye, D. & Lallemant, R. (1980). Evolution de la microflore d'un substrat avant et pendant sa colonisation par les lichens. I Le cas de toitures en amiante-ciment en zone urbain. *Cryptogamie Bryologie Lichénologie*, Vol. 1, pp. 21-31, ISSN 0181-1576

Tiano, P. & Tomaselli, L. (1989). Un caso di biodeterioramento del marmo. Arkos, Vol.6, (June 1989), pp. 12-18

Tiano, P. (1993). Biodegradation of cultural heritage : decay mechanisms and control methods, In: *Conservation of stone and other materials.* Vol. 2. *Prevention and treatment,* Thiel M.J. (ed), RILEM/UNESCO Paris, pp. 573-580, ISBN 0419188509 , E & FN Spon Press, London

Tiano, P.; Accolla, P. & Tomaselli, L. (1995). Phototrophic biodeteriogens on lithoid surfaces : an ecological study. *Microbial Ecology*, Vol.29, No.3, pp. 299-309, ISSN 0095-3628 (print) 1432-184X (online)

Tiano, P. (1998). Biodeterioration of monumental rocks: decay mechanisms and control methods. *Science and Technology for Cultural Heritage*, Vol.7, No.2, pp. 19-38, ISSN 1121-9122

Tomaselli, L.; Lamenti, G.; Bosco, M. & Tiano, P. (2000). Biodiversity of photosynthetic micro-organisms dwelling on stone monuments. *International Biodeterioration & Biodegradation*, Vol.46, No.3, (October 2000), pp. 251-258, ISSN 0964-8305

Tomaselli, L.; Lamenti, G. & Tiano, P. (2002). Chlorophyll fluorescence for evaluating biocide treatments against phototrophic biodeteriogens. *Annals of Microbiology*, Vol.52, No.3, pp. 197-206, ISSN 1590-4261

Turtura, G.C.; Perfetto, A. & Lorenzelli, P. (2000). Microbiological investigations on black crusts from open air stone monuments of Bologna (Italy). *New Microbiology*, Vol. 23, No.2, (April 2000), pp. 207-228, ISSN 1121 - 7138

Underwood, A.J. (1984). Vertical distribution and seasonal abundance of intertidal microalgae on a rocky shore in New South Wales. *Journal of Experimental Marine Biology and Ecology*, Vol.78, No.3, (June 1984), pp. 199-220, ISSN 0022-0981

Urzì, C.; De Leo, F.; De Hoog, S. & Sterflinger, K. (2000). Recent avances in the molecular biology and ecophysiology of meristematic stone-inhabiting fungi, In: *Of microbes and art - The role of microbial communities in the degradation and protection of cultural heritage*, pp. 3-19, Cifferi O., Tiano. P, Mastromei G., ISBN 0-306-46377-6, Kluwer Academic/Plenum Publishers, New York, Boston, Dordrecht, London, Moscow

Warscheid, Th. (1996). Impacts of microbial biofilms in the deterioration of inorganic building materials and their relevence for the conservation practice. *International Journal for Restoration of Buildings and Monuments*, Vol.2, pp. 493-503, ISSN 1864-7251 (print) 1864-7022 (online)

Warscheid, Th. & Braams, J. (2000). Biodeterioration of stone: a review. *International Biodeterioration & Biodegradation*, Vol.46, No.4, (December 2000), pp. 343-368, ISSN 0964-8305

Warscheid, Th. (2008). Heritage research and practice: towards a better understanding ? In: *Heritage, microbiology and science - microbes monuments and maritime materials*, E. May, M. Jones & J. Mitchell (Eds.), Springer Verlag, pp. 11-25, ISBN 978-0-85404-141-1, Great Britain

Wimpenny, J.; Manz, W. & Szewzyk, U. (2000). Heterogeneity in biofilms. *FEMS Microbiology Reviews*, Vol.24, No.5, (December 2000), pp. 661-671, ISSN 0168-6445 (print)1574-6976 (online)